Palgrave Studies in Minority Languages and Communities

Series Editor
Gabrielle Hogan-Brun
Vytautas Magnus University
Kaunas, Lithuania

Worldwide migration and unprecedented economic, political and social integration present serious challenges to the nature and position of language minorities. Some communities receive protective legislation and active support from states through policies that promote and sustain cultural and linguistic diversity; others succumb to global homogenisation and assimilation. At the same time, discourses on diversity and emancipation have produced greater demands for the management of difference.

This series publishes new research based on single or comparative case studies on minority languages worldwide. We focus on their use, status and prospects, and on linguistic pluralism in areas with immigrant or traditional minority communities or with shifting borders. Each volume is written in an accessible style for researchers and students in linguistics, education, politics and anthropology, and for practitioners interested in language minorities and diversity. We welcome submissions in either monograph or Pivot format.

More information about this series at
http://www.palgrave.com/gp/series/14611

Peter Jordan • Přemysl Mácha
Marika Balode • Luděk Krtička
Uršula Obrusník • Pavel Pilch
Alexis Sancho Reinoso

Place-Name Politics in Multilingual Areas

A Comparative Study of Southern Carinthia (Austria) and the Těšín/ Cieszyn Region (Czechia)

palgrave
macmillan

Peter Jordan
Institute of Urban and Regional Research
Austrian Academy of Sciences
Vienna, Austria

Faculty of Humanities
University of the Free State
Bloemfontein, South Africa

Marika Balode
Dept of Geography and Regional Studies
University of Klagenfurt
Klagenfurt am Wörthersee, Kärnten, Austria

Uršula Obrusník
Department of Anthropology
University of Aberdeen
Aberdeen, UK

Alexis Sancho Reinoso
Department of Social and Cultural Anthropology
University of Vienna
Vienna, Austria

Přemysl Mácha
Department of Memory Studies
at the Institute of Ethnology
Czech Academy of Sciences
Brno, Czech Republic

Luděk Krtička
Department of Human Geography and Regional Development
University of Ostrava
Brno, Czech Republic

Pavel Pilch
Department of Philosophy
Masaryk University
Brno, Czech Republic

Palgrave Studies in Minority Languages and Communities
ISBN 978-3-030-69487-6 ISBN 978-3-030-69488-3 (eBook)
https://doi.org/10.1007/978-3-030-69488-3

Cover illustration: Marika Balode

This Palgrave Macmillan imprint is published by the registered company Springer Nature Switzerland AG.
The registered company address is: Gewerbestrasse 11, 6330 Cham, Switzerland

This book was supported by the Austrian Science Fund [Fonds zur Förderung der wissenschaftlichen Forschung, FWF] with the project number I 2366-G23 and the Czech Grant Agency [Grantová agentura České republiky, GAČR] with the project number 16-34841L.

Contents

List of Figures

List of Tables

1

Introduction

Contributed by Peter Jordan and Přemysl Mácha

In our research, we analyzed the politics of toponymy, identity, and landscape in two multilingual areas—the Těšín/Cieszyn region in Czechia and southern Carinthia in Austria. In both areas, linguistic and ethnic composition is the result of population movements in Central Europe's vivid history and shares a lot of similar characteristics. For this reason, we saw a close cooperation between Czech and Austrian researchers highly advantageous. Also, a comparative study allows for broader generalizations based on shared similarities and differences.

Both regions have gone through a series of conflicts over bilingual signs in recent decades. We were interested in what such conflicts reveal about the relationship between name, place, and identity, what were the principal causes of these conflicts, and which factors most influenced people's attitudes toward bilingual signs. We drew on anthropological and geographical approaches to place names to gain a complex understanding of the issues. This presented many conceptual, methodological, and practical challenges, but significantly enriched our analysis. We hope that our findings not only shed light on processes and problems in our study areas but also offer general observations which can inform research and policy making in other areas.

P. Jordan et al., *Place-Name Politics in Multilingual Areas*, Palgrave Studies in Minority Languages and Communities, https://doi.org/10.1007/978-3-030-69488-3_1

1.1 Study Areas

There is, up to this day, no systematic study of bilingual/multilingual toponymy of the Těšín/Cieszyn region that would put place names in association with landscape, identity, and minority rights. It is true that several historical texts have appeared that provide excellent information about the historical development of the region (see, e.g., Korbelářová 2008 or Collective of authors 1997–2003). Nevertheless, virtually no in-depth toponymical analysis exists save a recent brief study on the linguistic landscape of the area in a wider comparative perspective by Szabó-Gilinger et al. (2012). This is rather surprising given the unique position of the area in Czechia. It is the only area with a high concentration of an ethnic minority (Polish) where according to Czech legislation the minority has the right to use their maternal language in official dealings with the government, in schools and newspapers, and on public signs. Over the past few years significant conflicts have arisen regarding the use of bilingual (Czech-Polish) public signs and place names, the Polish names being vandalized. The situation is even more complicated, because the real maternal language of both Poles and Czechs is neither Polish nor Czech but a unique local dialect called *po naszymu* ('in our language'). Traditional toponymy exists in this dialect. The region therefore is trilingual while public signs are bilingual, written in languages which few people use in everyday life. The pronunciation of place names thus becomes a deeply public and political matter (for the politics of place-name pronunciation and toponymic resistance see, e.g., Kearns and Berg 2002).

The Carinthian minority situation is up to the present day—albeit with declining intensity—marked by the fact that a Slavonic population present since the sixth–seventh centuries has later been socially overlaid by Bavarian settlers. The newcomers, supported by political powers, formed the upper strata of the society including traders and craftsmen while the Slavonic population remained the rural ground layer. Up to the end of the Middle Ages an ethnically/linguistically mixed situation persisted. Assimilation toward local majorities resulted in an ethnic/linguistic patchwork. This shapes Carinthian culture in many respects also today. It is also reflected by the namescape, which is a mixture of Slavonic and German names all over the province. In general, however, linguistic assimilation toward German-speakers, the upper strata of the society, proceeded. Social ascent was only possible by using the German language—very similar to the situation of Slovenes under Venetian rule in what is today Italy, where Venetian, later Italian were the languages of the dominant group. By the end of the Middle Ages a distinct language boundary within Carinthia had developed—very much coinciding with ecclesiastical boundaries between Salzburg and Aquileia established in 811. This boundary still exists, but also at the Slovenian side of the boundary the Slovenian population has decreased substantially. The strong decline is not only due to social stratification as mentioned before, but also to societal change in general (conversion of rural societies by industrialization and tertiarization) and the peripheric situation of the Slovenes in Austria in socio-economic terms. In addition, political events and forces had their strong impact: the rise of national ideas during the nineteenth century and national homogenization after World War I almost all over Europe and also in Austria; the fact that Austria had to cede some parts of Carinthia populated predominantly by Slovenes to the first Yugoslavian state (with the effect that the rest of the area populated by Slovenes was regarded as 'ours' and subjected to Germanization); repeated attempts of the Yugoslavian states (after World War I and II) to occupy at least larger parts of Carinthia; the atrocities of the National-Socialist regime between 1938 and 1945 and of Yugoslavian partisans at the end of World War II. Efforts to establish and improve minority rights after World War II met already a very small and further declining Slovenian group. As regards minority toponymy, the Austrian State Treaty as of 1955 included a principal statement, which needed to be specified by additional federal legislation. Unsuccessful attempts in 1972 and 1976/1977 were only in 2011 followed by a compromise that seems to have satisfied all parties and calmed down the at times delicate situation.

The Carinthian minority situation resembles the situation in the Těšín/Cieszyn region as regards place names in actual use frequently deviate from standardized names or names in the standard languages and as regards political conflict on place names in recent times. In contrast to the Těšín/Cieszyn region toponymy is linguistically well-investigated in

Carinthia (see a.o. Kranzmayer 1956, 1958; Kronsteiner 1974; Pohl 1981–1983, 1982, 1984, 1985a, b, 1986, 1987–1988, 2000, 2008, 2009a, b, 2010; Zdovc 1974, 1993, 2010). Not too much has been done, however, also in Carinthia on highlighting the roles of place names from a cultural-geographical perspective, that is, as mediators between human community and space including aspects such as space-related identity building by place names, place names as supporters of emotional ties between human beings and place, and the ambivalent attitudes also of some minority members toward minority place names in public space.

In view of this, our principal research questions are as follows:

1. What do place names mean for the identity of human communities in general and more specifically for linguistic minorities? What is the relationship between language, place, and identity, and how do we make ourselves at home through place names?
2. What toponymic strategies have been employed by different actors in establishing, maintaining, and subverting ethnic/national boundaries, and what are the principal social forces structuring the contemporary toponymic landscape and everyday toponymic practice?
3. How are the multilingual/multiethnic city-text and linguistic landscape produced, performed, interpreted, and contested?
4. When we speak of minority rights and cultural preservation, what role do place names play in this discussion? Why, how, by what means and procedures, by whom, and for whom should place names be protected?

1.2 Structure of the Book

In the second chapter we outline the theoretical framework of our study. We draw on several disciplines (namely anthropology, geography, history, and linguistics) as well as on different topics associated with names (etymology, socio-toponomastics, politics of place naming, emotional geography of names, cultural heritage value of names, linguistic landscape studies, etc.). Our discussion is therefore necessarily superficial, and we only select those themes which are most directly relevant for our research.

In the third chapter we summarize the methods we chose for our study describing how we collected and analyzed our data. We also report on the practical challenges we encountered in the field. We try to be maximally transparent, so that the reader can competently judge the relevance of our results and future researchers can learn from our experiences and mistakes.

In the fourth and fifth chapters, we provide a comprehensive survey on linguistic minorities and their political, legal, and societal embedding in our two countries (Chap. 4) as well as a detailed description of our research areas, including their historical, cultural, political, legal, socio-economic, and geographical background and situation (Chap. 5), since it is our conviction that a study in critical toponomastics is to make all these backgrounds and conditions transparent to the reader. The two regions share a part of their history, as both belonged to the Habsburg dominions for hundreds of years. After World War I they departed on their own ways, and for this reason we consider our comparative analysis very valuable since we can study how the changing context impacted ethnic minorities and minority names. After the common framework we therefore present the changes in the national contexts during the twentieth century up to the present.

In Chap. 6 we present the results of our research in the two situations in an identical structure, so that it is easy to compare them. The structure of this chapter follows that of the methodological chapter—place-name use on maps, linguistic landscape, media analysis, interviews, questionnaires.

In the final two chapters, we compare the results from both situations (Chap. 7) and offer our interpretation and conclusions (Chap. 8) critically reflecting on our work and suggesting possibilities for further research.

The text is accompanied by a comprehensive bibliography and by lists of figures and tables as well as appendices containing basic materials and documents.

In this book geographical features are primarily named by their English exonym, if such an exonym exists, or by the contemporary endonym(s). With the exception of country names the current official name is added in rectangular brackets, should it deviate from the name(s) outside brackets. It is, however, not consequently repeated, just when the feature is addressed in a new context. The same practice is applied to names of institutions, documents, and publications (e.g., books, newspapers), historical figures or events, where also at first an English translation or exonym is offered and the contemporary official and currently conventional name in the local official language is added in rectangular brackets.

Italics are used, when a name or term is addressed as a label, that is in its designating function, and not the feature or person designed by the name or term, the designatum.

1.3 Research Teams, Division of Work, Acknowledgments

The research project was conducted by Peter Jordan on the Austrian and Přemysl Mácha on the Czech side with Přemysl taking the initiative and defining most of the research design. Both of them formulated most of the book texts with Peter doing the final editing. Members of the Austrian team were Marika Balode and Alexis Sancho Reinoso, members of the Czech team Luděk Krtička, Uršula Obrusník, and Pavel Pilch (see also Fig. 1.1).

The Austrian team:

Peter Jordan, born 1949 in Hermagor, Carinthia, Austria, close to the bilingual area and up to 1961 living there, was educated in geography, cartography, and ethnology at the University of Vienna (PhD), habilitated at the University of Klagenfurt, worked 1977–2006 at the Austrian Institute of East and Southeast European Studies in Vienna [Wien] (2002–2005 as its director), since 2007 at the Austrian Academy of Sciences, Institute of Urban and Regional Research, in Vienna. He is research fellow of the University of the Free State, Faculty of the Humanities, South Africa. He was editor-in-chief of the map series "Atlas of Eastern and Southeastern Europe" (30 installments published) and authored 388 scientific publications, 147 of them on regional and cultural geography and 84 on toponomastics, the fields closest to the theme of this book. He was/is teaching regularly at the universities of Vienna, Klagenfurt, and Cluj-Napoca (Romania) in the fields of cultural and

Fig. 1.1 The two research teams (missing: Luděk Krtička) during their excursion in the Těšín/Cieszyn region in 2017. From left to right: Přemysl Mácha, Marika Balode, Alexis Sancho Reinoso, Peter Jordan, Pavel Pilch, Uršula Obrusník. (Photo by Luděk Krtička)

tourism geography, cartography, and toponomastics. From 2006 to 2017, he functioned as Convenor of the United Nations Group of Experts on Geographical Names' (UNGEGN's) Working Group on Exonyms, and 2007–2017 as Chair, Austrian Board on Geographical Names, and 2009–2017 as Editor, Annals of the Austrian Geographical Society [Mitteilungen der Österreichischen Geographischen Gesellschaft]. From 2017 he has been the chair on behalf of the International Cartographic Association (ICA) of the Joint ICA/IGU Commission on Toponymy. Together with Paul Woodman he edits the toponymic book series "Name & Place" (so far nine volumes published). Within the project presented in this book he coordinated the Austrian research, contributed to the theoretical and methodological framework, and organized a symposium at the Austrian Academy of Sciences. In this book he authored or co-authored all parts describing the Austrian situation as well as general chapters.

Marika Balode, BA, born 1976 in Liepāja, Latvia, and residing in Leppen/Lepena in the core area of the minority in Carinthia, is Slavist and geographer, is pursuing an MA at the Institute of Geography and Regional Research, University of Klagenfurt. She speaks five languages including German and Slovene and is interested in space-related identities, intercultural relations, and minority situations. Besides contributing to the concept of the research project she conducted about half of the interviews in Carinthia, did near to all the photo documentation of the linguistic landscape there and engaged herself in the analysis of the media discourse on the Carinthian situation.

Alexis Sancho Reinoso, BA, MSc, PhD, born 1983 in Barcelona, Catalonia [Cataluña/Catalunya], Spain, residing in Lower Austria [Niederösterreich], geographer with a special interest in intercultural relations, regional development, and rural space, acquired his academic grades at the University of Barcelona, and works at the Institute of Cultural and Social Anthropology, University of Vienna. He participated in several research projects, a.o. "Current developments in regional economy and community building in the rural space of Romania and Austria". Besides contributing to the concept of this research project he conducted about half of the interviews in Carinthia, engaged himself in the analysis of the media discourse on the Carinthian situation, surveyed and analyzed the linguistic landscape of Ferlach, initiated the questionnaire survey in Carinthia, and formulated draft versions of the subchapters on media discourse on minority place names (6.1.3), interviews (6.1.4), and questionnaires (6.1.5).

The Czech team:

Přemysl Mácha, born 1975 in Frýdek-Místek, Czechia, has lived most of his life in the historic Těšín/Cieszyn region. He obtained his PhD from the Institute of Ethnology at Charles University, Prague [Praha], Czechia. During the project, he worked as an assistant professor of anthropology and cultural geography at the Department of Human Geography, University of Ostrava, Czechia. He now works in the Department of Memory Studies at the Institute of Ethnology of the Czech Academy of Sciences in Brno, Czechia. He is interested in identity politics, place names, landscape, inclusive communities, and sustainability. He is the author, co-author, and editor of over 60 publications, including several books and many articles on place names—for example, "Place Names: Memory, Identity, Cultural Heritage" [Názvy míst: paměť, identita, kulturní dědictví] or "Secrets Hidden in Place Names" [Tajemství v názvech ukrytá]. He previously coordinated (together with Jaroslav David) a major research project titled "Place Names as Cultural Heritage" (2011–2014) financed by the Czech Ministry of Culture. Within the project presented in this book he coordinated the Czech research team, conducted interviews, organized the survey, documented the linguistic landscape, and contributed to archival research and media analysis of the Czech part. He also evaluated the research results and authored and co-authored all chapters pertaining to Czechia as well as parts of the general theoretical and all of the methodological framework.

Luděk Krtička, born 1978 in Rapotín, Moravia [Morava], Czechia, where he still lives, works as an assistant professor at the Department of Human Geography and Regional Development, University of Ostrava. He is finishing his PhD in physical geography, geo-ecology, and geoinformatics at the Comenius University in Bratislava, Slovakia. His interests include landscape change assessment, urban sustainability, GIS analysis and data collection, cartography, and landscape toponymy. In these fields he published more than ten peer-reviewed articles. He is also co-author of three textbooks on GIS and cartography. Within the project he was in charge of technology for mobile data collection, data processing in GIS, and creation of map outputs.

Uršula Obrusník, born 1987 in Ostrava, lives in Aberdeen, United Kingdom. She is currently pursuing a PhD in anthropology at the University of Aberdeen. She is interested in issues of belonging, minority rights, politics of memory, and contemporary urban spaces. She is a member of the Polish minority in Czechia. Within our project, she conducted several interviews with Polish minority members, assisted in setting up the questionnaire survey, and co-authored the media discourse analysis. She formulated draft versions of the subchapters on media discourse on minority place names, legislation regarding place names, and political conflict as well as subchapters dealing with history and the current situation of the Polish minority.

Pavel Pilch, born 1986 in Třinec, Czechia, that is right in the research area, and living in Brno, is bohemist, Slavist, and comics studies scholar, pursuing a PhD and teaching at the Institute of Slavonic Languages, Faculty of Arts, Masaryk University in Brno. As a native of the research area he speaks the local dialect. He contributed to the project by interviewing locals in rural areas and analyzing interviews gathered in the whole researched area. He also provided valuable insight into the local specifics of the Těšín/Cieszyn region.

This book is based on a research project conducted between 2016 and 2018 (with a cost-neutral extension by 2019) and bilaterally funded by the Austrian Science Fund [Fonds zur Förderung der wissenschaftlichen Forschung, FWF] with the project number I 2366-G23 and the Czech Grant Agency [Grantová agentura České republiky, GAČR] with the project number 16-34841L. The research team likes to express its utmost gratitude for this funding and extends it also to the Fund's officials responsible for communication and accounting. In the Austrian case, they were Martina Kunzmann, Georg Rücklinger, and Elisabeth Thoernblom, as well as to the anonymous international reviewers.

Very special thanks go to experts in the research field or to important actors in the research area essentially supporting us with their expertise and/or being instrumental with our activities like finding interview partners, disseminating questionnaires, organizing meetings and excursions. This refers on the Austrian side to Martina Piko-Rustia, head of the Urban Jarnik Institute of Ethnography in Klagenfurt am Wörthersee, Martin Pandel, editor with the Hermagoras Verlag/Mohorjeva založba in Klagenfurt, Heinz-Dieter Pohl, Professor Emeritus for General and Diachrone Linguistics at the University of Klagenfurt, Andreas Schuller, director of the high school in Hermagor, and Igor Roblek, geographer and minority activist from the 90% Slovene-speaking commune Zell (Sele).

We are equally and deeply thankful to the political representatives of the Polish minority (Congress of Poles, PZKO) and the Czech majority (current and former mayors and vice-mayors) as well as all the inhabitants of the Těšín/Cieszyn region who kindly shared with us their thoughts and experiences related to bilingualism, minority signs, and Czech-Polish relations. Special thanks go to Dr. Józef Szymeczek from the Congress of Poles and the University of Ostrava, not only for his practical help with resources and contacts but also for his politically engaged, yet academically critical, reflections of the Polish minority situation. Prof. Tadeusz Siwek from the University of Ostrava was also helpful with his consultations on the political background of the Polish minority organizations and social life. Emilia Świder provided valuable contacts as did Martyna Radłowska-Obrusník who also offered ongoing moral support. Finally, we would also like to express thanks to the students of the University of Ostrava who contributed to parts of the archival research (Horst Lassak, Blanka Pohorská), media analysis (Tereza Kuchařová, Jakub Lukáš), and interviews (Michaela Czeczotková).

Before a book like this goes to print, it undergoes a reviewing process and needs many organizational and technical inputs and efforts, for which we are very grateful to our anonymous peer-reviewers as well as to Palgrave Macmillan Publishers, notably to Alice Green and Cathy Scott.

References

Collective of authors. (1997–2003). Těšínsko. Díl 1–5. Český Těšín—Rožnov pod Radhoštěm.

Kearns, R. A., & Berg, L. D. (2002). Proclaiming place: Towards a geography of place name pronunciation. *Social and Cultural Geography,* 3(3), 283–302.

Korbelářová, I. (2008). *Těšínsko—země Koruny české = Ducatus Tessinensis—terra Coronae Regni Bohemiae (k dějinám knížectví od počátků do 18. století).* Český Těšín: Muzeum Těšínska.

Kranzmayer, E. (1956). *Ortsnamenbuch von Kärnten. I. Teil: Die Siedlungsgeschichte Kärntens von der Urzeit bis zur Gegenwart im Spiegel der Namen.* Klagenfurt: Verlag des Geschichtsvereines für Kärnten.

Kranzmayer, E. (1958). *Ortsnamenbuch von Kärnten. II. Teil: Alphabetisches Kärntner Siedlungsnamenbuch.* Klagenfurt: Verlag des Geschichtsvereines für Kärnten.

Kronsteiner, O. (1974). *Die slowenischen Namen Kärntens in Geschichte und Gegenwart (= Österreichische Namenforschung, Sonderreihe 1).* Wien: Österreichische Gesellschaft für Namenforschung.

Pohl, H.-D. (1981–1983). Kärntner Bergnamen I (1–4). *Österreichische Namenforschung,* 9–11, 55–82.

Pohl, H.-D. (1982). Linguistische Aspekte der Zweisprachigkeit in Kärnten. In F. Dotter (Ed.), *Kein einig Volk von Brüdern. Studien zum Mehrheiten-/Minderheitenproblem am Beispiel Kärntens* (pp. 35–53). Klagenfurt: Verlag für Gesellschaftskritik.

Pohl, H.-D. (1984). Kärntner Bergnamen II (5). *Österreichische Namenforschung*, 12, 25–45.

Pohl, H.-D. (1985a). Kärntner Bergnamen III (6–8). *Österreichische Namenforschung*, 13, 39–73.

Pohl, H.-D. (1985b). Slavische und romanische Oronyme in den Karawanken (und umliegenden Gebirgsgruppen). *Slavisticna revija*, 33, 177–184.

Pohl, H.-D. (1986). Die Ortsnamen des zweisprachigen Gebietes Kärntens—Eine Bestandsaufnahme. *Onomastica Slavogermanica*, XV, 103–133.

Pohl, H.-D. (1987–1988). Zu den slowenischen Ortsnamen Kärntens deutscher Herkunft. *Österreichische Namenforschung*, 15–16, 91–102.

Pohl, H.-D. (2000). Kärnten—deutsche und slowenische Namen. Koroška—slovenska in nemška imena. Kommentiertes zweisprachiges Verzeichnis der Siedlungs-, Berg- und Gewässernamen. *Österreichische Namenforschung*, 28(2–3), 1–148.

Pohl, H.-D. (2008). Sekundäre Umformungen von (Kärntner) Ortsnamen im deutsch-slowenischen Sprachkontaktgebiet. In P. Ernst (Ed.), *Namenarten in Österreich und Bayern. Vorträge der 4. Tagung des Arbeitskreises für bayerisch-österreichische Namenforschung am 28. und 29. September 2006 in Wien* (pp. 95–105). Wien: Praesens Verlag.

Pohl, H.-D. (2009a). Ortsnamen in einer historisch gewachsenen Kulturlandschaft unter besonderer Berücksichtigung Kärntens und Osttirols. *Europäisches Journal für Minderheitenfragen*, 2(2), 72–88.

Pohl, H.-D. (2009b). Sprachkontakt in Kärnten. In M. Elmentaler (Ed.), *Deutsch und seine Nachbarn* (pp. 117–132). Frankfurt a. M. et al.: Peter Lang.

Pohl, H.-D. (2010). *Unsere slowenischen Ortsnamen—Naša slovenska krajevna imena*. Klagenfurt: Hermagoras/Mohorjeva.

Szabó-Gilinger, E., et al. (2012). Discourse coalitions for and against minority languages on signs: Linguistic landscape as a social issue. In D. Gorter, H. Marten, & L. Van Mensel (Eds.), *Minority languages in the linguistic landscape* (pp. 263–280). New York: Palgrave Macmillan.

Zdovc, P. (1974). Einige Aspekte zu Ortsnamenfragen in Kärnten. *Carinthia I*, 164, 283–303.

Zdovc, P. (1993). *Die slowenischen Ortsnamen in Kärnten/Slovenska krajevna imena na avstrijskem Koroškem*. Klagenfurt/Celovec: Slowenisches Wissenschaftliches Institut in Klagenfurt/Slovenski znanstveni inštitut v Celovcu.

Zdovc, P. (2010). *Slovenska krajevna imena na avstrijskem Koroškem/Die slowenischen Ortsnamen in Kärnten* (= Razprave, 21). Ljubljana: Slovenska akadenija znanosti in umjetnosti.

2

The Wider Onomastic Scope of the Research Topic

Contributed by Peter Jordan and Přemysl Mácha

Place names have always been studied by several disciplines and for diverse reasons (Taylor 1998). Until recently, however, they were approached predominantly as 'windows into the past'. Research in these areas has been very productive and continues to be developed by several investigators (see, e.g., Conedera et al. 2007; Kathrein 2009).

In recent years, however, we can observe an important development in place-names research across social sciences (namely in cultural anthropology and geography) that represents a break with the past. Rather, the new approaches emphasize the contemporaneity of place names (while not ignoring their historical roots) and study them in relation to the political constitution and contestation of place, landscape, and identity.

Initial propositions of these new perspectives on place names were put forth by Yi-Fu Tuan (1977, 1990 [1974], 1991) within geography and Keith Basso (1988, 1996) in cultural anthropology. Both argue that naming is a fundamental social and existential practice, whereby people establish their relationship with the space they occupy. Tuan showed that human spatial perception is structured by language, place names playing an important role in our perception and representation of the environment. Basso specified place naming as a way of writing/making history

P. Jordan et al., *Place-Name Politics in Multilingual Areas*, Palgrave Studies in Minority Languages and Communities, https://doi.org/10.1007/978-3-030-69488-3_2

and relating to the world at a fundamental, existential level, place names being closely tied to our identity.

A number of researchers have taken up these propositions and applied them to case studies all over the world (see, e.g., Jett 1997; Thornton 1997a; Oliveira and Wahi 2009, or essays edited by Jordan et al. 2009). They all point out the importance of place names for understanding and establishing our relations with the space we occupy. At the same time, they also acknowledge that the relationship between place names and identities is by no means simple. Toponymic inventories of individuals differ in relation to gender, age, occupation, social status, and other attributes of social differentiation (Thornton 1997b: 221). In addition, place names undergo a constant re-interpretation and transformation, and so do the associated identities. Therefore, toponymic inventories and toponymic practice are highly fluid, situational, and dynamic, and thorough ethnographic studies of these phenomena are yet to come in larger numbers (for rare examples of these see, e.g., Gabbert 2007; David and Mácha 2014).

However, there is yet another dimension to place names—and that is power. We are not, and never have been, in equal positions to name places, individually or collectively. Place names may constitute cultural heritage and may be important for establishing and reproducing social identities, but they are also loaded with emotions, alternative interpretations, and contested histories, and as such they are not politically innocent. As Tuan shows, they must be understood in the context of the present relations of power which (strive to) reproduce themselves in space by various material and non-material practices, one of them being toponymy:

> Normally, only a sociopolitical revolution would bring about a change of name in a city or a nation. The idea behind taking such a step is not only that a correct label should be affixed to a new entity, but also that, somehow, the new name itself has the power to wipe out the past and call forth the new. (Tuan 1991: 688)

It is precisely this focus on the politics of place naming that the so-called critical toponymy approach has developed (see, e.g., Rose-Redwood et al. 2010). Critical toponymy is a lively current within contemporary place-name research that critically examines the relationship between toponymy and power. It analyzes ways in which political regimes and movements use place names to claim territories, erase linguistic traces of original populations, gain political legitimacy, delegitimize other political forces, naturalize certain versions of history, and silence dissent. This is possible because of place names'

> ability to incorporate an official version of history into such spheres of human activity that seems to be entirely devoid of direct political manipulation. This transforms history into a feature of the 'natural order of things' and conceals its contrived character. (Azaryahu 1996: 481)

Several very interesting case studies have appeared in the last two decades that critically analyze the utilization of toponymy in (post-)colonial settings (e.g., Nash 1999; Herman 1999), as tools in nation-building (e.g., Cohen and Kliot 1992; Yeoh 1996; Azaryahu and Golan 2001) or as strategies of legitimizing political regimes (e.g., Horsman 2006). Particular attention within critical place-name studies has been paid to commemorative street names in diverse historical, geographical, and political contexts (e.g., Autengruber 2013; González Faraco and Murphy 1997; Alderman 2000; Light 2004; Gill 2005; Palonen 2008). An important concept in the last set of studies is the concept of the city-text (Azaryahu 1996) which is a useful way to think about the internally heterogenous, palimpsest-like toponymic ensemble of a given city, together with other, non-toponymic and even non-linguistic components which contribute to the expression and reproduction of specific political, territorial, religious, ethnic, national, or class identities (e.g., Šakaja and Stanić 2011; Bucher et al. 2013).

A new development within critical place-name research which is associated with the concept of the city-text but which has yet to bear fruit is a growing interest in toponymy not as a fait-accompli but as an ongoing and permanently contested social process and a set of performative actions (Rose-Redwood 2008). The focus changes from studying the result of toponymic processes (i.e., place names) to the toponymic praxis itself (i.e., place naming)—how are names given, what are the procedures and rituals of name giving, how participatory, representative, and just are these procedures, how is a name-giving authority established (Kerfoot 2003), what are the motivations and strategies in place naming, and how do place names operate to structure the everyday life and social relations (e.g., Rose-Redwood 2008; Withers 2000; Alderman and Inwood 2013).

Within the scope of these directions, this chapter presents at first from a cultural-geographical angle basics of the place-naming process, the concept of the "great toponymic divide" (Woodman 2012) between the endonym and the exonym as well as of the basic roles that place names play in relating people to geographical space. In its second part the chapter highlights the theory of linguistic landscape discourse analysis as another important fundament of our case studies.

As regards the first range of topics, the chapter is based on the seminal works of Yi-Fu Tuan (1977, 1990 [1974], 1991), Keith Basso (1988, 1996), and Don Mitchell (2000) as well as contributions by Botolv Helleland (2009), Paul Woodman (2012), and Peter Jordan (2009, 2012, 2014, 2015a, 2019a, b).

2.1 The Place-Naming Process and the Central Role of the (Local) Community with an Outlook at the Endonym/Exonym Divide (Peter Jordan)

Three factors are involved in the place-naming process, that is, when it comes to attribute a place name to a geographical feature (Fig. 2.1):

- The *local community* in the sense of a group of people, who feel to have a common identity. It can vary in size between family/partnership (related to a flat or house) via inhabitants of a village/town/city/landscape, a nation, a language community up to the global community ('community of global citizens'). It needs not to be a cohesion group in the sense of a group of people tied by personal relations and almost in permanent interaction. It can also just be an identity group—a group of people feeling to have something in common. They neither

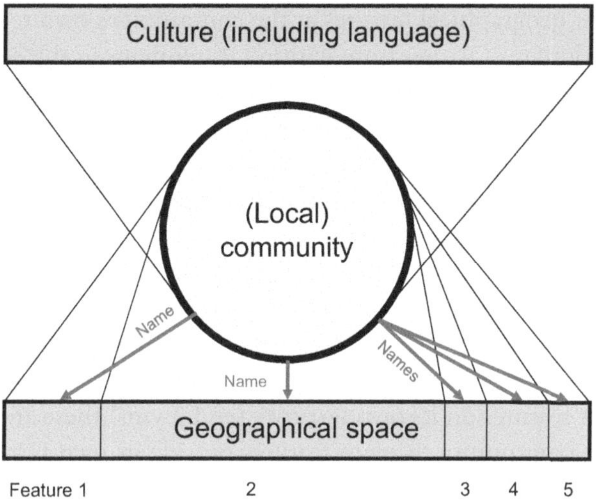

Fig. 2.1 Factors of the place-naming process. (Source: Jordan 2019a)

need not have personal relations nor to know each other. The community creating and applying place names is not necessarily the same. However, if a community applies place names created by another community, it appropriates these place names and makes them their own.

- The community's *culture including language*. Culture is understood here in the most comprehensive sense as the totality of all human expressions (Kroeber and Kluckhohn 1963; Lévi-Strauss 1946, 1949).
- *Geographical space* is subdivided into geographical features. Geographical space is understood here as a construct of immaterial relations, as "a logical structure, in which given elements can mentally be placed" (Weichhart 1999: 77).

The only actor in this process is the (local) community. It inhabits a certain section of geographical space (a flat, a house, a village, town/city, region, country), has developed a certain culture, speaks a certain language, and structures complex geographical space mentally into features on the background of its culture and led by its specific (e.g., economic) interests marking these features by place names in their language.

Names for geographical features at the community's own territory can be called endonyms ('names from within'). Endonyms in this sociological sense are symbols of appropriation: Who owns a feature usually has the right to name it. Who has the power to assign the name usually also has the power over this feature or at least responsibility for it (Jordan 2019b; Woodman 2012). When Genesis 2:20 says that "the man gave names to all the cattle, and to the birds of the sky, and to every beast of the field…" this function of naming (in general) is clearly addressed.

For geographical features outside its territory a community usually just adopts already existing place names, translates them into its own language, or adapts them morphologically or phonetically to it. In contrast to names for features on its own territory (endonyms) these are exonyms, needed by a community to mark features outside its own territory sufficiently important to it in a comfortable way (easy to be pronounced, to be communicated, to be memorized). In contrast to endonyms, exonyms are *not* symbols for appropriation and do *not* express claims, but indicate the importance of a feature for this community and the relations it has with it (Jordan 2009, 2015a, b; Woodman 2012). Exonyms just help to integrate this foreign feature into the cultural sphere of a community and help to avoid exclusion and alienation (Back 2002). But it is also true that the use of exonyms is sometimes conceived as expressing claims, especially when exonyms correspond to historical endonyms. This, however, is a misunderstanding, which should be erased, also by a politically sensitive use of exonyms (Jordan 2000).

Place naming is executed either by convention between the members of a community or by an institution charged and legitimized by the community for this purpose. Of course, also an individual can attribute a place name to a feature, but such a name will not get into use, assume communicative value and persist, if it is not accepted at least by the smallest local community. So, it is at the end always the community, who acts in this process.

No community, however, is completely homogenous. It is always composed of a dominant portion and non-dominant subgroups. The dominant portion of a community is of course in the position to decree the use of a place name and to oblige other community members to use it whether they like it or not. So it may happen that the place name applied by the own community assumes for some community members an 'exonymic character' in the sense that they have to accept and to use it, but that it is not the name with which they can fully identify themselves and which they would use, if this was their own decision. Situations of this kind occur with many place-name changes after a new political regime or direction assumed power—as it was in the eastern part of Europe after World War II, when Communism came into power, and again after the fall of Communism in 1989/1990 (Jordan and Woodman 2016; see also Autengruber 2013; Bucher et al. 2013; Šakaja and Stanić 2011; Palonen 2008; Gill 2005; Light 2004; Alderman 2000; González Faraco and Murphy 1997; Azaryahu 1996; Yeoh 1996; Cohen and Kliot 1992).

It is also a fact that we usually do not belong to only one community, but rather to a multitude of them—we have in fact multiple group identities (Fig. 2.2). These various communities have usually also different relations to space and feel responsible for different sections of spatial reality. We are global citizens, when we engage ourselves in questions like climate change, global disparities in development, and so on. Global institutions

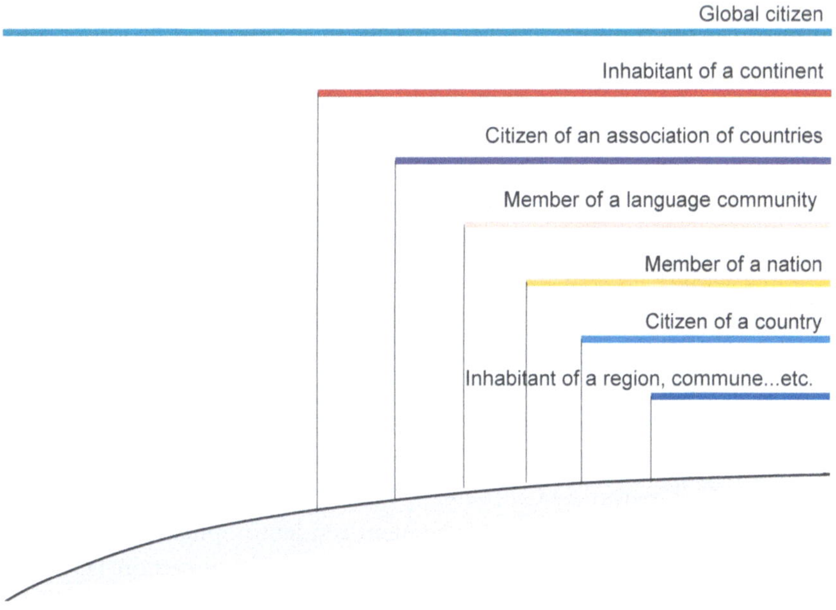

Fig. 2.2 Multiple space-related identities. (Source: Jordan 2019a)

and organizations support this community (e.g., the United Nations). We are inhabitants of our continent as far as we feel responsible and engage ourselves for this continent. We are citizens of an association of countries like the European Union, members of a language community (e.g., the English-speaking), members of a nation, citizens of a country, inhabitants of a region, a city, a commune, a village. Almost all these communities are in a way organized and feel responsibility for a section of space. All of these mentioned have certainly also a specific relation to space.

But there exist also communities with the same relation to space, because they reside in the same location, and differ just by their cultural identity (ethnic affiliation, language, religion, etc.)—as it is in minority situations, when a given territory is settled not only by one but by several communities.

All the community levels mentioned are also active in place naming. But they can attribute names in the status of endonyms (place names 'from within') only to geographical features at their very own level (scale), since the competence for attributing a place name (for applying the endonym) follows the principle of subsidiarity. It is always the smaller community, the community closer to the feature and responsible for it, who has the primary right to attribute a name. Thus, the name for the Earth is certainly an endonym in all languages spoken on Earth. Names for individual geographical features on Earth are, however, not anymore endonyms in all languages—even if we all feel to be global citizens. The name for a certain country, for example, is certainly an endonym in the language of all communities composing the permanent population of this country, while names in the languages of non-dominant communities for the capital of this country will be endonyms only, if these communities permanently reside in this capital—are a local community there. The Hawaiian name for the United States of America, *'Amelika Hui Pū 'ia*, for instance, is certainly an endonym, since Hawaiians are part of the United States' permanent resident population, but their name *Wakinekona, D.C.* for Washington, D.C., the U.S. capital, is just an exonym, because Hawaiians are not a part of its permanent resident population.

This principle of subsidiarity is also valid within a certain language community (so, not only when communities with different languages are involved). It frequently occurs that a local population calls a village differently from outsiders (speaking the same language). It happens also with other feature categories, for example, rivers. While, for example, the local German-speaking Saxons residing along the upper run of river Mureş in Transylvania [Ardeal] call this river *Mieresch*, the same river is addressed by the also German-speaking Danube Swabians and elsewhere by German-speakers *Marosch*. Thus, *Mieresch* is the German endonym for the upper run of the river, while *Marosch* is a German exonym for it.

Based on this concept, the endonym can be defined as the place name accepted and applied by the local community and the exonym as the place name not applied by the local community.[1]

If we accept these definitions, the endonym/exonym divide corresponds to the divide between space and place in the sense of Yi-Fu Tuan (1977), that is, the divide between (neutral) space and this section of space, to which a certain human community has assumed relations ('place'). "Naming turns space into place", as Bill Watt puts it (Watt 2009: 21), is very much to the point, although place naming is certainly not the only agent in this respect.

2.2 Four Basic Roles of Place Names in Relating People to Geographical Space (Peter Jordan)

Let us proceed now to the roles or functions of place names in relating humans to territory: Place names (can) have four main functions in relating humankind to territory (or communities to geographical space).

[1] These definitions deviate, however, from the definitions formulated by the United Nations Group of Experts on Geographical Names (UNGEGN) as documented in its Glossary of Toponymic Terms (UNGEGN 2007: 3): "*Endonym*: Name of a geographical feature in an official or well-established language occurring in that area where the feature is situated. *Examples*: Vārānasī (not Benares); Aachen (not Aix-la-Chapelle); Krung Thep (not Bangkok); Al-Uqşur (not Luxor). *Exonym*: Name used in a specific language for a geographical feature situated outside the area where that language is widely spoken, and differing in its form from the respective endonym(s) in the area where the geographical feature is situated. *Examples*: Warsaw is the English exonym for Warszawa (Polish); Mailand is German for Milano; Londres is French for London; Kūlūniyā is Arabic for Köln. The officially Romanized endonym Moskva for Москва is not an exonym, nor is the Pinyin form Beijing, while Peking is an exonym. The United Nations recommends minimizing the use of exonyms in international usage."

(1) They often highlight characteristics of space important for a certain community and reflect in this way a human community's perception of space.

They often describe location (*Upper Austria [Oberösterreich]*), morphology (*Rocky Mountains, Great Plains*), hydrology (*Lake District, Suhi dol*—'dry valley'), vegetation (*Dobrava*—'oak forest'), soil (*Steinfeld*—'gravel field') of a certain place or functions of a place within geographical space: bridge function (*Innsbruck*—'bridge across river Inn'), port function (*Newport*), pass function (*Cluj*—'narrowness'). They highlight in this way characteristics that seemed important to the people, who named the place on the background of their culture and their specific interests. Farmers had naming motives different from herdsmen, seafarers different from mountain dwellers. But even cultures less distinct in economic interests like the Slavonic and German settlers of Bohemia [Čechy] preferred different motives in place naming as it has been demonstrated by Sperling (2008). We, the people living today, have naming motives different from our ancestors.

Thus, place names can be called 'condensed narratives' in two directions: They tell us something about spatial reality at the time of naming and they tell us at the same time something about the cultural disposition of the name-giving community. For example, the village of Reifnitz in Carinthia has been named after a brook flowing there into Lake Wörth [Wörthersee]. The German name is derived from Slavonic *Ribnica* 'fish brook', and this name has obviously been given, because this brook is up to the present day known for the lots of salmonid fish (salvelinus) that hike it up for spawning. Obviously, this fact was the striking characteristic of this little river and obviously the name-giving community took notice of it, that is, it was in the scope of their (economic) interests.

For us the meaning of the place name can have lost its transparency, for example, when the name originates from a language spoken earlier and not anymore at a place (as it is with *Reifnitz*; and also with *Feistritz* derived from Slavonic *Bistrica* 'the quickly flowing' or *Graz* from *Gradec* 'the small castle' in parts of Austria, where Slovene is not spoken anymore). The meaning attributed to the feature needs not to have the same importance for us, since our culture and interests are different from the culture and interests of the people that named the feature or circumstances have changed. The name of the town Hranice in Moravia [Morava], for example, with its meaning of 'border' has lost its descriptive function, since there is no border anymore.

But it can be assumed that no name was meaningless for the people, who were the first to apply it. They are for this reason also keys to settlement and cultural history of a certain place. Since new settlers tend to adopt place names from the already resident population they meet in a certain place—as it was in all former colonies—toponymic research is a safe way to document the sequence of population and cultural layers. Gottfried Schramm (1994), for example, proved by toponymic research that—quite in contrast to their own historiography—Albanians were not the first to settle on the territory of modern Albania, but met a Slavonic population layer, when they arrived there in the eighth-century A.D. Eberhard Kranzmayer (1956) traced Carinthia's settlement history in the Early Middle Ages by an analysis of its place names.

Not all place names, however, exert this function. It is specific of descriptive place names, while commemorative place names in the sense of place names after persons, peoples, organizations, institutions, and events—very frequent in urban space, but also in former colonial lands—have been assigned on the ground of other motives.

(2) Place names mark the territory of a community.

Place names in the status of endonyms (as names given and/or applied by the local community) are markers of the community's own territory. This role is exerted by the display of place names in public space, for example, by town signs or plates, or in documents and publications, for example, on maps. Their function is similar to flags, coats of arms, labels, or logos. Since this role has already been discussed in the context of the place-naming process, it is not necessary to elaborate on it here. It should, however, be emphasized that it is this role that gives names in general, but place names in particular, always and inevitably a political, sociological, and juridical dimension and that makes them a potential source of conflict (see Horn 2004; Eller et al. 2008).

Place names mark the territories of all levels and scales of human communities. Already our workplace is usually marked by a name. It is the name of a person (Fig. 2.3), which in this function assumes the status of

Fig. 2.3 Label functioning as a place name at the entrance to an office. (Photo by Peter Jordan 2014)

Fig. 2.4 Westernmost bilingual (German/Slovene) town sign in Carinthia near Hermagor. (Photo by Peter Jordan 2015)

a place name[2] indicating that this is the room, where this individual has more rights (and responsibilities) than others. At the next level of human communities, a town sign marks a populated place (Fig. 2.4). When the

---

[2] The shift of a name between feature categories is termed *transonymy* in onomastics.

Fig. 2.5 Trilingual (Romanian/Hungarian/German) town sign in Romania. (Photo by Peter Jordan 2006)

town sign is bi- or multilingual (Fig. 2.5), it communicates that not just one community, but two or more reside there and share the identity of the place. This only works without conflict, when the dominant community is ready to accept sharing the identity of the place—which is not always the case (Fig. 2.6).

Exonyms have in principle not the role of marking territories, although it is sometimes attributed to them. This attribution is, as already explained, a misunderstanding and may—besides the intention to standardize according to the 'one name for one feature principle'—be a major reason, why the United Nations so far recommended by several resolutions the reduction of exonym use (UNGEGN 2020). It is, however, also true that sometimes exonyms are actually used in a demonstrative way to express territorial claims or remind in a nostalgic way of former 'possessions', for example, when territorial losses of a country are outlined on maps by exonyms (Tátrai and Erőss 2016).

(3) Place names structure space mentally.

Place names help to subdivide complex spatial reality into features. Every geographical feature (in the sense of a subunit of geographical space)

Fig. 2.6 Damaged bilingual (Italian/Resian) town sign in Valle di Resia, Italy. (Photo by Peter Jordan 2008)

is a mental construct. This is especially obvious with landscapes, cultural regions, or macro-regions lacking concrete or clear limits like current administrative boundaries, 'natural boundaries' like mountain ranges or rivers. Up to where Europe extends in the East is obviously just a convention. It is impossible to find clear boundaries of Central or Southeast Europe in reality. A place name is the vehicle, the instrument in this process of mental structuring of space. Without place names we would not be able to establish a system of space-related concepts, to communicate it, to maintain it. In many cases (e.g., cultural regions, landscapes) the place name is in fact the only identifier of a geographical feature.

Dalmatia [Dalmacija] as a cultural landscape is a case in point. It was a historical unit but is no current political entity anymore. Its boundaries are flexible and negotiable, conceived by different people in a different way. The concept of Dalmatia exists just by virtue of the name. But this name marks a concept rich in contents: rocky coast, Venetian-type towns, cold wind bora, tourism. Everybody has an imagination of it. The place name is a tourism brand and repeated by the names of newspapers, hotels and restaurants, ships, and so on.

Another case in point is the concept of Central Europe. It was introduced by August Zeune in 1815 (Zeune 1815) as a chorographic term without political connotations. In the later nineteenth century and toward World War I, however, it assumed various political meanings and was intensively discussed. Also in the interwar period and up to Europe's partition into two antagonistic political blocs it served as a political instrument. Already in the 1980s and after the fall of Communism, it enjoyed a certain renaissance. For practical reasons, the Permanent Committee on Geographical Names [Ständiger Ausschuss für geographische Namen, StAGN], the place-name committee of Germany with a coordinative function for all German-speaking countries and regions, found it necessary to define and outline it in terms of a cultural macroregion (Jordan 2005; StAGN 2015) based on historical factors that had left traces in the cultural landscape, have an influence on human attitudes and behavior up to the present day, are relevant for historical as well as current societal, political, and economic situations, and result in a spatial subdivision that sustains and is not subject to frequent changes. As factors in this sense the following were conceived: (1) synchronic or diachronic existence of Protestantism and Catholicism, while Orthodoxy and Islam play only marginal roles; (2) shaped—as a specific—by German and Jewish culture in addition to Slavonic, Romance, Hungarian, and other cultural layers (present also outside Central Europe); (3) early development of an urban system and an independent urban society in counterbalance to nobility, sovereign, and church compared to East and Southeast Europe; (4) early free farmers independent of feudal landlords; (5) traditions of local and regional self-government as a consequence of early political particularism; (6) politically and economically oriented toward the continent (and not toward overseas); (7) delayed industrialization compared to West Europe, but much earlier than in East and Southeast Europe. The territorial delimitation derived from these factors (Fig. 2.7, StAGN 2015) lacks, of course, empirical evidence and a study based on other factors may arrive at a very different result. The concept has nevertheless been recommended by StAGN, has assumed a kind of quasi-reality, but exists in fact just by virtue of its name.

(4) Place names support emotional ties between people and place and promote in this way space-related identity building.

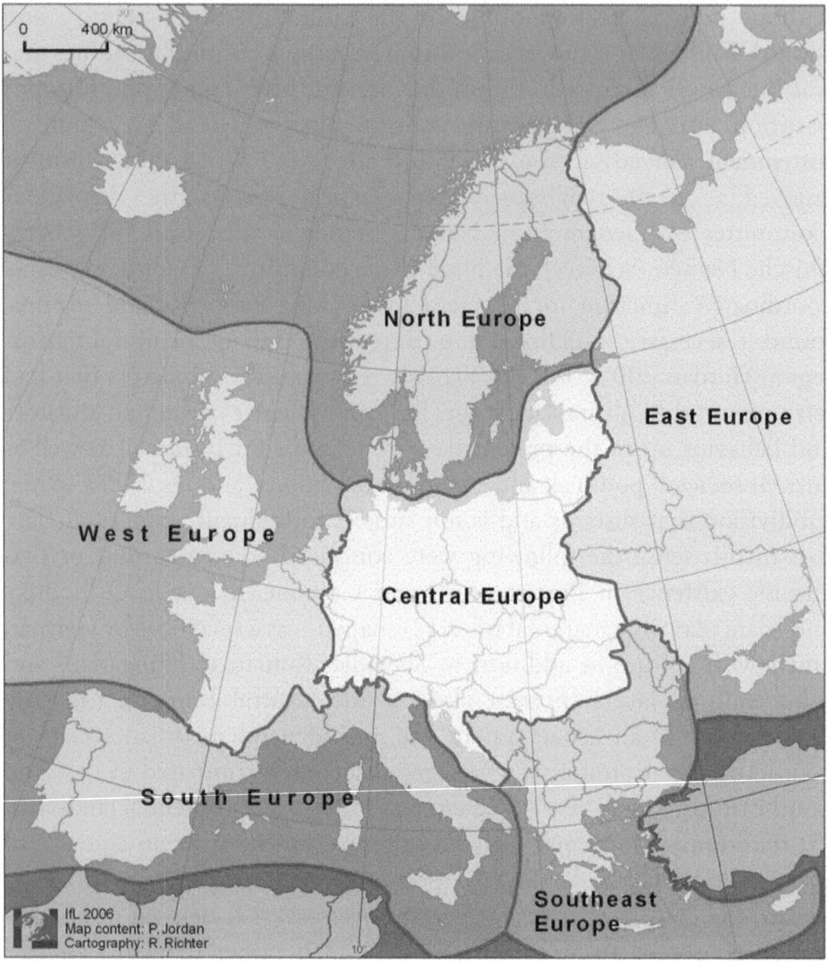

Fig. 2.7 Subdivision of Europe into macro-regions by cultural criteria. (Draft: Peter Jordan, cartography by R. Richter, Leibniz-Institut für Länderkunde)

If somebody acquainted with a place reads, mentions, or memorizes a place name, this recalls all the contents of a space-related concept with him/her, reminds her/him of sights, persons, events, smells, sounds associated with this place, and lets "the feel of a place" arise as Yi-Fu Tuan calls it (Tuan 1977).

Therefore, it is, for example, important to render minority place names on town signs (Jordan 2014). They give these communities the feeling of belonging, of being at home there. It is also a kind of an affirmative action, since non-dominant groups are in special need of being affirmed. For them group identity (including space-related identity as a prominent part of it) means a daily challenge—much more than for a dominant group.

The main task of minority place names on town signs is not information (assuming that somebody could not be able to read the name in the dominant language), but symbolic representation of the minority. By its own place name, the minority is to be affirmed that this is also its place and the minority has a share in its identity.

Therefore, it is also important that the town sign features the endonym as it is used and written by the minority group. A good example is the rendering of Ukrainian minority names in Cyrillic letters on town signs in Romania (Fig. 2.8) as it was introduced by the Romanian place name act of 2001 (Jordan 2005, 2006). Had the Ukrainian name been converted into Roman script, the local Ukrainian community would look at it as alienated and not recognize it anymore as 'theirs'.

Fig. 2.8 Town sign in Romania in two alphabets and scripts: Romanian-Latin, Ukrainian-Cyrillic. (Photo by Peter Jordan 2008)

How important place names are for space-related identity and emotional ties, can also be seen from emigrants, who frequently take the name of their home with them—as a last tie to their former home or to make the new place more familiar. *Nieuw Amsterdam*, the earlier Dutch name for New York is a prominent example. But there are many more, such as in the surroundings of Kitchener, Ontario, Canada, where, for example, an emigrant group from Breslau in former German Silesia, now Wrocław in Polish Silesia [Śląsk], has taken its name with them (Fig. 2.9).

We can conclude that if one looks at place names from a cultural-geographical point of view the (local) community is the essential factor (and only actor) in the place-naming process using place names as mediators between people and territory to highlight characteristics of a place important for this community, to mark its territory and distinguish between 'our own' and 'theirs', to structure space mentally, to support emotional ties to its territory (to turn space into place), in other words: to exert territoriality—an essential aspect of human life. Place names have for this reason always and inevitably sociological, political, and juridical implications. The community closer to the feature owning it or feeling responsible for it has the right on the primary name, the

Fig. 2.9 Town sign nearby Kitchener, Ontario, Canada. (Photo by Peter Jordan 2008)

endonym. The endonym/exonym divide thus reflects the difference between 'our own' and 'theirs', between place and space. This divide is frequently supported by difference in languages but exists also within a given language.

2.3 Linguistic Landscape Discourse Analysis (Přemysl Mácha)

In view of our research goals, we found it essential to include the linguistic landscape (LL) into our analysis. It is, after all, not a general disagreement over minority names but an actual conflict over bilingual signs which stands at the heart of our research. It would be impossible to understand this conflict without paying close attention to the physical appearance and development of the LL, the location and character of bilingual and minority signs, and the perceptions and attitudes associated with them.

In the last two decades, the LL has become a prominent object of study across several disciplines including linguistics, sociology, anthropology, and geography. It is now widely recognized that the LL is a powerful concept which makes it possible to analyze a whole range of interesting topics such as ethnolinguistic vitality, language ideologies, attitudes toward languages, language prestige, social and class relations, ethnic hierarchies, minority communities, economies of language, and public policies. They all become manifest in the ordinary spaces of everyday life whereby they are internalized and sometimes also contested.

During these years, LL studies have undergone significant theoretical and methodological transformations, following to a large extent comparable developments in social sciences and humanities. Namely, there has been a visible movement away from text, language, and discourse to performance, embodiment, and materiality (Barni and Bagna 2015). Also, we have seen a substantial critique of quantitative approaches and a strong push for more qualitative and ethnographic studies of the LL (Shohamy and Waksman 2009). In addition, the LL is no longer viewed as static and passive but rather as dynamic and interactive (Gorter and Cenoz 2015). Instead of *linguistic landscape* more and more authors now write of *linguistic landscaping*, stressing thus the processual, authored, unfinished, chaotic, and emergent character of the LL (Blommaert 2013). Finally, the LL has an implicit historic, diachronic dimension which makes it necessary to pay attention to the development of the LL, both in the past and into the future (ibidem).

It is not our ambition to offer a thorough theoretical treatment of the LL. Rather, the LL for us was one of several potential venues for exploring place-name conflicts in multilingual areas. Our goal was to understand the sources and dynamics of those conflicts, treating thus the LL as both an independent and a dependent variable—independent in the sense how the LL affects group relations and attitudes, and dependent in the sense of how minority politics and other factors shape the form of the LL. We draw in this on Puzey (2012) and Gorter (2013), to name just a few, who pointed out the dialectical nature of the LL:

> Linguistic landscape can both reflect and influence the relative power and status of different languages. (Puzey 2012: 127)

> The linguistic landscape not only reflects the status of different languages in society, but it also acts as a force shaping how languages are being perceived and used by the population. (Gorter 2013: 199)

Our interest in the LL is therefore somewhat limited—we see the LL as a product and an expression of ethnic relations in a given locality. The LL can certainly be analyzed for other purposes (e.g., commercial landscapes, political advertising, graffiti resistance, etc.). For us, however, only those aspects of the LL were of interest, which directly or indirectly reflect or arise from local ethnic relations. To this end we did our literature review and selectively chose those theories and methods which we considered most appropriate and insightful for our own research.

In terms of larger research paradigms, we find particularly useful the works of Fairclough (2003), Scollon and Scollon (2003), and Blommaert (2013). For Fairclough, a discourse cannot be reduced to text, but other, contextual aspects need to be taken into account as well. Specifically, he proposes to study discourses at three interrelated levels—as texts, practices, and events. At the most immediate level, a researcher needs to analyze the text itself, its form, structure, genre, visual appearance, and so on.

At the level of practices, one needs to investigate the processes which led to the production of the text—authorship, strategies, production, distribution, and reception. At the most general level, the researcher needs to approach discourses as instances of interdiscursively and intertextually connected social events closely tied in with larger and wider socio-cultural practices, structures, and ideologies. Only by studying all three levels can we obtain a complex understanding of discourse. The LL is an assemblage of texts on signs and can be considered a set of discourses, at the same time both chaotic and giving a wholesome, gestalt impression (Ben-Rafael 2009). While Fairclough did not develop his critical discourse analysis (CDA) for the LL, we think it can serve as a useful conceptual framework. As Grzech and Dohle (2018: 74) put it: "From the CDA perspective, the main aim of studying linguistic landscape is to uncover why and by whom certain content is presented in the public space, and what the rationale is behind presenting them in a certain way. The manner of presenting information is not limited to the choice of words used to convey it, but also includes the choice of certain languages or linguistic repertoires."

Scollon and Scollon (2003) build on the critical discourse analysis by adding spatial context. They argue that a sign (text) acquires its meaning only when it is placed in a specific socio-spatial context. They call their approach *geosemiotics* which they define as the "social meaning of the material placement of signs" (ibidem: 4). They identify four types of discourses that may be applied to the study of the LL. These are regulatory discourses, infrastructural discourses, commercial discourses, and transgressive discourses. Each may be characterized by a specific style, form, content, location, and process of creation. While the first two are mostly top-down, that is directed from the government to the local population, the last two are mostly bottom-up, that is produced by the local population either for themselves or for the government (e.g., in the case of graffiti). Scollon and Scollon argue that when we approach a sign, in addition to paying attention to the sign itself, we have to ask who the author is, who the presumed reader is, what the social situation is, and whether the location is relevant to this situation. That is, the socio-spatial context is crucial in interpreting the meaning of a sign. However, even specific features of a sign are important in its interpretation—for example, we need to notice the placement of multilingual texts (up vs. bottom, center vs. periphery, left vs. right, earlier vs. later), the use of fonts, colors, types of materials (permanence, newness, quality), layering, changes of state (neons), and pauses and absences. In totality then, the visual character of the sign and the socio-spatial context of its creation and placement combine to give the sign its real meaning.

Blommaert (2013) takes this approach a little further, adding an ethnographic dimension to sign interpretation. He not only is critical of purely quantitative studies, but also argues that even in qualitative studies greater attention should be paid to how signs are produced, how the decisions about their placement are made, what signs tell about community dynamics, and how are signs received by locals and visitors. In accordance with current trends in the social sciences, Blommaert does not see discourse as simple text, for him it is always mediated by social action and material objects. For the appropriate interpretation of the LL, individuals must acquire a certain capacity which he calls *enskillment*, that is knowledge, practice, performance, and bodily discipline necessary for the discourse-in-place to function (ibidem: 34–37). Communication via LL would not be possible without certain internalized regimes of truth, body politics, and fields of power. In consequence, the LL is always bound up with relations of power, produced and reproduced by everyday bodily practices: "communication in the public space…is communication in a field of power" (ibidem: 40). Blommaert therefore proposes to enrich Scollon and Scollon's perspective with additional methods and categories of analysis. First and foremost is ethnography, that is long-term participant observation and interviews with local actors to make clear the links between LL and community dynamics. And second, signs should be differentiated by purpose (permanent vs. event-related vs. noise), function (landmark vs. recruitment vs. inform vs. public statement vs. muted) and quality (shape, size, luxuriousness, visibility, etc.). These three theoretical-methodological approaches provide us with a fairly solid, albeit complex framework for LL analysis. However, translating them into a concrete research methodology presented a number of challenges on which we comment in detail in Chap. 3.

LL is both a qualitative and a quantitative phenomenon and a range of theories and methods can be applied to its analysis. Some researchers have proposed hypotheses related to the quantitative aspects of the LL which informed our research. Specifically, we reflected on the hypotheses proposed by Ben-Rafael (2009):

- Collective identity would be absent, at least in top-down flow, where power relations work against minorities.
- Where there is official policy promoting minority languages, the LL associated with minorities may be more visible in top-down signs than the bottom-up flow.
- Where multiculturalism is taken for granted, minority signs should be equally frequent in the top-down and bottom-up flows.

In addition, we also tested several hypotheses of our own:

- The location of bilingual and minority signs is a simple function of the location of all signs (i.e., where there are more signs, there will be more bilingual and minority signs).
- The location of bilingual and minority signs is associated with symbolic points in the populated place (schools, communal office, settlement boundaries).
- The number of bilingual and minority signs correlates with the number of minority members in the populated place (i.e., ethnolinguistic vitality hypothesis of Landry and Bourhis 1997).

Overall, we found the analysis of place-names conflicts viewed through the prism of the LL as a useful research paradigm which provided a physical reference point for attitudes and opinions voiced in the media, interviews, and questionnaires. The LL under conflict (and our research) was neither abstract nor hypothetical but it took specific local forms informing the observed debates and attitudes. It was therefore essential to obtain a good grasp of its concrete character as well as spatial and temporal dynamics.

2.4 Concluding Remarks (Peter Jordan and Přemysl Mácha)

Two roles of place names in relating people and space are specifically important for linguistic minorities (see Jordan 2004, 2014): the role of marking the territory of a community and the role of supporting emotional ties.

In areas that are inhabited by more than one socio-cultural community, these communities compete for the public, official designation of a geographical feature. With their strive for public recognition of their names both communities want this place to be designated as theirs, wish to relate their identities to it, and express that they feel responsible and accountable for this place (see, e.g., Weichhart 1990). Without conflict between the communities this is only possible if each of them accepts the claim of the other and feels comfortable with a shared or common identity of the place. A conflict—as it has occurred and occurs in many cases—indicates that such mutual acceptance is not (sufficiently) given and that the dominant community is not ready to give in or to share.

Dispute about the rendering of the place name in public space is always an expression of deeper conflict reasons (see, e.g., Horn 2004; Eller et al. 2008). For the non-dominant community it is usually more important than for the dominant to see its relationship to the place recognized by a name in public space right because it is the minority and non-dominant, and because it is not always obvious for the outside world that it is present there. A minority also requires a higher level of self-assurance. Members of a minority face almost daily the challenge to confess identity (see, e.g., Reiterer 2003). When non-dominant communities strive for the public recognition of their place names, they strive—abstractly formulated—for the symbolic function of marking their territory, for the opportunity of demonstrating their presence, but also for the visual support of their emotional attachment to the place. If a member of a non-dominant community reads the place name in his/her own language and script on a town sign or on a map, a sense of familiarity develops.

Since only communities established in a place for generations have developed own place names for the features in their surroundings, they regard the public presentation of their geographical names also as an acknowledgment of their presence for generations and recognition of the fact that their group has contributed to shape culture and cultural landscape. It is for this very reason also a wise decision on the side of the dominant community to grant the non-dominant group this right: It will satisfy the non-dominant group, it will promote its sense for cooperation and its loyalty. Having understood this, many countries grant this right to their minorities (see, e.g., Ormeling 1983).

For members of the non-dominant community the public rendering of minority place names means in the first line that they can regard this place as theirs, that it is (also) the place of *their* group. Information to the outside world is just a secondary function. The geographical name in the minority language should therefore strictly observe the orthography of the minority language, with all the diacritics and special characters. An alienated notation adapted to the pronunciation of the majority language does not satisfy this purpose. If linguistic minorities write their names in a script different from the majority, it is for the same reason also appropriate to use this other script and not to convert it.

With defining the importance of (their own) place names for the linguistic minorities in the focus of our research and the situation of these minorities in general the following items of the linguistic landscape may be regarded as specifically indicative: (1) Is minority legislation related to place names completely effectuated? (2) Are signs in minority language a matter of conflict; and if it is so, which kind of signs are in the focus of this conflict? (3) What is the proportion between mandatory, 'top-down', and voluntary, 'bottom-up' signage in the minority language? (4) Who is the main promotor of voluntary minority-language signage? (5) Are there spatial and socio-spatial variations in the share of minority-language signs and which are their reasons? (6) Is there a temporal development in frequency, share, and status of minority-language signs?

Answers to these questions can shed light on power relations between majority and minority, on status and prestige of the minority language (and the community behind it, since the prestige of a language is closely connected with the prestige of the group using this language), on who is/are the core group(s) of the minority, on which socio-economic factors have a positive or negative influence on the maintenance of the minority language, and on how all these aspects develop over time.

Sources and Literature

Alderman, D. H. (2000). Naming places and commemoration in the American South. *The Professional Geographer,* 52(4), 672–684.

Alderman, D. H., & Inwood, J. (2013). Street naming and the politics of belonging: Spatial injustices in the toponymic commemoration of Martin Luther King Jr. *Social & Cultural Geography,* 14(2), 211–233.

Autengruber, P. (2013). Straßennamen in Wien. Unter besonderer Berücksichtigung von Namen mit geographischem Bezug. *Mitteilungen der Österreichischen Geographischen Gesellschaft,* 155, 263–290.

Azaryahu, M. (1996). The power of commemorative street names. *Environment and Planning D: Society and Space,* 14, 311–330.

Azaryahu, M., & Golan, A. (2001). (Re) naming the landscape: The formation of the Hebrew map of Israel 1949–1960. *Journal of Historical Geography,* 27(2), 178–195.

Back, O. (2002). *Übersetzbare Eigennamen. Eine synchronische Untersuchung von interlingualer Allonymie und Exonymie.* Klagenfurt: Praesens Verlag.

Barni, B., & Bagna, C. (2015). The critical turn in LL: New methodologies and new items in LL. *Linguistic. Landscape,* 1(1–2), 6–18.

Basso, K. H. (1988). Speaking with names: Language and landscape among the Western Apache. *Cultural Anthropology,* III(2), 99–130.

Basso, K. H. (1996). *Wisdom sits in places: Landscape and language among the Western Apache.* Albuquerque: University of New Mexico Press.

Ben-Rafael, E. (2009). A sociological approach to the study of the linguistic landscapes. In E. Shohamy & D. Gorter (Eds.), *Linguistic landscape: Expanding the scenery* (pp. 40–54). New York: Routledge.

Blommaert, J. (2013). *Ethnography, superdiversity and linguistic landscapes: Chronicles of complexity.* Bristol, Buffalo, Toronto: Multilingual Matters.

Bucher, S., et al. (2013). The perception of identity through urban toponyms in the regional cities of Slovakia. *Anthropological Notebooks,* 19(3), 23–40.

Cohen, S. B., & Kliot, N. (1992). Place-names in Israel's ideological struggle over the administered territories. *Annals of the Association of American Geographers,* 82(4), 653–680.

Conedera, M., Vassere, S., Neff, C., Meurer, M., & Krebs, P. (2007). Using toponymy to reconstruct past land use: A case study of 'Brusada' (Burn) in Southern Switzerland. *Journal of Historical Geography,* XXXIII, 729–748.

David, J., & Mácha, P. (2014). *Názvy míst—paměť, identita, dědictví.* Brno: Host.

Eller, N., Hackl, S., & Ľupták, M. (Eds.). (2008). *Namen und ihr Konfliktpotential im europäischen Kontext*. Regensburg: Edition vulpes.
Fairclough, N. (2003). *Analysing discourse: Textual analysis for social research*. London and New York: Routledge.
Gabbert, L. (2007). Distanciation and the recontextualization of space: Finding one's way in a small western community. *Journal of American folklore*, 120(476), 178–203.
Gill, G. (2005). Changing symbols: The renovation of Moscow place names. *The Russian Review*, 64(3), 480–503.
González Faraco, J. C., & Murphy, M. D. (1997). Street names and political regimes in an Andalusian town. *Ethnology*, 36(2), 123–148.
Gorter, D. (2013). Linguistic landscapes in a multilingual world. *Annual Review of Applied Linguistics*, 33, 190–212.
Gorter, D., & Cenoz, J. (2015). Translanguaging and linguistic landscape. *Linguistic Landscape*, 1(1–2), 54–74.
Grzech, K., & Dohle, E. (2018). Language Landscape: An innovative tool for documenting and analysing linguistic landscapes. *Lingue Linguaggi*, 25, 65–80.
Helleland, B. (2009). Place names as means of landscape identity. In P. Jordan, H. Bergmann, C. Cheetham, & I. Hausner (Eds.), *Geographical names as a part of the cultural heritage* (pp. 25–31). Wien: Institut für Geographie und Regionalforschung der Universität Wien, Kartographie und Geoinformation.
Herman, D. (1999). The Aloha State: Place names and the anti-conquest of Hawai'i. *Annals of the Association of American Geographers*, 89(1), 76–102.
Horn, J. (2004). *Ortsnamenkonflikte. Lösungswege für mehrsprachige Gebiete*. St. Augustin: Asgard.
Horsman, S. (2006). The politics of toponyms in the Pamir mountains. *Area*, 38(3), 279–291.
Jett, S. (1997). Place-naming, environment, and perception among the Canyon de Chelly Navajo of Arizona. *Professional Geographer*, 49, 481–493.
Jordan, P. (2000). The importance of using exonyms—pleading for a moderate and politically sensitive use. In: J. Sievers (ed.), *Second International Symposium on Geographical Names "GeoNames 2000" Frankfurt am Main, 28–30 March 2000* (pp. 87–92). Frankfurt am Main: Bundesamt für Kartographie und Geodäsie.
Jordan, P. (2004). Ortsnamen als Kulturgut—Die symbolische Wirkung von Ortsnamen auf Ortstafeln und in Karten. In M. Pandel (Ed.), *Ortstafelkonflikt in Kärnten—Krise oder Chance?* (pp. 216–229). Wien: Braumüller.

Jordan, P. (2005). Großgliederung Europas nach kulturräumlichen Kriterien. *Europa Regional,* 13(4), 162–173.

Jordan, P. (2009). Place names as ingredients of space-related identity. In P. Jordan, H. Bergmann, C. Cheetham, & I. Hausner (Eds.), *Geographical names as a part of the cultural heritage* (pp. 33–39). Wien: Institut für Geographie und Regionalforschung der Universität Wien, Kartographie und Geoinformation.

Jordan, P. (2012). Geographische Namen als Ausdruck menschlicher Raumbindung. *Mitteilungen der Österreichischen Geographischen Gesellschaft,* 154, 67–88.

Jordan, P. (2014). The meaning of bilingual naming in public space for the cultural identity of linguistic minorities. *Review of Historical Geography and Toponomastics,* IX(17–18), 21–24.

Jordan, P. (2015a). The endonym/exonym divide: On the state of our discussion, and a final attempt towards new definitions. In: P. Jordan, & P. Woodman (eds.), *Confirmation of the Definitions. Proceedings of the 16th UNGEGN Working Group on Exonyms Meeting, Hermagor, 5–7 June 2014* (= Name & Place, 4) (pp. 9–17). Hamburg: Verlag Dr. Kovač.

Jordan, P. (2015b). The endonym/exonym divide from a cultural-geographical point of view. In J. Löfström & B. Schnabel-Le Corre (Eds.), *Challenges in synchronic toponymy* (pp. 163–179). Tübingen: Narr Francke Attempto Verlag.

Jordan, P. (2019a). Role of place names in relating people and space. In S. D. Brunn & R. Kehrein (Eds.), *Handbook of the changing world language map* (pp. 2037–2051). Cham: Springer Nature Switzerland AG.

Jordan, P. (2019b). The endonym/exonym divide from a cultural-geographical perspective. *Language and Society,* 10, 5–21.

Jordan, P. & Woodman, P. (Eds.) (2016). *Place-name changes. Proceedings of the Symposion in Rome, 17–18 November 2014* (= Name & Place, 5). Hamburg: Verlag Dr. Kovač.

Jordan, P., Bergmann, H., Cheetham, C., & Hausner, I. (Eds.). (2009). *Geographical names as a part of the cultural heritage.* Wien: Institut für Geographie und Regionalforschung der Universität Wien, Kartographie und Geoinformation.

Kathrein, Y. (2009). Onomastic research within a project on historical mining sites. The Research Program "HiMAT" in Austrian Tyrol. Working Paper 4, presented by the head of the Austrian UNGEGN delegation, Peter Jordan, at the Session of the United Nations Group of Experts on Geographical Names (UNGEGN) in Nairobi 05.-12.05.2009.

Kerfoot, H. (2003). National agencies, models, and procedures. In: F. Ormeling, J. Sievers, & H. Stabe (eds.), *Training course on toponymy* (= Mitteilungen des BKG, 28) (pp. 39–46). Frankfurt am Main: Bundesamt für Kartographie und Geodäsie.

Kranzmayer, E. (1956). *Ortsnamenbuch von Kärnten. I. Teil: Die Siedlungsgeschichte Kärntens von der Urzeit bis zur Gegenwart im Spiegel der Namen*. Klagenfurt: Verlag des Geschichtsvereines für Kärnten.

Kroeber, A. L., & Kluckhohn, C. (1963). *Culture, A critical review of concepts and definitions*. New York: Vintage Books.

Landry, R., & Bourhis, R. Y. (1997). Linguistic landscape and ethnolinguistic vitality: An empirical study. *Journal of Language and Social Psychology*, 16(1), 23–49.

Lévi-Strauss, C. (1946). Natur und Kultur. In W. E. Mühlmann (Ed.), *Kulturanthropologie* (pp. 80–107). Köln, Berlin: Kiepenheuer and Witsch.

Lévi-Strauss, C. (1949). *Les structures élémentaires de la parenté*. Paris: Mouton.

Light, D. (2004). Street names in Bucharest, 1990–1997: Exploring the modern historical geographies of post-socialist change. *Journal of Historical Geography*, 30(1), 154–172.

Mitchell, D. (2000). *Cultural geography. A critical introduction*. Oxford: Blackwell Publishing.

Nash, C. (1999). Irish placenames: Post-colonial locations. *Transactions of the Institute of British Geographers*, 24(4), 457–480.

Oliveira, K. R., & Wahi, A. K. (2009). Place names as vehicles of ancestral memory. *AlterNative*, 5(2), 101–115.

Ormeling, F. (1983). *Minority toponyms on maps. The rendering of linguistic minority toponyms on topographic maps of Western Europe*. Utrecht: University of Utrecht.

Palonen, E. (2008). The city-text in post-communist Budapest: Street names, memorials, and the politics of commemoration. *GeoJournal*, 73(3), 219–230.

Puzey, G. (2012). Two-way traffic: How linguistic landscapes reflect and influence the politics of language. In D. Gorter, H. F. Marten, & L. Van Mensel (Eds.), *Minority languages in the linguistic landscape* (pp. 127–147). Hampshire: Palgrave Macmillan.

Reiterer, A. F. (2003). Sprache ist auch Heimat. *Raum*, 51, 24–26.

Rose-Redwood, R. (2008). "Sixth Avenue is now a memory": Regimes of spatial inscription and the performative limits of the official city-text. *Political Geography*, 27(8), 875–894.

Rose-Redwood, R., Alderman, D., & Azaryahu, M. (2010). Geographies of toponymic inscription: New directions in critical place-name studies. *Progress in Human Geography*, 34(4), 453–470.

Šakaja, L., & Stanić, J. (2011). Other(ing), self(portraying), negotiating: The spatial codification of values in Zagreb's city-text. *Cultural Geographies*, 18(4), 495–516.

Schramm, G. (1994). *Anfänge des albanischen Christentums. Die frühe Bekehrung der Bessen und ihre langen Folgen*. Freiburg im Breisgau: Rombach.

Scollon, R., & Scollon, S. W. (2003). *Discourses in place. Language in the material world*. London, New York: Routledge.

Shohamy, E., & Waksman, S. (2009). Linguistic landscape as an ecological arena: Modalities, meanings, negotiations, education. In E. Shohamy & D. Gorter (Eds.), *Linguistic landscape: Expanding the scenery* (pp. 313–331). New York: Routledge.

Sperling, W. (2008). *Bäume und Wald in den geographischen Namen Mitteleuropas: Die böhmischen Länder*. Leipzig: Leipziger Universitätsverlag.

Ständiger Ausschuss für Geographische Namen (StAGN) (ed.). (2015). *Empfehlung zur Großgliederung Europas*. Retrieved December 30, 2015, from http://www.stagn.de/drupal/Publikationen%20und%20Downloads

Tátrai, P., & Erőss, Á. (2016). Toponymic silence in contested geographical spaces: The neglect of place-name changes in Cyprus and in Central and Eastern Europe. In: P. Jordan, & P. Woodman, (eds.), *Place-Name Changes. Proceedings of the Symposion in Rome, 17–18 November 2014* (pp. 385–401). Hamburg: Verlag Dr. Kovač.

Taylor, S. (Ed.). (1998). *The uses of place names*. Edinburgh: Scottish Cultural Press.

Thornton, T. (1997a). Know your place: The organization of Tlingit geographic knowledge. *Ethnology*, 36, 295–307.

Thornton, T. (1997b). Anthropological studies of Native American place naming. *American Indian Quarterly*, 21(2), 209–228.

Tuan, Y.-F. (1977). *Space and place: The perspective of experience*. Minneapolis: University of Minnesota Press.

Tuan, Y-F. (1990 [1974]). *Topophilia. A study of environmental perception, attitudes, and values*. New York: Columbia University Press.

Tuan, Y.-F. (1991). Language and the making of place: A narrative-descriptive approach. *Annals of the Association of American Geographers, LXXXI*, 684–696.

United Nations Group of Experts on Geographical Names (UNGEGN) (ed.) (2007). Glossary of Terms for the Standardization of Geographical Names. New York, United Nations, ST/ESA/STAT/SER.M/85Addendum.unstats.un.org/unsd/geoinfo/UNGEGN/docs/pdf/Glossary_of_terms_revised.pdf (accessed 12 March 2020).

United Nations Group of Experts on Geographical Names (UNGEGN) (2020). https://unstats.un.org/unsd/ungegn/ (accessed 12 March 2020).

Watt, B. (2009). Cultural aspects of place names with special regard to names in indigenous, minority and regional languages. In P. Jordan, H. Bergmann, C. Cheetham, & I. Hausner (Eds.), *Geographical names as a part of the cultural heritage* (pp. 21–24). Wien: Institut für Geographie und Regionalforschung der Universität Wien, Kartographie und Geoinformation.

Weichhart, P. (1990). Raumbezogene Identität. Bausteine zu einer Theorie räumlich-sozialer Kognition und Identifikation (= Erdkundliches Wissen, 102). Stuttgart: Steiner.

Weichhart, P. (1999). Die Räume zwischen den Welten und die Welt der Räume. In Meusburger, P. (Ed.), *Handlungszentrierte Sozialgeographie. Benno Werlens Entwurf in kritischer Diskussion* (= Erdkundliches Wissen, 130). Stuttgart: Franz Steiner Verlag.

Withers, C. W. J. (2000). Authorizing landscape: 'authority', naming and the Ordnance Survey's mapping of the Scottish Highlands in the nineteenth century. *Journal of Historical Geography,* 26(4), 532–554.

Woodman, P. (Ed.). (2012). *The great toponymic divide. Reflections on the definition and usage of endonyms and exonyms.* Warszawa: Head Office of Geodesy and Cartography.

Yeoh, B. S. A. (1996). Street-naming and nation-building: Toponymic inscriptions of nationhood in Singapore. *Area,* 28, 298–307.

Zeune, A. (1815). *Erdansichten oder Abriss einer Geschichte der Erdkunde.* München: Maurer.

3

The Challenges of Studying Place-Name Politics in Multilingual Areas

Contributed by Přemysl Mácha, Uršula Obrusník, Peter Jordan, and Alexis Sancho Reinoso

In this chapter we would like to offer a brief description and a critical reflection of our research approach. The international and multidisciplinary nature of our project, together with the politically sensitive character of the research topic and the geographical setting of the research in a complex border situation, generated numerous challenges. Most were addressed successfully, but certainly not all, and many doubts (and lessons) remain. In addition, some issues could not be addressed at all and simply had to be reckoned with as a given. Nevertheless, in spite of the inevitable limitations, we believe we achieved a balanced mix of theoretically grounded methods which helped us to navigate the perils of studying place-name politics in multilingual areas.

In order to address our principal research questions, we chose to use a wide range of anthropological and geographical methods and sources. By triangulation (see, e.g., Rothbauer 2008), we wanted to make sure that the data gathered from various sources can support and complement each other; but we also wanted to see how different factors and processes, both historical and present, influence and determine each other. Methods

P. Jordan et al., *Place-Name Politics in Multilingual Areas*, Palgrave Studies in Minority Languages and Communities, https://doi.org/10.1007/978-3-030-69488-3_3

chosen by the team were both qualitative and quantitative, synchronic and diachronic, and single-case study as well as comparative. In the following sections we will deal with each of the principal methods separately, always giving a summary of how the method was used, which previous investigations informed its design, what challenges we encountered, and how we tried to resolve them. Although we strived for direct comparability between the Těšín/Cieszyn region and southern Carinthia, this was not always possible due to local idiosyncrasies. Therefore, where relevant, divergence in methodologies is also mentioned. The summary is by no means exhaustive, but it does provide a sufficiently detailed context for the critical interpretation of our data which we present in Chaps. 6 and 7.

3.1 Archival Research and Map Analysis

Our archival research consisted of three types of analyses that we conducted in order to understand the normative cartographic context in which the struggle for the official recognition of minority names in official documents and in the public space takes place. Maps form an important component of the public discourse, and because of their authority and disciplinary character, they significantly shape people's understanding of toponymy in relation to their nationality and residence. Therefore, we adopted a critical approach to map analysis, following Harley (1988), Pickles (1992), and Crampton (2001). We approached maps as problematic texts with agendas and messages, communicating specific political visions and ideologies. The maps we studied, for example, clearly expressed implicit territorial claims just as Cohen and Kliot (1992) found for Israel.

The first method we selected for our project was archival research focused on the toponymic changes in the two regions. Our interest was in the ways in which minority names were represented on official maps throughout the centuries. We analyzed the linguistic forms of place names indicated on available large-scale maps of the Těšín/Cieszyn region and southern Carinthia ranging from the earliest regional maps through the Austrian land surveys to the most recent state and tourist maps. In all these maps, we noted the use of minority names, the types of features with minority names, the linguistic variants recorded, and the forms of names transcription. Since both regions were part of the Habsburg dominion from the 1500s until World War I, comparable maps could be found. Yet, in spite of this shared past, significant differences in the representation of minority names between the Těšín/Cieszyn region and southern Carinthia were found. The same, of course, holds true for the post-World War I era when Czechoslovakia and Austria embarked on separate routes.

In addition to this general comparative analysis, we carried out a case study of the street-name changes in the town of Teschen [Cieszyn] from the 1800s to the present. We studied how Austrian, Czechoslovak, Polish, German, and Silesian nationalism influenced the city text in competition with other political ideologies and different scales of territorial identification. In addition to old town maps we also analyzed municipal records and communications which contained a wealth of information on the character and motivation of different actors who intervened in the design and implementation of street-name changes from the Austrian period to the present. Since we published the results of this analysis elsewhere (see Mácha et al. 2018), we will not comment on them in detail in this book. However, our analysis uncovered historic layers in the city text and gave us points to be pursued during interviews, such as street names that were used in the past but are now invisible in the public space. We will comment on those.

The third component to the archival research was the analysis of selected historic and current local maps published by individual villages for their inhabitants and/or tourists. Most villages have no official street names, but they do have other kinds of names (field names, names of hills, streams, valleys, village parts, etc.). In Czechia communal assemblies, in Austria communal councils have to decide what should be represented and in what form. The decisions made by the communal assembly/council were not always well received by the local population and, as several interviews indicated, in some instances these local maps became an important component of people's ideas about bilingualism in the area.

In terms of methodology, we followed Jordan (2018), who carried out a similar analysis for a more limited sample of maps in the context of Austria. In the archival research itself, the greatest challenge was to find contextual sources which would help to interpret critically the otherwise generally easily available digitalized maps. Ironically, in the case of the town of Teschen [Cieszyn], significantly more documents were available for the older periods than for the more recent, post-World War II ones. The analyzed documents were in Slovene, German, Polish, and Czech. Where the archival context was missing as in the case of recent local maps, we tried to complement it by conducting interviews with public representatives and activists responsible for the printing of these maps.

Another challenge specific to the Těšín/Cieszyn region was the linguistic proximity of Polish, Czech, and the dialect. Some names are written identically in all three languages, and in such cases, there was no way to decide from the name itself in which language it was written. Only the context of the entire map could help to resolve this issue, but not always, since many maps contain a mixture of names in different languages. In addition, most pre-World War I maps use German orthography which means that most names are rendered either in German or more commonly in a corrupt, hybrid form making it impossible to judge the name's original language. This is further aggravated by the fact that Czech and Polish standard orthography took a long time to develop and even longer to enter daily use. In consequence, the use of what today would be seen as Czech or Polish orthography may not necessarily indicate a Czech or Polish origin of the name but simply an orthographic preference of the map maker. Overall, this proved to be a particularly intractable issue, and our interpretation of the representation of minority names on maps thus rests on arguably not very firm ground. Fortunately, the situation in southern Carinthia was different, and our analysis of Carinthian maps could compensate some of the downfalls of the Těšín/Cieszyn maps.

Also Carinthia was throughout history represented by a large variety of maps, mostly of private editors and publishers starting with the famous Atlas of Austria Interior [Atlas von Innerösterreich] by J.K. Kindermann published between 1789 and 1797 (Dörflinger 1984: 110f) and ending up with maps of the Austrian Alpine Club [Österreichischer Alpenverein], tourist maps, road maps, and maps of various institutes and associations of our days. We decided, however, not to include into our analysis private maps, since they pursue varying interests and purposes and apply in consequence various representation modes also related to place names. It would have needed extended efforts to analyze their rendering of minority place names in relation to their editorial intentions and purposes. Therefore, the Carinthian treatise confines itself to the use of minority place names on official Austrian topographical maps and related gazetteers. Their intention, purpose, and user group remained in principle the same over time, although with them, too, intention, purpose, and user group underwent some change, for example, from purely military purposes to the purpose of public information. However, much better than maps of private publishers they allow—ceteris paribus—to examine how changes of the legal status of the minority, their general embedding into the wider political community, and their recognition as a cultural and identity group express themselves in the rendering of minority place names or—looking at it the other way round—what the mode of rendering minority place names tells us about the legal, political, cultural, and societal status of the minority in the period in question.

3.2 Analyzing the Linguistic Landscape

As a counterweight to the representation of place names on maps we also analyzed their presence in the linguistic landscape (LL). Members of the research team traveled around the case study areas and documented existing bilingual and minority signs to understand their spatial distribution and character. Some settlements had many signs, others had few and still others had none at all, in spite of a strong minority presence. This helped us to identify key questions to be pursued in interviews. Almost 3000 signs were documented in southern Carinthia and approximately 1500 in the Těšín/Cieszyn region. Due to the different size, configuration, and structure of the two research areas, however, in southern Carinthia only two places were the target of an (almost) complete documentation (documentation of the totality of signs), and for the rest of the area only a selection of typical cases was documented, while in the Těšín/Cieszyn region four communes were documented completely. Thus, while it was by no means an exhaustive survey, it did provide important information about the physical appearance of bilingual signs in the case study areas which then served as a background and a context for the interpretation of interviews and other data.

Our analysis mostly focused on the current situation but, where it was possible, we also looked for historical photographs and postcards capturing parts of the local linguistic landscape in different periods of time to see to what extent what was represented on maps also appeared in the real world. Understandably, road signs, street signs, or other types of signs in the public space were not of interest to photographers of the past. However, in the background of some photos we could see some signs and, thus, put them into the context of other historical textual sources as well as memories of people we interviewed. Although the visual record of the historic LL was extremely fragmented and could only be used illustratively, it did provide useful insights into the character of the LL, the degree to which names represented on maps corresponded with their appearance on signs, and the occasional physical resistance of the LL to place-name changes implemented from above.

In addition, all signs (monolingual as well as multilingual) in the communes Bystřice/Bystrzyca, Hnojník/Gnojnik, Chotěbuz/Kocobęz, and Jablunkov/Jabłonków in the Těšín/Cieszyn region and in the town Ferlach (Borovlje) and the commune Zell (Sele) in southern Carinthia were documented to put the minority and bilingual signs into the larger context of the entire linguistic landscape. This survey yielded several thousand signs. All signs were categorized and spatially located with the help of the mobile app Collector for ArcGIS connected with a web-based geodatabase (ArcGIS Online). In this we followed the work of Barni and Bagna (2009). We documented official street and road signs as well as commercial and unofficial signs, such as posters and community announcements. The number, location, and visual aspects of the signs were of particular interest, in their quantitative as well as qualitative character (see Table 3.1 for details).

Finally, we carried out an in-depth study of selected signs to understand their particular social history and spatial context. This research was based primarily on interviews with sign creators, public officials, vandals,

Table 3.1 Attribute categories for linguistic landscape analysis

Attribute type	Values
Object ID	Unique numeric value
Location	GPS coordinates, cadastre commune
Sign type	Town sign, street sign, road sign, tourist sign, information table, welcome sign, shop sign, company sign, company billboard, communal office sign, school or kindergarten sign, sign of another public institution, train-stop sign, bus-stop sign, culture and sport sign, restaurant, inn, or hotel sign, text on a sacral building, grave inscription, monument inscription, poster, information board, orientation sign, landscape-feature sign, other signs
Ownership	Public (state, regional, or local government, religious communities, associations, NGOs)
	Private (individuals and companies)
Voluntarity	Mandatory (sign had to be introduced because of legislation)
	Voluntary (sign was introduced irrespective of legislation)
Languages used	Majority language only
	Minority language only
	Both majority and minority language
	Multilingual (includes other languages in addition to Czech/Polish or Slovene/German)
	Other (e.g., English only, dialect only, languages of migrant workers, etc.)
Order of languages	Majority language first
	Minority language first
	Other (e.g., English first, etc.)
Size	Small (up to 40 × 60 cm)
	Medium (between 40 × 60 cm and 100 × 200 cm)
	Large (from 100 × 200 cm)
Visibility	Small (either located on a side street or obscured by obstacles)
	Medium (normal visibility on a regular street)
	Great (exceptional visibility due to size or location)
Vandalization	Yes (disappearance, physical damage, graffiti)
	No

and local inhabitants, supplemented by visual documentation and the reception of signs in the local press. The results of the individual analyses helped us to identify the key factors influencing location and interpretation of signs in space which seem to be their visibility, symbolism, and role in social boundary marking.

Our primary sources of methodological inspiration were Fairclough (2003) on critical discourse analysis in general and Scollon and Scollon (2003) and Blommaert (2013) on critical LL analysis in particular. We adopted the combined quantitative-qualitative approach promoted by Blackwood (2015). However, while documenting and mapping the linguistic landscape we encountered several problems which are well-known in LL studies. As Gorter (2006) points out, these dilemmas include, for example, drawing a sample (Where do we select signs, how many do we need, and which ones do we choose?), defining the unit of analysis (What is a sign, do we count only permanent, static signs, or do we also include temporary and mobile signs, and do we count a shop window as one sign or do we count each text on a shop window as a separate sign?), and categorization of signs (What qualities of the sign do we follow?). In terms of sampling, in addition to documenting bilingual and minority signs in the region, we also recorded all signs in selected settlements. This was rather novel as most other researchers limited their sample to a single street or a shopping mall (e.g., Trumper-Hecht 2009) or a neighborhood only (e.g., Gorter and Cenoz 2015). Of course, this approach only moves the sampling problem one level up, but if the settlement is well chosen or a random sample of settlements is drawn, the sampling risks are diminished while a significantly more representative and complex picture of the LL emerges.

In terms of the unit of analysis we followed Cenoz and Gorter (2006) and classified a group of signs on a shop window all related to the same shop as one sign rather than several individual signs. Shop signs or temporary signs, such as posters, were considered relevant but signs on vehicles were considered what Blommaert (2013) referred to as 'noise'. However, we did account in our classification for 'toponymic warfare' (Kadmon 2004), that is, cases when the sign was vandalized. In terms of content, we mostly followed Blommaert (2013) to categorize and analyze signs by location, ownership, degree of voluntarity, size, visibility/prominence, languages used, order of languages, and vandalization (see Table 3.1 for details).

For the qualitative analysis (possible on a smaller sample) we also studied the visual quality, fonts, colors, materials, and the spatial and social contexts:

- the sign itself—material, color, size, quality, permanence, function (according to Blommaert 2013), discourse type (according to Scollon and Scollon 2003), text, fonts, order of languages, visual organization of the individual aspects of the sign, nominal message;
- the immediate socio-spatial context of the sign—location, visibility, visual competition, authorship, motivation, presumed audience, reception and perception (from interviews, questionnaires, media), history, local ethnic relations, cooperation/competition/dialog with other nearby signs, performative character of the sign, vandalization;
- larger political-legal, ethnic, and economic context (interdiscursivity, intertextuality)—connection between the sign and dominant discourses of identity, nation and the state, economic discourses (tourism), discourses of home.

Finally, probably the greatest challenge was to account for the dynamic character of the LL. As repeated visits to the same places clearly showed, the LL is in a continuous state of flux. Both official and unofficial signs might disappear and reappear. Cases of vandalism or disappearance of bilingual signs are not as common as they were ten years ago, but still occur. As a result, in LL analysis the question of 'when' is as important as the question of 'where'. Unfortunately, the temporal dimension of the LL is not yet sufficiently accounted for in literature and more work has to be done to better appreciate the extent and character of the constant changes. In our estimate, between 10% to 20% of signs as we defined them for the purpose of our research changed during any given year. This fluctuation does not concern only commercial billboards, posters, and community announcements which often are temporary by their nature. Shops and restaurants open and close, road signs appear and disappear (due to changes in traffic organization, vandalization, traffic accidents, new housing development, etc.) and buildings are constructed and torn down. A peculiar class of signs is what we might call 'zombie' signs—that is, signs referring to objects which no longer exist or function (i.e., shops that have closed, buildings which were demolished, services which moved elsewhere, etc.). According to our estimate, some 10–15% of signs in a locality may be of this type. The actual LL is thus in a constant state of flux and it is always a little behind its current functional and social character. Interesting questions then arise: Are some signs more prone to replacement or change than others? Which signs survive the longest? Is there a difference in permanence between majority and minority signs? While we will not be able to give definite answers to these questions yet, we strongly encourage researchers to investigate them. The temporal limitations for us, however, meant that we had to set a definite timeframe for data collection with the painful awareness that the LL we analyzed and will report on no longer exists today.

3.3 Media Analysis

In order to get a complete picture of the public discourse on minority names and bilingual signs, we were also interested in how the topic was represented in the media at both regional and national levels. Media form a fundamental component and shaping factor of the public sphere, and although the actual influence of 'old' and 'new' media on public opinion is still debated, there is a fairly strong evidence that the power of the media is growing with all of its corresponding positive and negative aspects (see, e.g., Fuchs 2014; Galpin and Trenz 2017). Media, just as the LL, shape the socio-spatial context for the formation of opinions, including attitudes toward bilingual signs and minority rights. While we do not overestimate the power of the media, we do believe they both reflect and influence public debates and agendas (see, e.g., Dołowy-Rybińska 2017).

In our analysis we focused on printed and online newspapers and journals, radio and TV broadcasting, and a sample of internet debates. We covered a vast range of sources from local to national, from broadsheets to tabloids. In the case of the Těšín/Cieszyn region, we searched the Anopress database which contains the vast majority of Czech newspaper, journal, radio, and TV outputs for the past three decades. Keyword search made it relatively straightforward to identify relevant articles. However, some important media outlets were not included in the database, namely the two most important Polish language sources—the semi-daily *Głos* and the monthly *Zwrot*—as well as two regional newspapers—*Třinecký Hutník* and *Horizont*—which are widely read and followed by the local population in the Těšín/Cieszyn region. For a lack of electronic database, these four sources had to be manually searched in the library which was one of the main practical challenges of the media analysis. We did not draw a sample but simply worked with the entire population since the numbers were manageable. The period covered in our analysis was between 2000 and 2018, and the search yielded 323 Czech-language media outputs and 237 Polish-language media outputs.

In Carinthia, a comparable approach was taken, although the available online digital database maintained by the Austria Press Agency [Austria Presse Agentur, APA] contained only textual media. A keyword search (in German and Slovene) using the terms *Carinthia* and *town sign* was conducted. The resulting articles were selected using stratified random sampling so that the number of analyzed articles was proportional to the years in which they appeared. In addition, we selected some relevant articles from a private press archive devoted to this topic. Local, regional, national, and international media are represented in our sample. The period covered was between 2000 and 2017 and the search yielded 199 media outputs altogether (87 in German language and 112 in Slovene).

In the case of the Těšín/Cieszyn region only, we also analyzed selected internet debates from the same period (2000–2018) as a counterpoint to the official media discourse. We used a regular search engine to find them and unfortunately have no way of estimating their representativeness. We are well aware of the perils of internet debate analyses, although as Albrecht (2006) found out, internet debate participants are not necessarily radically different from public debate participants in the 'real' world. We entirely excluded social networks for a lack of access and personnel. Social-network debates on bilingual signs (or place names more generally) is a topic for future researchers. However, since most of the analyzed debates took place under standard online articles published by national media outlets, they did provide a very important comparison in terms of their content, argumentation, and prevalent position on bilingual signs—that is, they offered us insights into the reception of media outputs by their consumers.

All of the recorded media and discussion items were coded for a set of predefined criteria (e.g., date, media type, media name, authorship, reach, topic, length, position on bilingual signs, voiced arguments—see Table 3.2 for details) and submitted for a content analysis (see, e.g.,

Table 3.2 List of argument types identified in the media and internet debates

Argument type	Typical statements
Pragmatic	
plan (language)	Against: They speak German/Czech anyway.
	For: It will strengthen the Slovene/Polish language.
pnum (number)	Against: There are few Slovenes/Poles here.
	For: There are many Slovenes/Poles here.
pmon (money)	Against: It will cost a lot of money.
	For: It will not cost us anything, the state is going to pay for it.
pout (outside)	Against: It should not be decided from above.
	For: The EU/national government requires it.
psig (signs)	Against: There are no Czech/German signs in Poland/ Slovenia either.
	For: Minorities in other countries also have signs.
poth (other)	Other pragmatic arguments.
Symbolic	
sinc (inclusion)	Against: Every other minority should have their own signs, as well, then.
	For: It is a way of including Slovenes/Poles into the society.
sasi (assimilation)	Against: We are Czechs/Austrians and they have to adapt.
	For: The Slovenes/Poles have to defend themselves from Germanization/Czechization.
srep (representation)	Against: It is just a provocation and increasing tensions.
	For: It is a way of recognizing the existence of the minority.
shis (history)	Against: This is a historically Czech/German-speaking region.
	For: Poles/Slovenes have always lived here.
spow (power)	Against: Let us move beyond partisan politics.
	For: Bilingual signs are a symbol of the cultural diversity of the region.
soth (other)	Other symbolic arguments.

Weber 1990; Schulz et al. 2004). We used Szabó-Gilinger et al. (2012) to establish basic argument categories (i.e., pragmatic/symbolic); however, trial categorization of the data revealed the need for more fine-tuned argument categories, both for and against bilingual signs. We used 16 principal categories (see Table 3.3) and compared their importance by media type, language, and other factors. Certain relationships and observed differences were tested for statistical significance using IBM

Table 3.3 Categories followed in the media analysis

Category	Values
Item ID	Unique numeric values
Date	Exact date (day, month, year) of publication
Media type	Printed newspaper, printed regional newspaper, printed journal, online newspaper, radio, TV, internet debate
Media name	Official name
Title	Name of article/report
Author	Name of author
Length	Number of characters for texts, length of broadcast (in minutes) for radio and TV
Page	For printed media only
News category	National, regional, economy, politics, culture, sports, other
Order of news	Order of appearance, for broadcasted news only
Language	Czech, Polish, German, Slovene
Topic	Politics, economy, culture, sports, tourism, other
Sub-topic	Introduction of signs, vandalization
Importance	Importance of the topic of bilingual signs in the media item—main, secondary
Overall position	For, against (bilingual signs)
Arguments used	See Table 3.2 for details
Sample arguments	Verbatim citation of arguments voiced in the article/report
Brief summary	Brief summary of the content of the article/report
Full text	Full text of the article and radio/TV report (where available)

SPSS Statistics. We also performed critical discourse analysis (Fairclough 2003) of the coverage of selected events in order to understand how the discourse on bilingual signs relates to other discourses and the overarching social context. It is important to note that the media from the other side of the border, that is Poland and Slovenia, were not included in our media analysis.

3.4 Interviews

After gaining an overall understanding of the larger historical, social, political, and media context of our research areas, including the physical form of the bilingual linguistic landscape, we began interviewing local people in order to see how this context reflected on their perceptions, attitudes, and behavior in relation to bilingual signs and place-naming practices. We were interested in personal stories connected to the local linguistic landscape and the usage of place names in daily life, all in the context of their personal life stories. Between 2016 and 2018 we conducted semi-structured in-depth interviews with 105 people of Czech and Polish nationality living in urban and rural areas in the Těšín/Cieszyn region and with 132 people with majority or minority background living in southern Carinthia (see Tables 6.6 and 6.24). In selecting people, we strove to have different subgroups of the population roughly proportionately represented (age, gender, residence, education, ethnicity). Snowball and opportunistic sampling proved most useful in this respect. In spite of our efforts, however, our sample overrepresented university-educated and minority members in comparison with their percentage in the general population.

Due to the politically sensitive nature of the topic, requests for interviews were sometimes rejected. The rejection rate was especially high among German-Carinthians[1] which is a clear indication of the degree to which this topic has been politicized even at the everyday level. Nevertheless, we succeeded to interview a sufficient number of people representing the different positions in the debate, including several individuals directly involved in the vandalization of bilingual signs. In addition to 'ordinary' inhabitants of the region, we also interviewed political representatives of the minority and majority (mayors, directors of political and cultural organizations, activists). These 'elite' interviews presented specific challenges which had to be addressed in addition to regular issues accompanying all interviews (Mikecz 2012). Finally, in the Těšín/Cieszyn region, we talked to a limited number of people from the other side of the border, namely from Cieszyn, on their perception of the bilingual signs in Český Těšín.

[1] The term *German-Carinthians* is in this book just used as a substitute for the majority population in Carinthia, when it is unavoidable to address the other cultural component of Carinthia's population besides Carinthian Slovenes. Hardly any member of this component would use it as a self-designation (see Chap. 6, Sects. 6.1.4 and 6.1.5). It seems, however, preferable to *German-speaking Carinthians*, since near to all Carinthian Slovenes have also a command of German, many of them regard German as their mother tongue and language is just one of many potential identity markers of a cultural group.

Interviews usually took place either in the respondents' homes or in public cafes. We encouraged the respondent to choose the preferred space, so that the interview occurs in a familiar, comfortable setting. At the beginning of each interview, the interviewer introduced him/herself and the research project briefly in order to provide the respondent with the context of the interview. If the respondent had any questions, they were answered. The interview itself was semi-structured and drawing on the traditions of ethnographic interviewing (Spradley 1979). We had a list of questions/topics we wanted to cover during the interview (see Appendix 4), but the order in which they were addressed differed from interview to interview because we wanted to give the respondent freedom to express his/her ideas and draw links between different issues. When the interview became too side-tracked, the interviewer led it back to topics of interest.

With the permission of the respondent we recorded the vast majority of the interviews which were later transcribed to allow for an in-depth analysis. In those few cases where recording was not available, detailed notes were taken. Interviews lasted anywhere between 20 minutes and 2 hours. The vast majority of questions used in the Těšín/Cieszyn region and southern Carinthia were identical; however, some diverged due to differences in the local legislative, political, and cultural context. In several instances, we also carried out walking interviews (see, e.g., Evans and Jones 2011); however, these proved rather difficult to record and they tended to get significantly more off-track as they progressed, so we did not pursue them further.

The transcribed interviews were then read, hand-coded, and analyzed using the critical discourse analysis approach (Fairclough 2003). Because interviews were analyzed by several researchers, coding reliability also had to be taken into account (Campbell et al. 2013). Qualitative analysis of our interviews was based on identifying links with theory on the one hand and the everyday reality lived by the respondent on the other. More importantly, interviews helped us understand better the intimate and emotional role place names play in home- and place-making and in mediating relations between different social groups. They also provided a good foundation for the formulation of hypotheses and questions for the questionnaire survey (see Chap. 3, Sect. 3.5) with which we completed our research.

The principal challenge associated with interviewing (and, to a certain extent, also with some of the other methods) was that we worked in a border situation which is inherently political (Johnson et al. 2011). The memories of historical grievances are still alive and continue to shape public debates about current issues as well as the presentation and interpretation of historical facts. In addition, the topic of our research was itself highly politically charged and people generally held rather strong opinions. As a consequence, 'neutrality' was a principle that was extremely difficult to maintain as our respondents constantly sought to bring us over to 'their' side. Similar situations have been noted elsewhere in the context of research interviews on the border (see, e.g., Galasińska 2006). We had to be very careful to phrase our questions in a way that avoided signaling our political stance, emphasizing open-ended explorations of the topic. Where possible, we tried to benefit from the position of 'outsiders' (see, e.g., Balogh 2013). This was possible in southern Carinthia, since—as we explain below—the interviewers were originally from other countries than Austria and Slovenia. In the case of Těšín/Cieszyn, it was more difficult to achieve, since all of the members of the Czech team were originally from the case study area. On the other hand, these ties often opened doors which would have otherwise remained closed, so balancing our role as outsiders and insiders was a delicate and strategic endeavor. To a certain extent, the Czech team engaged in what has been referred to as autoethnography (Anderson 2006).

The best illustration of the way politics challenged our research, however, was the choice of language of the interview. It is well-documented that the choice of language influences responses (see, e.g., Lee and Pérez 2014). The local population has various preferences of usage of Czech, Polish, and the local dialect (po naszymu) in the Těšín/Cieszyn region and the Carinthian-Slovene dialects, Carinthian-German dialects, standard Slovene, and standard German in Carinthia. To a significant extent, the language preference correlates with residence and ethnicity and, thus, the choice of language with which a respondent was approached could be read as signaling group membership (local vs. outsider, Czech- vs. Polish-, Slovene- vs. German-speaker) and a presumed associated political position. This could introduce a systematic (ethnic) bias (see, e.g., Adida et al. 2016) which we had to critically reflect.

In the Těšín/Cieszyn region, it was particularly difficult to navigate the perils of 'incorrect' language choice. While members of the Polish minority are usually fluent in both Czech and Polish, for the vast majority of them the local dialect is the real native language. We assumed that a Polish minority speaker might react differently to a researcher approaching them in Polish than to a researcher approaching them in Czech or in the dialect. In hopes of making our respondents as comfortable as possible, we decided to approach them in the dialect and Polish in cases where we knew that was their preferred language. In cases where we had no indication of a preferred language, we approached them in Czech and clearly stated the possibility to switch to Polish or the dialect at any point (which often happened). On the other hand, having a large sample of interviews in three languages and mixtures of them, we have been able to compare responses in relation to the preferred language and we did not find significant or systematic differences in respondents' position on bilingual signs. The choice of language, however, certainly did affect the choice of place-name variants that were mentioned during the interview (see also Meinhof and Galasiński 2005). Because the interview was an artificial setting, the comparison of the place-name variant preference in everyday use was not possible. Overall, it is difficult to assess how much the language of the interview influenced the comparability of the final responses, but it would be reasonable to assume it has—at least to a certain extent.

In southern Carinthia, the research situation was somewhat different, though no less challenging. Neither of the two interviewers was from Austria: one was Spanish with no previous ties to the region and with no (or little) knowledge of local German and Slovene dialects; the other was Latvian, yet living in Carinthia for years, and developing close ties to the people in the region during the course of the project, including learning the local Slovene dialects. Interviews thus took place either in standard German or in a mixture of Slovene and German. This inadvertently predetermined the place-name variants mentioned in the interviews as well as the details provided by the respondents knowing the interviewer perhaps would not appreciate them sufficiently. Some details may have also been lost in translation. On the other hand, the great advantage to this situation was that the (relative) absence of ties to the region made it possible to approach the research field without any pre-existing local ethnic, social, or political group links which would influence respondents' answers. Also, given the interviewers' neutrality, the respondents may have felt less need to strategize in their responses.

One of the most important findings of our interviews, however, has been that regardless of the choice of language and the possible ties of the interviewer to the region, the fundamental character of respondents' answers was consistent and differed only little. Or, rather, as the questionnaire survey later clearly demonstrated, the observed differences in attitudes and opinions expressed in the interviews could most likely be attributed to other factors than the language of the interview and the personality of the interviewer.

3.5 Questionnaires

In order to gain a broad picture of how the issue of bilingual place names and signs is perceived by the local population, we carried out a questionnaire survey in early to mid-2018. Specifically, we wanted to identify the key factors which influenced people's attitudes toward bilingual signs and minority/majority relations. The tested factors included nationality, age, education, gender, place of residence, usage of the local dialect, and preferred versions of local place names. The contents of the Czech and Austrian questionnaire and the methods of their distribution differed in some important aspects. So we will comment on them separately.

In the Czech case, the questionnaire contained 19 questions (see Appendix 5 for the English translation). Most of them were closed-ended multiple-choice or five-point Likert scale, only four were purely open-ended (including age and residence) and two were semi-open. Questionnaires were in Czech and Polish and they were distributed through local schools and other venues (e.g., cultural centers, businesses, etc.). We did not use an online questionnaire. Of the approximately 4000 questionnaires which were distributed nearly 2000 were returned and 1804 were complete and included in the analysis. The resulting sample was fairly representative of the larger population and evenly distributed across the region (see Table 42) with a slight overrepresentation of young people, university-educated, and women. Interestingly, although respondents were given neutral response options in most questions ("I do not know" or "I cannot evaluate"), the vast majority chose to express their opinions clearly. Actually, many added on different emphatic messages in their own handwriting, underlining their feelings toward bilingual signs. This clearly shows that the sentiments we registered in our sample are very strong and, in view of the size and representativeness of the sample, also probably fairly widespread.

Language issues were important not only during in-depth interviews but also during our questionnaire survey. We decided at the beginning that it was important that questionnaires were available in both the majority and minority languages. We distributed the Czech-language questionnaires through the schools with Czech as the language of tuition and Polish-language questionnaires in schools with Polish as the main language of tuition. In public situations we made both versions available. The questions were originally formulated in Czech and then translated to Polish by a native speaker. Both versions of the questionnaire were tested on a small sample of both minority and majority members to check for possible issues with understandability. Since standardized questionnaires, unlike semi-, and non-standardized interviews, do not allow for explanations and follow-up questions, the issue of validity is a central one. All respondents must understand all questions equally, and in our case, in both language versions. As Pérez (2009) found, surveys carried out in different languages yield different results also because of problems of translation. This issue partially surfaced in our survey in the question asking for the respondent's affiliation with the region when we struggled with an adequate term for the region in Czech and a sufficiently analogical translation in Polish. Our decision—*Těšínsko* in Czech and *Śląsk Cieszyński* in Polish—turned out wrong since the Czech respondents most commonly associated the term with the immediate surroundings of the town of Český Těšín rather than with the historical Těšín/Cieszyn region. As a consequence, we had to disregard the responses given to this question entirely and we were not able to test the influence of one's ties to the region on his/her attitude toward bilingual signs.

In the Austrian case, the questionnaire contained 18 questions, including 7 related to personal data at the end of the questionnaire. Most questions were closed-ended multiple-choice or five-point Likert scale, only three were purely open-ended (including residence). The form and content of some questions differed significantly from the Czech questionnaire. Namely, the Austrian questionnaire had more nuanced questions on languages used in different situations in everyday life and, more importantly, on space-related identity which proved to be the decisive factor in explaining differences in attitudes toward bilingual signs in the Těšín/Cieszyn sample. Also, in the Austrian case, responses to the identity question were not mutually exclusive and respondents were asked to rank their choices by importance. Responses thus yielded a different kind of data than the Czech questionnaire, and we had to devise a way to interpret them in a manner that would be comparable to the Těšín/Cieszyn data. As a counterpoint, we may mention that respondents in the Těšín/Cieszyn region also could indicate more identity options, but almost none chose to do that which only confirms how strong and clearly defined ethnic boundaries in the region are.

The differences in our questionnaires reflected the realities in the two regions. We provide more details on the linguistic and ethnic situation in the two regions in the following chapters, so in this chapter we only briefly explain the motivation for the questionnaire differences. While in the Těšín/Cieszyn region, Czech, Polish, and the dialect are all Slavic languages which are mutually intelligible to a high degree and the dialect is shared by most, in southern Carinthia, Slovene and German, being from different language groups, lack any mutual intelligibility, so the particular choices people have to make in everyday situations are somewhat more complex than in the Těšín/Cieszyn region. As far as nationality is concerned, again the situation in the Těšín/Cieszyn region is rather straightforward as national boundaries are clearly drawn, everyone knows who is a Pole and who is a Czech and these simple categories are, for the most part, seen as mutually exclusive. In Carinthia, by contrast, Slovene-speakers rarely consider themselves as Slovenes in the national sense; rather, the Austrian cultural-national identity is shared by most, and regional identity (Carinthian) or ethnic-group identity (Carinthian Slovene) play a crucial role.

The Austrian questionnaire was distributed in printed as well as electronic format. The paper version was distributed through a high school located in the Gail Valley [Gailtal/Ziljska dolina], that is, only in one of the territorial subunits of the case study area, while the electronic version was sent by email to various groups and organizations representing different positions on bilingual signs in the region, as well as to private individuals. In the latter case, we asked our respondents to forward the questionnaire to their contacts (e.g., relatives, friends, neighbors). Comparable to the Těšín/Cieszyn case, the questionnaires were made available in both German and Slovene. The original German version was translated by a native speaker of Slovene. A total of 589 usable questionnaires were returned. The Carinthian sample is thus significantly smaller than the Těšín/Cieszyn one, but it was still large enough to allow for statistical testing. Reaching sufficient respondents was the main challenge of the Austrian team. Several reasons should be mentioned: first, the channel used. A digital questionnaire is easy to fill in and send back, but respondents' willingness and trust are needed to succeed. Also, the timing was probably not ideal (during the last weeks of the school season with all the associated stress).

As with the interviews, certain German-speaking people with no ties to the Slovene-speaking minority were reluctant to answer and/or to forward the questionnaire. The sample is, therefore, biased toward minority-friendly opinions. In retrospect, we also see problems with the wording and form of some questions which did not bring the answers expected. For example, the question concerning the use of language in everyday life had a wide range of possible answers which were sometimes not properly understood, while the question asking respondents to rank their identities in order of importance was skipped by many altogether. Finally, and this holds true for both regions, the questions asking respondents to name places in their surroundings could not be fruitfully analyzed in a quantitative or a qualitative way. Future cartographic analysis may, nevertheless, prove useful.

Questionnaires were subsequently coded and analyzed using IBM SPSS Statistics to test for the significance of the various factors. Because most questions were either nominal or ordinal, Chi Square test was used, and significance was set at p=.05. Only where ratio and ordinal data were correlated, ANOVA was used (e.g., age x feelings toward signs measured on a 5-point Likert scale) and where two ordinal variables were correlated, Kendall Tau-b and Tau-c were used. In Chap. 6, we report on the results which proved significant on the selected significance level but also on those which did not prove significant. We believe that a lack of significant relationship is also an important result that should be reported in order to avoid outcome reporting bias.

In spite of the aforementioned challenges, we consider both surveys to have been successful, bringing very interesting and in some cases also surprising results. They also completed the mosaic which we have pieced together from the various research methods to gain a complex and complete understanding of the region.

3.6 Conclusion and Recommendations

As many researchers will confirm, there is no one good way to collect or interpret data. We hardly choose from ideal options and all we can hope for is to find ways to minimize the damage. One possible solution to this dilemma that we chose to adopt was to triangulate our methods to complement the shortcomings of each. We believe we have been quite successful in this regard, although the sampling and coding limitations are clear. The other solution was to be maximally transparent about our decisions and not to be overly confident about the generalizability of our findings. We hope our project can help others to prepare research proposals which build on our experiences and avoid repeating some of our mistakes.

Our principal recommendations to future researchers of place-name politics in bilingual regions would be these:

1. triangulate methods
2. make sure your samples are representative
3. test the validity of questionnaires in all language versions
4. be sensitive to respondents' language preferences but also reflect critically on the effect of language choice on responses
5. where possible, work as an outsider and use the insider's role carefully

6. maintain neutrality even when it does not seem advantageous at the moment
7. account for the temporal and spatial dynamic of the linguistic landscape
8. notice both the politics and poetics of place names (only some names are politicized at any given time, do not overpoliticize your research)

References

Adida, C. L., Ferree, K. E., Posner, D. N., & Robinson, A. L. (2016). Who's asking? Interviewer coethnicity effects in African survey data. *Comparative Political Studies*, 49(12), 1630–1660.

Albrecht, S. (2006). Whose voice is heard in online deliberation? A study of participation and representation in political debates on the internet. *Information, Community and Society*, 9(1), 62–82.

Anderson, L. (2006). Analytic autoethnography. *Journal of Contemporary Ethnography*, 35(4), 373–395.

Balogh, P. (2013). The outsider advantage: Interviewing planners and other elites in the Polish-German borderland. *Journal of Settlements and Spatial Planning*, 4(1), 101–108.

Barni, B., & Bagna, C. (2009). A mapping technique and the linguistic landscape. In E. Shohamy & D. Gorter (Eds.), *Linguistic landscape: Expanding the scenery* (pp. 126–140). New York: Routledge.

Blackwood, R. (2015). LL explorations and methodological challenges. *Linguistic Landscape*, 1(1–2), 38–53.

Blommaert, J. (2013). *Ethnography, superdiversity and linguistic landscapes: Chronicles of complexity*. Bristol; Buffalo; Toronto: Multilingual Matters.

Campbell, J. L., Quincy, C., Osserman, J., & Pedersen, O. K. (2013). Coding in-depth semistructured interviews: Problems of unitization and intercoder reliability and agreement. *Sociological Methods & Research*, 42(3), 294–320.

Cenoz, J., & Gorter, D. (2006). Linguistic landscape and minority languages. *International Journal of Multilingualism*, 3(1), 67–80.

Cohen, S. B., & Kliot, N. (1992). Place-names in Israel's ideological struggle over the administered territories. *Annals of the Association of American Geographers*, 82(4), 653–680.

Crampton, J. W. (2001). Maps as social constructions: Power, communication and visualization. *Progress in Human Geography*, 25(2), 235–252.

Dołowy-Rybińska, N. (2017). Krajobrazy językowe mniejszości językowych Europy: Znaczenie, rola, wyzwania. *Kultura Współczesna*, 97(4), 79–91.

Dörflinger, J. (1984). *Österreichische Karten des 18. Jahrhunderts*. Wien: Verlag der Österreichischen Akademie der Wissenschaften.

Evans, J., & Jones, P. (2011). The walking interview: Methodology, mobility and place. *Applied Geography*, 31(2), 849–858.

Fairclough, N. (2003). *Analysing discourse: Textual analysis for social research*. London and New York: Routledge.

Fuchs, C. (2014). Social media and the public sphere. tripleC: Communication, Capitalism & Critique. *Open Access Journal for a Global Sustainable Information Society*, 12(1), 57–101.

Galasińska, A. (2006). Border ethnography and post-communist discourses of nationality in Poland. *Discourse & Society*, 17(5), 609–626.

Galpin, C., & Trenz, H. (2017). The spiral of Euroscepticism: Media negativity, framing and opposition to the EU. In M. Caiani & S. Guerra (Eds.), *Euroscepticism, democracy and the media* (pp. 49–72). London: Palgrave Macmillan.

Gorter, D. (2006). Introduction: The study of the linguistic landscape as a new approach to multilingualism. In D. Gorter (Ed.), *Linguistic landscape: A new approach to multilingualism* (pp. 1–6). Clevedon; Buffalo; Toronto: Multilingual Matters.

Gorter, D., & Cenoz, J. (2015). Translanguaging and linguistic landscape. *Linguistic Landscape*, 1(1–2), 54–74.

Harley, J. B. (1988). Maps, knowledge, and power. In D. Cosgrove & S. Daniels (Eds.), *The Iconography of Landscape* (pp. 277–312). Cambridge: Cambridge University Press.

Johnson, C., Jones, R., Paasi, A., Amoore, L., Mountz, A., Salter, M., & Rumford, C. (2011). Interventions on rethinking 'the border' in border studies. *Political Geography*, 30(2), 61–69.

Jordan, P. (2018). The use of minority place names on maps of the Third Austrian Military Survey—With a regional focus on Slovenian and Croatian Lands. *The Cartographic Journal*, 55(2), 178–186.

Kadmon, N. (2004). Toponymy and geopolitics: The political use—And mis-use-of geographical names. *The Cartographic Journal*, 41(2), 85–87.

Lee, T., & Pérez, E. O. (2014). The persistent connection between language-of-interview and Latino political opinion. *Political Behavior*, 36(2), 401–425.

Mácha, P., Lassak, H., & Krtička, L. (2018). City divided: Place names and nationalism in the Czech-Polish Borderlands. *Mitteilungen der Österreichischen Geographischen Gesellschaft*, 160, 303–329.
Meinhof, U. H., & Galasiński, D. (2005). *The language of belonging*. London: Palgrave Macmillan.
Mikecz, R. (2012). Interviewing elites: Addressing methodological issues. *Qualitative Inquiry*, 18(6), 482–493.
Pérez, E. O. (2009). Lost in translation? Item validity in bilingual political surveys. *The Journal of Politics*, 71(4), 1530–1548.
Pickles, J. (1992). Texts, hermeneutics and propaganda maps. In T. J. Barnes & J. S. Duncan (Eds.), *Writing worlds: Discourse, text and metaphor in the representation of landscape* (pp. 193–230). London and New York: Routledge.
Rothbauer, P. (2008). Triangulation. In L. Given (Ed.), *The SAGE encyclopedia of qualitative research methods*. Thousand Oaks, CA: Sage Publications.
Schulz, W., et al. (2004). *Analýza obsahu mediálních sdělení*. Prague: Karolinum.
Scollon, R., & Scollon, S. W. (2003). *Discourses in place. Language in the material world*. London and New York: Routledge.
Spradley, J. P. (1979). *The ethnographic interview*. New York: Holt, Rinehart and Winston.
Szabó-Gilinger, E., et al. (2012). Discourse coalitions for and against minority languages on signs: Linguistic landscape as a social issue. In D. Gorter, H. Marten, & L. Van Mensel (Eds.), *Minority languages in the linguistic landscape* (pp. 263–280). New York: Palgrave Macmillan.
Trumper-Hecht, N. (2009). Constructing national identity in mixed cities in Israel. In E. Shohamy & D. Gorter (Eds.), *Linguistic landscape: Expanding the scenery* (pp. 238–252). New York: Routledge.
Weber, R. (1990). *Basic content analysis* (= Quantitative Applications in the Social Sciences. A Sage University Paper, 49). Newbury Park; London; New Delhi: Sage Publications.

4

Linguistic Minorities in Austria and Czechia: Historical, Political, and Cultural Contexts

4.1 Austria (Peter Jordan)

4.1.1 Ethnic and Linguistic Structure: Historical Development and Current State

The current cultural landscape and structure of Austria are the result of several cultural layers accumulated throughout history. These layers can very well be traced and documented by toponymy confirming the value of place names as keys to cultural and settlement history.

Austria's namescape contains pre-Bavarian and pre-Alemannic linguistic layers, since Bavarian-Alemannic settlement of most of modern Austria did not occur before the eighth and ninth centuries. Most of the names of larger rivers in modern Austria—like in other parts of Europe—are of Celtic (or pre-Celtic) origin: *Donau* (Danube) can be derived from Indo-Germanic *déh-nu* 'river' (such as *Don, Dniester* [Nistru]) (Pohl and Schwaner 2007: 183); *Drau/Drava* from Indo-Germanic *drowo* 'river course' (ibidem: 184); *Inn* from Indo-Germanic *penios, enios* approximately 'muddy river' (ibidem: 186); *Enns* from Indo-Germanic

P. Jordan et al., *Place-Name Politics in Multilingual Areas*, Palgrave Studies in Minority Languages and Communities, https://doi.org/10.1007/978-3-030-69488-3_4

pen-, pon- 'swamp' (ibidem: 184); *Mur* from Indo-Germanic *mori-* 'standing water' (ibidem: 189); *Gail* from late Illyric *gelias* 'the frothing' (ibidem: 185).

Some mountain names also originate in Celtic or pre-Celtic. However, many were later forgotten and newly shaped in the course of tourism development. Examples for old names include *Tauern*, derived from Indo-Germanic *tûr* 'mountain, pass' (Pohl and Schwaner 2007: 218); or *Karawanken* from Celtic *karvos* 'deer'. This name, however, underwent a later Celtic re-interpretation, while it originally came from pre-Celtic *kar* 'rock, stone' (ibidem: 212). As it can be seen, many of these names are the result of onymization in the sense of general terms of a language assuming the character of names.

Also some region and country names are part of the Celtic cultural layer. *Kärnten* (Carinthia), for example, can either be derived from Celtic *karantāna* 'place of friends' as Eberhard Kranzmayer saw it (a later tourism slogan "Carinthia—Vaccations with friends" could have taken up this meaning as a suggestion) or from Celtic *karant* 'stone, rock' as Pohl and Schwaner conceive it (Pohl and Schwaner 2007: 157).

Roman settlement affected Austria up to the Danube in the North (Danube Limes). Carnuntum (nearby Bratislava), Vindobona [Wien], Lauriacum (nearby Enns) were Roman municipia ('cities', at the same time centers of larger administrative units) along the Danube, Ovilava [Wels], Iuvavum [Salzburg], Flavia Solva (nearby Leibnitz), Virunum (north of Klagenfurt), Teurnia (nearby Spittal an der Drau), Aguntum (nearby Lienz), and Brigantium [Bregenz] held this function in the interior of the Roman provinces Pannonia (superior), Noricum, and Raetia covering modern Austria and areas beyond. Roman urban and road networks laid the foundations of modern structures, and their continuity in location from Roman times is remarkable.

Roman rule promoted, however, mainly Romanization of the local Celtic population in linguistic and other cultural terms, much less the immigration of Romans from central and other parts of the Empire which remained more or less confined to the fields of administration, the military, and trade. In the phase of the late Roman Empire—after

Emperor Theodosius had elevated the Christian Church to the rank of the state church in 391 AD—modern Austrian lands received a first wave of Christianization, although Germanic and Slavonic invasions from the fourth to the seventh centuries let Christians almost disappear.

Roman names, most of them adapted to German, are still preserved up today. They are, however, much more frequent in the western parts of Austria, where Roman settlement lasted longer locally and was not overlaid by Slavonic settlement. Examples are *Fontanella* from the Latin diminutive of *fontana* 'source'; *Kuchl* from Latin *cucullus* 'hood' in the sense of hilltop. The old High German ethnonym *walah* for 'welsh, romance' indicates longer Roman or Romance settlement as it can be found with *Seewalchen* (Upper Austria), *Straßwalchen* (Salzburg), *Walchsee* (Tyrol), *Walchen* (Salzburg, Styria [Steiermark]—upper Enns Valley [Ennstal]). It also occurs as a family name: *Walcher, Walker*. It can be assumed that the invasions of Germanic tribes from the late fourth century onward caused most Romans to re-migrate to the safer central parts of the then West Roman Empire (after the Empire's split in 395) and reduced also the Romanized Celtic population to a minimum.

The Slavonic settlement of the eastern part of modern Austria from about 600 onward as a facet of the great migration of Slavs from their earlier homes between Vistula [Wisła] and Dnjepr [Dnìpro] may thus have met only an economically and politically weak local population. Slavonic settlers left many toponymic traces not only in Carinthia [Kärnten] and Styria [Steiermark] (also outside of today's minority areas), but also in Burgenland, Lower Austria [Niederösterreich], Upper Austria [Oberösterreich], Salzburg, and East Tyrol [Osttirol]. *Feistritz* as a river and a settlement name derived from it, from Slavonic *bistrica* 'the fast, clear', occurs more than 30 times in Austria (Pohl and Schwaner 2007: 185). The common suffix in German orthography *-ach* is derived from the Slavic locative like in *Göriach, Dellach, Görtschach, Amlach, Leisach, Tainach, Kainach*. Although written in German orthography, it is completely Slavic (Slavic *gora* 'mountain') and means 'where the mountain dwellers live'. The diminutive *-itz* can be found in many river names (and settlement names derived from them): *Türnitz, Rabnitz, Lafnitz, Lassnitz, Görtschitz*, or *Mürz* derived from *Murica* 'the little Mur'.

Names of Slavic origin have been preserved far outside of today's 'suspect areas' (such as Carinthia, Styria, Lungau, East Tyrol): *Türnitz* in the Lower Austrian Limestone Alps [Niederösterreichische Kalkalpen]; *Ötscher*, the name of a mountain that can be seen from afar in the Lower Austrian Alpine foothills, from Slavic *očan* 'godfather' (Pohl and Schwaner 2007: 214); *Jauerling*, the name of a mountain in the Waldviertel at the edge of the Wachau, from Slavic *javornik* 'maple mountain'. Analogous to the ethnonym *walch* for Romans, Middle High German coined *windisch* for Slavic. Place names with this adjective indicate a longer Slavic settlement: *Windischgarsten, Windisch Minihof*, formerly: *Windisch Matrei* (today *Matrei in Osttirol*).

The Alpine Slavs—as they are called in modern historiography—came as pagans, but were soon Christianized, north of river Drau/Drava by Salzburg as the main Bavarian ecclesiastical center and south of it by Aquileia as the main ecclesiastical center along the northern Adriatic. Both meant to be Christianized in the Latin, Western rite and not to become part of the Byzantine realm as it happened in the ninth century temporarily to the Slavonic population of the Moravian Empire (including larger parts of the Pannonian Basin). These were missioned by the 'Slavs apostles' Cyril and Method on behalf of the Moravian rulers and Byzantium, although the Christian Church was at that time not yet officially split into the Church of Rome and the Byzantine Church. It meant also to practice Latin as the sacral language and to write in Roman script, while for the Slavs of the Moravian Empire Old Church Slavonic was the sacral language and Glagolitic their script. Thus, in contrast to the Pannonian Slavs as well as to Slavs in the later lands of the Bohemian Crown, Alpine Slavs were also immediately affected by and later integrated into the Western political and societal system, created and represented at that time by the Franconian Empire and its eastern outpost Bavaria [Bayern].

Christian mission by Salzburg as well as political affiliation to Bavaria promoted also Bavarian settlement down to at least river Drau/Drava and gradual acculturation of Alpine Slavs. Bavarian settlement proceeded from the already old-Bavarian settlement areas in western Upper Austria [Oberösterreich], around the city of Salzburg and larger parts of Tyrol [Tirol] from the eighth to the thirteenth century. Modern Vorarlberg and western parts of Tyrol belonging to the Duchy of Swabia [Schwaben] had been settled by Alemanni. As the Duchy of Carinthia developed into a political power independent of Bavaria in the Early Middle Ages and maintained a bilingual (Bavarian/Alpine Slavonic) cultural structure up to the late Middle Ages, Alpine Slavonic has been better preserved in its dominions. Up to the end of the Middle Ages, however, the language boundary shifted also there toward the South as will in more detail be shown in Chap. 5.

Hungarians invaded the Pannonian Basin and larger parts of Europe from the late ninth century onward and caused thus the fall of the Moravian Empire. Beaten by the Franconian Empire in 955 they founded a state centered on the Pannonian Basin around 1000, politically associated to the Franconian Empire and missioned in the Latin, Western Christian sense from Bavaria (Passau). They were also present in the western areas of the Pannonian Basin today forming the Austrian province of Burgenland that was only in the thirteenth century settled by Bavarians. Thus, Hungarians were the first and Bavarians the latecomers—an important issue in later national historiography. As a part of the Hungarian Kingdom, Hungarians constituted the dominant group there up to 1918, although they were by numbers in minority.

In the sixteenth century, the non-dominant (not by numbers, but political and societal status) Bavarian/German West Hungarians were joined by Croats evacuated by their Hungarian landlords from the Austrian frontier against the Ottoman Empire. Like elsewhere in contemporary West Hungary they settled places destroyed by the Ottoman invasions in an island-like manner (Breu 1970).

Later immigration to the territory of modern Austria from the nineteenth century onward, in the Communist period and during the Yugoslavian secession wars, was in some cases numerically not less important than some of the flows mentioned. But it left—with the limited exception of Czechs—due to its character of labor migration or of refuge much fewer sustaining traces in the cultural landscape. Most of these migrants had been assimilated in the third generation at the latest and did not settle down compactly creating new cultural regions. Even immigration of recent years, although partly large in numbers, is—due to its city-bound and dispersed character—not creating new cultural areas with specific namescapes.

The limited exception is Czech immigration in the later nineteenth century and up to World War I. Czechs were attracted by the booming metropolis of Vienna and came prevailingly as industrial and construction workers (male) as well as catering workers and helping hands in better-off households (females). A part of them formed an ethnic cluster in Vienna's 10th District (Favoriten) preserving their cultural identity supported by a Czech high school (Österreichische Rektorenkonferenz 1988).

Tables 4.1, 4.2, 4.3, and 4.4 show the numerical development of autochthonous ethnic minorities in Austria as documented by the population censuses 1971–2001 not only for Austria in total, but also by federal provinces, where they have their traditional homes. The criterion of the colloquial language used by these censuses does not refer to a language in the sense of a 'softer', less stricter standard language in between standard language and dialect, but in the sense of the language usually applied in the private sphere (family, relatives, friends, etc.) conforming to all Austrian population censuses since 1880 except the three censuses in the interwar period (1923, 1934, 1939). It is seemingly an objective criterion, but in practice respondents conceived it as a question for their subjective ethnic affiliation. Thus, it may very well serve as a characteristic portraying ethnic consciousness as it was conceived by the Constitutional Court of Justice in 1987 (Öhlinger 2006: 128).[1] This is also the reason why citizens in Carinthia declaring to speak 'Windisch' were not included into the figures. They recognize their Slavonic origin and culture, but do not feel affiliated to the ethnic group of Carinthian Slovenes. The population census of 2001 was the last in Austria to be conducted in the traditional way as a complete survey of the population by asking every household 'on the ground', while already the census of 2011 was taken as a so-called register census combining all available data files and for the first time not taking into account language or ethnicity. The data as of

[1] The linguist Heinz-Dieter Pohl, a distinguished expert of Carinthian culture, holds the opinion that "the declaration of colloquial language does not allow safe conclusions on nationality" (Pohl 2002: 183).

Table 4.1 Colloquial language of Austrian citizen resident population in 1971

Colloquial language	Austria in total	Austria %	Burgenland abs.	Burgenland %	Carinthia abs.	Carinthia %	Vienna abs.	Vienna %
all languages	7,279,630	100	270,539	100	517,586	100	1,558,316	100
German	7,195,090	98.8	240,462	88.9	495,795	95.8	1,538,578	98.8
Croatian	28,084	0.4	24,332	9.0	143	0.0	2316	0.1
Slovene[a]	19,604	0.3	43	0.0	17,011	3.3	500	0.0
Czech	7967	0.1	45	0.0	35	0.0	6528	0.4
Hungarian	14,815	0.2	5447	2.0	141	0.0	6099	0.4

Source: Statistik Austria (2020)

[a]Not including 'Windisch'

Table 4.2 Colloquial language of Austrian citizen resident population in 1981

Colloquial language	Austria in total	Austria %	Burgenland abs.	Burgenland %	Carinthia abs.	Carinthia %	Vienna abs.	Vienna %
All languages	7,263,890	100	267,750	100	528,023	100	1,417,929	100
German	7,150,623	98.4	244,493	91.3	509,392	96.5	1,380,579	97.4
Croatian	22,113	0.3	18,648	7.0	168	0.0	2557	0.1
Slovene[a]	16,290	0.2	29	0.0	14,204	2.7	624	0.0
Czech	5101	0.1	22	0.0	29	0.0	4104	0.3
Hungarian	12,043	0.2	4025	1.5	121	0.0	5683	0.4

Source: Statistik Austria (2020)

[a]Not including 'Windisch'

Table 4.3 Colloquial language of Austrian citizen resident population in 1991

Colloquial language	Austria in total	Austria %	Burgenland abs.	Burgenland %	Carinthia abs.	Carinthia %	Vienna abs.	Vienna %
All languages	7,278,096	100	263,092	100	530,726	100	1,343,196	100
German	7,107,411	97.7	237,516	90.3	512,122	96.5	1,269,417	94.5
Croatian	29,596	0.4	19,109	7.3	295	0.0	6604	0.5
Slovene[a]	19,289	0.3	65	0.0	13,962	2.6	1825	0.1
Czech	9822	0.1	81	0.0	106	0.0	6429	0.5
Hungarian	19,638	0.3	4973	1.9	247	0.0	8930	0.7

Source: Statistik Austria (2020)

[a]Not including 'Windisch'

Table 4.4 Colloquial language of Austrian citizen resident population in 2001

Colloquial language	Austria in total	Austria %	Burgenland abs.	Burgenland %	Carinthia abs.	Carinthia %	Vienna abs.	Vienna %
All languages	7,322,000	100	265,005	100	527,333	100	1,301,859	100
German	6,991,388	95.5	240,228	90.6	508,543	96.4	1,139,196	87.5
Burgenland-Croatian	19,374	0.3	16,245	6.1	25	0.0	2456	0.2
Slovene[a]	17,953	0.2	70	0.0	12,554	2.4	1412	0.1
Czech	11,035	0.2	189	0.1	192	0.0	5778	0.4
Hungarian	25,884	0.4	4704	1.8	313	0.0	10,686	0.8

Source: Statistik Austria (2020)

[a]Not including 'Windisch'

2001 are therefore the last conveying language or ethnic consciousness and have to suffice as the most reliable basis for portraying current Austrian citizens' ethnic consciousness.

Table 4.1 shows that Croatian-speakers reside predominantly in Burgenland, Slovene-speakers in Carinthia, and Czech-speakers in Vienna [Wien] but are even there only small minorities. Most Hungarian-speakers reside outside Burgenland. They, however, do not belong to the traditional Hungarian minority of former Western Hungary, but are Hungarian refugees after the Hungarian uprising against the Communist regime in 1956 residing mainly in Vienna and adjacent parts of Lower Austria.

Table 4.2 shows a significant decline of all minorities compared to 1971 with Hungarian-speakers suffering only from a smaller decline due to the recognition of Hungarian refugees of 1956 as an ethnic group according to the Federal Act on Ethnic Groups 1976.

The decade 1981–1991 (Table 4.3) is—very likely due to the Act on Ethnic Groups 1976, its recognition of minority rights, and its creating a better atmosphere for minorities—marked by a numerical increase of all linguistic/ethnic minorities, but much less in their traditional areas of residence than in other parts of Austria. This latter fact is with Croatian-speakers due to Croatian labor migration (from Yugoslavian Croatia and Bosnia-Herzegovina) mainly to Austrian cities and industrial centers. With speakers of Slovene, Czech, and Hungarian it is due to the fact that their educational and labor migration is directed to Austrian cities and that people sympathizing with minorities declared minority affiliation without any factual background.

In 1982, Burgenland-Croatian had been codified as a standard language different from Croatian and in 1987 been declared an official language in some communes of Burgenland (besides German). This resulted in a strong decline from 'Croatian-'speakers in 1991 to 'Burgenland-Croatian-'speakers in 2001 and their concentration on Burgenland as the Burgenland-Croats' traditional homeland (Table 4.4). Besides Burgenland, only Vienna as an educational and labor center hosts a significant number of Burgenland-Croats. With other minorities the long-term decline continues.

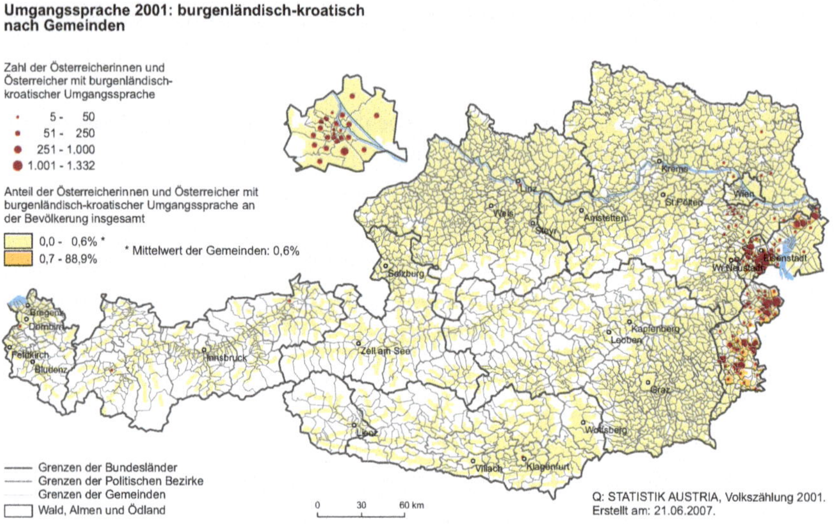

Fig. 4.1 Speakers of Burgenland-Croatian by communes in 2001. (Source: Statistik Austria 2020)

Figures 4.1, 4.2, and 4.3 reflect cartographically the results of the 2001 population census and visualize the homelands of Burgenland-Croats (Fig. 4.1), Slovenes (Fig. 4.2), and Hungarians (Fig. 4.3). They clearly reveal the island-like structure of Burgenland-Croatian residences, the comparatively compact settlement area of Carinthian Slovenes, while Slovenes are also spread over cities (Vienna!), and a small Slovenian minority is visible along the Styrian border with Slovenia. The distribution of Hungarian-speakers demonstrates that the number of Hungarian refugees of 1956 concentrated in Vienna and its (southern) surroundings is much higher than that of the traditional West-Hungarian autochthonous community.

4.1.2 State Formation

The modern Republic of Austria [Republik Österreich] has formally not been founded earlier than in 1945, after the end of World War II, but is in respect to territory, law, and administrative-territorial system as well as

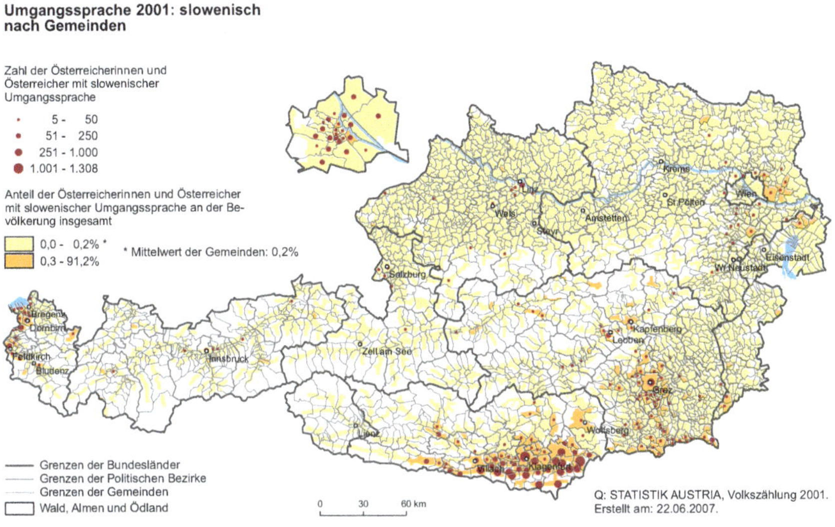

Fig. 4.2 Speakers of Slovene by communes in 2001. (Source: Statistik Austria 2020)

space-related identities based on earlier political entities that cannot be disregarded, when the modern state is to be analyzed properly.

As a predecessor of modern Austria may be regarded the conglomerate of lands (Monarchische Union) ruled by the Habsburg dynasty. In the middle of the seventeenth century, this conglomerate became not only the dominion of one and the same ruler (personal union), but it also received a common parliament (Generallandtag) and some common law (real union) in addition to law valid for the individual lands (Brauneder and Lachmayer 1992; Brauneder 2009). This applied to a territorial configuration that comprised apart from Austria [Österreich] in the contemporary sense the Alpine lands Styria [Steiermark, Štajerska], Carniola [Kranjska], Carinthia [Kärnten, Koroška], Tyrol [Tirol] (including South Tyrol [Südtirol/AltoAdige] and Welsh Tyrol [Trentino]), 'Vorderösterreich' (later Vorarlberg), the Bohemian lands Bohemia [Čechy], Moravia [Morava], Silesia [Śląsk, Slezsko], and Lusatia [Lausitz/Lužice] as well as the Pannonian lands Upper Hungary (today roughly corresponding to Slovakia), Western Hungary (modern Burgenland and the western fringe of modern Hungary), and the parts of Croatia [Hrvatska] not occupied

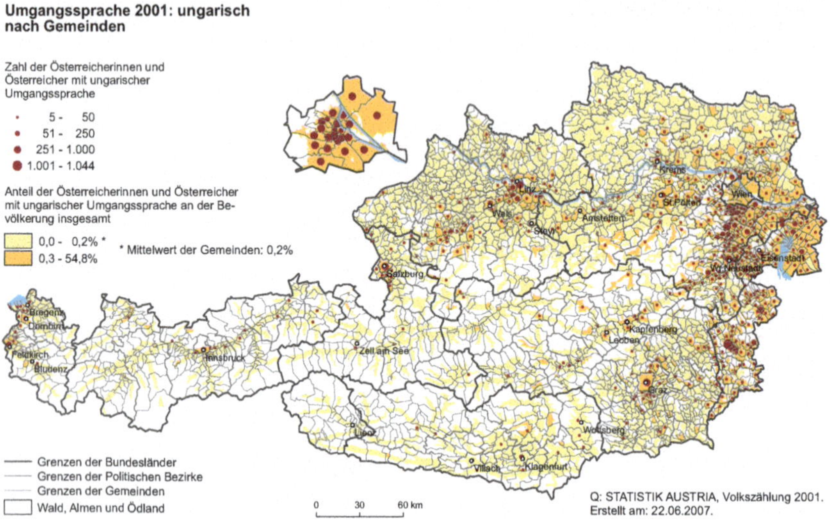

Fig. 4.3 Speakers of Hungarian by communes in 2001. (Source: Statistik Austria 2020)

by the Ottoman Empire. This configuration combined territories of rather different historical shaping, cultural characteristics (language, after the Reformation also religion), and distinct identities. What was titled *Austria* above was at that time composed of the former Babenberg Duchy as the germ of the later Habsburg lands—today roughly corresponding to the federal state of Lower Austria [Niederösterreich], with Vienna [Wien] as its center, and what is today Upper Austria [Oberösterreich].

This first step of state formation occurred under (and was driven by) a macropolitical situation characterized by a devastated, politically weak, and internally estranged Central Europe after the Thirty-Years War (1618–1648) and an expansive Ottoman Empire that had conquered large parts of the Pannonian Basin and two times (1529, 1683) sieged Vienna. Very helpful with establishing a central power was the fact that the Habsburg ruler was at the same time the Emperor of the Holy Roman Empire with all his authority and central institutions at hands. This first stage of an Austrian state was called 'Habsburg Lands' [Habsburgische Erblande] or 'House of Austria' [Haus Österreich, Casa d'Austria].

After the Austrian armies had driven the Ottoman Empire out of the Pannonian Basin and most of it had been integrated into the Habsburg realm, the Pragmatic Sanction of 1713 provided the Habsburg Lands with a first constitution ruling that they were undividable. The power of individual lands was further reduced but remained nevertheless strong, and the state was heterogenous in all respects (Brauneder and Lachmayer 1992; Brauneder 2009).

From 1749 onward under Mary Theresa [Maria Theresia] (1740–1780) and her son Joseph II (1780–1790), this consolidated conglomerate of lands was transformed into a state conforming to the ideas of enlightened absolutism. Additional central authorities, predecessors of ministries, were established and made responsible for foreign affairs, the interior, defense, and finances. Most essential and a good basis for further development was the introduction of civil officers with a solid remuneration by the state. The constitution and additional legislation were further developed resulting in a rather homogenous legal system all over the country (Brauneder and Lachmayer 1992; Brauneder 2009). A common official language (German) was proclaimed—promoting coherence, but from the angle of cultural identity and the (later) rising national ideas not unproblematic in a truly multilingual environment. Austria was also only second to France in conducting a complete topographic survey (starting 1764). It was now justified to speak of a state in the modern sense, and the name 'Austrian Empire' [Kaisertum Österreich], used from 1804 onward, was a symbol of it. Following the dissolution of the Holy Roman Empire in 1806, Austria was the only empire in Central Europe. The lands remained nevertheless political entities and strong references of personal and group identities, although governments [Gubernien] comprising several lands at the upper regional level as well as counties [Kreise] at the lower regional level, both executing deconcentrated state administration, controlled them and confined their political power (Brauneder and Lachmayer 1992; Brauneder 2009). It was nevertheless a federal system, albeit with varying federal state competences. All these developments referred to a relatively stable territory, although gains and losses marked its fringe, especially in the North and South.

The year 1848 was of civil revolutions in most parts of Europe demanding the control of power and legitimation of rule by the citizens, and so it was also in the Austrian Empire. They had a strong national component due to the spread of the national idea originating in Enlightenment and the French Revolution at the end of the eighteenth century. The national idea assumed in the central (and other parts of Europe) in contrast to the francophone and anglophone sphere not the variant of the civic nation, but (inspired by Johann Gottfried Herder) of the cultural nation in the sense of a community of people bound together by elements of culture like language or religion and striving for a genuine political self-government (Herder 1813; Symmons-Symonolewicz 1985). In a multicultural state like Austria it had an inherent potential for conflict. Indeed, many national identities had emerged, most explicitly among Hungarians, but not much less so also among the Slavonic and Romance peoples of the Empire. Many of them were striving for self-government, if not for establishing their own nation state or joining their 'mother nation' in the way of irredentism.

The response of the Emperor was a constitution still imposed by him, but granting the citizens participation in exerting power by electing one chamber of the parliament (House of Deputies [Abgeordnetenhaus]), while the members of the other chamber—the Senate [Senat]—were nominated by the ruler. The parliament had to be convened by the ruler, could only pass acts submitted by the government, and was subject to a potential veto by the ruler. This constitution, implemented in 1848 and referred to as the Pillersdorf Constitution [Pillersdorf'sche Verfassung] after the Minister who elaborated it, laid thus the first foundations of the democratic principle. In addition, it implemented a hierarchy of independent courts and proclaimed basic rights strengthening the rule of law. At the sub-national administrative levels self-government was enforced by indirect voting (voting electors, not delegates) emphasizing the federal structure of the state (Brauneder and Lachmayer 1992; Brauneder 2009).

After the loss of most Italian possessions and having been defeated by Prussia [Preußen] in the battle of Hradec Kralové in 1866, the Empire found itself in the position of being only the second in Central Europe and German lands and of competing with a new power in the Adriatic space due to the proceeding unification of Italy (1860–1870). Furthermore, it was confronted with Hungarian aspirations of establishing a nation state. The way out was the Austro-Hungarian Compromise of 1867. It was interpreted by Austria and Hungary in different ways: While from the Austrian perspective, it was a decentralized federation with two federal entities, from the Hungarian perspective it was a confederation of two states with some common authorities. In practice, the Hungarian perspective was effectuated (Brauneder and Lachmayer 1992; Brauneder 2009).

While the Hungarian part of the Austro-Hungarian Monarchy [Österreichisch-ungarische Monarchie/Osztrák-magyar monarchia], as it was called now, developed into a centralized Hungarian nation state, the Austrian part, as defined by the Act on Basic Rights [Staatsgrundgesetz] of 1867, was a federation in itself, composed of self-governing 'crownlands' [Kronländer] having their elected parliaments and executive authorities as well as up to (dependent on their linguistic structure) four official languages (Brauneder and Lachmayer 1992; Brauneder 2009). In Carinthia, for example, German and Slovene were both official, in Austrian Silesia German, Polish, and Czech (Staatsgrundgesetz RGBl. 142/1867). After the Berlin Congress of 1878 had charged Austria-Hungary with the occupation of former Ottoman Bosnia-Herzegovina and this acquisition was administered as a condominium of the two confederative units, as well as after having been awarded by the same Congress a small extension along the Montenegrin coast, the Monarchy had reached its territorial shape and administrative structure that existed up to its dissolution in 1918 (Fig. 4.4).

Below the level of crownlands, the Austrian part of the Monarchy was subdivided into 'political districts' [Politische Bezirke] (Fig. 4.4) that were already in the 1850s established to complement the juridical districts in the non-juridical sphere and represented deconcentrated state administration. They persist up to the present day or were after the Communist period re-established (under varying designations) in almost all successor states, certainly in the Republic of Austria—there even only with minor territorial modifications. At the local administrative level, the Austrian part of the Monarchy was composed of self-governing communes with an elected council that in turn elected the mayor and the communal government.

World War I (1914–1918) and the dissolution of the Austro-Hungarian Monarchy in 1918 had a profound impact on the political and

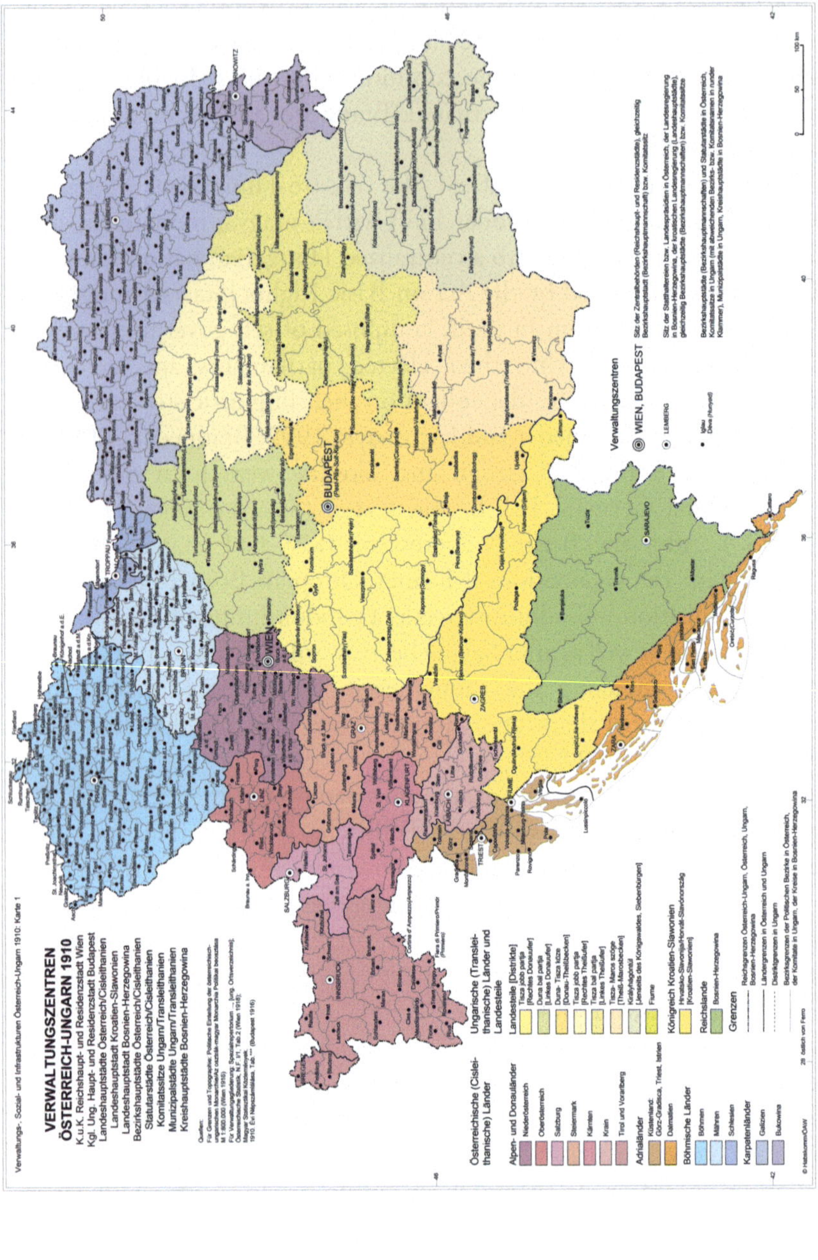

Fig. 4.4 Austro-Hungarian Monarchy in 1910—Administrative-territorial subdivision. (Source: Rumpler and Seger 2010)

socio-economic development of the territory of modern Austria. However, elements of continuity overshadowed fractures, especially as regards the administrative-territorial system at the sub-national levels, political culture, and space-related identities.

The Republic of Austria was initially founded in 1918 under the name *Deutsch-Österreich* ('German-Austria'), since German-speaking Austrians conceived themselves as Germans in the sense of a cultural nation and the new political elite was aiming at a unification with Germany (which was then forbidden by the Peace Treaty of Saint Germain in 1919). Austria saw itself (in contrast to Hungary) as a new state and not as the legal successor of the Austro-Hungarian Monarchy or even only of the Austrian part of it. It nevertheless took over most of the acquis of the Monarchy below the constitutional level.

Significant elements of continuity are also the sub-national administrative-territorial structures. Thus, the traditional lands with their constitutions as of 1861 remained, although some of them had lost larger (Tyrol, Styria) or smaller (Carinthia, Lower Austria) parts of their territory to other successor states. As already mentioned, also the political districts persisted as units of deconcentrated state administration—institutionally related to the federal states, as the lands can also be called. The communes, too, remained self-governing units, and the Act on Communes [Reichsgemeindegesetz] as of 1862 was—as regards its contents—integrated into the Republic's constitution in 1925 (Brauneder and Lachmayer 1992; Brauneder 2009).

The Republic adopted also the Act on Basic Rights as well as the act on supreme courts from the constitution of 1867. The act on general male voting, implemented in 1907, was extended to females on 18 December 1918 (Brauneder and Lachmayer 1992; Brauneder 2009).

An important element of continuity was also the persistence of the major political parties not only in terms of content, but also formally, that is, the Christian Socials, the Social Democrats, and the German Nationals. On the background of an initially complete break with the old political system and the fact that all its political authorities had lost their functions and competences it was on them to establish a new political system—even if taking advantage of most of the acquis and the institutions of the former one.

A lasting new achievement of the young Republic is the constitution of 1920, essentially elaborated by Hans Kelsen and significantly amended in 1929 by strengthening the position of the Federal President [Bundespräsident]. It defines with minor modifications the political system of modern Austria and will be presented under this item. Later in the interwar period, in 1934, and in the course of a growing societal fractioning it was replaced by an authoritarian constitution before Austria discontinued its existence as an independent state in 1938, when it was occupied by the German Reich [Deutsches Reich] and dragged into World War II.

After the German Reich's capitulation in May 1945, based on the Moscow Declaration of the later victorious allies of World War II as of 1943 and a proclamation of the "antifascist" Austrian political parties (Austrian People's Party [Österreichische Volkspartei], Socialist Party of Austria [Sozialistische Partei Österreichs], Communist Party of Austria [Kommunistische Partei Österreichs]) as of 1945, the Republic of Austria was "re-established" as a federation in its borders of 1937, that is, before the occupation ['Anschluss'] by the German Reich. Thus, the victorious allied powers as well as the relevant Austrian parties of the immediate postwar period held the view that Austria had been occupied by the German Reich against its own will and that the postwar state continued the state of the interwar period. Accordingly, the latter's constitution of 1920 and 1929 was fully adopted (Brauneder and Lachmayer 1992; Brauneder 2009; Vocelka 2002). This would not allow to speak of a First and a Second Republic. It is, however, also true that the constitutional continuity of the Republic had already been interrupted in 1934. For this reason and due to the fact that the narrative of a First and a Second Republic has become conventional in Austrian historiography it seems to be preferable.

After a decade of occupation by the victorious powers (France, Soviet Union, United Kingdom, United States) Austria regained its full sovereignty by the State Treaty of Vienna [Staatsvertrag von Wien] as of 1955. The Act on Neutrality [Neutralitätsgesetz], passed immediately after the State Treaty and after the last occupation forces had left the country (so demonstratively not under external pressure, but certainly as the result of a political deal with the Soviet Union), as well as accession to the European

Union (EU) in 1995 still modified Austria's international position. While Austria's neutrality—always just understood in the military, not political or ideological sense—was of great importance in times of an antagonistic Europe split into two political and military blocs, its importance declined following the end of this antagonism in 1989 to the level of non-participation in military alliances and denying the stationing of foreign military forces. Membership in the EU meant in the phase of a progressing European integration (up to the later 2000s) the ceasing of competences to EU authorities—such as renouncing to conduct a currency and a financial policy of one's own or acknowledging the primacy of the European Court of Justice over national jurisdiction.

4.1.3 Political System

The Federal Constitution defines as state authorities two chambers of the parliament, that is National Council [Nationalrat] and Federal Council [Bundesrat], as well as the Federal Government [Bundesregierung] and the Federal President [Bundespräsident]. The personal composition of the National Council with its 183 members results from nationwide general, free and secret elections according to the principles of proportional representation in intervals of five years at the maximum. The Federal Council is composed of 61 delegates of provincial parliaments and corresponds in its political structure to the results of elections to these parliaments. It is thus a body of delegated self-administration. When both chambers of the parliament meet, they constitute the Federal Assembly [Bundesversammlung]. According to the 1920 Constitution and up to 1929 it elected the Federal President. The amendment of 1929 ruled direct election of the Federal President by the population (in intervals of six years at the maximum) thus augmenting his/her democratic legitimation and in consequence political power (Brauneder and Lachmayer 1992; Brauneder 2009).

The Federal President is in principle free to charge anybody with constituting a federal government. For pragmatic reasons, however, this is usually the candidate of the largest fraction in the National Council, since a government needs to be supported by a majority there. A federal government without a majority in the National Council runs the risk of being swept out of office by a vote of non-confidence. In contrast to, for example, Germany, where a vote of non-confidence needs to be constructive, that is to present an alternative government, the Austrian constitution opens the way also for a non-constructive vote of non-confidence. The Federal President would then have to define a next candidate to constitute a government. This shows that the Federal President can in times of political turbulences play the role of a stability anchor and as a focal point of the political business at the federal level. In times of a stable government, however, his/her role is confined to representative functions.

Anyway, the Austrian political system at the federal level is far from a presidential or even only semi-presidential. The most important role is attributed to the federal government supported by the National Council and responding to it. It is the federal government that usually submits bills to the National Council (although also the National Council is entitled to do this) and gets them passed—since it can rely on a majority there. The Federal Council can object to acts passed by the National Council, but only with a suspensive effect (Brauneder and Lachmayer 1992; Brauneder 2009). This occurs anyway only exceptionally, since voting in the Federal Council follows not so much the interests of the lands, but party loyalties—and party majorities in the Federal Council usually resemble majorities in the National Council. Thus, the role of a chamber of the lands allotted to the Federal Council is hardly fulfilled.

The political system at the federal level is by constitution in the field of jurisdiction complemented by the Constitutional Court of Justice [Verfassungsgerichtshof], the Administrative Court of Justice [Verwaltungsgerichtshof], and the Supreme Court [Oberster Gerichtshof], the latter as the final of all levels of courts (district courts [Bezirksgerichte], provincial courts [Landesgerichte]). Jurisdiction is—also with reference to subordinate levels—a prerogative of the Federation (Brauneder and Lachmayer 1992; Brauneder 2009).

The Federal Constitution of 1920 defines the Republic of Austria also as a federation of self-governing provinces, states, or lands [Länder]. Since 1921 their number amounts to nine: Burgenland, Carinthia [Kärnten], Lower Austria [Niederösterreich], Salzburg, Styria [Steiermark], Tyrol [Tirol], Upper Austria [Oberösterreich], Vienna [Wien], and Vorarlberg

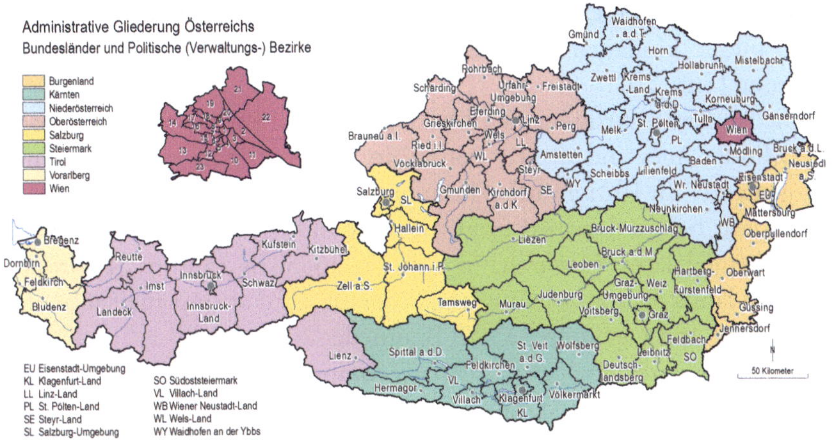

Fig. 4.5 Administrative-territorial subdivision of Austria into federal states, political districts, and (in Vienna [Wien]) communal districts (Source: Seger 2019)

(Fig. 4.5). They dispose over parliaments [Landtage], elected in intervals of five or six years at the maximum by the provincial population, as well as executive bodies (provincial governments [Landesregierungen] and governors [Landeshauptleute]) elected by the parliaments. The provincial parliaments have legislative power and are thus in the position to pass their own constitution and additional legislation, but provincial legislation can maneuver only within the limits of federal law. Governors are not only heads of their provincial government and thus bodies of provincial self-government, but at the same time authorities of delegated federal administration when they execute tasks on behalf of the federation assisted by the heads of the political districts [Bezirkshauptleute]. In exerting this function, they are, however, obliged to respect directives of the relevant federal minister (Brauneder and Lachmayer 1992; Brauneder 2009).

The lands dispose over their own budget, which is, however, prevailingly nourished by transfer payments from the Federation ('financial equalization' [Finanzausgleich]). This means that the Federation levies most taxes, while the lands have small tax revenues compared to their expenses. Periodical negotiations on this 'financial equalization' are a delicate item in the Austrian political calendar. Land competences extend to social and health care, nature protection, agriculture, fishing and forestry, public kindergartens and primary schools, tourism, and spatial planning (Sitte 2020).

The lands can partly be traced back to political entities that exist from the Middle Ages and display continuity and tradition ever since. Just Burgenland—in 1921 awarded from Hungary to Austria—and Vienna—in the year before detached from the surrounding Lower Austria—have been formed later. However, in contrast to Swiss cantons [Kantone] Austrian lands do not constitute the Federation but are just their subunits (Brauneder and Lachmayer 1992; Brauneder 2009).

At the lower regional level Austria is composed of 79 political districts as units of deconcentrated state administration (Fig. 4.5). Their heads are not elected, but civil officers installed by the land. The political districts can be traced back to the 1850s, when they were established complementary to juridical districts to cover the tasks of general administration. Ever since, they have modified their territorial configuration only in rare cases.

Local administration is performed by 2096 self-governing communes (as of 1 January 2019) with a communal council [Gemeinderat, Gemeindevertretung], in an interval of at the maximum five years elected by the local population, and an executive board [Gemeindevorstand] as well as a mayor [Bürgermeister] elected by the communal council. Later, in most lands except Vienna, Lower Austria, and Styria, a direct electing of the mayor by the population has been introduced. In total 770 out of the 2096 communes are 'market communes' [Marktgemeinden] and 201 'city/town communes' [Stadtgemeinden]. These are status categories that have something to do with (but are not bound to) numbers of inhabitants, historical traditions, and central place functions. The competences of communes rest mainly in the fields of public space management, emergency provision, the drafting of a surface management plan and of a development plan as well as in construction permits (Sitte 2020). There is no further territorial-administrative subdivision of communes except in Vienna.

Cities, but also some towns with a long history, are classified as 'cities/towns by statute' [Statutarstädte, Städte mit eigenem Statut] resulting in an administrative body, the magistrate [Magistrat], that combines the competences of a commune and a political district. The 15 cities/towns by statute (Fig. 4.5) are for this reason not part of the political district surrounding them (but frequently seat of its office [Bezirkshauptmannschaft]). Reforms have repeatedly changed size and territorial configuration of communes.

Vienna, Austria's federal capital, is at the same time a land and commune. Its communal council convenes for some agenda items in its capacity as provincial parliament and its city councilors and the mayor are at the same time provincial councilors and governor, respectively, although they are usually called city councilors and mayor. Vienna is subdivided into 23 communal districts [Gemeindebezirke] with a 'district representation' [Bezirksvertretung] elected in intervals of five years at the maximum by the district population. The district representation elects the 'head of the district' [Bezirksvorsteher]. The communal districts are no juridical persons, but components of the Commune of Vienna. Their competences are limited to the maintenance of schools and kindergartens, traffic, and public space management.

The Austrian administrative-territorial system is thus characterized by federalism and two (in Vienna three) sub-national levels of self-governing units (lands, communes, in Vienna also communal districts). In between them (except in Vienna) figures a level of deconcentrated state administration (political districts) that is to provide citizens with better access to services of the federal and provincial administration. The system fits well into Central European traditions that are shaped by subsidiarity instead of centralism from the High Middle Ages onward.

4.1.4 National Identity and Cultural Characteristics

Contrary to the interwar period, when German-speaking Austrians used to affiliate themselves to the German cultural nation, most Austrians today feel to be part of an Austrian cultural nation in the sense of a community bound together by common cultural characteristics (see Herder 1813; Symmons-Symonolewicz 1985). After World War II, this new consciousness has been supported by all major political forces—not the least to mitigate the consequences of the war defeat and to escape the stigmatization of all that was German by National Socialism (Bruckmüller

1984, 1996; Pelinka 1990; Pohl 1999; Thaler 2001). As an essential characteristic Austria's specific history was highlighted, since by language or religion the vast majority of Austrians (except linguistic minorities) is not different enough from Germans (see, e.g., Österreichische Rektorenkonferenz 1988: 7ff). The Austrian cultural nation can thus also function as a haven for the autochthonous linguistic minorities in Austria, who indeed use to regard themselves as parts of the Austrian cultural nation, although their 'mother nations' look at them usually as 'theirs'. Immediately after World War II, the contemporary federal minister of education Felix Hurdes (in office 1945–1952) considered even codifying an Austrian standard language distinct from German. This idea was, however, discarded later, mainly because this would have meant a break with Austria's German-speaking cultural traditions.

The politically promoted consciousness of an Austrian cultural nation spread quickly. Thus, the share of people regarding themselves as Austrians in the cultural-national sense rose from 47% in 1964 to 80% in 1993 (Bruckmüller 1996). Besides the activities of the postwar political actors this was due to "experiences of difference in the context of rule" (Bruckmüller as quoted by Stourzh 2004) that could be made between 1938 and 1945, the prosperous socio-economic development since 1945 (Thaler 2001) that was also very much based on the European Recovery Program (ERP) of the Marshall Plan granted by the United States as well as social peace that in turn may be regarded as a merit of the political and economic elites, who had learned their lessons from the interwar period. Austrian nation-building, even by success and velocity, is, however, not at all singular as the examples of Macedonians, Moldavians, or Montenegrins show. This confirms the ironic statement of Paul Robert Magocsi (related to the Rusins in Transcarpathia [Zakarpattja]): "Of the making of nationalities there is no end" (Magocsi 2000).

Austrian national consciousness, certainly less emphasized than French, Italian, or Croatian, but today perhaps more proudly presented than German, meets nevertheless two limitations. It is firstly second to regional consciousness, especially related to the very traditional lands, but also to smaller historical-cultural landscapes. Carinthia is a case in point with the strongest regional consciousness of all Austrian lands according to opinion polls (Seger 2005: 245).[2] It is secondly also limited by the consciousness of Austrian German-speakers to be part of a larger German-speaking area connected not only by traditional and modern literature, but also by the media. While Austrians play rather a more than proportional role in co-shaping German-speaking culture, it is simply the proportion of 1 to 10 in population numbers and economic power that matters and results in a much stronger impact of German culture on Austria than vice versa. This manifests itself in media consumption as well as 'brain migration', that is, the influx of German scientists and professors to Austria with the effect of a 'Germanization', for example, of Austrian universities (see Jordan and Seger 2019).

The emergence of a cultural-national Austrian identity after World War II based on Austria's specific history and its swift gaining ground in the vast majority of the population eased also essentially the minority situation in Carinthia by replacing the former German-Slovenian national antagonism. Majority as well as minority could now feel parts of the same cultural nation. As an ethnic group of the Austrian nation the Slovenian minority had now not only regional but also national identity in common with the majority in Carinthia.

Although it is true that language is not a constituent marker of Austrian cultural-national identity, since (1) also non-German speakers can be part of the Austrian nation and (2) the standard version of German spoken in Austria (like the German spoken in other parts of the German-speaking area including Switzerland) is not different from that spoken in Germany, German is a polycentric language with several varieties—and the so-called Austrian German is one of them. There exists quite a comprehensive Austrian dictionary. Many of its words are, however, neither used in all parts of Austria, nor is their use confined to Austria, but extends also to neighboring Bavaria [Bayern]. Thus, the existence of an 'Austrian German' is disputable (Pohl 1999, 2007).

[2] If the commercial use of place names (for names of hotels, restaurants, foodstuff, newspapers, music groups, etc.) is regarded as indicating prominence of regional identity, Carinthia ranks second among Austrian federal provinces behind Tyrol [Tirol]—provided that Vienna [Wien] and Salzburg are not taken into account, where city and province coincide (Vienna) or have the same name and can thus not be discerned (Salzburg) (see Stummvoll 2013; Jordan 2019).

It is, however, a fact that while the standard German written in Austria integrates Austria very well into the German-speaking area at the cognitive level, melody and intonation of the spoken language have an emotionally distancing effect. This is not so obvious opposite the southern parts of Germany, but very much so opposite the German North. A North German understands an Austrian speaking standard German without any problems and perceives him/her as a part of the same linguistic community, but emotionally as an 'other' (and vice versa). Austria and Germany are thus indeed "separated by the common language more than by anything else" as the Austrian cabaret artist Karl Farkas formulated it. It is, however, also true that the all-German television offer, mainly produced in (northern) Germany, exports (north) German expressions and phrases also to Austria and has a strong impact especially on the younger generation (see Jordan and Seger 2019).

The other standard languages spoken in Austria by autochthonous communities and residing in compact areas are Slovene, Burgenland-Croatian, and Hungarian. Slovene is the modern standard language of the approximately 18,000 (by colloquial language according to the population census 2001[3]) Slovenes living predominantly in southern Carinthia and minor parts of southern Styria. The Slovenian standard language is, however, not based on Carinthian-Slovenian dialects, but on the dialect of Carniola [Kranjska] in modern Slovenia, and Carinthian Slovenes learn it in the first line as an expression of their sense of community and almost like a foreign language.

Burgenland-Croatian is the modern standard language of about 19,000 Burgenland-Croats (Census 2001, Statistik Austria 2020) scattered over several 'linguistic islands' in the easternmost Austrian province of Burgenland (plus Vienna). It has been codified in 1982 as a standard language to escape the contemporary Yugoslavian language dispute (Serbo-Croatian or Croatian?) and to detach it from the odium of a clerical idiom. In 1987 it was elevated to the status of an official language in some communes of Burgenland (besides German).

[3] The 2001 Census was the last to document language or ethnic affiliation. The number does not include the category 'Windisch'.

The Hungarian with 4700 speakers of an autochthonous minority in Burgenland[4] (Census 2001, Statistik Austria 2020) conforms to the Hungarian standard language also spoken in Hungary. In some communes of Burgenland, it is an official language besides German.

"The German language is, irrespective of the rights granted to linguistic minorities the language of the Austrian state" (Article 8, Austrian Constitution, BKA 2020). Standard German is therefore also the only nationwide official language in Austria. Official languages with regional reference are in addition Slovene in Carinthia as well as Burgenland-Croatian and Hungarian in Burgenland. They hold this status in communes with a minimum share of the relevant minority of 17.5% in the total population according to recent population censuses (1991 and 2001) and enjoy it at courts, offices, and with names of towns and villages (not of communes and political districts) as well as at the army headquarters of these provinces, but not in provincial parliaments and provincial legislation.

4.1.5 Federal Provinces

The nine federal states, provinces, or lands correspond to the NUTS-2 level of the EU statistical classification system and have an average area of 9320 sq.km and an average population number of 981,967 (Statistik Austria 2020). By area they vary between 415 sq.km (Vienna) and 19,186 sq.km (Lower Austria), by population between 293,433 (Burgenland) and 1,897,491 (Vienna) (Statistik Austria 2020; see also Table 4.5). They roughly correspond to the central-place system at the upper level—with the significant exception of Vienna, whose catchment extends far beyond its boundaries into Lower Austria and even to Northern [Nordburgenland] and Central Burgenland [Mittelburgenland], and with the minor exceptions of Southern Burgenland [Südburgenland] gravitating toward Graz in Styria, East Tyrol [Osttirol], Political District Lienz, gravitating toward Carinthia, southwestern parts of Upper Austria

[4] The larger number of Hungarian-speakers outside Burgenland is not included here, since it is mainly composed of Hungarian refugees after the 1956 uprise and their descendants.

Table 4.5 Federal states, provinces, or lands [Länder] (see also Fig. 4.5)

Land	Area (in sq.km)	Population number in 1000 (2019)	Capital
Burgenland	3962	293	Eisenstadt
Carinthia [Kärnten]	9538	561	Klagenfurt a.W.
Lower Austria [Niederösterreich]	19,186	1678	Sankt Pölten
Salzburg	7156	555	Salzburg
Styria [Steiermark]	16,401	1243	Graz
Tyrol [Tirol]	12,640	755	Innsbruck
Upper Austria [Oberösterreich]	11,980	1482	Linz
Vienna [Wien]	415	1897	–
Vorarlberg	2601	394	Bregenz

Source: Statistik Austria (2020)

gravitating toward Salzburg, and western parts of Lower Austria gravitating toward Linz in Upper Austria.

Besides Carinthia [Kärnten] that will be highlighted in detail in Chap. 5 Burgenland is the other Austrian land with autochthonous minorities and the latest Austrian 'territorial acquisition'. The Treaty of Saint Germain (with Austria) as of 1919 and the Treaty of Trianon (with Hungary) as of 1920 ruled that (south)western parts of the former West-Hungarian counties Pressburg (Hungarian *Pozsony* [Bratislava]), Wieselburg [Moson], Ödenburg [Sopron/Ödenburg],[5] and Eisenburg [Vas] had to be awarded to the First Austrian Republic. Due to partisan resistance this could not be effectuated immediately, but only on the basis of the Venetian Protocols of 1921 signed between Hungary and Austria upon Italian mediation and convening on a swift surrender of the territory to Austria except the region of Sopron/Ödenburg, where a plebiscite had to be held (see, e.g., Suppan 2019). As this plebiscite resulted in a vote against Austria, the urban center of the region and its 'natural' capital remained with Hungary. After a provisional capital had been

[5] Sopron/Ödenburg and Mosonmagyaróvár/Wieselburg-Ungarisch Altenburg are according to Hungarian minority legislation since 1993 officially bilingual and therefore adequately represented here.

established in Mattersdorf, later renamed to *Mattersburg*, Eisenstadt became Burgenland's capital.

Austria's latest territorial acquisition is founded on the argument that it had a German population from the twelfth and thirteenth centuries, very likely Bavarian migrants from neighboring Lower Austria and Styria, perhaps also from Danubian regions west of Lower Austria [Niederösterreich] up to Passau, speaking a Middle Bavarian dialect (Fig. 4.6). This population extended even beyond Burgenland's current boundaries into modern Hungary, especially into the areas of Mosonmagyaróvár/Wieselburg-Ungarisch Altenburg and Sopron/Ödenburg. When the Hungarian national idea gained ground in the nineteenth century and after the Austro-Hungarian Compromise of 1867 turned the Hungarian Kingdom into a Hungarian nation state with Hungarian as the only official language, German-speakers began to feel some sympathy with the adjacent Austrian lands, although irredentism found no local support in German West-Hungary (in German

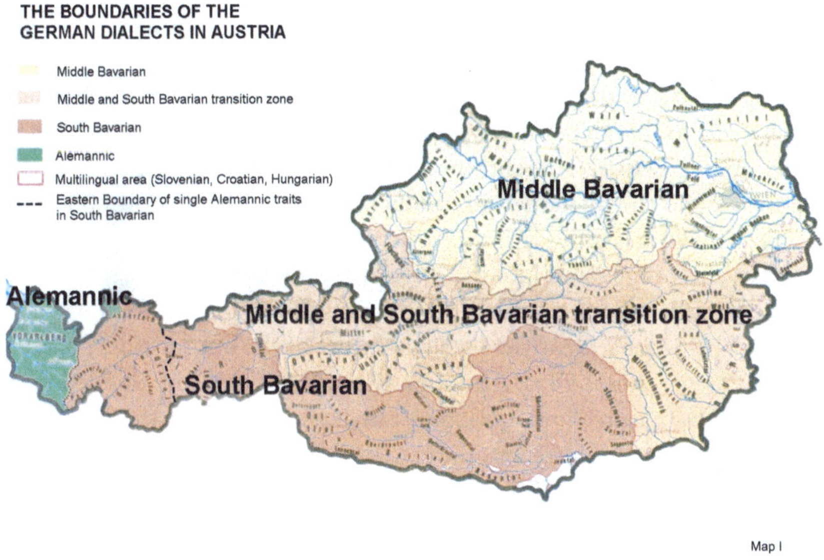

Fig. 4.6 German dialects spoken in Austria. (Source: AKO 2012)

Deutsch-Westungarn)—as roughly the region of modern Burgenland was then called.

A second community making modern Burgenland specific, but in smaller numbers also present in western parts of modern Hungary and southwest Slovakia, are Croats having been evacuated by their Hungarian landlords from the Austrian frontier against the Ottoman Empire in the sixteenth century to areas in contemporary West-Hungary earlier devastated by Ottoman invasions. They were not settled in a compact area, but in a kind of islands scattered over various parts of Burgenland (see Breu 1970).

Except a German ("German West-Hungary") Burgenland had no common identity up to its awarding to Austria as it had been composed of parts of four Hungarian counties. Also a name still had to be found. *Burgenland* ('land of the castles') proved to be the most appropriate, since all Hungarian counties out of which it had been composed had the generic *burg* ('castle') in their German name (Pressburg, Wieselburg, Ödenburg, Eisenburg). Identity building was thus a difficult and initially also much disputed process (Haslinger 2001). While Social Democrats and (of course) German Nationals favored in the interwar period a German identity, Christian Socials advocated a 'Pannonian', within Austria rather exotic identity. Following World War II, the latter idea gained ground and proved to be a real success, also in tourism marketing. It is also due to this fact that Burgenland styles itself today a multicultural region augmented by autochthonous Croatian, Hungarian, and Roma minorities, and the introduction of official bilingual naming in 2000 did not meet (in contrast to Carinthia) any problems (Richnovsky 2013).

Today, the Burgenland identity is distinct and even more distinct than the respective land identity in the traditional Lower Austria, since it is less contested by quarter identities—the identities of the four distinct parts of this land (Seger 2005: 245). Although Burgenland is clearly divided into three parts by nature (low mountain ranges) in the North-South direction, that is, into Northern Burgenland [Nordburgenland], Central Burgenland [Mittelburgenland], and Southern Burgenland [Südburgenland], these are no significant anchors of space-related identity. It is rather the Seewinkel east of Lake Neusiedl [Neusiedler See] that has this capacity thanks to its truly Pannonian character (close to that of the Hortobágy Puszta in Hungary). The three sections of the province are, however, rather distinct by external orientation. While Northern Burgenland and—somewhat less—also Central Burgenland are oriented toward Vienna in terms of commuting to labor, longer term supply, higher education (the three in the direction to Vienna), and recreation (in the opposite direction), Southern Burgenland has similar relations with the Styrian capital Graz.

Due to the traditional socio-economic gap between the Austrian lands and the Hungarian Kingdom, Burgenland was poor and peripheric in the interwar period and could not gain ground at this time, because the First Republic in its entirety was economically deteriorated. Burgenland was thus the main source of emigration from the territory of modern Austria to overseas (see Dujmovits 2012). Also in the postwar period up to 1989 Burgenland remained clearly the last of all Austrian lands, although it had profited from some tourism from the 1970s onward. Being Austria's outpost at the Iron Curtain to Hungary and Czechoslovakia and having been part of the Soviet occupation zone between 1945 and 1955 prevented any essential step forward. This was only possible later by placing European pre-accession funds and—after Austria's EU accession in 1995—EU regional funding in this region at a socio-economic level below the EU average. Appropriate investment based on these funds, together with the fall of the Iron Curtain and the effects of a restored metropolis function of Vienna (mainly on Northern Burgenland) caused the socio-economic rise of Burgenland, also in the ranking of Austrian lands. While yokes on Burgenland and the Burgenlanders were popular some decades ago, they have lost their factual background today.

4.1.6 Minority Legislation

As the starting point of minority legislation in Austria can be regarded the Act on Basic Rights as of 1867 [Staatsgrundgesetz RGBl. 142/1867, BKA 2020] passed with establishing the dual monarchy and valid for the kingdoms and lands represented in the Imperial Council, that is the 'Austrian lands' or 'Cisleithania', not for the Hungarian Kingdom that conducted a policy of Magyarization (Brauneder and Lachmayer 1992;

Brauneder 2009). It rules by its Article 19 that "all peoples of the state have equal rights, and every people has an inviolable right to preserve and maintain its nationality and language. The state recognizes the equality of all national languages in school, office and public life. In the lands in which several peoples live, public educational institutions should be set up in such a way that each of these peoples receives the necessary means of training in their own language without the need to learn a second national language."[6] This was progressive compared to contemporary European standards and generously effectuated at least related to toponymy as Chap. 6, Sect. 6.1.1 will show. Yet, it could not calm down the rising nationalism of that time.

In the interwar period nationalism culminated all over Europe, and the first Austrian republic was no exception. German-speaking Austrians conceived themselves as Germans and Austria as a second German state, as a nation state of Germans with all the implications for ethnic and linguistic minorities. In the first years of its existence, a majority of its political elite (certainly German Nationals and Social Democrats) was even determined to attach it to Germany. Austria nevertheless adopted by content the Act of Basic Rights as of 1867 also related to its minority regulations and had furthermore to accept the minority regulations of the Peace Treaty of Saint Germain as of 1919 that were to be controlled by the League of Nations (Brauneder and Lachmayer 1992; Brauneder 2009). They are summed up in Articles 67 and 68 with the following wording:

> Article 67: "Austrian nationals belonging to a minority by race, religion or language enjoy the same treatment and the same guarantees, legally and factually as the other Austrian nationals; in particular, they have the same right to set up, manage and supervise charities, religious or social institutions, schools and other educational institutions at their own expense, with

[6] Article 19 in its original wording: "Alle Volksstämme des Staates sind gleichberechtigt, und jeder Volksstamm hat ein unverletzliches Recht auf Wahrung und Pflege seiner Nationalität und Sprache. Die Gleichberechtigung aller landesüblichen Sprachen in Schule, Amt und öffentlichem Leben wird vom Staate anerkannt. In den Ländern, in welchen mehrere Volksstämme wohnen, sollen die öffentlichen Unterrichtsanstalten derart eingerichtet sein, daß ohne Anwendung eines Zwanges zur Erlernung einer zweiten Landessprache jeder dieser Volksstämme die erforderlichen Mittel zur Ausbildung in seiner Sprache erhält" (BKA 2020).

the right to use their own language in them at will and to practice their religion freely."[7]

Article 68: "As far as public education is concerned, the Austrian government in the cities and districts where a comparatively large number of non-German-speaking citizens of Austria live will provide reasonable facilities to ensure that in primary schools children of these Austrian citizens are taught in their own language. This provision will not prevent the Austrian government from making teaching the German language in the schools in question a compulsory subject.

In cities and districts, where a relatively large number of Austrian citizens live who belong to a minority by race, religion or language, these minorities are secured a fair share of benefits and uses from all amounts, for example for education, religious or charitable purposes, from public funds in state, community or other budgets."[8]

As much as the Peace Treaty in total, its minority regulations were at least by Austria's political elite and especially by the elites in Carinthia conceived as imposed and effectuated reluctantly, if at all (see, e.g.,

[7] Article 67 in its original wording: "Österreichische Staatsangehörige, die einer Minderheit nach Rasse, Religion oder Sprache angehören, genießen dieselbe Behandlung und dieselben Garantien, rechtlich und faktisch, wie die anderen österreichischen Staatsangehörigen; insbesondere haben sie dasselbe Recht, auf ihre eigenen Kosten Wohlthätigkeits-, religiöse oder soziale Einrichtungen, Schulen und andere Erziehungsanstalten zu errichten, zu verwalten und zu beaufsichtigen mit der Berechtigung, in denselben ihre eigene Sprache nach Belieben zu gebrauchen und ihre Religion frei zu üben" (BKA 2020).

[8] Article 68 in its original wording: "Was das öffentliche Unterrichtswesen anlangt, wird die österreichische Regierung in den Städten und Bezirken, wo eine verhältnismäßig beträchtliche Zahl anderssprachiger als deutscher österreichischer Staatsangehöriger wohnt, angemessene Erleichterungen gewähren, um sicherzustellen, daß in den Volksschulen den Kindern dieser österreichischen Staatsangehörigen der Unterricht in ihrer eigenen Sprache erteilt werde. Diese Bestimmung wird die österreichische Regierung nicht hindern, den Unterricht der deutschen Sprache in den besagten Schulen zu einem Pflichtgegenstande zu machen. In Städten und Bezirken, wo eine verhältnismäßig beträchtliche Anzahl österreichischer Staatsangehöriger wohnt, die einer Minderheit nach Rasse, Religion oder Sprache angehören, wird diesen Minderheiten von allen Beträgen, die etwa für Erziehung, Religions- oder Wohltätigkeitszwecke aus öffentlichen Mitteln in Staats-, Gemeinde- oder anderen Budgets ausgeworfen werden, ein angemessener Teil zu Nutzen und Verwendung gesichert" (BKA 2020).

Suppan 2019).[9] This refers also to toponymy: No Slovenian name was elevated to official status and Slovenian names were eradicated from official topographical maps (see Chap. 6, Sect. 6.1.1).

Styled as a kind of reward for the plebiscite of 1920, when also a considerable share of Carinthian Slovenes (about 40%) had voted for Austria and against the State of the Serbs, Croats, and Slovenes (SHS) (for more details, see Chap. 5, Sect. 5.1), and as an attempt to provide the Slovenian ethnic group with a definite status in Austria, a 'cultural autonomy' was offered to them. The negotiations between 1925 and 1929, however, failed, since the German-speaking side did not accept the Carinthian-Slovenian demand to nominate the Slovenian school inspectorate (Reimann 2004: 73). Other sources mention a German-Carinthian demand that every Carinthian Slovene had to register in an 'ethnic cadaster' and that this had not been accepted by the Slovenian side (Österreichische Rektorenkonferenz 1988: 17). These other sources are confirmed by the historian Helmut Rumpler, who qualifies the offer of a cultural autonomy as a tactical maneuver of its initiators, the Carinthian Social Democrats, demanding from the Carinthian Slovenes ethnic confession, while they knew that their representatives would never accept this (Rumpler 2005: 40f).

After the National-Socialist period and World War II (1938–1945), when Austria was incorporated into the German Reich, with all its atrocities having affected all Austrian minorities to a special extent,[10] the situation in Austria was characterized by occupation by the victorious allies, limited sovereignty up to 1955, and being located at the frontier to the Communist bloc. This bloc included up to 1948, the break between Tito and Stalin, also Yugoslavia. The latter had (like the first Yugoslavian state after World War I) occupied parts of Carinthia and held a claim on them.

---

[9] The same is true for all other successor states of the Austrian-Hungarian Monarchy, in particular for Czechoslovakia as regards the Language Act of 1920 and the Brno Treaty between Czechoslovakia and Austria of the same year (Suppan 2019: 180).

[10] The National-Socialist regime pursued in Carinthia the "extinction of the Slovenes' ethnic individuality" (Ferenc 1980: 5), liquidated, expelled, and deported many Carinthian Slovenes and replaced them mainly by German optants from Italian South Tyrol [Südtirol/Alto Adige] and the Kanal Valley [Val Canale/Val Cjanâl/Kanalska dolina/Kanaltal]. In reaction, local and Yugoslavian partisans liquidated German-Carinthians, and many of them were deported to Yugoslavia with no return.

After Yugoslavia had left the 'Moscovite camp' and started its own way toward Socialism in 1949, this claim was not maintained anymore. It nevertheless joined the Treaty of Vienna 1955, colloquially called 'Austrian State Treaty', on the side of the victorious powers and felt thus to be entitled to watch Austrian minority legislation—at least as regards South-Slavonic minorities. Consequently, Austria enacted measures to comfort minorities and to prevent interventions by the victorious, later by the Treaty's signatory powers. In Carinthia, an act on tuition implemented in 1945 ruled that in all elementary schools of the so-called bilingual area with an overall majority of German-speakers and several areas with near to zero Slovene-speakers all pupils had to learn both provincial languages, that is German and Slovene, whether they were members of the minority or not. The act theoretically affected 107 schools in 62 communes but was actually practiced only in 96–98 schools according to Claudia Fräss-Ehrfeld (2005: 123).[11]

The Treaty of Vienna as of May 1955 re-establishing Austria's full sovereignty, bringing occupation by the victorious allies to an end and elevated to the ranks of Austria's constitutional law, included extensive regulations for ethnic/linguistic minorities by Article 7 that was essentially shaped by the Soviet Union (Veiter 1980; Matscher 2006; Fräss-Ehrfeld 2005: 120) and became the baseline of all later minority legislation in Austria. This article titled "Rights of the Slovenian and Croatian minorities" runs as follows:

1. Austrian citizens of the Slovenian and Croatian minorities in Carinthia, Burgenland and Styria enjoy the same rights on the basis of the same

[11] It needs to be remarked that compulsory bilingual tuition in the bilingual area (following the territorial and not the personal principle) has a different meaning for majority and minority. While for the minority it is in the public and professional sphere necessary to have a command of the majority language and actually every minority member in Carinthia has a command of it, it is for majority-language speakers just a matter of humanism to learn it. Thus, necessity and motivation are asymmetric. When modern Slovenian and Croatian Istria [Istra] with small Italian minorities are quoted as examples of good practice in this respect, it should not be disregarded that Italian was up to 1945 the dominant and official language there with all its prestige and that every Slovene and Croatian native speaker residing in the minority area is anyway bilingual. It has, however, also to be mentioned that Slovenia practices in the Prekmurje a similar language regime, where at least today the minority language Hungarian has not this same prestige.

conditions as all other Austrian citizens, including the right to their own organizations, assemblies and press in their own language.
2. They are entitled to elementary tuition in Slovenian or Croatian and to a proportionate number of their own secondary schools; in this context, school curricula will be reviewed and a department of the School Inspectorate will be set up for Slovenian and Croatian schools.
3. In the administrative and juridical districts of Carinthia, Burgenland and Styria with Slovenian, Croatian or mixed population, the Slovenian or Croatian language is permitted as an official language in addition to German. In such districts, the names and inscriptions of a topographical nature are written both in Slovenian or Croatian and in German.
4. Austrian citizens of the Slovenian and Croatian minorities in Carinthia, Burgenland and Styria take part in the cultural, administrative and judicial institutions in these areas based on the same conditions as other Austrian citizens.
5. The activity of organizations which aim to deprive the Croatian or Slovenian population of their property and their rights as a minority must be prohibited.[12]

This Article was, however, too generally formulated to be effectuated directly, it was not "self-executing" and required more detailed legislation (Veiter 1980; Unkart et al. 1984; Matscher 2006; Öhlinger 2006).[13] As

[12] Article 7 in its original wording: "1. Österreichische Staatsangehörige der slowenischen und kroatischen Minderheiten in Kärnten, Burgenland und Steiermark genießen dieselben Rechte auf Grund gleicher Bedingungen wie alle anderen österreichischen Staatsangehörigen einschließlich des Rechtes auf ihre eigenen Organisationen, Versammlungen und Presse in ihrer eigenen Sprache. 2. Sie haben Anspruch auf Elementarunterricht in slowenischer oder kroatischer Sprache und auf eine verhältnismäßige Anzahl eigener Mittelschulen; in diesem Zusammenhang werden Schullehrpläne überprüft und eine Abteilung der Schulaufsichtsbehörde wird für slowenische und kroatische Schulen errichtet werden. 3. In den Verwaltungs- und Gerichtsbezirken Kärntens, des Burgenlandes und der Steiermark mit slowenischer, kroatischer oder gemischter Bevölkerung wird die slowenische oder kroatische Sprache zusätzlich zum Deutschen als Amtssprache zugelassen. In solchen Bezirken werden die Bezeichnungen und Aufschriften topographischer Natur sowohl in slowenischer oder kroatischer Sprache wie in Deutsch verfaßt. 4. Österreichische Staatsangehörige der slowenischen und kroatischen Minderheiten in Kärnten, Burgenland und Steiermark nehmen an den kulturellen, Verwaltungs- und Gerichtseinrichtungen in diesen Gebieten auf Grund gleicher Bedingungen wie andere österreichische Staatsangehörige teil. 5. Die Tätigkeit von Organisationen, die darauf abzielen, der kroatischen oder slowenischen Bevölkerung ihre Eigenschaft und ihre Rechte als Minderheit zu nehmen, ist zu verbieten" (BKA 2020).

[13] According to Fräss-Ehrfeld the rather general wording of Article 7 as proposed by the Soviet Union was appreciated by the Western allies (as well as by the Austrian government) and accepted by them, because it offered a wide range for interpretation (Fräss-Ehrfeld 2005: 120).

regards Paragraph 3, relevant for toponymy, for example, at least five items had to be clarified (see also Matscher 2006): (1) Who is a minority member? (Somebody declaring to speak Slovene as the colloquial language? Also somebody declaring to speak 'Windisch' as the colloquial language? Somebody affiliating himself/herself to the minority in a secret ballot? Families enrolling their children for bilingual tuition?) (2) Reference area of minority regulations. Should it really be "administrative and juridical districts" in their entirety including in most cases areas without any minority population; should it be the so-called bilingual area to which also the act on bilingual tuition was applied; or should it be communes or even only populated places with a minority population irrespective of their spatial context?[14] (3) Which share/percentage of minority population is required to speak of a "mixed population"? (4) What means "officiality" of a language? Is it to be official at the communal level, at the level of juridical and administrative districts, at the level of the province, or even at the level of the Federation? In which communicative situations and for which purposes is it to be official? (5) What are "names and inscriptions of a topographical nature" and where have they to appear in both languages? Do they comprise only names of populated places or perhaps also names of streets, communes, and political districts, perhaps even names of natural features like mountains and water bodies, of technical infrastructure like railroad stations, railroad lines, roads, and motorways? And have they to be represented only by local town signs or also on road signs, on maps or in official texts on these places?

Already in October 1955, the two contemporary umbrella organizations of the Carinthian Slovenes, the Christian-conservative Council of Carinthian Slovenes [Rat der Kärntner Slowenen/Narodni svet koroških Slovencev], and the rather leftist Central Association of Slovenian Organizations in Carinthia [Zentralverband slowenischer Organisationen in Kärnten/Zveza slovenskih organizacij na Koroškem] presented a memorandum interpreting Article 7 of the Treaty and defining their position related to it. In respect to its Paragraph 3 highlighting "names and

[14] Representatives of Carinthian Slovenes used to conceive the reference area as "clearly defined" by the reference area of the act on bilingual tuition as of 1945. The negotiators of the Treaty of Vienna, however, were not even aware of this act on tuition and its reference area (Fräss-Ehrfeld 2005: 121).

inscriptions of a topographical nature" they subsumed under this heading not only town signs, but also street signs, plates on all public buildings including schools as well as post service and railroad facilities (Hren 2004: 100).

In an atmosphere of diminished external threat (e.g., by Yugoslavia), growing welfare, and self-consciousness in Austria, Austrian political leaders tended to postpone solving these questions, partly neglecting the minority problem, partly not willing to touch this delicate issue. A telling indicator of the feeling not to be under external pressure anymore is the withdrawal of the tuition act of 1945 and its replacement in 1959 by a regulation ruling that parents had to enroll their children for tuition in the minority language, after a decree of the Carinthian governor as of 1958 had already allowed parents to sign out their children from bilingual tuition. It had resulted in 83% of pupils signed out—a motion called "negative minority assessment" by Ralf Unkart, the former head of the Carinthian provincial office (quoted by Fräss-Ehrfeld 2005: 131 and 133). The new act on bilingual tuition in Carinthia affecting 98 elementary schools was accompanied by an act on language at courts in Carinthia. Both dissatisfied the Slovenian minority.

It has to be noted at this point that in this period the Socialist Party [Sozialistische Partei Österreichs, SPÖ][15] had the absolute majority in the Carinthian provincial parliament and held consequently also the position of the Carinthian governor. One could assume that Socialists, elsewhere known for rather internationalist and a-national attitudes would have tried to promote measures of minority protection or at least to make efforts of promoting an understanding for them among their own party officials and electorate. This was, however, not the case in Carinthia (Hren 2004) and has its primary reason in a historical fact that may at first sight be regarded as far-fetched: After most of the Habsburg lands had been affected by Reformation, Habsburg-supported Counter-Reformation had succeeded also in Carinthia, but to a minor extent than in other parts of the Habsburg realm. A larger Protestant minority remained among German-Carinthians up to the present day than in

[15] The Austrian Social Democrats named themselves *Socialists* (Socialist Party of Austria [Sozialistische Partei Österreichs, SPÖ]) between 1945 and 1991.

other parts of modern Austria, and Protestantism had a lasting impact on lifestyle and attitudes also on re-converted Catholics expressing itself inter alia in a certain inclination toward a-clerical positions and in a political landscape characterized (on the German-Carinthian side) by an oscillation between (a-clerical) Social Democrats, at times calling themselves Socialists, and the (a-clerical) German Nationals, in the Second Republic the Freedom Party and its offsprings, while Conservatives (closer to Catholicism) were well into the postwar period associated with Carinthian Slovenes.

Slovenes in Carinthia had—as Slovenes in their entirety—also been reformed, but later completely re-converted to Catholicism.[16] During their 'national awakening' in the nineteenth century, the Roman-Catholic clergy—replacing a missing bourgeoisie—had even assumed the role of their political leaders associating them with the Roman-Catholic Church even more. Thus, Social Democrats, in the postwar period Socialists, conceived themselves—like the later Freedom Party—as representatives of the German-Carinthian majority and were competing for the same electorate, contrasting themselves from 'clerical' Conservatives and attributing them in a pejorative way 'minority-friendliness'. It can thus also be said that the later shift of political dominance in Carinthia from Socialists/ Social Democrats to the Freedom Party led by Jörg Haider—obtaining international prominence as a right-wing populist—did mean all but a fundamental change in Carinthian minority politics.

A second reason for a certain German-Carinthian bias of more or less all major political parties in Carinthia is the historically rooted fear of the socially dominating majority to lose a part of its realm and influence (see Moritsch 2000)—conventionally called 'Carinthian primeval fear' [Kärntner Urangst]. It was nourished by the rise of national ideas that had among Slovenes the effect of defining a "Slovenian ethnic territory" including southern Carinthia. Up to the end of World War I, it figured as the territorial reference of a "Slovenian kingdom" to be realized. This would have divided several pre-national cultural-historical units, among

[16] With the tiny exception of the Slovenes in the Prekmurje, a part of modern Slovenia that belonged up to 1918 to Hungary proper, where Catholic Counter-Reformation was executed much less energetically than in the Austrian lands.

them Carinthia (Moritsch 2000: 20). A first document of this attitude is Peter Kozler's map of the "Slovenian lands and regions" as of 1853 (Kozler 1853, see Fig. 5.15), a next, already very physical manifestation of the occupation of southern Carinthia by SHS forces after World War I[17] (see, e.g., Suppan 2019: 68ff), another the claims of Communist Yugoslavia on the "Slovenian ethnic territory" in Carinthia after World War II as documented, for example, by an atlas published by an official Yugoslavian authority (recognizable by the coat of arms of the Socialist Federative Republic of Yugoslavia at its cover) but without an explicit author or editor (Anonymous s.a. 1945).[18] For the sake of scientific correctness, it also has to be mentioned that this external claim was supported locally at least by a fraction of the Carinthian minority, that is, the Liberation Front for a Slovenian Carinthia [Osvobodilna Fronta za Slovensko Koroško, OF] figuring after World War II as a political party admitted by the Republic of Austria as well as by the British occupation forces and initially even represented in the Carinthian provincial parliament (Fräss-Ehrfeld 2005: 90). It also has to be taken into account that Carinthian Slovenes were conceived not only by German-Carinthians, but also by 'official Austria' at least between 1945 and 1955 as supported by the victorious allies and thus as enjoying a powerful position. The Treaty of Vienna as of 1955 undoubtedly took away from this scenario much of its sharpness and realistic content, but Yugoslavia remained a powerful player in this wider region up to its dissolution. Only the independence of (a small and not at all 'dangerous') Slovenia in 1991 and her joining the European Union in 2004 deprived these fears of any realistic background.

What also has to be taken into account is that Austrian federal politics and Austrian mainstream media outside Carinthia used to rank good

[17] Already the national council of the ephemeric first South-Slavonic state formation, the State of Slovenes, Croats, and Serbs, held in Zagreb 19 October 1918, proclaimed the "unification of the entire people of Slovenes, Croats and Serbs on its entire ethnographic territory" (Rumpler 2005: 13). Even earlier, since 19 September 1918, the Ljubljana national council maintained a branch in Klagenfurt (titled "Slovenian National Council for Carinthia") that denied to take part in re-establishing Carinthia as a federal state of German-Austria (Rumpler 2005: 13).

[18] The introduction to this atlas ends with the sentence: "Consequently, the right of the Carinthian Slovenes to be united to Yugoslavia is founded on ethnical facts which the Austrian Germans of Carinthia tried to dissimulate and to misrepresent by all means at their disposal" (Anonymous s.a. 1945).

relations with Yugoslavia and Austria's favorable image at the international scene as prior to "those irrational fears" of German-Carinthians. This, in turn, nourished German-Carinthian suspicions, Vienna would be ready to cede parts of Carinthia, which may not always have been unrealistic as the historian Helmut Rumpler confirms at least for 1918 ("Oktobermanifest" of 16 October 1918, Rumpler 2005: 13).

Returning after this longer deviation to historical backgrounds to the flow of events in the later 1950s, it can—in face of the delicate character of minority legislation and the rather limited prospects of gaining popularity by promoting it—be supposed that Carinthian Socialists felt relaxed by a decision of the Constitutional Court of Justice as of 1958 ruling that the execution of the Treaty of Vienna as well as all minority legislation in Austria is a prerogative of the federal level. It was now the federal parliament (composed of National Council and Federal Council) that had the power to pass acts with relevance for ethnic/linguistic minorities, as usual proposed and submitted by the federal government. Relevant provincial parliaments (Burgenland, Carinthia) had no other choice than to accept them and convert them literally to provincial legislation. This had and has the advantage that not the local or regional majority—in a case like Carinthia heavily involved into local disputes and burdened by historical events—but the more distanced, 'neutral' authorities in Vienna decide about minority rights. This bears, however, also the danger of passing decisions without sufficiently exploring local/regional acceptance.

The minority place-name act of 1972 is a striking example for that. Following a growing impatience of the Slovenian minority to execute Article 7 of the Treaty of Vienna and in particular its Paragraph 3 expressing itself in vandal actions of "supplementing" (Hren 2004: 102) monolingual town signs by Slovenian names, it was submitted by the Socialist federal government led by Federal Chancellor Bruno Kreisky and passed by the federal parliament with an absolute Socialist majority. The federal government had the support of Carinthia's Governor Hans Sima, also from the Socialist Party, and relied on his assessing local acceptance well. The Act ruled the placement of 205 bilingual town signs in Carinthia with a minimum share of 20% Slovene-speakers according to the population census of 1961 (Matscher 2006: 115). These town signs were immediately placed and prompted local and regional resistance going as far as damaging or tearing out these town signs (titled "Carinthian place-name storm"). Many of these actions had certainly been orchestrated by 'homeland defenders'[19]—as the traditional Carinthian associations not in favor of minority rights and at least at that time regarding the Slovenian ethnic group as a threat to Carinthian unity conceive themselves. Others, however, were certainly also spontaneous, since inhabitants of some villages and towns, not only German-Carinthians, did not wish to be 'stigmatized' by bilingual town signs.

What can be said for certain is that the political approach characterized by taking advantage of an absolute political majority at the federal and provincial levels, not taking other political directions 'on board' combined with a missionary attitude (not untypical for Socialists or Social Democrats), was insufficient. Facing the magnitude of problems arising and in a mode of political opportunism, first Federal Chancellor Bruno Kreisky gave in, followed by a majority of Carinthian Socialist Party officials being afraid of reducing their chances with upcoming communal elections, leaving Governor Hans Sima 'standing in the rain'. The Act was finally withdrawn, the bilingual town signs disappeared, and Hans Sima had to resign first as the chair of the Carinthian Socialist Party (1973), in 1974 also as the governor of Carinthia (Hren 2004).

The following years saw intensive efforts to 'repair' the political and societal damages caused by the events of 1972. After an unsuccessful nationwide mother tongue census of 14 November 1976[20] they resulted in the Federal Act on Ethnic Groups [Volksgruppengesetz] of 1976[21] and its implementing legislation by decrees in 1977. It established ethnic-group advisory councils [Volksgruppenbeiräte] for the Croatian, Slovenian, Hungarian, Czech, Slovakian, and Roma ethnic groups with the Federal Chancellery [Bundeskanzleramt]. The Decree of Topography for Carinthia [Topographieverordnung für Kärnten] as of 1977 (BGBl. 306/1977, BKA 2020) ruled that 91 populated places [Ortschaften] in

[19] Josef Feldner, chair of the Carinthian Homeland Association [Kärntner Heimatdienst], denies that his organization had orchestrated any action of this kind (Feldner and Sturm 2007).

[20] The census was boycotted by the minority organizations and resulted in a number of only 2535 speakers of Slovene as a mother tongue in Carinthia. In Vienna [Wien] a higher number was registered (Matscher 2006: 122).

[21] A revised version of this Act has been published in 2013 (BGBl. 84/2013, see BKA 2020).

Carinthia had to have officially bilingual names. The threshold relevant for this number was 25% of Slovene-speaking population in communes [Gemeinden] as of 1955 and according to the population census of 1951. Reference to 1955 and 1951 was made, because the Decree was conceived to be the execution of Article 7 of the Treaty of Vienna (1955).

This Decree was, however, only hesitatingly and gradually effectuated. Neither were all 91 bilingual town signs located on the spot, nor were the officially bilingual names of populated places represented on official Austrian topographical maps. This prompted the author of this subchapter to demonstrate by a book (Jordan 1988) that rendering official minority place names on topographical maps would not only comply to Austrian traditions before World War I and to modern European standards, but also not overburden maps cartographically—a pragmatic argument that was frequently brought forward against bilingual names. Based on this book, Karel Smolle, a Carinthian-Slovenian delegate to the Austrian National Council, initiated a parliamentary process resulting in a directive of the relevant federal ministry to the Federal Office of Metrology and Surveying ruling that officially bilingual names of populated places in Carinthia had to be represented on their topographical maps. The Federal Office reacted by gradually introducing them on their maps with every cartographic revision of the individual map sheet.

The low number of officially bilingual names and their reluctant implementation by town signs dissatisfied, however, the minority in Carinthia and stirred up the political climate. Surprisingly enough, a federal government coalition of the conservative Austrian People's Party [Österreichische Volkspartei, ÖVP] and the Austrian Freedom Party [Freiheitliche Partei Österreichs, FPÖ] under Federal Chancellor Wolfgang Schüssel, the Freedom Party known for their all but minority-friendly attitudes, passed in 2000 a Decree on Topography in Burgenland [Topographieverordnung Burgenland] (BGBl. 170/2000, BKA 2020) ruling that 47 populated places with at least 10% minority-language speakers according to the population census 1991 in this federal province must have bilingual German/Croatian town signs and 4 bilingual German/Hungarian. Much in contrast to Carinthia in 1977, this Decree was effectuated immediately and without any exception. It met—for reasons explained in Sect. 4.1.5—also no local opposition.

In 2001, in reaction to the intervention of a Carinthian-Slovenian advocate, the Constitutional Court of Justice [Verfassungsgerichtshof] referring to the Decree of 1977 concerning Carinthia passed a decision stating that a percentage of 25% minority speakers was too high and recommended to reduce it to 10% based on the average results of more recent population censuses (1961, 1981, 1991, 2001, and with reference to the criterion 'colloquial language'), but recommended also to regard this threshold of 10% just as a guideline and basis for comprehensive political negotiations. When regarding censuses of the colloquial language as relevant, the Court however contradicted an earlier finding of 1987 "that proof of ethnic group membership based on objective criteria such as origin or colloquial language should not be required due to the associated discriminatory effects" (VfGH Slg. 11.585/1987 quoted after Öhlinger 2006: 128).[22] Long-lasting negotiations between all relevant political forces and several unsuccessful attempts to find a solution followed. In the meantime, the number of officially bilingual populated places in Carinthia had been raised by two from 91 to 93.

In 2011, finally a compromise could be achieved that was accepted by all relevant political forces both at the federal and Carinthian provincial level and legally implemented accordingly. Most important for this success was the intention of the contemporary federal government led by the Social Democrats and the Carinthian provincial government led by Gerhard Dörfler representing Alliance Future Austria, the successor of the Freedom Party, to have all relevant political parties and directions involved—including the Carinthian homeland associations (Carinthian Homeland Association [Kärntner Heimatdienst] and Union of Carinthian Homeland Defenders [Kärntner Abwehrkämpferbund]). A strong symbol with some impact was also the remarkable personal reconciliation between former exponents of the two antagonistic ethnic 'camps', Josef Feldner, chair of the Carinthian Homeland Association, and Marjan Sturm, chair of the Central Association of Slovenian Organizations in Carinthia [Zentralverband slowenischer Organisationen in Kärnten/

[22] When it is, however, taken into account that answering the question for 'colloquial language' is by most census respondents rather understood as subjectively declaring one's ethnic affiliation (as the author of this subchapter thinks that it is) the two findings of the Court can be reconciled.

Zveza slovenskih organizacij na Koroškem]. They had established a so-called consensus group [Konsensgruppe] advocating and promoting more tolerance, a better understanding of the historically rooted prejudices and the abandoning of extreme positions inside both 'camps' (Feldner and Sturm 2007). On the German-Carinthian side, it was mainly motivated by a negative image of Carinthia that had gained ground through Austrian and international media due to the continued and seemingly 'irrational' place-name conflict and that began to negatively affect even tourism. While the consensus group did not succeed in including all regional players and was notably not supported by the political parties (Feldner and Sturm 2007), was perhaps also regarded as an elitist undertaking, it paved the way toward an atmosphere of understanding and laid thus the basis for a political compromise.

This compromise, although confined to names of populated places and the official language of communes, seems to have satisfied all parties in Carinthia and has essentially calmed down at times politically delicate situation (see Hren and Pandel 2012). Up to 2020 it laid the ground for initiatives and activities in favor of the minority and its culture that would have been hardly imaginable just ten years ago. Thus, it has, for example, become possible to publish hiking and tourist maps with Slovenian field and house names financially supported by (in majority German-Carinthian) communes (see Österreichische UNESCO-Kommission 2020). In 2020, a quadrilingual photo volume on Carinthia was published with the Slovenian name of Carinthia (*Koroška*) appearing in the second (!) and very visible position on the book cover with Italian and English names just following (see Stranner and Lehner 2020). In the same year the Commune Sankt Jakob im Rosental with 17.5% share of minority population (as of 2001) decided as the first in Carinthia to introduce bilingual street naming.

Federal Act No. 46/2011 (BKA 2020) and its Annex 1 rule that 164 populated places [Ortschaften] in Carinthia have officially bilingual names (see also Appendix 1). Town signs in front of them and road signs hinting at them, if located at the territory of the same place, have to show their names bilingually. The number of 164 is based on a share of 17.5% Slovene-speaking population according to the 2001 population census (as a compromise between the earlier share of 25% and the share of 10% proposed by the

Constitutional Court of Justice). Minor deviations from this percentage have been possible when this was found locally acceptable. The number of 164 includes all 93 populated places with officially bilingual names already before the new act. The 164 places are dispersed over four political districts [Politische Bezirke], that is Hermagor, Villach-Land, Klagenfurt-Land, and Völkermarkt, and thus roughly reflect the so-called bilingual area, but do not cover this area completely. Appendix 2 presents a comparison of the regulations by the Act of 2011 with the three earlier regulations for official bilingual names of populated places in Carinthia as of 1972, 1977, and 2001. Act No. 46/2011, Annex 1 also defines 47 populated places in the federal province Burgenland as bilingual German-Croatian and 4 as bilingual German-Hungarian—just confirming the Decree on Topography in Burgenland [Topographieverordnung Burgenland] (BGBl. 170/2000, BKA 2020) as of 2000.

Accompanying the Act and actually signed already before (26 April 2011) the Act was passed by all relevant parties, a "Memorandum concerning bilingual topographical signs" had been published mentioning inter alia the possibility that, in addition to the 164 populated places defined by the Act, Carinthian communes may use their autonomy to place further bilingual signs for populated places and other features. This possibility has in the meantime (in 2020) been used by the Commune Sankt Jakob im Rosental for all its populated places not yet covered by Act No. 46/2011 as well as by the Commune Sittersdorf to name its village Sielach/Sele bilingually. It has, however, to be taken into account that this memorandum is just a convention between relevant parties lacking the force of law. Thus, it can rather be attributed to the (new) political climate in Carinthia that additional town signs can be placed, and nobody opposes.

The Act of 2011 suffers (like the Act of 1976/1977) from the weakness that it does not make the place name as such and in every context bilingually official, but refers explicitly just to town signs and road signs (the latter at the same place! See also Chap. 6, Sect. 6.1.2), while several issues remain subject to interpretation: Has the place to be reflected bilingually also on maps? Has it to be reflected bilingually also on road signs in some distance, in official communication, and so on?

Neither the names of political districts nor of communes were made officially bilingual. The Act does also not refer to names of streets (that remain within the responsibility of communes), of railroad stations (that remain within the responsibility of railroad companies), and of natural features of all kinds, for the rendering of which in spatial reality tourism associations, the Austrian Alpine Club and similar associations and the hydrographic office in the relevant federal ministry are responsible, while for their rendering on official topographical maps the Federal Office of Metrology and Surveying takes the decision based on empirical search of their topographers for the local use of these names. In contrast to earlier legislation, the 164 Slovenian names of populated places conform to standard Slovene according to the principles mentioned in Chap. 5, Sect. 5.1.2.5. This is usually in line with the demands of minority speakers but meets locally also opposition as Fig. 4.7 shows, where some local Slovenes would obviously prefer the dialect version.

The Act and its Annex 2 also rule that in the federal provinces Burgenland and Carinthia certain communes, juridical districts, and political districts as well as other offices of the Federation and the Province responsible for these administrative units have to regard the minority language as a second official language of communal offices, police inspectorates, district courts, offices of the political district, offices of these federal and provincial institutions as well as the military commands in these provinces. This does not mean that these communes, juridical districts,

Fig. 4.7 In addition to the official town sign according to Federal Act No. 46/2011 naming a village in the Commune Hermagor *Dellach/Dole*, a local initiative placed a 'private' town sign showing the name *Dule* in the local Slovenian dialect. The 'private' sign has in the meantime been removed. (Photo by Maciej Zych 2014)

and political districts have officially bilingual names, but have an impact on the linguistic landscape insofar as the related public buildings, very often also school buildings, are named bilingually including a bilingual place name (see Chap. 6, Sect. 6.1.2).

In Carinthia this regulation applies to the communes Ebenthal in Kärnten, Feistritz im Rosental, Ferlach, Ludmannsdorf, Sankt Margareten im Rosental in the Political District Klagenfurt-Land, to the communes Rosegg and Sankt Jakob im Rosental in the Political District Villach-Land, to the communes Bleiburg, Eisenkappel-Vellach, Feistritz ob Bleiburg, Globasnitz, Neuhaus, Sittersdorf in the Political District Völkermarkt, in addition to the communes Eberndorf and Sankt Kanzian am Klopeinersee in communication with residents just of some of its populated places (individually addressed in the Act's Annex 2), further to the district courts Ferlach, Eisenkappel, Bleiburg as well as to the political district offices Villach-Land, Klagenfurt-Land, and Völkermarkt (BKA 2020).

Minority languages have, however, neither official status in provincial parliaments [Landtage], nor in one of the chambers of the federal parliament. There is also—in contrast to some other European countries, for example, Romania, Slovenia, or Croatia—no regulation ruling that an ethnic group has the right on a minimum number of delegates in any body of self-government from the communal to the federal level irrespective of votes.

4.2 Czechia (Přemysl Mácha and Uršula Obrusník)

4.2.1 Linguistic and Ethnic Structure: Historical and Current Context

4.2.1.1 Early History

As in the case of Austria, the ethnic and linguistic structure and toponymy of contemporary Czechia is the result of complex and sometimes dramatic demographic changes over the course of the past two millennia.

Due to its proximity to Austria, the regions conforming Czechia today—Bohemia [Čechy], Moravia [Morava], and Silesia [Slezsko]—often experienced similar population movements with similar effects. Unlike Austria, however, they lied north of the Roman border line (*limes romanus*) and were never subject to direct Roman control and cultural influence (save occasional military incursions). In the history of the Czech lands, the Romans established no permanent populations and left no linguistic or toponymic traces. From the perspective of Rome, the Czech lands remained Barbaric.

The earliest ethnically identifiable toponymic traces (and historical comments) indicate the presence of the Celts during most of the first millennium B.C. They coined the names of some rivers (e.g., *Jizera*) and gave name to one of the Bohemian crown lands. *Bohemia* is derived from the name of one of the Celtic groups living in the territory, the Boii (via the Germanic *Boiohaemum*, i.e., 'the home of the Boii').[23] This name, however, is not used in the Czech language, it has survived only as an exonym in other European languages. In Czech, Bohemia is called *Čechy*, a Slavic name derived from *Čech*, the personal name of the mythical founder of the Slavic population which arrived at the territory in the sixth century.

Before the arrival of the Slavs, however, Germanic groups settled in the region, replacing the Celts most of whom appear to have left the territory by the turn of the first millennium B.C. In historical records, these Germanic groups are referred to as *Marcomanni* (in Bohemia) and *Quadi* (in Moravia). They left their traces in a number of toponyms, namely rivers, either directly (such as *Vltava* from the presumed Germanic *Wiltahwa*, 'wild water') or indirectly by suffixes (such as *Opava* from the Indoeuropean *apa*, 'water', and Germanic *-ahwa*, 'water'). The majority of rivers and streams in Czechia carry the Germanic suffix *-ava* today even when their names were coined after the arrival of the Slavs who adopted this toponymic marker from the Germanic tribes. It is not clear how many Germanic groups stayed in the region after the Slavs arrived. However, the testimony of geographical names indicates that some undoubtedly did, sharing their names with the newcomers.

---

[23] Incidentally, the Boii also left their legacy in the names of *Bavaria* (*Baiowaria*) and *Bologna* (*Bononia*).

4.2.1.2 Middle Ages

Still, because significant depopulation always accompanied the aforementioned demographic changes, and new settlements were frequently moved, there is no continuity in settlement names from the Celtic period to the present. Contemporary settlement and field names in Czechia are mostly the result of later internal colonization and international migration intensifying after the gradual consolidation of the Bohemian state in the 900s.

This state was built on the ashes of the Moravian Empire, a Slavic political unit centered in modern southeastern Czechia and southwestern Slovakia and controlling extensive surroundings of its core areas (see, e.g., Galuška 1991). It existed during the larger part of the 800s and was brought to its end by 907 by the Hungarians, who had invaded the Pannonian Basin in the later ninth century. The Moravian Empire laid institutional and economic foundations for Bohemian statehood. As the only previous historical instance of a joint Czech-Slovak state, it was also used as one of the justifications for the establishment of Czechoslovakia in 1918 and played an important role in the national mythology of the country in spite of its short historical duration.

From the toponymic and archival records it appears that in the Early Middle Ages the Bohemian Lands were inhabited principally by Slavic populations differentiated into various tribal groups and divided into autonomous 'Lands of the Bohemian Crown' [země Koruny české], that is Bohemia [Čechy], Moravia [Morava], Silesia [Slezsko, Śląsk], and Lusatia [Lausitz/Lužice]. In total—and like the major part of modern Austria—the Bohemian Lands were part of the Holy Roman Empire.

In the 1200s, during the apogee of the Bohemian Kingdom under Přemysl Otakar II, when a large portion of modern Austria including Carinthia [Kärnten] was temporarily integrated into the Bohemian Kingdom, extensive areas of Bohemia, Moravia, and Silesia were settled by Bavarians, Saxons, Swabians, and other Germanic groups. They were invited by the monarch to settle the border regions (to substantiate the Bohemian claim to the territory) and to bring the latest agricultural, technological, and legal know-how. Many new villages and towns were founded, extensive areas of forests came under cultivation, and gold, silver, and metal-ore prospection and mining brought prosperity to the entire Kingdom.

Over the course of many decades, this migration dramatically transformed the ethnic structure of the Kingdom. The German population prospered in the border regions (later to be called *Sudetenland*) and it also played a significant demographic, economic, and political role in cities while gradually dominating state administration. Only the expulsion of the German population after World War II brought an end to this ethnic plurality which otherwise characterized the entire history of the Bohemian Lands.

German colonization and the influence of German law on settlement foundation imprinted itself strongly on toponymy. As a consequence, a number of cities, villages, castles, and other objects bear Slavicized German names such as Šenov (from *Schönau*), Petřvald (from *Peterswald*), Frenštát (from *Frankstadt*), or Pernštejn (from *Bärenstein*) well outside the historic German-speaking areas. The long linguistic contact with German also heavily influenced colloquial Czech in general and the Těšín/Cieszyn dialect in particular, displaying many loan words and phrases borrowed from German, which survived even the extensive linguistic purification efforts of the 'Czech national revival movement' in the 1800s.

4.2.1.3 Austrian Period (1526–1918)

With the exception of the German population in the border regions and in cities, Jews in cities and the Polish population concentrated in the Těšín/Cieszyn region (see Chap. 5), there was no other significant ethnic or national minority. Even though between 1526 and 1918 the Bohemian Lands formed part of the Habsburg dominions, they were not a preferred destination for settlement from other parts of these dominions, and immigration from other countries was not encouraged. There were minor migrations of small groups from different parts of Europe (e.g., Croatia, Bulgaria) which left occasional traces in toponymy (e.g., *Charváty*,

Bulhary) and many foreigners settled individually, but nothing even remotely comparable to the German colonization occurred again.

Tensions between Czechs and Germans are documented in historical records throughout the entire period, especially in Prague [Praha] and other cities with a substantial German population. There were also repeated pogroms of Jews. However, by far the most dramatic conflicts in the medieval and Austrian periods had religious motives (which were later often re-interpreted in ethnic terms). The Bohemian Lands have been Christian since the time of the Moravian Empire and Christian in the Western, Latin sense from the moment when the Bohemian ruling family decided to turn to the West for geopolitical reasons. (The Moravian Empire was associated with the eastern rite and the Byzantine Church.) The first attempt to reform the Western, later Catholic Church in Bohemia, was led by Jan Hus in the early 1400s. After his execution for heresy, the movement was transformed by his followers, the Hussites, into a holy war which decimated the population of the Bohemian Lands. Although the movement was eventually crushed, it prepared the ground for the rise of local reformed churches and the spread of the Lutheran reformation in the 1500s.

The conflict between the mostly Protestant Bohemian nobility and the staunchly Roman-Catholic supreme authorities in Vienna [Wien] over self-governance and autonomy escalated in the early 1600s and resulted in the Thirty-Years War (1618–1648). This war involving many European states once again hit the Bohemian Lands both demographically and economically. In its aftermath, Vienna outlawed Protestantism and Protestants were severely persecuted until the late 1700s. The state confiscated the landholdings of the Protestant nobility and executed many of its representatives. Loyal noblemen mostly of German origin replaced them as both landholders and state administrators. In consequence, Austrian rule was associated with Catholicism and both were associated with the German element.

In the mythology of the 'Czech national revival', the Hussites and the Bohemian and Moravian Protestant nobility were the true heroes of the eternal Czech resistance to German oppression. They were portrayed so also later during Czechoslovak times. It therefore comes as no surprise that on 3 November 1918, just a few days after the declaration of independent Czechoslovakia, the extravagant statue of the Virgin Mary on Old Town Square in Prague was torn down by an angry mob as a symbol of this dual, political, and religious Austrian oppression. Its recent renewal was met with strong resistance and criticism indicating that the memory of Austrian rule in present-day Czechia is still alive (see, e.g., Novinky.cz 2020).

Until the late 1800s, there are no reliable census data which would make it possible to estimate the ethnic composition of the Bohemian Lands. Using the colloquial language (*Umgangssprache*) as the criterion for assigning ethnicity, the first Austrian census carried out in 1880 counted a total of 8.22 million inhabitants on the territory of modern Czechia. Of those, 62.1% declared to speak Czech, 35.9% German, 1.9% Polish, and 0.1% other languages. However, there were significant differences both within and between the three lands because the ethnicities were not necessarily equally mixed. In Bohemia, the proportion of Germans was higher than in Moravia (37.2% to 29.4%) but still remained an overall minority. In the border regions of both lands, however, the German population represented the overwhelming majority, while in the interior it often was just a small minority. Only in Silesia did the German population constitute the largest ethnic group (48.9%), followed by Poles (28.1%) and Czechs (23.0%). But even there, the ethnic distribution was not uniform. Due to territorial concentration, German-speakers dominated in cities and northeastern Silesia while the countryside in the Těšín/Cieszyn region was predominantly Slavic (Czech and Polish). These proportions did not change significantly before World War I (Kladiwa et al. 2016: 231).

Under Austrian rule, the language of state administration in the Bohemian Lands gradually changed from Czech to German, and Emperor Joseph II (1780–1790) declared German the official language of the entire Austrian Empire. From the Austrian-Hungarian Compromise of 1867 and by the Act on Basic Rights (*Staatsgrundgesetz*) of 1867 for the Austrian Lands ('Cisleithania', in German *Zisleithanien*) including all Bohemian Lands under Austrian rule, Czech achieved the status of a second "external"[24] official language in Bohemia (in German *Böhmen*) and Moravia (in German *Mähren*), in Silesia (in German *Schlesien*) besides Polish. Between 1880 and 1897, Czech also became the de facto internal official language in many public institutions but was formally elevated to this status only by the Language Decree of the Administration Badeni as of 1897. This Decree, however, prompted heavy protests by the Germans of the Bohemian Lands, resulting in its withdrawal in 1899.

The affair marks just the peak of an intensive conflict over official language between the Germans of the Bohemian Lands and many Czech intellectuals, who, although fluent in German, fiercely resisted German domination. The Czech 'national revival movement' thus crystallized in opposition to the German language and culture. In its imaginary, Germans were painted as the eternal enemy of the Czech nation. The problematic nature of the distinction between ethnic and civic nationalism notwithstanding (see, e.g., Brubaker 1999), we can characterize the two nationalisms—German and Czech—as ethnic, pitying themselves against each other at the expense of building a common Austrian or later Czechoslovak identity. As a consequence, during the 1800s, ethnic identification gradually began to replace regional identities (Bohemian, Moravian, Silesian) with a long historical tradition. This led to mostly unsuccessful linguistic demands of Czech intellectuals and politicians to give the Czech language an equal standing on par with German (Soukupová 2015). Paradoxically, while Czechs constituted the demographic majority, they remained a political minority in the context of the Austrian state.

This feeling of historical inequality impacted heavily on the creation of Czechoslovakia in 1918, a process in which, from the perspective of Czech politicians, the balancing out of the German element was of utmost importance. As a consequence, the relationship between the Czechoslovak state and minorities living on its territory was strained from the beginning and eventually led to tragic outcomes.

We will only focus on national minorities as these had played an important role in the history of modern Czechia, unlike religious and

[24] 'External' in the sense of admitted in communication between citizens and a public authority, not for communication inside the authority.

racial minorities, whose presence had been marginal (Petráš 2009: 136). Our discussion will focus predominantly on the Czech part of the Republic, adding the reference to the Slovak part of the country as needed to clarify certain points.

When looking at the historical and contemporary position of national minorities within the Czech state (and the Czechoslovak state prior to 1993), we have to pay attention to the position of researchers whose work we use. Many of the issues presented in this chapter remain a sensitive topic for parts of the Czech society as well as for the members of national minorities. In the case of the Polish minority specifically, Czech and Polish historians might present a different narrative of the historical events; historians from the minority itself might also offer a unique point of view (Siwek et al. 2001). While we do not aim to provide a comparative analysis of these approaches or of the many historical events, we hope to provide readers with an understanding of the policies of the state regarding its various minorities. We will present the material chronologically, starting from the birth of Czechoslovakia as an independent country after World War I. We will proceed through important regime changes and finish with a short discussion on the position of national minorities in the modern Czech state.

4.2.1.4 First Czechoslovak Republic (1918–1938)

From the beginning of its existence the Czechoslovak state had to come to grips with the numerous minorities living within its borders. Following World War I and Wilson's doctrine of national self-determination, the representatives of both Czechs and Slovaks realized that, if taken as two separate nations, Slovaks would only be the third-largest group in the new state, after Czechs and Germans (Prokop 1998). In the light of these numbers, Czech and Slovak politicians decided to 'create' a new 'Czechoslovak' nation based on presumed close linguistic and historical proximity. A new state, Czechoslovakia, was established on 28 October 1918, with a total population of 13,374,364 (Tóth et al. 2012). Of course, they were only able to do this because of favorable geopolitical considerations of world powers intending to weaken Germany and the

Austro-Hungarian Empire—in the latter case not all its successor states, but just German-Austria and Hungary, who were regarded only responsible for the war. The treaties of Saint Germain (1919) and Trianon (1920) thus gave rise to not only Czechoslovakia but also several other countries, while sowing seeds of future discontent. Wilson's doctrine of national self-determination was not applied to Germans and Hungarians, leaving large portions of German and Hungarian populations as minorities in the newly established countries (Suppan 2019).

The new state had to finalize its borders as one of its first tasks. The internal and international challenges to its territorial integrity were grave and their relatively rapid resolution proved short-lived. At the international level, Czechoslovakia had to resolve territorial disputes with Poland, Hungary, and Austria. For this text the conflict over the Těšín/Cieszyn region with Poland was particularly important as it left a lasting mark on the historical consciousness of the local population while continuing to shape the public debate about bilingualism (see Chap. 5 for more information about the region).

Although the situation in this region was the most contentious, a large part of the borderland of the new state was inhabited by minorities in general. Germans settled along the northern, western, and southern borders of the Czech part of Czechoslovakia and in a number of large towns. Poles settled in the northeast, where the new border cut through the older historical land of Silesia. Jews were also an important minority during this time and settled in many towns throughout the country. Ruthenians, living in the easternmost part of the country, were also recognized as a minority. In the Slovak part (up to 1918 called "Upper Hungary", in Hungarian *Felvidék*) was also a numerous Hungarian minority along the southern border with strong ties to Hungary, in addition to Germans living in towns and villages everywhere. The location and strength of various minorities were a concern for the state, especially since some of the minorities located in the borderlands (Germans and Hungarians) did not agree with the establishment of the independent Czechoslovak state to begin with (Prokop 1998; Petráš 2009).

The heaviest initial threat came from the German-speaking areas of Bohemia, Moravia, and Silesia, which was a symptom of the long-strained relations between Czech- and German-speaking inhabitants of the new country. The Sudetenland areas expressed a wish to be annexed to the First Austrian Republic that titled itself initially *German-Austria*. Their delegates insisted to remain in the Viennese parliament. The stand-off between Czechoslovak authorities and Sudetenland representatives lasted for months and only the Peace Treaty of Saint Germain (1919) temporarily quelled their resistance. However, their dissatisfaction with the new political arrangement never abided and resurfaced dramatically in 1938, when these areas became part of the Third Reich, a decision strongly supported by the majority of Czechoslovak Germans.

All national minorities continued to strive for some degree of autonomy throughout the interwar period, which was viewed as a challenge to the integrity of the state. Instead of promoting autonomy, the internal administrative organization of the state tried to limit the influence of minorities arising from their territorial concentration. Initially, the original land system continued to exist (Bohemia, Moravia, Silesia) and two new lands were created—Ruthenia (in Czech *země Podkarpatská*) [Zakarpattja] and most notably Slovakia (in Czech *země Slovenská*) which emerged on the political map of Europe for the first time, albeit as a substate unit. In 1928, the land system was reorganized presumably in order to rationalize state administration. Moravia and Silesia were merged into a single land Moravia-Silesia (*země Moravskoslezská*). This was strongly opposed by German and Polish representatives who saw the reform as a poorly veiled attempt to dilute extensive political influence of minorities in Silesia (see, e.g., Gawrecká 2004).

The new state also had to navigate the international agreements it had signed that guaranteed the protection of minorities under the framework imposed by the League of Nations (Petráš 2009: 65). These became part of the postwar international order attempting to protect newly created minority populations in Central Europe. An additional factor of the postwar establishment of state borders in Central and Eastern Europe was the fact that the minority in one state was at the same time a majority in a neighboring state, such as was the case with the Polish, German, and Hungarian minorities in Czechoslovakia. In its foreign policy then, Czechoslovakia had to take into account the position of minorities especially in its relationship with Germany, which continued to be a dominant force in regional politics (Tóth et al. 2012).

Minority population in the First Republic formed a large segment of the society. The first Czechoslovak census of 1921 indicated that almost one-third of the population belonged to a minority (Żelazny 2016). The same holds true for the Czech part. As Tables 4.6 and 4.7 indicate, between the last Austrian census in 1910 and the Czechoslovak census in 1921, there was, however, a significant decrease in the number of Germans and Poles and a rise in the number of Czechs, although it has to be taken into account that the census of 1910 had asked for colloquial language and not ethnicity. This change was most dramatic in Silesia where the German majority became a minority after many people previously classifying themselves as German-speakers (and Polish-speakers) declared Czech nationality. In other words, the change in the political context together with the change in the methodology of the census (from colloquial language to declared nationality) resulted in the change of the ethnic composition of the territory. This was strongly criticized by

Table 4.6 Ethnic composition (by colloquial language) of the Czech lands (see also Fig. 4.8)

	Bohemia		Moravia		Silesia	
Nationality	Abs.	Rel. (%)	Abs.	Rel. (%)	Abs.	Rel. (%)
Czechs	4,241,918	63.19	1,868,971	71.75	180,348	24.32
Germans	2,467,724	36.76	719,435	27.62	325,523	43.90
Poles	1541	0.02	14,924	0.57	235,224	31.72
Total	6,712,944	100.0	2,604,857	100.0	741,456	100.0

Source: Kladiwa et al. (2016: 305)

Note: Silesia is without the Hlučín region belonging to Germany at that time, but includes the entire Těšín/Cieszyn region before its division between Czechoslovakia and Poland

Table 4.7 Ethnic composition of the Czech lands in 1921 (see also Fig. 4.8)

	Bohemia		Moravia		Silesia	
Nationality	Abs.	Rel. (%)	Abs.	Rel. (%)	Abs.	Rel. (%)
Czechs	4,382,788	66.64	2,048,426	78.29	296,194	47.56
Germans	2,173,239	33.04	547,604	20.93	252,365	40.53
Jews	11,251	0.17	15,335	0.58	3681	0.59
Poles	973	0.02	2080	0.08	69,967	11.24
Total	6,576,825	100.0	2,616,436	100.0	622,738	100.0

Source: Kladiwa et al. (2016: 305)

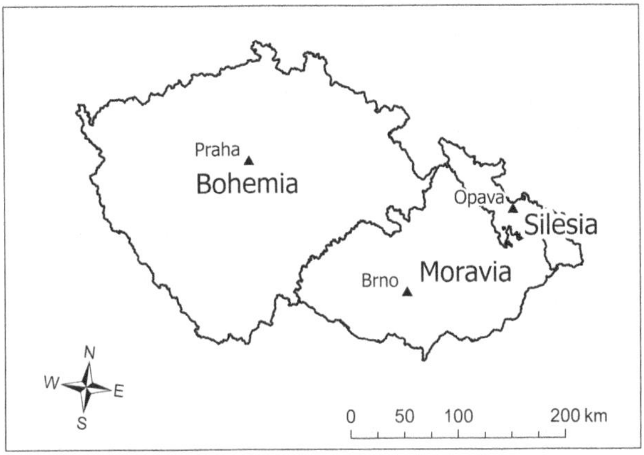

Fig. 4.8 Historical lands and their original administrative centers in the current territory of Czechia. (Cartography: Přemysl Mácha)

minority representatives who feared statistical extinction. Publishing the final results of the census thus became a delicate political matter and numerous 'adjustments' had to be made to accommodate the demands of the minorities (Kladiwa et al. 2016: 82–168).

Considering the numbers as well as the location of minorities, coupled with constraints of international law and demands of foreign policy, the state took a measured approach toward the minority question. There were legal provisions ensuring state support of education in minority languages, support for minority culture, and provisions for the usc of minority language in official settings (see Sect. 4.2.4 Minority legislation). However, while legally the situation of minorities was relatively good, the implementation of the laws in daily life was not as straightforward. There was a strong mutual distrust between Czechs, Germans, and Poles. Numerous provocations and public obstructions in language use took place (see, e.g., Soukupová 2015).

The minorities were not represented in the first Czechoslovak parliament, the Revolutionary National Assembly [Revoluční národní shromáždění] which worked between 1918 and 1920. Only Czech and Slovak deputies were nominated from parties which ran candidates in the Imperial Council [Reichsrat] elections in 1911. There was no popular election. In spite of its limited mandate, however, this improvised

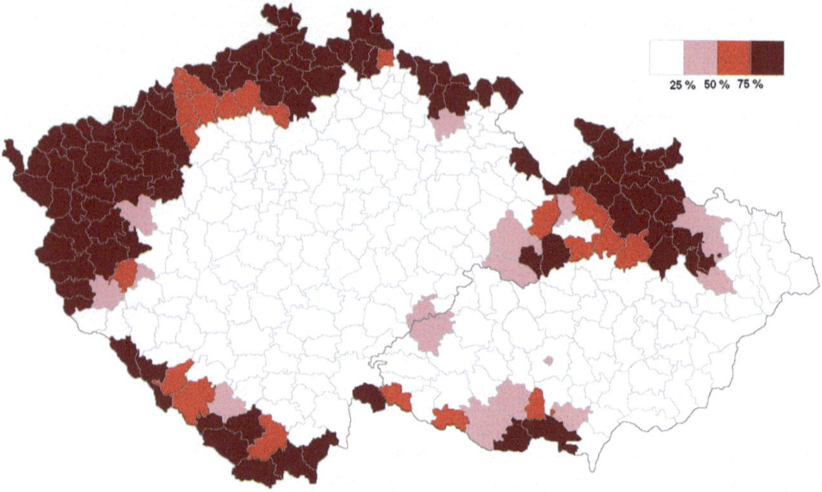

Fig. 4.9 Share of German-speaking population in the Czech lands in 1930. (Source: https://en.wikipedia.org/wiki/Sudetenland)

parliament passed important legislation with immediate consequences for the minorities (constitution, land reform, language law) (Petráš 2009). Although minorities had their representatives in the subsequent parliaments and especially German parties played an important role in national politics, by this time the majority of important legislature had already been passed, so the effect of their presence was limited, contributing thus to their growing frustration with the state (Fig. 4.9).

Minorities under Czechoslovak rule were thus confronted with the everyday ethnic reality of state administration defined by Czech- and Slovak-speaking authorities. The gradual assimilation of minorities became the implicit goal of state policies which had a profound impact on minority life as well (Tóth et al. 2012: 593). Czechs and Slovaks were considered 'state-forming' nations, which was one of the more significant changes in the legal framework compared to the Austro-Hungarian system, where all nations of the monarchy were, at least in theory, equal (Petráš 2009: 63).

The second important shift from the Austrian period to independent Czechoslovakia was that in the newly formed state Czechs and Slovaks became the majority nations, while Germans became a minority. This reversal of power balance caused both short-term and long-term tensions. One of the most crucial issues of the time was the existence of the state language. While in the late Habsburg Empire (from the late eighteenth century until 1867) German had been the exclusive official language and after the Austrian-Hungarian Compromise of 1867 Czech had achieved the status of a second "external" official language in Bohemia and Moravia and together with Polish in Silesia, in Czechoslovakia Czech became recognized as the exclusive official language by Czechoslovak law which was heavily criticized among the numerous well-educated members of the German-speaking population.

Overall, the situation of minorities and the state approach to them in the First Czechoslovak Republic were everything but favorable. There was little legislative support for minority rights and a lack of trust on the basic social level, resulting from the initial resistance of minorities to the creation of Czechoslovakia. Language conflicts and economic disparities further exacerbated the already strained relations. The World Economic Crisis of 1929–1932 affected German and Polish minorities concentrated in the heavily industrialized regions of the country particularly severely and only confirmed their suspicion that the Czechoslovak state had no interest in its minorities (Prokop 1998). Finally, as a clear legislature was lacking, decisions were undertaken by local administration which led to stark differences in the treatment of minorities and a sense of unpredictability and insecurity on their part (Tóth et al. 2012: 591). In general, the situation of minorities in the First Republic was complicated and remained de facto unresolved until World War II.

4.2.1.5 World War II (1938–1945)

The signing of the Munich Agreement in September 1938 led to a significant loss of territory and population of Czechoslovakia after the so-called Sudetenland was ceded to Nazi Germany and Slovakia declared independence. The First Czechoslovak Republic ceased to exist. While this was a result of a complex geopolitical situation, the presence of a strong German minority was seen as largely responsible for such a turn of events. The majority of the German population overtly supported the annexation after having repeatedly voted for political parties which had the annexation as their long-term goal. Similarly, in the northeast of the country, the Polish minority was seen as being responsible for the entry of the Polish army in October 1938, which led to the temporary annexation of a part of the Czech Těšín region to Poland. While this was not technically part of the Munich treaty and was negotiated on a separate occasion (Baran 2009), in the collective memory of the country these two events remained an example of dangers of minorities inhabiting the borderlands. The sense of betrayal shaped the thinking about minorities in the Czechoslovak state for years to come.

The Second Republic ended in March 1939, when the German Protectorate of Bohemia and Moravia was established. The state approach to minorities clearly fell out of line with any previous development (Petráš 2009). Jews and Roma were severely persecuted as elsewhere under the Nazi regime. The Polish minority in the Těšín/Cieszyn region was also specifically targeted after Poland itself was occupied by Germany in September 1939. Because the Těšín/Cieszyn region at the time was a part of the occupied Polish territory and not the Protectorate, the legal provisions differed slightly from the ones established in the Protectorate. The German population in Sudetenland was incorporated directly into Nazi Germany.

4.2.1.6 Early Postwar Years (1945–1948)

After World War II, Czechoslovakia was restored but without Ruthenia (Subcarpathian Rus (in Czech: *Podkarpatská Rus*) [Zakarpattja]), which was ceded to the Soviet Union. As a result, the number of Ruthenians living in Czechoslovakia decreased dramatically. Even more dramatic events were to occur internally, however. The situation of the German minority was particularly difficult because Czechs at large demanded some sort of retribution if not downright revenge for the suffering caused during the war. In consequence, the state stripped the Czechoslovak Germans of citizenship and civil rights (Petráš 2009). By 1948, nearly three million Czechoslovak Germans were expelled from the country and their property was confiscated on the basis of decrees issued by President Beneš. While this was done in agreement with the allied powers, the process itself was often violent and chaotic, especially in its early stages. Ultimately, the German minority, residing in the Czech lands from the Middle Ages, almost disappeared.

The transfer of the German minority out of the country had a profound effect on the social and economic makeup of postwar Czechoslovakia (see Table 4.8). The border areas were suddenly depopulated and in need of new settlers. This led to population shifts in which people from the interior and Slovakia (were) moved to the borderlands and took over the confiscated German property. The settler population, however, never approached the pre-expulsion numbers and large parts of the Sudetenland remained permanently depopulated, eventually turning into militarized 'Iron-Curtain' zones. Echoes of these events are felt up to this day as the former Sudetenland continues to differ from the rest of the country in electoral behavior, unemployment, and other aspects (see, e.g., Šimon 2015).

The state wanted to conduct a similar process when it came to the Hungarian minority in Slovakia but that was not approved by the allied powers. In turn, Czechoslovakia agreed with Hungary on a bilateral population 'exchange' when an equal number of Hungarians from Czechoslovakia were moved to Hungary while a similar number of Slovaks living in Hungary were moved to Czechoslovakia. This, however, was a much smaller number of people than the Hungarian minority in

Table 4.8 Ethnic composition of the Czech lands before and after World War II in %

Nationality	1930	1950	1970	1991
Czech	68.4	93.8	94.5	81.2
Moravian	–	–	–	13.2
Silesian	–	–	–	0.4
Slovak	0.4	2.9	3.3	3.1
Polish	0.8	0.8	0.6	0.6
German	29.5	1.8	0.8	0.5
Hungarian	0.1	0.2	0.2	0.2

Source: Czech Statistical Office (2014)

Czechoslovakia, so there was still a large Hungarian minority left within the borders of the country (Žáček 2001; Spurný 2011).

While the approach to non-Slavic minorities was openly hostile, the approach to Slavic minorities was only slightly more favorable (Petráš 2009: 91; Žáček 2001: 88). The Poles were seen as one of the factors of dissolution of the First Republic, even though they organized a strong resistance movement in the Těšín/Cieszyn region and were mercilessly persecuted by the Nazis. Also, the Czech-Polish conflict over the Těšín/Cieszyn region came back to life. Czechoslovakia made a twofold claim pertaining to the area. Firstly, it hoped to move the members of the Polish minority to Poland. As Poland was going through its own wave of internal migration caused by a significant shift in its borders, the hope was that it would support the transfer of population from Těšín/Cieszyn Silesia to its western borderlands. Secondly, Czechoslovakia made claims to parts of Silesia [Śląsk] around Raciborz, Kłodzko, and Głubczyce (Žáček 2001), which had historically belonged to the Bohemian Kingdom. Poland, on the other hand, did not agree to the population transfer and hoped for the restoration of the borders back to the situation just before the start of the war, that is, to its largest territorial gain in the region. The regional tensions grew till 1947, when Soviet diplomacy intervened and for its own political reasons facilitated a border agreement between the two countries, with added provisions for the rights of the Polish minority in Czechoslovakia.

4.2.1.7 Communist Era (1948–1989)

During the first half of the Communist regime in Czechoslovakia, the treatment of minorities was ambivalent at best. Until 1968, there were no consistent legal provisions for minorities in the constitution. Many proposals emerged but these were rarely translated into executable law (Petráš 2009: 106). The minorities often did not demand their rights simply because they did not know about them. As there was no clear state policy, it was up to the authorities of local governance to institute the rights and regulations as they saw fit. This meant that the conditions of minority populations varied widely depending on the region. Concurrently, the lack of clear state policy meant that on the local level the practice often followed the procedures set up during the First Republic (ibidem: 105).

After the displacement of the German minority in the late 1940s, Hungarians became the largest minority in the country. The regime's minority policy was therefore often based on the situation of the Hungarian minority in Slovakia. While some aspects of minority life were supported by the state (especially education in minority languages and cultural life), minority issues were largely ignored by the state in favor of the narrative of an 'internationalist' state. The argument of the time was that any possible conflict between majority and minority was caused by different living standards across the various regions of the country and not by ethnic or national differences per se. Once the economic differences were equalized by the Socialist system, the tensions between minority and majority would disappear.

In 1948, the system of the historical lands was abolished, and regions [kraje] began to play the role of intermediaries between the state, district [okres], and communal [obec] levels. Regions had existed more or less continuously since medieval times, but their number and size had changed frequently. None of the sub-state administrative units, however, enjoyed true autonomy. The Communist regime was heavily centralized, and regions, districts, and communes often only carried out decisions made at the state and Communist party level. The newly established regions, however, further suppressed regional identities as they failed to respect the historic boundaries between Bohemia, Moravia, and Silesia. The impact of this reorganization on the Polish minority in the Czech part of the country was minimal and minorities were not given any special representation either at the regional, at the district, or at the communal level.

The political changes and temporary loosening of the party grip on Czechoslovak society following the Prague Spring of 1968 were visible also in the minority question. The period of 1968 was characterized by increased social participation, which was also the case with minority organizations. The 1968 Act was the first comprehensive law dealing with the position of the minorities in Czechoslovakia since 1920 and as such was of historical importance, despite its shortcomings. It allowed individuals the right for self-determination and prohibited any forms of forced assimilation. Formally, minorities were conceived as state-forming subjects, together with Czechs and Slovaks, rather than the state being seen as the state of Czechs and Slovaks only, as it had been until then (Petráš 2009: 120).

The reforms of 1968 also led to the official federalization of the country, which gave Slovakia statehood within the federation. The Slovak representatives had fought for wider autonomy for decades, and the reform finally admitted that the idea of the Czechoslovak nation never had a future. This, however, meant that Czech-Slovak relations now became more important than relations between Czech and Slovaks on the one hand and other groups on the other. Choosing federal officials and national Communist party representatives thus turned into a delicate balancing game between Czechs and Slovaks, in which there was little room for minorities.[25] Those were relegated to the local level with little impact on state and federal politics.

The so-called normalization process after the invasion of the Warsaw Pact armies in 1968 brought to a definite halt all attempts at the broadening of minority rights. The Communist hardliners and internationalists strengthened their position again. While the 1968 Act made important changes for minorities, the subsequent lack of implementation meant that these were rarely reflected in practice. Some aspects of minority life were well maintained—such as minority education or presence of bilingual signs. However, other concerns remained unanswered. The minority issues were pushed aside once again and remained largely unresolved until the end of the Communist era.

4.2.1.8 Post-Communist Development (1989–)

The regime shift in 1989 from Communism to democracy brought a renewed interest in the minority question. This surprised a lot of people as for many the question seemed largely resolved by simply being absent in public discourse for so long (practically since the late 1940s). Minorities, however, saw the regime change as an opportunity to bring

[25] For an overview of Czech-Slovak relations, see, for example, Leff (2014, 2018).

their concerns back into the public sphere. By the early 1990s, Hungarians in Slovakia and Poles in the Těšín/Cieszyn region had become the best-organized minorities. Poles had especially well-developed social networks and community life. Minority representatives were even present in the first post-revolutionary state parliament (Czech National Council [Česká národní rada]). The main issues were the ongoing support for minority education, return of properties confiscated by the Nazis, and financial support in the new economic reality of capitalism (Borák 1998). There were secondary issues such as the presence of bilingual signs in the minority areas, which used to be on all state enterprises during Communism but started to disappear as the enterprises were privatized. While the political changes of 1989 are generally seen as positive, from the minorities' point of view they brought both positives and negatives. A large part of Czechoslovak society had only a minimal understanding of the situation of minorities, and this has not changed significantly since then (Małysz and Kaszper 2009).

While changes in the constitution and conception of the minority law were discussed in parliament, and a project of setting up a special government committee that would deal with minority issues was suggested, the differences in approaches between the Czech and Slovak side of the country made this essentially unfeasible. Between 1989 and 1993, there were a series of ad hoc steps ensuring the continuation of minority protection already in place, but no systematic solution was developed (Sulitka 2009).

In addition to a renewed attention to minority issues, the early 1990s also saw a renaissance of Moravian identity. In the 1991 census, 1.3 million people, that is, 13% of the entire Czech population and approximately 37% of the inhabitants of Moravia, declared Moravian nationality (see Table 4.8). Several political parties promoting the renewal of the historical land system even had their representatives in the federal and Czech parliaments. Their efforts, however, failed, as the public debate was entirely overshadowed by growing tensions between Czech and Slovak politicians. Although the number of people identifying with Moravian nationality decreased in the subsequent censuses (see Table 4.9), it still continues to be statistically substantial while having a strong presence in daily life of Moravia's inhabitants even if they do not declare this identity in the census. Interestingly, Silesian identity is extremely weak in

Table 4.9 Ethnic composition of Czechia according to the censuses of 2001 and 2011

	2001		2011		% change
Nationality	Abs.	%	Abs.	%	2011/2001
One nationality	10,044,255	98.2	7,630,246	73.1	76.0
Czech	9,249,777	90.4	6,711,624	64.3	72.6
Moravian	380,474	3.7	521,801	5.0	137.1
Silesian	10,878	0.1	12,214	0.1	112.3
Slovak	193,190	1.9	147,152	1.4	76.2
Polish	51,968	0.5	39,096	0.4	75.2
German	39,106	0.4	18,658	0.2	47.7
Roma	11,746	0.1	5135	0.0	43.7
Hungarian	14,672	0.1	8920	0.1	60.8
Vietnamese	17,462	0.2	29,660	0.3	169.9
Ukrainian	22,112	0.2	53,253	0.5	240.8
Russian	12,369	0.1	17,872	0.2	144.5
Other	40,501	0.4	58,289	0.6	143.9
Dual nationalities	12,978	0.1	163,648	1.6	1,261.0
Czech and Moravian	–	–	99,028	0.9	–
Czech and Slovak	2783	0.0	17,666	0.2	634.8
Czech and Roma	698	0.0	7026	0.1	1006.6
Czech and German	–	–	6158	0.1	–
Other	9497	0.1	33,770	0.3	355.6
Not indicated	172,827	1.7	2,642,666	25.3	1529.1
Total	10,230,060	100.0	10,436,560	100.0	102.0

Source: Czech Statistical Office (2014)

comparison with Moravian, while Bohemian identity is practically dead after *Bohemian* became a synonym for *Czech*.

In 1993, the conflict between Czech and Slovak politicians resulted in the peaceful break-up of Czechoslovakia and the formation of two new countries—the Czech Republic and the Slovak Republic. This was a significant move for minorities in the Czech Republic as their previous efforts were often based on their population numbers, which were bolstered by a significant Hungarian minority in the Slovak part of the Republic. After the split, the Polish minority in Czechia saw its negotiating position weaken significantly, while Slovaks in Czechia turned from a state-forming nation into the (by official data) largest national minority. While the relationship between the two countries has been excellent, because of the strong presence of Slovaks in the Těšín/Cieszyn region, competition for limited resources destined for the minorities has often led to tensions between minority representatives in the region, including the conflict over bilingual signs.

After the dissolution of Czechoslovakia, Czech policy toward minorities was shaped by its foreign policy and plans of joining the European Union (EU). This required that Czech laws were aligned with EU laws, including minority legislation. The main change for the relationship between the state and the minorities came with the ratification of two legal documents—the Framework Convention for the Protection of National Minorities and the European Charter for Regional and Minority Languages. The Charter gave a legal backing to a number of changes, including the introduction of bilingual place names in the Těšín/Cieszyn region. One of the issues that the members of minorities often pointed out with these two legal documents was that they lacked implementation measures and were as such open to interpretation by the state. This can also be seen in the time delay between the ratification of the documents and their actual implementation in the Czech legal system. (The Framework was signed in 1995 but ratified in 1997, the Charter was signed in 2000 but ratified in 2006 (Sulitka 2009).) While members of minorities were hopeful the changes required to join the EU will be beneficial to them, they were subsequently disillusioned by the delays in implementation, lack of effective companion legislature, and negative public reactions that went hand in hand with these changes. As we mentioned before, during Communism conversations about minority issues were largely non-existent due to larger ideological pressures. After the change in the political system, the unresolved issues and grievances of the past came back to the public sphere. This was especially the case in the Těšín/Cieszyn region, where a sense of resentment still existed on both sides.

One of the most contentious issues of the past 15 years has been the introduction of bilingual place-name signs in communes where the minority population reaches 10% according to the two most recent census data. While 30 communes fulfill this requirement, the number of places with bilingual signs is far lower. The introduction of bilingual place-name signs in the region is analyzed in detail in Chap. 5. This process, however, shows that while legislation and general political mood in the country might be favorably inclined toward minorities, local practice, interpretation, and implementation of these laws might often lag behind.

State policies regarding national minorities of the 1990s and 2000s concentrated on fulfilling the legal requirements for joining the EU and on continuing the support for minorities in the new economic reality. There were also efforts to create a separate comprehensive minority act that would unify and bring together various existing regulations. This, however, proved to be problematic, and eventually only a limited version of the law was passed in 2001 (Sulitka 2009).

Czechia joined the EU in 2004 and the Schengen area three years later. This marked a significant change in relationship between the Czech state and national minorities living in the country. The disappearance of borders led to an increased international cooperation, especially through the EU Interreg programs and creation of cross-border regions. Minorities became an important partner in this process through their networks of relationships in the kin countries. Minorities also came to be appreciated as an interesting local feature that signals the historical richness of regions and can increase the interest in the country overall.

The current ethnic structure of the country can only be estimated based on the most recent population census carried out in 2011. This census is summarized in Table 4.9 and showed several trends. First and foremost, it is the continuing assimilation of historic minorities such as Poles and Germans. In the context of this book, the case of Poles deserves particular attention because they suffered a great statistical loss in spite of a strong campaign "Bet on Polishness" [Postaw na Polskość] persuading Poles to declare Polish nationality (see, e.g., Motýl 2011). Because many minority rights are tied to the 10% threshold, Polish representatives feared that the decreasing numbers of Poles could lead, for example, to the disappearance of the hard-won bilingual signs in some communes. These fears proved warranted, and if the next census in 2021 confirms this trend, this book may quickly become outdated. As an aside we should draw attention to the number of people declaring German nationality. As is clear, the current German population in Czechia is roughly 0.5% of its pre-World War II size.

Second, we see a significant rise of new minorities—namely Vietnamese and Ukrainian. This process is tied to labor migration and family reunifications and reflects the growing number of foreigners settling in the country. Third, we see a gradual renaissance of older territorial identities—Moravian and Silesian, either as exclusive or dual identities. Finally, we also see a decrease in the number of people willing to declare any national identity. Once again, if this trend continues in the next census, the census will become a useless tool for estimating the ethnic structure of the country, and other tools will have to be devised.

The geography of national identification is also interesting. Figures 4.10, 4.11, 4.12, 4.13, 4.14, 4.15, 4.16, 4.17, 4.18, 4.19, and 4.20 show the spatial distribution of all minorities combined and then individual minorities identified in the census of 2011 and aggregated at the district [okres] level. Poles are clearly concentrated in the Těšín/Cieszyn region, while other census nationalities are primarily located in big cities (Prague [Praha], Brno, Ostrava, Plzeň) and the former Sudetenland where they

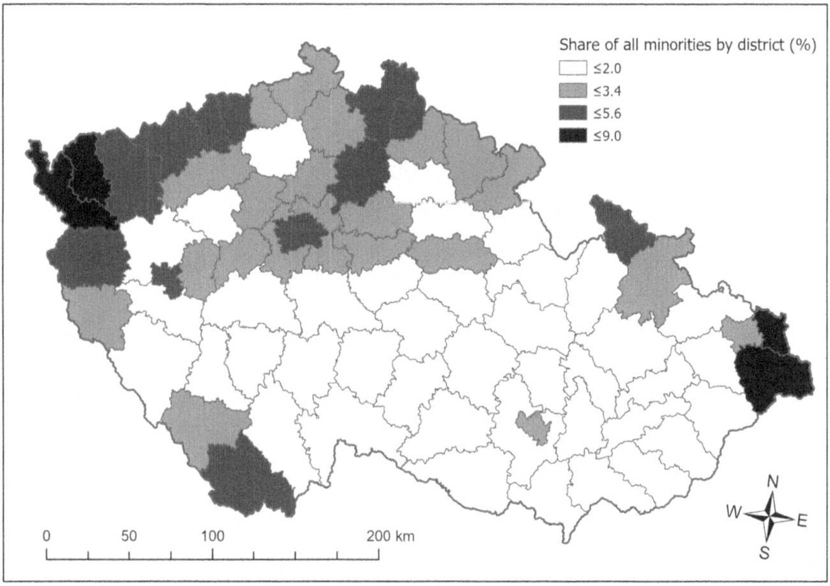

Fig. 4.10 Nationalities other than Czech in the 2011 census by districts [okresy]. (Source: Czech Statistical Office 2020, cartography: Přemysl Mácha)

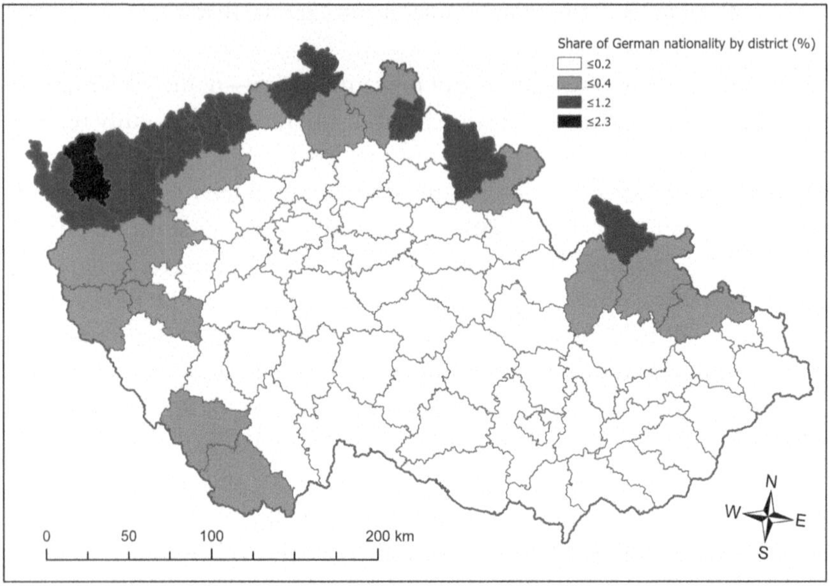

Fig. 4.11 Germans in the 2011 census by districts [okresy]. (Source: Czech Statistical Office 2020, cartography: Přemysl Mácha)

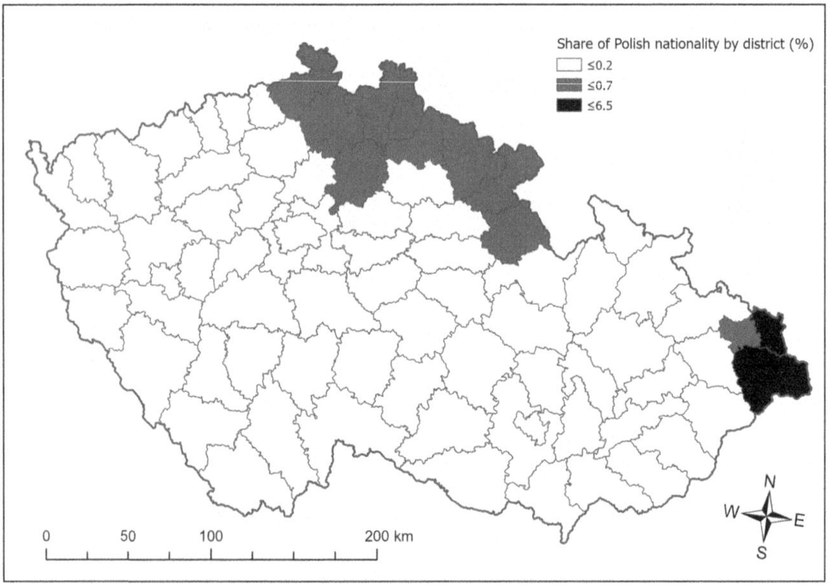

Fig. 4.12 Poles in the 2011 census by districts [okresy]. (Source: Czech Statistical Office 2020, cartography: Přemysl Mácha)

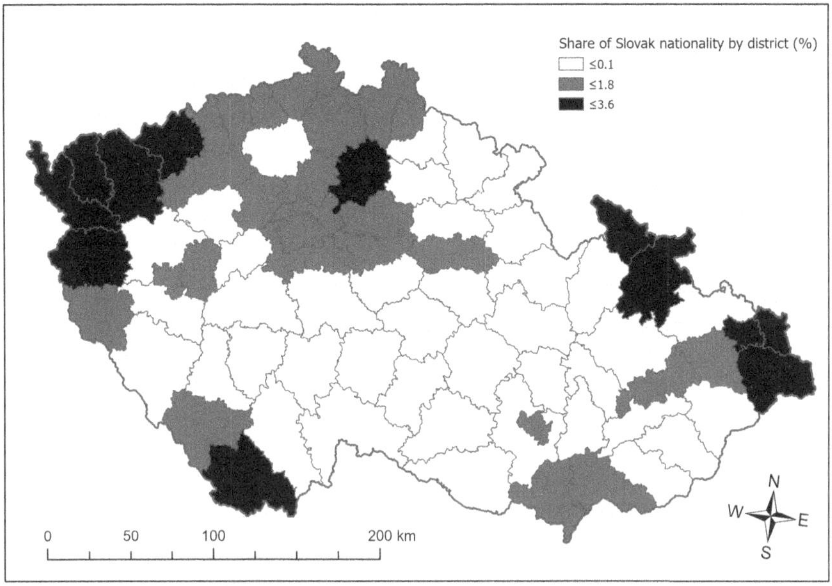

Fig. 4.13 Slovaks in the 2011 census by districts [okresy]. (Source: Czech Statistical Office 2020, cartography: Přemysl Mácha)

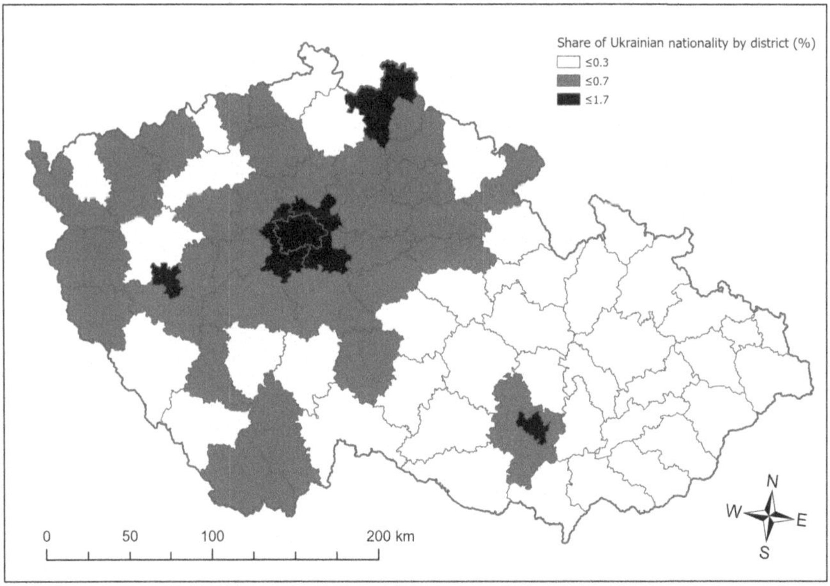

Fig. 4.14 Ukrainians in the 2011 census by districts [okresy]. (Source: Czech Statistical Office 2020, cartography: Přemysl Mácha)

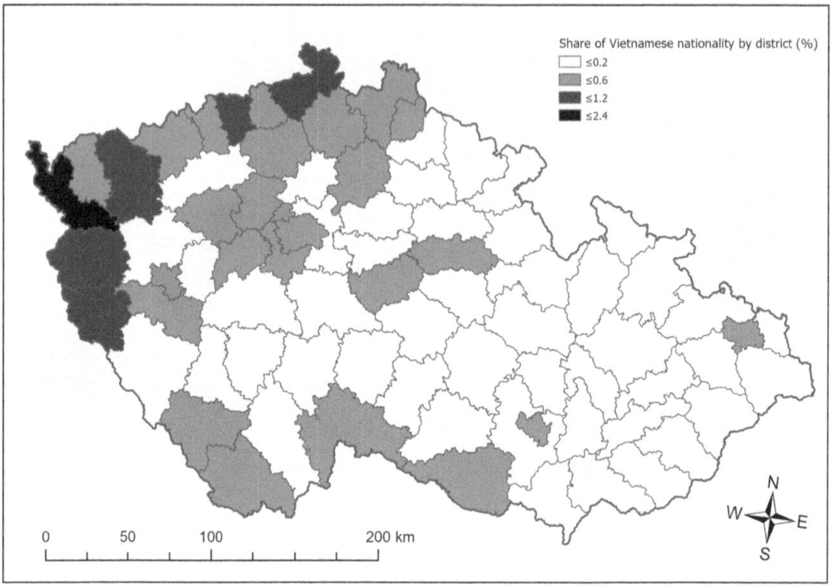

Fig. 4.15 Vietnamese in the 2011 census by districts [okresy]. (Source: Czech Statistical Office 2020, cartography: Přemysl Mácha)

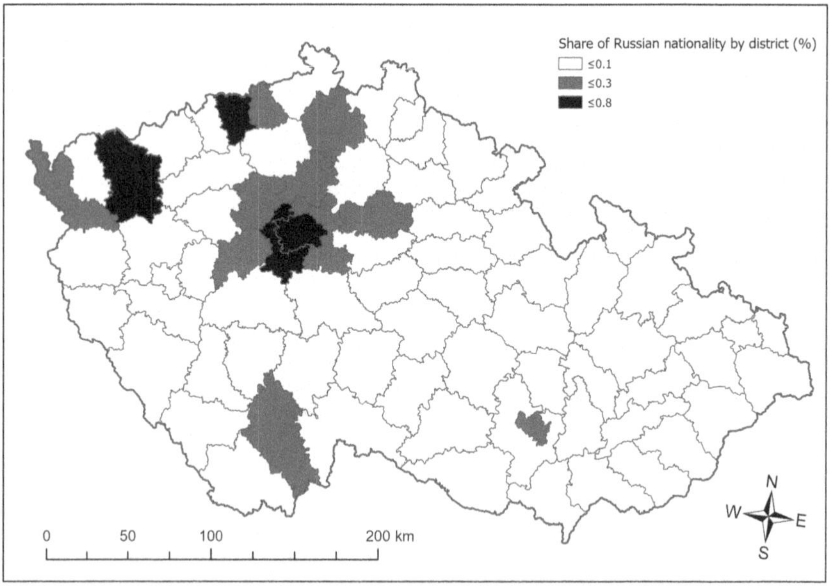

Fig. 4.16 Russians in the 2011 census by districts [okresy]. (Source: Czech Statistical Office 2020, cartography: Přemysl Mácha)

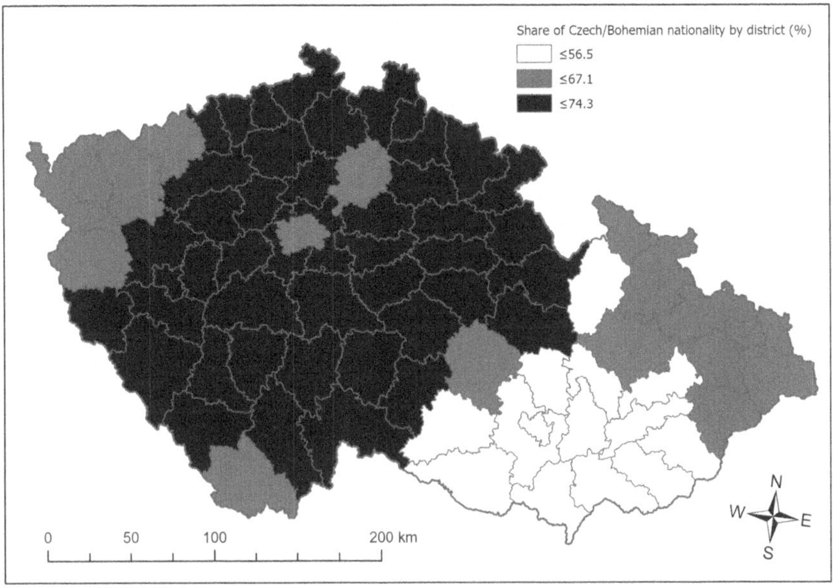

Fig. 4.17 Czechs/Bohemians in the 2011 census by districts [okresy]. (Source: Czech Statistical Office 2020, cartography: Přemysl Mácha)

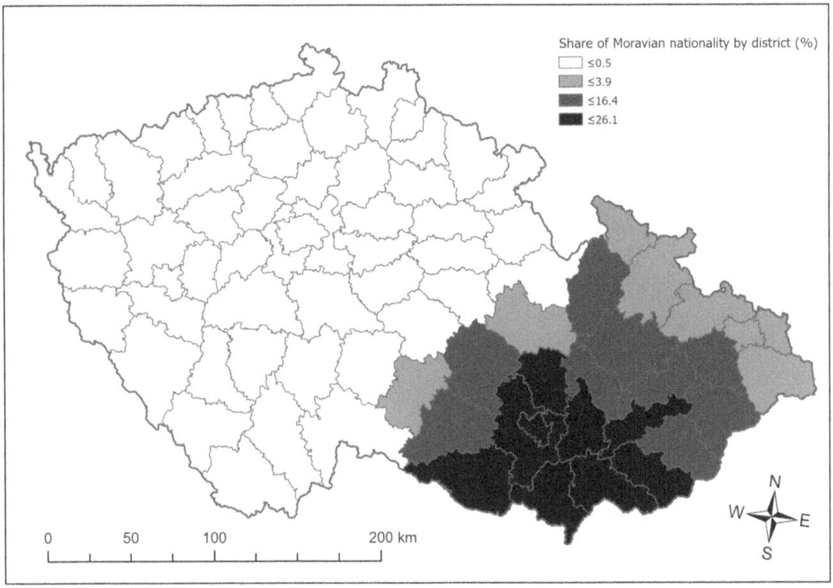

Fig. 4.18 Moravians in the 2011 census by districts [okresy]. (Source: Czech Statistical Office 2020, cartography: Přemysl Mácha)

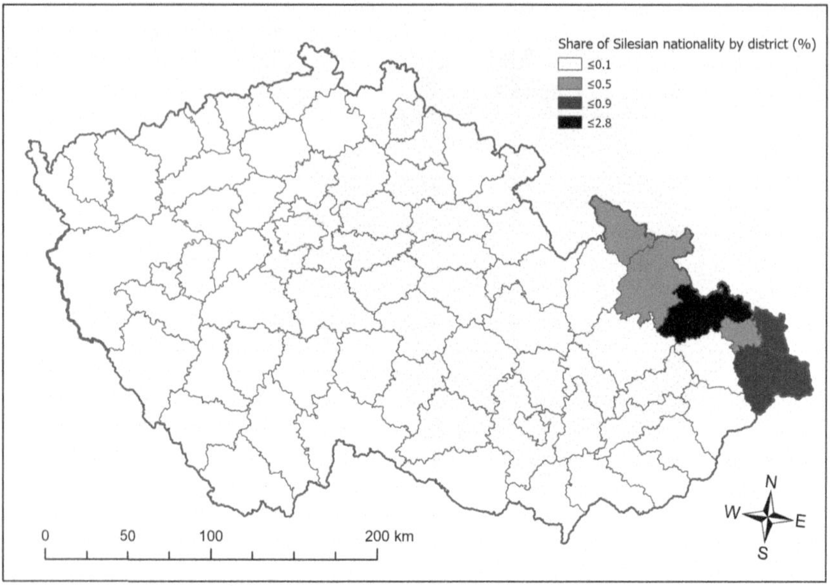

Fig. 4.19 Silesians in the 2011 census by districts [okresy]. (Source: Czech Statistical Office 2020, cartography: Přemysl Mácha)

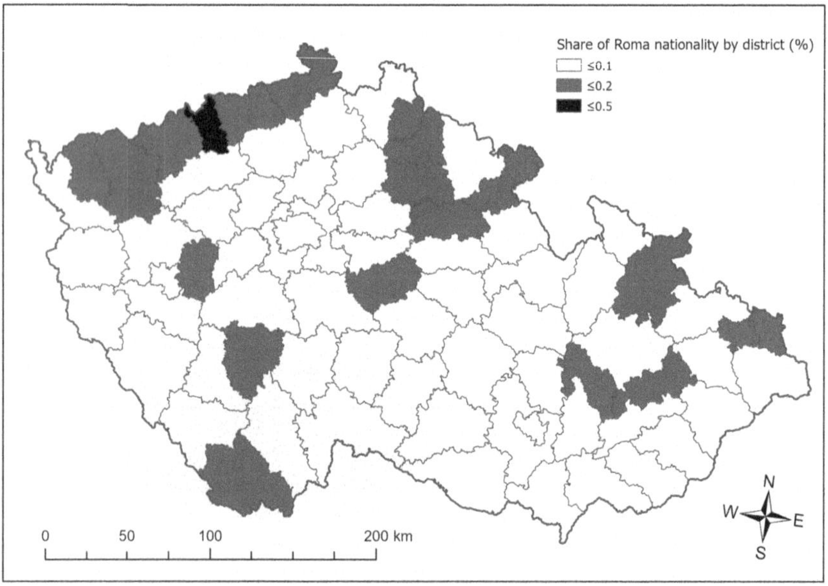

Fig. 4.20 Roma in the 2011 census by districts [okresy]. (Source: Czech Statistical Office 2020, cartography: Přemysl Mácha)

replaced the largely (but not entirely) relocated German population. In the context of minority rights, it is important to note that with the exception of Poles and Roma (see below) national minorities in Czechia are not sufficiently territorially concentrated to benefit from territorially based minority rights such as the use of their language in official communication with the state or the placement of bilingual signs. Everywhere, they represent individually and even collectively only a small proportion of all inhabitants.

The former land identities—Moravian, Silesian, and Bohemian—appeared in the census as national categories as well. Figures 4.17, 4.18, and 4.19 show the spatial distribution of people declaring Bohemian, Moravian, and Silesian nationality. As becomes clear, they generally follow the former land borders. However, these statistics are rather problematic for several reasons. First, the census question specifically asked about nationality, not land or regional identification. This means that people were forced to choose between Czech and Moravian, as if the two were on an equal level. In fact, Moravian identity is very strong as are anti-Bohemian and anti-Prague sentiments in Moravia [Morava], but most people consider Moravian identity to be a sub-identity of the Czech nationality rather than a nationality in its own right. The same holds true for Silesian identity with which even fewer people identified themselves in the census. Thus, the census data do not reveal the salience of either Moravian or Silesian identities as such but rather their perception as national identities. And second, in Czech language no distinction is made between 'Czech' and 'Bohemian' (both are just *Čech* and *český*). This means that there is no way to know whether people identifying with "česká národnost" ('Czech/Bohemian nationality') in the census meant Czech or Bohemian.

However, there is yet another important metric observable in the census—the negligible amount of people declaring Roma nationality. It is difficult to establish their actual number because 'Roma' is a highly fluid category and its definitions differ widely (see, e.g., Jakoubek 2004). The government has a special Inter-Ministerial Commission for Roma Community Affairs [Rada vlády pro záležitosti romské menšiny] (see Government of the Czech Republic 2020) which attends to the special needs of the Roma community. In addition, the Roma are also represented in the Government Council for National Minorities [Rada pro národnostní menšiny] which we describe in more detail below. None of them, however, has exact data on the size and spatial location of the Roma community.

In the last census, only a few thousands declared Roma nationality (see Table 4.9). The Czech government, however, estimates their number at 240,000 (Office of the Government 2018). This would make them by far the largest ethnic minority in the country. What is more, they are often spatially concentrated, usually in low-income, socially excluded localities. Nevertheless, because they do not declare Roma nationality in the census, they do not enjoy any linguistic rights, nor do they seem to claim them. The threat of repression from members of the majority society is real and they prefer to keep low profile, save occasional cultural and music festivals.

After the expulsion of Germans, it was the Roma who became the quintessential Other (see, e.g., Fawn 2001). Easily differentiated by their darker complexion and despised for their ostentatious disregard for settled life, they become an easy target for venting national frustrations and frequent racially motivated neo-Nazi and skinhead attacks. Only in recent years have they been allowed to breathe a little, their place on the pedestal of hate having been taken up by Muslims and refugees during the recent refugee crisis (although practically no refugees arrived in the country).

It is difficult to create a realistic map of the Roma population in Czechia based on census data. Figure 4.20 shows census results but only roughly corresponds with what is known about the Roma's primary areas of concentration—large cities, northern Bohemia, and the Moravian-Silesian Region [Moravskoslezský kraj]. However, even the 2011 census results are now outdated in the case of the Roma because in recent years, the Roma population has been moved out into internal and external peripheries by real-estate speculators and 'entrepreneurs with poverty' who capitalized on state accommodation subsidies for the poor. A better indication of their actual spatial distribution are socially excluded localities (see, e.g., Agency for Social Inclusion 2020). But even that provides only for an insufficient picture because many Roma live in regular residential neighborhoods while many non-Roma live in socially excluded localities. Finally, not only is mapping the Roma scientifically difficult, if achievable at all, it is also extremely problematic ethically in view of the aforementioned racist attacks. We therefore prefer to follow the census results and map only those who voluntarily declared Roma identity.

In addition to Czech citizens declaring minority membership, approximately 564,000 foreigners lived on a long-term basis in the country as of December 2018. Of those, 41% were from EU countries, half of them from Slovakia representing one-fifth of all foreigners. The rest came from other parts of the world. Most commonly, they were from Ukraine and Vietnam, the two countries making up for more than a half of all non-EU nationals and one-third of all foreigners (Czech Statistical Office 2019). As they gradually acquire Czech citizenship, they will strengthen the existing Ukrainian and Vietnamese minority, rewriting thus fundamentally the ethnic map of the country.

4.2.2 Political system

Present-day Czechia is a unitary parliamentary republic. The legislative branch is formed by the Parliament [parlament] divided into two chambers—the House of Deputies [poslanecká sněmovna] and the Senate [senát]. The House has 200 members, the Senate 81. The executive branch is formed by the president [prezident] and the cabinet [vláda] led by the prime minister [předseda vlády]. The president is elected directly but presidential powers are severely limited, the real executive power resting with the prime minister and his/her cabinet recruited from the parliament and appointed by the president. The judicial branch has many levels and is independent of both the executive and the legislative branches of government.

The electoral system relies on proportional representation for the House of Deputies and two-round majority vote for the Senate and the presidential office. The term of service is four years for the House of Deputies, six years for the Senate, and five years for the president. These temporal and electoral differences make it difficult for any single party or coalition to control all offices. The system of proportional representation in the House of Deputies requires that a party gains at least 5% of the popular vote to obtain seats in the House. This has at least two important consequences. First, there usually are between five and eight political parties represented in the House which leads to the inevitability of coalition cabinets. And second, the downside to the 5% threshold designed to prevent excessive parliamentary fragmentation is that in any given elections a substantial number of votes given to parties which fail to pass the threshold is wasted. Voting for smaller parties thus becomes a strategic decision between principles and pragmatic chances for success. From the perspective of the minorities, it means that given their small numbers, they stand no chance of having their own representation in the House. They can only run on other party tickets.

Administratively, the country is divided into 13 regions [kraje] and Prague [Praha] which is also a commune [obec]. The 13 regions are further divided into 6258 communes [obce]. Each region has its regional assembly [zastupitelstvo kraje] composed of 45–65 members elected for four years by proportional representation. The 5% threshold also applies. The assembly elects the regional council [rada kraje], an executive organ which oversees the everyday operation of regional offices and services including hospitals, high schools, road maintenance, and others. The regional governor [hejtman] presides over the council and represents the region. The Těšín/Cieszyn region is part of the Moravian-Silesian Region [Moravskoslezský kraj], one of the largest and most populous regions of Czechia. The Těšín/Cieszyn region forms its easternmost part (Fig. 4.21).

The communes are organized similarly. They have an assembly [zastupitelstvo obce] elected by proportional representation. Larger communes including all towns create executive councils [rada obce, rada města] led by a mayor [starosta, primatór] and vice-mayor [místostarosta, zástupce primátora] elected by the assembly. In smaller communes, the everyday operation of the commune is overseen only by mayors. The size of the communes ranges widely from very small villages (the smallest independent commune has only 17 inhabitants) to Prague, the country's capital, with a population of 1.3 million. Large cities are usually divided into self-governing districts [městský obvod] with their own assemblies [zastupitelstvo městského obvodu] elected by proportional representation. These assemblies also elect their own councils [rada městského obvodu] and mayors [starosta] who oversee basic services in the district. The city-wide assembly controls the operation of the city as a whole.

Fig. 4.21 Regions [kraje] and their administrative centers. (Cartography: Přemysl Mácha)

In practice, it is at the communal level that minorities can enjoy their legally guaranteed rights such as support for bilingual schooling or the introduction of bilingual signs, provided their proportion exceeds 10% in the last two population censuses. In theory, a minority can enjoy similar rights at the regional level if its proportion exceeds 5% in the region [kraj]. However, no minority does, so this provision of the law remains on paper only.

The political landscape of Czechia has transformed significantly over the last ten years, moving from a relatively stable system of political parties located on the left-right political spectrum toward a great prominence of populist and ideologically opaque groupings. After 1989, a broad coalition of political forces called Civic Forum [Občanské forum] in Czechia and Public against Violence [Verejnost proti násiliu] in Slovakia dominated elections with the primary goal of turning the Socialist country into a Western-style liberal democracy. As soon as this step was constitutionally completed, both movements disintegrated into a number of political parties which, together with renewed historical parties such as Social Democrats or Christian Democrats, formed a multiparty parliamentary system in which left-leaning coalitions took turns with right-leaning coalitions. All the while, the Communist party has continued to exist on the margins and obtain seats in the parliament and at lower levels.

The most prominent political parties were the Czech Social-Democratic Party [Česká strana sociálně-demokratická] on the left and the Civic Democratic Party [Občanská demokratická strana] on the right. These were seconded by the Christian Democratic Union—Czechoslovak Popular Party [Křesťansko-demokratická strana—Československá strana lidová] and TOP 09 (an acronym for tradition, responsibility and prosperity, and the year of creation), both on the right side of the spectrum but occasionally cooperating with Social Democrats. Numerous other parties emerged and vanished, sometimes leaving a strong imprint on the political scene, without, however, changing fundamentally the basic party system.

In the last decade, however, this system has undergone a profound restructuring with the appearance of several populist and activist groupings which have pushed the traditional parties to political margins. The most influential one is ANO (acronym for Action of Dissatisfied Citizens [Akce nespokojených občanů]) founded, financed, and tightly controlled by Andrej Babiš, the second-wealthiest Czech (actually Slovak—see the next section on who can be recognized as a true Czech) and owner of a vast agricultural, chemical, and food production empire. ANO has no ideology or program, its steps are unpredictable and chaotic, but it has gained the support of the elderly and left-leaning voters, essentially draining the Communists and Social Democrats of their traditional support. It is by far the most influential political party on the scene, and even the excessive scandals associated with Babiš seem to do it no harm. The center- and liberal right-leaning voters have, in turn, been lured by the Pirate Party, ideologically also very poorly defined and very unpredictable, though mostly liberal and green. The more conservative, nationalistic, and neo-Nazi voters have been captured by another populist project called Freedom and Direct Democracy [Svoboda a přímá demokracie] with an overt anti-Roma, anti-Muslim, and anti-EU rhetoric. These three groupings together receive the support of over 50% of the population. The traditional parties are left to compete for the rest. Overall, Czech democracy is not at its best currently, and authoritarian and nationalistic tendencies are becoming more prominent, though not reaching the levels witnessed in Hungary or Poland yet.

In the last three decades, minority rights have never been a topic for national or regional politics and no major party, with the exception of the Greens, has had them as one of its principal electoral program points. Minorities themselves, however, have often been targeted in elections, rhetorically and visually. This is especially true for the Roma community which has played the unfortunate role of a scapegoat for all the problems of the world. The Roma formed a political party called Roma Civic Initiative [Romská občanská iniciativa], but because of its limited electoral support it never succeeded in having a strong political representation. The Polish community in the Těšín/Cieszyn region formed a small party called Coexistentia/Soužití, but it also never achieved national prominence, gradually losing even local electoral gains.

Minority rights including bilingual signs always remained a local topic where it impacted significantly local politics. Interestingly, however, bilingual signs never became an explicit electoral topic as such. Rather, they turned into a point of contention in public debates after the elections. All of this is an important difference from Austria in general, and Carinthia in particular, described above.

4.2.3 National Identity and Cultural Characteristics

As mentioned above, Czech national identity was defined in its early stages in opposition to Germans. By this negative differentiation, it adopted many of the traits of its imaginary enemies, including ethnic and cultural conceptualization of nationhood. This had several important consequences. First and foremost, one must be born Czech, it is not possible to become one. One can acquire Czech citizenship but not Czech nationality. At least it is so in popular imagination. Politicians repeatedly capitalize on this fact during electoral campaigns targeting refugees, immigrants, the EU, and the Roma.

Second, from the perspective of the minorities, this concept of nationhood puts an extreme pressure on assimilation. Hyphenated identities are not acceptable. One has to decide if he/she is a Czech or a Pole or a Roma because these identities are seen as mutually exclusive. This pressure is responsible for the statistical reduction of traditional minorities visible in the population census. It is also responsible for the weakening of historic land identities—Bohemian, Moravian, and Silesian. Even though they manifest some renaissance, the proportion of people officially declaring them in the census is relatively small compared to their role in everyday discourse.

Regional (kraj) identities are also weak. Regional (kraj) boundaries have been redrawn many times, making them historically the least stable administrative unit in Czechia. There were also periods when regions [kraje] were abolished altogether. Regions [kraje] in their present spatial extent have existed for only two decades and they fail to respect historical land boundaries. As a consequence, regional [kraj] identities are only in the process of birth, somewhat boosted by license plate letters and by efforts by regional governments to promote tourism and encourage identification with regions by local inhabitants.

The pressure on assimilation has also seriously impacted sub-regional identities associated with traditional dialectal and cultural areas. With a few historical exceptions, these never competed politically with national or regional identities but coexisted with them as a differentiating factor in ordinary social relations. In Bohemia [Čechy], dialectal and cultural identities and differences have mostly disappeared, while eastern Moravia [Morava] and eastern Silesia [Slezsko] still demonstrate a strong dialectal and cultural pride. In regionalist movements, these traditional Moravian and Silesian dialects and folklore often serve as symbols for the cultural uniqueness of Moravia and Silesia. In reality, however, they are representative only of small parts of these regions while the majority of their inhabitants have already thoroughly assimilated into mainstream culture and colloquial Czech.

Third, the expulsion of Germans after World War II together with the ideology of the Czechoslovak nation accepted by the majority of Czechs (not so much by Slovaks) created a semblance of an ethnically homogenous territory with a continuous historical occupation, self-justified in the restoration of an independent state building on the tradition of the Bohemian Kingdom. The seat of the president is in the Prague Castle [Pražský hrad], the historic seat of Bohemian kings. There is a great distrust of all major powers (e.g., Germany, United States, France, Russia, China) with a strong sense that Czechs do not really need anyone and that if they were just left alone, they would prosper and be role models for the entire world thanks to their "golden hands" and ability to improvise under the most unfavorable circumstances.

Fourth, Czech nationhood was constructed in opposition to the Austrian state associated with the German element. For a long time, it was not state-forming but rather state-resisting. Even after the creation of Czechoslovakia, this skepticism and distrust of the state continued and were definitely sealed by the Nazi invasion and subsequent Communist rule. As a result, the state was never really perceived as 'ours' and civil society developed in a parallel, deformed, ethnic manner (Müller 2002). Public patriotism is therefore rarely seen, and it is usually cynically doubted for its authenticity. The role model in this sense is The Good Soldier Švejk[26] written by Jaroslav Hašek and translated into many languages. Excellent scientific studies of the Czech national imaginary were published, for example, by Holý (1996) and Hroch and Malecková (1999).

Although the increase in the number of foreigners living in the country and the opening of borders after the fall of Communism allowed for more intensive contact with the Other, the prevalent conceptualization of Czech nationhood remains fairly conservative. It is based on birth with a strong emphasis on language as a differentiating trait vis-à-vis neighboring countries, and on appearance vis-à-vis foreigners from more exotic places, and, internally speaking, vis-à-vis the Roma.

Class and ethnicity overlap strongly among the Roma as they do among, for example, the Ukrainians (construction, services) or the Vietnamese (restaurants, nail studios, cheap clothing shops). This is not the case with the Polish minority which is economically and socially as stratified as the majority population. Overall, however, regardless of socio-economic standing, due to the extremely low tolerance for alterity

[26] Its full Czech title: Osudy dobrého vojáka Švejka za světové války (The fates of the good soldier Švejk during the World War).

and a great pressure for assimilation in Czech society, it is not easy to be a member of any minority in Czechia. It is doubly difficult when this otherwise visually inconspicuous minority becomes visible by means of bilingual signs as was the case with the Polish minority in the Těšín/Cieszyn region.

4.2.4 Minority Legislation

The issue of minority legislation played an important role in the legal system of the Czechoslovak and later the Czech Republic throughout its many changes. A certain continuity in treatment of minorities can be traced from Austro-Hungarian legislation until contemporary legal solutions (with the exception of the period of Nazi occupation). This section provides an overview of the position of minorities in the state as well as outlining the legal background for the introduction of bilingual signs in the Těšín/Cieszyn region.

Treatment of minorities was one of the key elements of the legal system of the First Czechoslovak Republic (1918–1938). Minorities in the new state formed up to one-third of the population. The German and Hungarian minorities especially found themselves in a new position as no longer the dominant segments of the society. This shift caused a lot of tensions in the political landscape of the First Republic.

International law had a crucial influence on national legislation (Kuklík and Petráš 2017: 78). Specifically, the Minorities Treaty of Saint Germain (1919) played an important role in shaping the minority policies of the First Republic. The Constitutional Charter (Act No. 121/1920 Sb) was adopted on 29 February 1920. A number of provisions were incorporated to align the constitution with the aforementioned treaty in matters concerning minorities. For example, Article 128 in Section VI of the Charter called Protection of National, Religious and Racial Minorities guaranteed the equality of citizens and free use of all languages in private and business spheres in much the same way as Article 7 of the Treaty: "State citizens of the Czechoslovak Republic can freely use any language in private and business contacts, in religious matters, in newspapers and in any publications or public meetings provided these are within the limits of general laws."[27] Article 130 of the Charter guaranteed the freedom of association including the use of minority languages which corresponds to Article 8 of the Treaty. Article 131 of the Charter guaranteed the right to education in minority languages pursuant to Article 9 of the Treaty. Linguistic rights in general were also mentioned in the Charter (Article 129) but only in the sense that these will be addressed by a special constitutional act (see Language Act below). As there were some inconsistencies between the Constitutional Charter of the First Czechoslovak Republic of 1920 and the Treaty of Saint Germain, one of the main questions of the time was whether the international treaty takes legal precedence over national law. The Czechoslovak government also had difficulties in fulfilling promises it gave early on, especially when it came to the autonomy of the Ruthenian minority in the eastern part of the Republic (ibidem).

The conflict between the German minority and the Czechoslovak majority had a particularly significant impact on the development of policies that in effect related to all minorities living in the First Republic. The tensions were most visible in the issues surrounding the creation of the Language Act (Act No. 122/ 1920 Sb) and its implementation orders through Government Order No. 17/1926 Sb. The Language Act was a brief text which mostly recapitulated minority rights already mentioned in the Charter and specified conditions for their implementation. Most notably, however, Article 1 of the Act recognized the Czechoslovak language as the official language of the Czechoslovak Republic and made its use mandatory in all areas of the country: "the Czechoslovak language is a state, official language of the republic... It is, above all, a language (1) in which ... all courts, bureaus, institutes, companies and organs of the Republic conduct their business, issue their orders and write their public signs; (2) in which official state banknotes are issued; (3) which is used by the military in its command

[27] Part 3 of Article 128 in the original: "Státní občané republiky Československé mohou v mezích všeobecných zákonů volně užívati jakéhokoli jazyka ve stycích soukromých a obchodních, ve věcech týkajících se náboženství, v tisku a jakýchkoli publikacích nebo ve veřejných shromážděních lidu."

chain and as its service language, only when soldiers do not know this language is it allowed to use their maternal language."[28]

This was a clear attempt to break the resistance of German- and Hungarian-speaking areas to the use of Czech and Slovak. At the same time, it admitted to the reality of an army largely composed of German-speakers. Interestingly, while Article 1 defined Czechoslovak as the official language, subsequent articles of the Act treated Czech and Slovak as distinct languages and made their use legitimate everywhere. This proves that the lawmakers were aware of significant linguistic differences between Czech and Slovak, but to maintain the semblance of a homogenous nation, the official Czechoslovak language was created.

In addition, Article 2 of the Act set a 20% minority threshold at the juridical district level[29] based on the last population census. In such districts, minority members were entitled to the use of their language in official government business while governmental institutions were required to respond in minority language and have bilingual signs. This was a crucial stipulation because German-, Hungarian-, and Polish-speaking populations were highly concentrated, often accounting for more than 90% of all inhabitants, which would make the mandatory exclusive use of Czechoslovak a nonsense. On the other hand, the 20% threshold became fiercely contested in areas where minorities barely met the threshold or stayed slightly below it. As it was already mentioned, this often led to political interventions into the publication of population census results in order to decrease the tensions (Soukupová 2015).

The language used among the various parts of the state apparatus was one of the most significant issues of minority legislation and politics in the First Republic because of its great symbolic power (Petráš 2009).

[28] Article 1 of the Language Act No. 122/1920 in the original: "Jazyk československý jest státním, oficelním jazykem republiky… Jest tedy zejména jazykem, 1. v němž … děje se úřadování všech soudů, úřadů, ústavů, podniků a orgánů republiky, konají se jich vyhlášky a zevní označení, 2. v němž upraven jest hlavní text státovek a bankovek, 3. jehož používá branná moc při velení a jako jazyka služebního; ve styku s mužstvem jazyka toho neznalým užíti lze take jeho jazyka mateřského."

[29] At this time, the country was divided into lands [země], political districts [politické okresy], juridical districts [soudní okresy], and communes [obce]. The juridical districts were the smallest supra-commune administrative units. For example, the Czechoslovak part of the historic Těšín/Cieszyn region consisted of six juridical districts: Bohumín, Slezská Ostrava, Frýdek, Fryštát, Český Těšín, and Jablunkov. These six districts were grouped into three political districts: Frýdek, Fryštát, and Český Těšín. All three were part of the Silesian Land.

While the legal provisions of the First Czechoslovak Republic roughly followed the Austro-Hungarian system, Czechoslovak became the official state language of the new country. This added to the already existing tensions, especially in regions with numerous minority populations (Petráš 2009; Kuklík and Petráš 2017).

While the Language Act was passed in 1920, the implementation orders took another six years, during which the tensions between the majority and minority populations grew. In effect, a number of ambiguities that were present in the original 1920 formulation of the Act were eradicated by 1926 when a much stricter reading was implemented (Kuklík and Petráš 2017). In theory, minority members could use their mother tongue in contact with the state in districts where they reached the 20% threshold, but it was a complex and lengthy procedure. The state language had to be used for all internal procedures, which was especially complicated in regions with a strong German minority with large numbers of German-speakers in the official posts. The Language Act is seen as one of the most cumbersome pieces of legislation of the First Republic pertaining to minorities and one that in effect caused more tensions while bringing little benefit to the majority population.

There were other pieces of legislation in the interwar period that affected the situation of minorities in the Czechoslovak Republic. The Land Reform of 1919 nationalizing land over 150 ha, while not a specifically minority law, did affect minorities to a large degree as many of the wealthy landowners in the country were from either the German or Hungarian minority. Similarly, the Administrative Reform of 1927 (Act No. 125/ 1927 Sb) worsened the situation between the minorities (especially the German minority) and the majority. According to this reform, the administrative powers were to be redistributed into new regional units and strengthen thus the position of the state in local administration. However, fears that the German minority would in fact have a dominant position in several of the newly established districts and thus could use them as a base for an irredentist movement, led to gerrymandering on the Czechoslovak side and a sense of betrayal on the German side (Kuklík and Petráš 2017). It is also important to repeat that minorities were excluded from the first independent parliament and so several important legislation measures, including the Constitutional Charter, were passed without their input (Petráš 2009).

World War II presented a break in the continuity of legal provisions for minorities in Czechoslovakia as the country was weakened and many laws were imposed by the Nazi regime. The Těšín/Cieszyn region itself became part of the Third Reich, as did vast areas of the Czechoslovak borderlands inhabited by the German population (Sudetenland). The German minority was often blamed for the demise of the First Czechoslovak Republic. In light of this, following World War II there was a much stronger push to rid the country of minorities completely rather than to engage with them in a constructive dialogue (ibidem: 91).

As was explained above, in the years immediately after the war the policy of the Czechoslovak state was inimical toward minorities. Postwar Czechoslovakia was to become a national state of Czechs and Slovaks. Minority members were given individual rights (e.g., education in the mother tongue) but the wider minority rights of the First Republic were rejected as obsolete. The consensus of the time was that minorities would eventually assimilate into the Czech and Slovak nations.

There were no consistent legal provisions for minorities in the constitution during the first 20 years of the Communist rule (Nowak 2009; Petráš 2009; Friedl 2011; Spurný 2011). The new Constitution adopted in 1948 (No. 150/1948) did not regulate the question of rights of national minorities in any significant way (Żelazny 2016). At the same time, however, it stated that the Czechoslovak Republic is a united country of two equal Slavic nations—Czechs and Slovaks. The legal provisions from the First Republic were deemed obsolete and effectively canceled by the new Constitution. A number of regulation proposals were introduced ensuring certain rights (minority schooling, use of the minority language in contact with the state, etc.), yet these were circulated on an intra-party level and were never made available to the public (Petráš 2009). For example, in 1952 the government set out language resolutions, which outlined the principles of bilingual signs in minority areas (on public and industrial buildings) as well as principles of representation of minorities in national councils and other institutions. These resolutions, however, were not published or indeed included into any of the existing legislation (ibidem: 107). The situation on the ground varied depending on local administrative practice. The regime's treatment of minorities was often based on the situation of the Hungarian minority in the Slovak part of the country, which had become the largest minority in Czechoslovakia after the expulsion of Germans.

State's impact on minorities was not only through direct legislation but also through other legal changes and policies that impacted the minorities. One such example is the Administrative Reform of 1960 (No. 36/1960) which redrew the maps of the units of local government and occasionally led to ethnic gerrymandering. In the new system, for example, the district of Český Těšín with a traditionally high concentration of the Polish minority was split between the districts of Frýdek-Místek and Karviná, which meant that the minority now had relatively fewer people in the newly formed units (Petráš 2009: 115; Nowak 2009: 69). While this particular change was not originally aimed at minorities, it had a profound effect on their ability to gain political power and representation. Furthermore, increased immigration from the interior to the Těšín/ Cieszyn region motivated by employment in the local industries and mines 'diluted' the concentration of the local minority population and weakened its position (Friedl 2009).

A new constitution was passed in 1960, offering specific protection for three minorities: Poles, Hungarians, and Ukrainians (Constitution 1960 no. 100/1960). Article 25 of the new constitution provided backing for cultural development and education in the minority language for these three groups: "The state guarantees the citizens of Hungarian, Ukrainian and Polish nationality all possibilities and means for education in the mother tongue and for cultural development."[30] The Constitution was further amended in 1968 by the Constitutional Act no. 144/1968 on the Position of Nationalities in the Czechoslovak Socialist Republic. This important piece of legislation recognized collective rights of minorities, rather than just individual rights of citizens as members of these nationalities (Kuklík and Petráš 2017: 251). It declared the equality of all Czechoslovak citizens regardless of their language and nationality

[30] Article 25 of the 1960 Constitution in the original: "Občanům maďarské, ukrajinské a polské národnosti zabezpečuje stát všechny možnosti a prostředky ke vzdělání v mateřském jazyce a ke kulturnímu rozvoji."

(ibidem: 250). Article 1 offered protection, for the first time, to the German minority (Nowak 2009). Article 2 guaranteed proportional representation of minorities in political bodies. Article 3 then specified minority rights: "(1) Citizens of Hungarian, German, Polish and Ukrainian (Ruthenian) nationality are guaranteed proportioned to the interest in their national development and within the limits set by laws (a) the right to education in their language, (b) the right to a comprehensive cultural development, (c) the right to the use of their language in official government business in the respective nationality areas, (d) the right to organize in national cultural social organizations, (e) the right to media and information in their language."[31]

While the 1968 Act created an important framework for the existence of minority rights, the subsequent lack of implementing regulations meant that these were limited to local and regional level only. Some aspects of minority life were well maintained—such as minority education or presence of bilingual signs, other concerns, however, remained unanswered throughout the remaining years of the Communist rule.

The sweeping political and social changes of 1989 also shifted the position of minorities in Czechoslovakia. Initially a special government committee for minorities was to be set up; however, the differences in approaches between the Czech and Slovak part of the country made a unified approach essentially unfeasible. During the period between 1989 and 1993, there were a series of ad hoc steps ensuring the continuation of minority protections already in place, but no systematic solution was developed (Sulitka 2009) due to the growing tensions between Czech and Slovak politicians.

In 1990, the constitution was amended to eliminate the Socialist character of the legal framework. In 1991, the constitutional order was extended to include the Charter of Fundamental Rights and Freedoms (no. 23/1991) which contained several articles directly addressing

[31] Article 3 of the 1968 Constitutional Act: "(1) Občanům maďarské, německé, polské a ukrajinské (rusínské) národnosti se v rozsahu přiměřeném zájmům jejich národního rozvoje a za podmínek stanovených zákony zabezpečuje: (a) právo na vzdělání v jejich jazyku, (b) právo na všestranný kulturní rozvoj, (c) právo užívat jejich jazyka v úředním styku v oblastech obývaných příslušnou národností, (d) právo spolčovat se v národnostních kulturních společenských organizacích, (e) právo na tisk a informace v jejich jazyku."

minority concerns. Article 2 of the Charter prohibited discrimination based on nationality and provided protection against forced assimilation. Article 25 in Section 3 on the Rights of National and Ethnic Minorities specified minority rights:

(1) Citizens belonging to national or ethnic minorities are guaranteed comprehensive development, namely the right to develop together with other members from the minority their own culture, the right to distribute and receive information in their mother tongue and to form minority associations...
(2) Citizens belonging to national or ethnic minorities are also guaranteed under the conditions defined by law

(a) the right to education in their language
(b) the right to use their language in official government business
(c) the right to participate in matters pertaining to national and ethnic minorities.[32]

These provisions mostly only confirmed those already valid. However, two important differences should be noted. First, no specific nationalities were mentioned, unlike in previous legislations. And second, inclusion of minority representatives in decisions related to minorities was an important step in ensuring that minority voices be truly heard.

With the dissolution of Czechoslovakia in 1992 and creation of Czechia in January 1993, the Charter was adopted into the Czech constitutional order without any change in the aforementioned section on minorities. No implementing legislation was introduced, so minority rights remained a constitutional promise depending on good will of local and national politicians. The main change for the relationship between state and minorities came only with the ratification of two international

[32] Article 25 of the Charter in the original: "(1) Občanům tvořícím národnostní nebo etnické menšiny se zaručuje všestranný rozvoj, zejména právo společně s jinými příslušníky menšiny rozvíjet vlastní kulturu, právo rozšiřovat a přijímat informace v jejich mateřském jazyku a sdružovat se v národnostních sdruženích...

(2) Občanům příslušejícím k národnostním a etnickým menšinám se za podmínek stanovených zákonem zaručuje též (a) právo na vzdělání v jejich jazyku, (b) právo užívat jejich jazyka v úředním styku, (c) právo účasti na řešení věcí týkajících se národnostních a etnických menšin."

legal documents—the Framework Convention for the Protection of National Minorities and the European Charter for Regional or Minority Languages. The Framework Convention was signed by the Czech Republic in April 1995 and came into effect in 1998 (as Act no. 96/1998), the Charter was signed in 2000 but ratified only in 2006 (Sulitka 2009; Forejtová 2015).

Once the Framework Convention was signed, the parliament asked for a legal analysis of the existing provisions and for changes to be suggested that would align the legal provisions with the requirements of the Framework Convention. This analysis concluded that the existing regulations pertaining to education, use of minority languages, and administration were insufficient. This was the first impulse to begin working on the Minority Act, which was eventually passed in July 2001 as Act no. 273/2001 (Sulitka 2009; Szymeczek 2009). The Minority Act states the right to education in minority language as well as the right to use the minority language in contact with courts. Members of the minority have the right to maintain and develop their culture and traditions, to which the state must create suitable conditions. The Minority Act also established the Council for National Minorities as a consultative body for the government. The Council consists of representatives from twelve national minorities and delegates from the ministries, the parliament, and the Office of the President (Vermeersch 2003; Sulitka 2009). Its role is to connect local and national politics and ensure that minority concerns are properly addressed. The Council also distributes funding for cultural activities of the minorities. Although the Minority Act was an important step in the protection of minorities in Czechia, it received a mixed reception especially from members of the minorities themselves. Discussions surrounding the work on the Act meant that only its limited version was eventually adopted.

The European Charter for Regional or Minority Languages played a crucial role in the introduction of bilingual signs in Czechia. After the ratification of the Charter in 2006, it became possible to change the existing legislation in order to introduce bilingual signs to their full extent. One of the issues that the members of minorities often pointed out with both the Framework Convention and the Charter was that they lacked implementation measures and were as such open to interpretation by the state. The extensive delays between signing and ratification were also a point of debate (Sulitka 2009).

Multilingual signs in the Těšín/Cieszyn region had been present throughout all the historical periods. During Communism some public buildings as well as state-owned enterprises carried names in both Czech and Polish, although town signs, that is signs with the name of the settlement, located at the entry to and exit from the settlement, were always only Czech. According to some authors, this led to an association of the bilingual signs with the former regime (Szymeczek 2009). This in turn often meant that when the regime changed in the early 1990s, many of the bilingual signs disappeared either immediately or gradually during the following decade. The initiative to bring the bilingual signs back to the region was brought by the ratification of the Charter which gave a legal backing to the introduction of bilingual place names in the Těšín/Cieszyn region.

The process was set out by the Commune Act No. 128/ 2000 Sb (Section 29). Article 29 of the Act stated that minority names should be added to signs indicating the name of the commune, its parts, streets, and other public spaces as well as governmental administrative buildings provided that at least 20% of the population in the commune declared minority nationality in the last census and at least 50% of adults from the minority signed a petition requesting the introduction of signs (Szymeczek 2009). The petition requirement caused major concerns among minority members who feared being singled out while it also proved to be impossible to fulfill. The Act was therefore later amended, the petition requirement was taken out, and the threshold was lowered to 10%. The Congress of Poles [Kongres Poláků/Kongres Polaków], the principal political body of Czech Poles (see Chap. 5), played an instrumental role in safe-guarding and amending minority rights at the national level.

The state tried to resolve the dissonance between state policies and their local implementation by establishing National Minority Committees [výbory pro národnostní menšiny] in 2006. These committees are mandatorily set up in every commune with at least 10% and in every region [kraj] with at least 5% minority population. The committees consist of representatives of all recognized minorities in the commune or region and representatives of the Czech majority. These committees have been instrumental in the introduction of the bilingual place-name signs and contribute to the visibility of minorities at the communal level.

Currently, bilingual signs can be introduced in communes, where the minority reaches the 10% threshold based on the last two available censuses and where members of the minority request them either through the local committee for national minorities, which then recommends it to the communal assembly, or through a minority organization that has been active in the commune for at least five years. The threshold for minority population is generally within the reach of the Polish minority and the process itself as laid-out in the Act is relatively straightforward. In practice, however, it became apparent that there was still space for complications and reluctance, which occasionally led to extensive delays in the process (for more details, see Chaps. 5 and 6).

Because the right to bilingual signage is not fixed as in Austria but rather is dependent on the latest census results, it is feared that the next census in 2021 will confirm the continuing decrease in the proportion of Poles and threaten the position of the minority in several communes. Minority representatives therefore attempt to negotiate a more permanent solution by way of a legislative amendment to safeguard their rights and be able to concentrate on other issues of concern for the minority. It is, however, too early to say whether such an amendment has a chance of succeeding, and if so, in what form it will be. We can thus conclude that the changing ethnic demography of Czechia together with the inconclusive minority legislation leaves the minority question still open. This existential and legislative insecurity will continue to generate tensions and conflicts.

References

Agency for Social Inclusion. (Ed.). (2020). Retrieved July 7, 2020, from https://www.socialni-zaclenovani.cz/lokality/.

Anonymous. (s.a. [post 1945]). *Maps relating to the ethnical structure of Slovenian Carinthia*. s.l.

Austrian Board on Geographical Names (AKO). (Ed.). (2012). *Toponymic Guidelines for Map and Other Editors for International Use*. Retrieved April

25, 2020, from http://ortsnamen.at/wp-content/uploads/2017/12/Toponymic_Guidelines_AT_2012_07.pdf.

Baran, I. (2009). Polská menšina v československo-polských vztazích mezi světovými válkami. In B. Małysz & R. Kaszper (Eds.), *Poláci na Těšínsku* (pp. 36–42). Český Těšín: Kongres Poláků v České republice.

Borák, M. (1998). Problémy polské menšiny v ČR v 90. letech. In O. Šrajerová & G. Sokolová (Eds.), *Národnostní menšiny a majoritní společnost v České republice a v zemích střední Evropy v 90. letech XX. století* (pp. 230–241). Opava: Slezský ústav Slezského zemského muzea v Opavě.

Brauneder, W. (2009). *Österreichische Verfassungsgeschichte*. Wien: Manz Verlag.

Brauneder, W., & Lachmayer, F. (1992). *Österreichische Verfassungsgeschichte*. Wien: Manz Verlag.

Breu, J. (1970). *Die Kroatensiedlung im Burgenland und in den anschließenden Gebieten*. Wien: Deuticke.

Brubaker, R. (1999). The Manichean myth: Rethinking the distinction between 'civic' and 'ethnic' nationalism. In H. Kriesl et al. (Eds.), *Nation and national identity: The European experience in perspective* (pp. 55–71). Zürich: Verlag Rüegger.

Bruckmüller, E. (1984). *Nation Österreich. Sozialhistorische Aspekte ihrer Entwicklung*. Wien, Köln, Graz: Böhlau.

Bruckmüller, E. (1996). *Nation Österreich. Kulturelles Bewusstsein und gesellschaftlich-politische Prozesse*. Wien: Böhlau.

Bundeskanzleramt der Republik Österreich (BKA). (2020). Retrieved March 17, 2020, from https://www.ris.bka.gv.at/.

Czech Statistical Office (CZSO). (Ed.). (2014). *Národnostní struktura obyvatel*. Retrieved from https://www.czso.cz/documents/10180/20551765/170223-14.pdf.

Czech Statistical Office (CZSO). (Ed.). (2019). *Cizinci v České republice / Foreigners in the Czech Republic*. Praha: Czech statistical office. https://www.czso.cz/csu/czso/cizinci-v-cr-xf62lf2yaw.

Czech Statistical Office (CZSO). (Ed.). (2020). *Results of the 1980, 1991, 2001 and 2011 population census*. Retrieved July 10, 2020, from https://www.czso.cz/csu/sldb/historie_scitani.

Dujmovits, W. (2012). *Die Amerika-Wanderung der Burgenländer*. Berlin: Epubli.

Fawn, R. (2001). Czech attitudes towards the Roma: 'Expecting more of Havel's country'? *Europe-Asia Studies*, 53(8), 1193–1219.

Feldner, J., & Sturm, M. (2007). *Kärnten neu denken. Zwei Kontrahenten im Dialog*. Klagenfurt: Drava Verlag, Verlag Johannes Heyn.

Ferenc, T. (1980). *Quellen zur national-sozialistischen Entnationalisierungspolitik in Slowenien 1941–1945*. Maribor: Založba Obzorja.

Forejtová, M. (2015). Dvojjazyčné označování měst a obcí dle Evropské charty regionálních či menšinových jazyků a Rámcové úmluvy o ochraně národnostních menšin. *Acta Universitatis Carolinae Iuridica*, 4, 37–44.

Fräss-Ehrfeld, C. (2005). Zwischen Bundeskompetenz und Kärntner Realität. Die Kärntner Minderheitenproblematik in der Zweiten Republik 1945–1976. In C. Fräss-Ehrfeld & H. Rumpler (Eds.), *Kärnten und die nationale Frage. Kärnten und Wien. Zwischen Staatsidee und Landesbewusstsein* (pp. 83–162). Klagenfurt/Celovec, Ljubljana/Laibach, Wien/Dunaj: Hermagoras/ Mohorjeva.

Friedl, J. (2009). Polská menšina v letech 1945–1956. In B. Małysz & R. Kaszper (Eds.), *Poláci na Těšínsku* (pp. 60–68). Český Těšín: Kongres Poláků v České republice/Kongres Polaków w Republice Czeskiej.

Friedl, J. (2011). *Státní politika vůči polské menšině na Těšínsku v letech 1945–1949. Výběrová edice dokumentů*. Praha: Historický ústav Akademie věd České Republiky.

Galuška, L. (1991). *Great Moravia*. Brno: Moravian Museum.

Gawrecká, M. (2004). *Československé Slezsko mezi světovými válkami 1918–1938*. Opava: Slezská univerzita v Opavě.

Government of the Czech Republic. (Ed.). (2020). Retrieved July 7, 2020, from https://www.vlada.cz/en/ppov/zalezitosti-romske-komunity/the-council-for-roma-community-affairs%2D%2D50634/.

Haslinger, P. (2001). Building a Region: Identity Politics and the Formation of Identities in the Federal Province of Burgenland 1921–1938. *Austrian History Yearbook*, 32, 105–123.

Herder, J. G. (1813). *Ideen zur Philosophie der Geschichte der Menschheit*. Wien: Johann von Müller, Haasische Buchhandlung.

Holý, L. (1996). *The Little Czech and the Great Czech Nation: National Identity and the Post-communist Social Transformation* (Cambridge Studies in Social and Cultural Anthropology). Cambridge: Cambridge University Press.

Hren, K. (2004). Die SPÖ und der Kärntner Ortstafelsturm. In *Kärntner Jahrbuch für Politik 2004* (pp. 100–110). Klagenfurt: Kärntner Druck- und Verlagsgesellschaft.

Hren, K., & Pandel, M. (Eds.). (2012). *Ein Jahr danach. Die Ortstafelregelung 2011 und was daraus wurde*. Klagenfurt/Celovec, Ljubljana/Laibach, Wien/ Dunaj: Hermagoras/Mohorjeva.

Hroch, M., & Malecková, J. (1999). The construction of Czech national history. *Historein*, 1: 103–112. Retrieved March 20, 2020, from http://ortsnamen.at/wp-content/uploads/2017/12/Toponymic_Guidelines_AT_2012_07.pdf.

Jakoubek, M. (2004). *Romové—konec (ne)jednoho mýtu*. Praha: Socioklub.

Jordan, P. (1988). *Möglichkeiten einer stärkeren Berücksichtigung slowenischer Ortsnamen in den heutigen amtlichen topgraphischen Karten Österreichs*. Wien: Österreichische Akademie der Wissenschaften, Institut für Kartographie.

Jordan, P. (2019). Three types of commercial place-name use—With examples from Austria. In H. Beyer, M. Brenzinger, T. Du Plessis, & P. E. Raper (Eds.), *Critical toponymy. Place names in political, historical and commercial landscape. Proceedings of the 4th International Symposium on Place Names 2017, Windhoek, Namibia, 18–20 September 2017* (pp. 111–122). Bloemfontein: UFS, Department of South African Sign Language and Deaf Studies.

Jordan, P., & Seger, M. (2019). Österreich als Staatswesen. In M. Seger, *Österreich. Raum und Gesellschaft* (pp. 12–25). Klagenfurt am Wörthersee: Naturwissenschaftlicher Verein für Kärnten.

Kladiwa, P. et al. (2016). *Národnostní statistika v česklých zemích 1880–1930,* 2 volumes. Praha, Ostrava: Nakladatelství Lidové noviny, Ostravská univerzita.

Kozler, P. (1853). *Zemljovid Slovenske dežele in pokrajin* [1:576,000]. S.l.

Kuklík, J., & Petráš, R. (2017). *Minorities and Law in Czechoslovakia, 1918–1992*. Praha: Karolinum.

Leff, C. (2018). *The Czech and Slovak republics: nation versus state*. London and New York: Routledge.

Leff, C. S. (2014). *National conflict in Czechoslovakia: The making and remaking of a state, 1918–1987*. Princeton: Princeton University Press.

Magocsi, P. R. (2000). *Of the Making of Nationalities There Is No End* (= East European Monographs, 540). Boulder, Colorado: Columbia University Press.

Małysz, B., & Kaszper, R. (Eds.). (2009). *Poláci na Těšínsku*. Český Těšín: Kongres Poláků v České republice/Kongres Polaków w Republice Czeskiej.

Matscher, F. (2006). Die Ortstafelfrage aus der Sicht der Ortstafelkommission. In: Amt der Kärntner Landesregierung, Abteilung 1—Landesamtsdirektion/Volksgruppenbüro, Dokumentationsstelle für Zeitgeschichte (Ed.), *Die Ortstafelfrage aus Expertensicht. Eine kritische Beleuchtung* (= Kärnten Dokumentation, Sonderband 1) (pp. 111–123). Klagenfurt: Verlag Land Kärnten.

Moritsch, A. (2000). Nationale Ideologien in Kärnten. In A. Moritsch (Ed.), *Kärntner Slovenen/Koroški Slovenci 1900–2000* (pp. 9–28). Klagenfurt/Celovec, Ljubljana, Wien: Verlag Hermagoras/Mohorjeva.

Motýl, I. (2011). Sčítání lidu bude i bitvou o česko-polské cedule. *Týden*. Retrieved February 7, 2011, from https://www.tyden.cz/rubriky/domaci/scitani-lidu-bude-i-bitvou-o-cesko-polske-cedule_193179.html?showTab=nejnovejsi.

Müller, K. (2002). *Češi a občanská společnost. Pojem, problémy, východiska*. Praha: Triton.

Novinky.cz. (2020). Na pražské Staroměstské náměstí se vrátil mariánský sloup. Retrieved June 4, 2020, from https://www.novinky.cz/domaci/clanek/na-prazskem-staromestskem-namesti-vztycuji-mariansky-sloup-40326441.

Nowak, K. (2009). Polská menšina v letech 1956–1989. In B. Małysz & R. Kaszper (Eds.), *Poláci na Těšínsku* (pp. 68–82). Český Těšín: Kongres Poláků v České republice/Kongres Polaków w Republice Czeskiej.

Office of the Government. (2018). *Zpráva o stavu romské menšiny v České republice za rok 2017*. Praha. Retrieved June 23, 2020, from https://www.vlada.cz/assets/ppov/zalezitosti-romske-komunity/dokumenty/Zprava-o-stavu-romske-mensiny-2017.pdf.

Öhlinger, Th. (2006). Verfassungsrecht und Volksgruppenschutz. In: Amt der Kärntner Landesregierung, Abteilung 1—Landesamtsdirektion/Volksgruppenbüro, Dokumentationsstelle für Zeitgeschichte (Ed.), *Die Ortstafelfrage aus Expertensicht. Eine kritische Beleuchtung* (= Kärnten Dokumentation, Sonderband 1) (pp. 124–132). Klagenfurt: Verlag Land Kärnten.

Österreichische Rektorenkonferenz. (Ed.). (1988). Bericht der Arbeitsgruppe "Lage und Perspektiven der österreichischen Volksgruppen". Wien.

Österreichische UNESCO-Kommission. (Ed.). (2020). Retrieved March 22, 2020, from https://www.unesco.at/kultur/immaterielles-kulturerbe/oesterreichisches-verzeichnis/detail/article/slowenische-flur-und-hofnamen-in-kaernten.

Pelinka, A. (1990). *Zur österreichischen Identität. Zwischen deutscher Vereinigung und Mitteleuropa*. Wien: Ueberreuter.

Petráš, R. (2009). Historie právního postavení menšin v českých zemích. In R. Petráš, H. Petrův, & H. C. Scheu (Eds.), *Menšiny a právo v České republice* (pp. 31–147). Praha: Auditorium.

Pohl, H.-D. (1999). Österreichische Identität und österreichisches Deutsch. *Kärntner Jahrbuch für Politik 1999*, 71–103.

Pohl, H.-D. (2002). Die ethnisch-sprachlichen Voraussetzungen der Volksabstimmung. In H. Valentin, S. Haiden, & B. Maier (Eds.). *Die Kärntner Volksabstimmung 1920 und die Geschichtsforschung. Leistungen, Defizite, Perspektiven* (pp. 181–188). Klagenfurt: Verlag Johannes Heyn.

Pohl, H.-D. (2007). Zur Diskussion ums österreichische Deutsch. *Kärntner Jahrbuch für Politik 2007*: 157–180.

Pohl, H.-D., & Schwaner, B. (2007). *Das Buch der österreichischen Namen*. Wien, Graz, Klagenfurt: Pichler Verlag.

Prokop, R. (1998). Geopolitické problémy národnostních struktur České i Československé republiky v procesu vývoje. In O. Šrajerová & G. Sokolová (Eds.), *Národnostní menšiny a majoritní společnost v České republice a v zemích střední Evropy v 90. letech XX. století* (pp. 39–46). Opava: Slezský ústav Slezského zemského muzea v Opavě.

Reimann, R. (2004). Minderheiten—Brücke oder Bedrohung? *Zeitschrift des Historischen Vereins für Steiermark*, 95, 69–82.

Richnovsky, R. (2013). *Die Einbettung der burgenländischen Kroaten in das Landesimage und die Landesidentität des Burgenlandes sowie ihr Verhältnis zu den anderen Nationalitäten*. Diploma thesis, University of Vienna.

Rumpler, H. (2005). Die nationale Frage im Spannungsfeld von kärntnerischem Landespatriotismus, österreichischem Staatsbewusstsein und völkischem Nationalismus 1918–1938. In C. Fräss-Ehrfeld & H. Rumpler (Eds.), *Kärnten und Wien. Zwischen Staatsidee und Landesbewusstsein* (= Kärnten und die nationale Frage, 4) (pp. 9–82). Klagenfurt/Celovec, Ljubljana/Laibach, Wien/Dunaj: Hermagoras/Mohorjeva.

Rumpler, H., & Seger, M. (2010). *Die Gesellschaft der Habsburgermonarchie im Kartenbild. Die Habsburgermonarchie 1848–1910, Band IX/2: Soziale Strukturen. Verwaltungs- Sozial- und Infrastrukturen. Nach dem Zensus von 1910. Atlas thematischer Karten*. Wien: Verlag der Österreichischen Akademie der Wissenschaften.

Seger, M. (2005). Sympathie und Freizeitmobilität. Der Kärntner Fremdenverkehr unter Berücksichtigung der Beziehungen zu Wien. In C. Fräss-Ehrfeld & H. Rumpler (Eds.), *Kärnten und Wien. Zwischen Staatsidee und Landesbewusstsein* (= Kärnten und die nationale Frage, 4) (pp. 241–264). Klagenfurt/Celovec, Ljubljana/Laibach, Wien/Dunaj: Hermagoras/Mohorjeva.

Seger, M. (2019). *Österreich. Raum und Gesellschaft*. Klagenfurt am Wörthersee: Naturwissenschaftlicher Verein für Kärnten.

Šimon, M. (2015). Measuring phantom borders: The case of Czech/Czechoslovakian electoral geography. *Erdkunde*, 69(2), 139–150. https://doi.org/10.3112/erdkunde.2015.02.04.

Sitte, Ch. (2020). Kompetenzaufteilung zwischen Bund, Ländern und Gemeiden in Österreich. Retrieved May 20, 2020, from https://homepage.univie.ac.at/Christian.Sitte/erasmus/kompetenzaufteilung.htm.

Siwek, T., Zahradník, S., & Szymeczek, J. (2001). *Polská národnostní menšina v Československu 1945–1954*. Praha: Ústav pro soudobé dějiny AV ČR.

Soukupová, B. (2015). Jazykový zákon (1920): česká odplata, upevňování národního charakteru státu, hospodářsko-sociální pragmatismus, nebo kultivace tradice bilingvismu? K názorům české veřejnosti na významy státního jazyka. *Acta Universitatis Carolinae Iuridica*, 4, 23–35.

Spurný, M. (2011). *Nejsou jako my: česká společnost a menšiny v pohraničí (1945–1960)*. Praha: Antikomplex.

Statistik Austria. (Ed.). (2020). Retrieved March 21, 2020, from http://www.statistik.at.

Stourzh, G. (2004). Österreichs Weg zum Staatsvertrag und zur Neutralität. In: Forum Politische Bildung (Ed.), *Frei—Souverän—Neutral—Europäisch. 1945–1955–1995–2005* (= Informationen zur Politischen Bildung, 22) (pp. 7–20). Innsbruck, Wien: Studien Verlag.

Stranner, E., & Lehner, C. (2020). *Kärnten vielseitig/Pestra Koroška/Carinzia versatile/Carinthia diverse*. Klagenfurt/Celovec: Verlag Johannes Heyn.

Stummvoll, O. (2013). *Geographische Namen als Ausdruck raumbezogener Identität. Dargestellt am Beispiel der kommerziellen Verwendung von Landschafts- und Gebietsnamen in Österreich*. MA thesis, University of Vienna.

Sulitka, A. (2009). Národnostní menšiny v České republice po roce 1989 a národnostněmenšinová politika. In R. Petráš, H. Petrův, & H. C. Scheu (Eds.), *Menšiny a právo v České republice* (pp. 148–185). Praha: Auditorium.

Suppan, A. (2019). *Imperialist Peace Order in Central Europe. Saint-Germain and Trianon 1919–1920*. Wien: Austrian Academy of Sciences Press.

Symmons-Symonolewicz, K. (1985). The concept of nationhood: Toward a theoretical clarification. *Canadian Review of Studies in Nationalism*, XII(2), 215–222.

Szymeczek, J. (2009). Prosazování práv polské menšiny v České republice na příkladu dvojjazyčnosti a dvojjazyčných názvů. In B. Małysz & R. Kaszper (Eds.), *Poláci na Těšínsku* (pp. 90–93). Český Těšín: Kongres Poláků v České republice.

Thaler, P. (2001). *The ambivalence of identity. The Austrian experience of nation-building in a modern society*. Purdue, Indiana: Purdue University Press.

Tóth, A., Novotný, L., & Stehlík, M. (2012). *Národnostní menšiny v Československu 1918–1938*. Praha: Univerzita Karlova.

Unkart, R., Glantschnig, G., & Ogris, A. (1984). *Zur Lage der Slowenen in Kärnten. Die slowenische Volksgruppe und die Wahlkreiseinteilung 1979—eine Dokumentation*. Klagenfurt: Verlag des Kärntner Landesarchivs.

Veiter, Th. (1980). *Die Kärntner Ortstafelkommission* (= Das gemeinsame Kärnten/Skupna Koroška, 8). Klagenfurt: Hermagoras/Mohorjeva.

Vermeersch, P. (2003). EU enlargement and minority rights policies in Central Europe: Explaining policy shifts in the Czech Republic, Hungary and Poland. *JEMIE*, 4(1), 1–32.

Vocelka, K. (2002). *Geschichte Österreichs*. Graz, Wien: Styria Verlag.

Žáček, R. (2001). Politika československého státu vůči menšinám v letech 1945–1948. In I. Kufová (Ed.), *Minulost a současnost národnostních menšin na Těšínsku* (pp. 83–93). Český Těšín: Pedagogické centrum pro polské národní školství.

Żelazny, K. (2016). Status prawny mniejszości narodowych na obszarze Czech w byłej Republice Czechosłowacji w latach 1918–1992. *Miscellanea Historico-Iuridica*, 15(1), 117–128.

5

The Two Minority Situations Compared

5.1 Carinthia (Peter Jordan)

5.1.1 The Wider Geographical and Historical Setting (including Neighborhoods)

Carinthia [Kärnten] is the southernmost federal state of Austria. By area it is the fifth largest (9538 sq.km) and by population the sixth (560,931 as of 1 January 2019). However, it is the component of modern Austria with the longest tradition as a political entity. It borders on Italy and Slovenia in the South and has boundaries with three other Austrian federal states in the West, North, and East, that is, Tyrol [Tirol], Salzburg, and Styria [Steiermark] (Fig. 5.1).

It is framed by partly very high mountain ranges rising up to 3798 m in the case of the Hohe Tauern with Großglockner, the highest peak of Austria. It can be accessed at the valley level only at four spots, in fact rather narrow gates: from Tyrol, where the Drau/Drava enters Carinthia, along River Gailitz/Ziljica [Slizza] toward Tarvisio in Italy, between Bleiburg/Pliberk and the Mießtal [Mežiška dolina], and nearby Lavamünd toward Slovenia, where River Drau/Drava leaves Carinthia

P. Jordan et al., *Place-Name Politics in Multilingual Areas*, Palgrave Studies in Minority Languages and Communities, https://doi.org/10.1007/978-3-030-69488-3_5

Fig. 5.1 Austria's territorial-administrative subdivision into federal states and political districts. (Source: Hölzel 2011: 16)

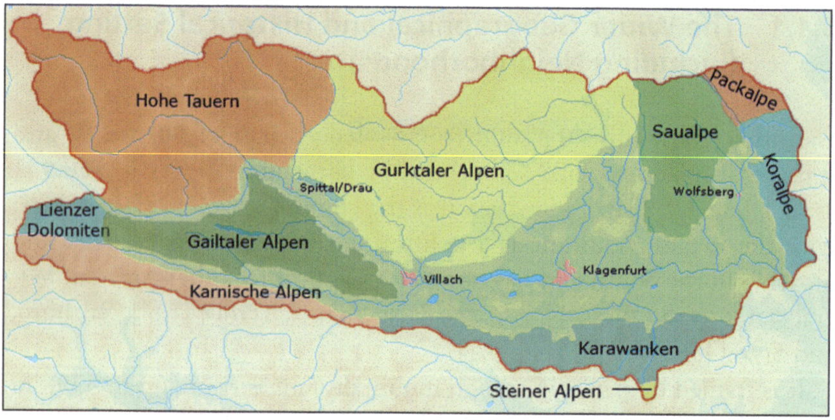

Fig. 5.2 Carinthia's natural-geographical structure. (Source: Wikipedia)

(Fig. 5.2). This geomorphological structure is the result of glaciers in the last Ice Age (Fig. 5.3).

Figure 5.4 shows the mountain range of the Carnic Alps [Karnische Alpen/Alpi Carniche] marking Carinthia's border with Italy in the South.

This geomorphological setting is the basis of a traditional and widespread narrative of Carinthia (in this respect similar to Bohemia [Čechy]

Fig. 5.3 Carinthia's glaciation in the last ice age, glaciers in blue. (Source: Seger 2010: 104). (Color figure online)

Fig. 5.4 Carnic Alps [Karnische Alpen/Alpi Carniche]. (Photo by Peter Jordan 2008)

and Transylvania [Ardeal]) as a 'natural political unit' bound together by nature. A division would be 'against nature'—an argument frequently used by the German-Carinthian majority in defending its dominion against all aspirations to attach the Slovenian or bilingual part to other political entities. What cannot be denied is that the wide intra-Alpine basin surrounded by mountain ranges and accessible only via mountain passes and narrow gates offers well-protected spaces for human settlement and political structure-building and favors internal networking compared to trans-boundary contacts.

Thus, the Celtic kingdom Noricum with its wealth of (not the least Carinthian) iron ore and salt mines covering the largest part of modern Austria had its political and economic center there—at Magdalensberg north of Klagenfurt (Fig. 5.5).

Also the Roman province of Noricum (Fig. 5.6), at least its Mediterranean part (Noricum mediterraneum), was focused on Carinthia

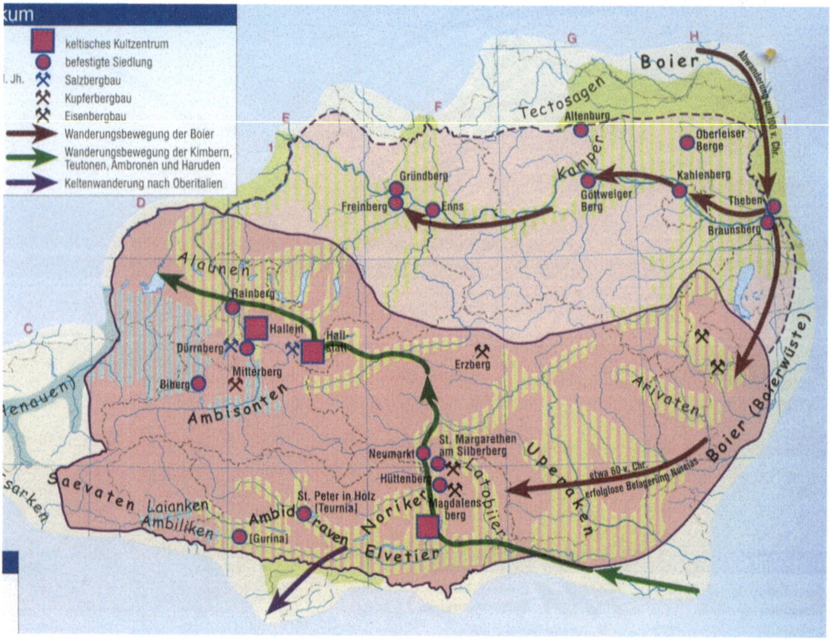

Fig. 5.5 Celtic kingdom Noricum. (Source: Hölzel 2004: 55)

Fig. 5.6 Roman province Noricum. (Source: Lendl et al. 1972: 35)

(see Glaser 2001). Its capital Virunum (near Klagenfurt) was located there and shifted later to Teurnia (near Spittal an der Drau) in upper Carinthia (Glaser 2001: 70). Iuenna [Hemmaberg/Gora svete Heme] right in the bilingual area (in the subregion of Jauntal/Podjuna, named after this place) was a transregional place of pilgrimage. From the density of ancient Christian churches, it can be concluded that in the later Roman period Christianization gained more ground in Noricum inside the Alpine arc (Noricum mediterraneum) than in Noricum outside (Noricum ripense). This may also have been caused by a stronger influx of Romance population. The later arrival of the Aryan-Christian Ostrogoths frequently resulted in two Christian churches standing side by side (Glaser 2001: 82).

In the sixth and seventh centuries the Alpine-Slavonic principality of Carantania (Fig. 5.7) had its center also north of Klagenfurt and Carinthian places and monuments are markers of all-Slovenian (comprising Slovenes in Slovenia and elsewhere) identity up to the present day. In the middle of the eighth century, however, this principality, threatened by

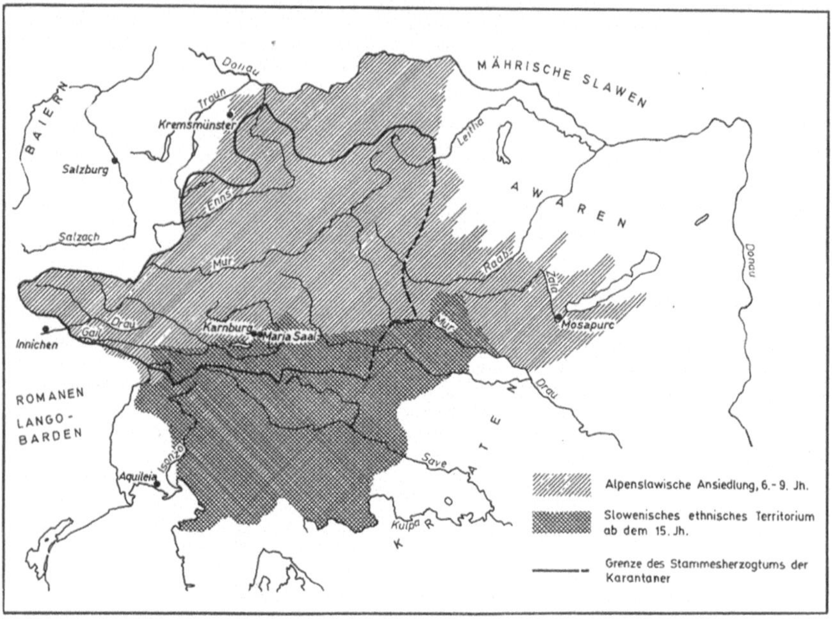

Fig. 5.7 Alpine-Slavonic settlement in the sixth to ninth centuries (black screens) and principality of Carantania (surrounded by a black line). (Source: Lukan and Moritsch 1988)

the Avars in the East and having asked for protection, had to acknowledge Bavarian/Franconian domination and was fully integrated into the political system of the Franconian Empire by the end of the eighth century (see Krahwinkler and Wolfram 2001).

In the Early Middle Ages (since 976), in the same year, when a first Austrian political unit (Ostarrichi) emerged in the surroundings of Vienna [Wien], the dukes of Carinthia having become independent of Bavaria [Bayern] and the Franconian Empire, ruled over territories and marches down to the Adriatic Sea (Friuli, Verona, Istria [Istra]) and included most of what is today Styria [Steiermark] (Carantanian March) and Slovenia (March Carniola [Kranjska], March at the Sann [Savinja]), even some parts of modern Lower Austria [Niederösterreich], that is the March of Pitten [Pittener Mark], and Upper Austria [Oberösterreich], that is the Enns Valley south of Steyr (Fig. 5.8).

Fig. 5.8 Duchy of Carinthia [Kärnten] with marches (dark orange) and Ostarrichi (paler orange) around 1000. (Source: Lendl et al. 1972: 47). (Color figure online)

The lack of a strong and continuous dynasty as well as a shift of major trade flows toward the Danube axis with Vienna as a rising hub resulted in economic decline and a loss in political power. Thus, in 1335, Carinthia, albeit only its eastern part,[1] together with Carniola [Kranjska] had to accept integration into the conglomerate of lands ruled by the Habsburg dynasty residing mainly in Vienna and composed of what was

[1] Most western parts belonged at that time to the County of Gorizia and to the ecclesiastical dominions of Salzburg and Bamberg (see, e.g., Moro 1957).

at that time called 'Austria' [Herzogtum Österreich] (Lower Austria [Niederösterreich] plus eastern parts of Upper Austria [Oberösterreich]), Styria [Steiermark], and Sundgau in the Rhine Valley—all of them also parts of the Holy Roman Empire. This meant, however, just subordination to a common ruler, not integration into common state structures as they emerged only much later. Carinthia remained in this way a rather self-governing unit with the local nobility and independent cities having a strong saying. However, as the historian Helmut Rumpler states, the narrative of Carinthia's political and economic decline "since Habsburg took over in 1335" as well as a skeptic stance toward Vienna are constants of Carinthian historiography (Rumpler 2005: 10f).

In the ecclesiastical sphere, Carinthia remained divided between the archbishopric of Salzburg and the patriarchate of Aquileia with River Drau/Drava forming the boundary since 811. This division was to gain great importance as regards the minority situation as it will be shown later.

Relative self-governance as well as the existence of independent cities and mining (for iron ore, gold, silver, lead, and zinc) as a still important economic activity provided also a fertile soil for Protestantism in the course of the Lutheran Reformation. Especially citizens of independent cities like Klagenfurt (see Baum 2002) and miners with their strong economic standing converted to Protestantism—many more than in other Habsburg lands, like Tyrol or Austria (at that time in the sense of Lower Austria and the largest parts of Upper Austria [Oberösterreich]), but certainly less than in Bohemia with its double (Hussite and Lutheran) Reformation.

As already mentioned in Chap. 4, Sect. 4.1.6, Habsburg-supported Counter-Reformation succeeded in Carinthia to a minor extent than in other parts of the Habsburg realm, although some Protestants (the so-called *Landler* from central parts of Carinthia[2]) were even expelled to Transylvania [Ardeal] in what is today Romania. Nevertheless, a larger Protestant minority remained up to the present day (see Fig. 5.9). One of the lasting impacts of Protestant lifestyle and attitudes influencing also

[2] Besides Carinthia, also the Ausseerland in Styria [Steiermark] and the Traun Quarter [Traunviertel] in Upper Austria [Oberösterreich] were source regions of Landler emigration to Transylvania (see Schabus 1996; Bottesch et al. 2002).

Fig. 5.9 Share of protestants in Carinthia in 1923. (Source: Wutte et al. 1925: Map 53). (Color figure online)

re-converted Catholics is an inclination toward a-clerical positions and a political landscape characterized by an oscillation between (a-clerical) Social Democrats and the (a-clerical) Freedom Party, while Conservatives (closer to Catholicism) are—much in contrast to other parts of modern Austria—in a rather weak position.

The emergence of an Austrian statehood with central supreme authorities and institutions as well as centralized legislation and military forces under Mary Theresa (1740–1780) and her son Joseph II (1780–1790), who introduced German as the common official language of the multilingual empire, found Carinthia as an integrated part of one of the major European powers, albeit within this large empire, after the defeat of the Ottoman Empire now also comprising the Pannonian Basin, in an economically weak and peripheric position. Carinthia was not anymore the core region of a rule conglomerate surrounded by marches like in the Early Middle Ages, but far off the centers of the new empire and even second to Styria, its former march, where Graz played now the role of the center of Innerösterreich ('Austria interior'), a rather autonomous unit with Styria, Carinthia, and Carniola as its components.

This development continued in the nineteenth century and up to World War I with railroad construction, industrialization, and urbanization as its main formative factors. Carinthia was bypassed by the main track of the Southern Railroad [Südbahn] connecting the imperial capital Vienna with the Empire's main port, Trieste. Also in respect to industrialization, a factor fundamentally changing lifestyle, demographic behavior, social relations, and settlement structure, it fell behind the Mur-Mürz axis in Styria, the southern Vienna Basin [Wiener Becken], and even more so behind Bohemia [Čechy] and Silesia [Slezsko, Śląsk] remaining essentially an agricultural region.

This also had an effect on the position of Carinthia's cities in the central place system of the Empire. By 1910, due to its location at the main track of the Southern Railroad and its industrialization, Graz became the Empire's sixth largest city (152,000 inhabitants and preceded only by Vienna, Budapest, Prague [Praha], Trieste, and Lemberg [L'vìv]), Trieste (230,000 inhabitants), in turn, due to its port function grew even to the fourth place. By contrast, Klagenfurt and Villach—the bipolar centers of Carinthia—remained truly provincial cities (Jordan 2016b: 7).

The interwar period, which saw a politically fragile and economically weak Austrian federal republic conceiving itself as a second German nation state with tendencies toward unification with Germany, did not modify Carinthia's position very much. It had lost some territories in the South to Italy (Kanal Valley [Val Canale/Val Cjanâl/Kanalska dolina/Kanaltal]) and the State of the Serbs, Croats, and Slovenes [Država Srba, Hrvata i Slovenaca/Država Srbov, Hrvatov in Slovencev, SHS], later Yugoslavia (Mießtal [Mežiška dolina], Seeland [Jezersko]), and suffered from national antagonism between Germans and Slovenes.

The much more prosperous development of Austria after World War II affected also Carinthia but could not make it escape the rather backward position among Austrian federal states. Although a remarkable development of international tourism could partly compensate the weakness of the industrial sector, gross national product per capita is still below the Austrian average (€37,200 compared to €43,640 in 2018) putting Carinthia at the seventh rank among nine Austrian federal states (Statistik Austria 2020b). As regards demographic development, Carinthia ranks even last due to significant rural out-migration and much less immigration from other countries, which is usually directed to larger cities. Transnational regional cooperation already in the period before the fall of the Iron Curtain in the framework of the Alpine-Adriatic Working

Community, a construct in configuration not unsimilar to the early medieval Duchy of Carinthia (see Lausegger 1992; Nećak 2001), and the process of European integration after 1989 ending in 2004/2007[3] up in the situation that Carinthia found itself together with its southern neighbors, Italy and Slovenia, in the common internal European market as well as in the European currency and Schengen union, certainly opened up new possibilities. They have been used at the micro-level and may have contributed to Carinthia not falling back further compared to other parts of Austria but have certainly not yet meant the essential 'jump forward'.

5.1.1.1 Settlement and Cultural History

When we look at settlement and cultural history, we still find remnants of Celtic settlement all over Carinthia—also in place names, especially in names of rivers: all major Carinthian rivers bear names derived from Celtic. The archeological excavations on Magdalensberg north of Klagenfurt reveal the political and economic center of the Celtic kingdom of Noricum. Roman settlement has also left its traces, especially in and around urban centers in Roman times, Virunum north of Klagenfurt and Teurnia near Spittal an der Drau, as well as along the rather dense Roman road network and at Roman mining sites (see Glaser 2001).

The great migration of Slavs in the sixth-seventh centuries brought a Slavic population (Alpine Slavs, later called 'Slovenes') also to the eastern Alps. They were the dominant group in their own principality, Carantania, of which modern Carinthia was the core region, but they had to acknowledge Bavarian/Franconian political domination by the middle of the eighth century. This resulted also in their Christianization in the Latin sense and in Bavarian (German-speaking) colonization. Christianization proceeded on the one hand from Salzburg with Maria Saal, north of Klagenfurt (Fig. 5.10), as its Carinthian outpost, on the other hand from Aquileia (Fig. 5.11) up to River Drau/Drava, since 811 the boundary between the two ecclesiastical realms. While Aquileian mission was not connected with colonization, Salzburg combined it with the settlement

[3] Slovenia joined the EU in 2004, but became part of the currency and the Schengen union only in 2007.

Fig. 5.10 Maria Saal

of Bavarians. The Bavarian settlers, backed by the political supremacy of Bavaria and the Franconian Empire, entered Carinthia in the kind of a social upper layer founding cities and towns. When they came as farmers, they preferred elevated locations, since valley bottoms had already been occupied by Alpine Slavs (Moritsch 2001). The Bavarian newcomers became politically and socially dominant, while the Slavic population remained the rural ground layer. In effect, up to the end of the Middle

Fig. 5.11 Aquileia. (Photos by Peter Jordan 2005 and 2012)

Ages an ethnically/linguistically mixed situation persisted shaping Carinthian culture in many respects up to the present day. This is also true for the namescape, which is a mixture of Slavonic and German elements all over the state as Eberhard Kranzmayer (1956) and Heinz-Dieter Pohl (2008, 2009a, 2009b, 2010) have impressively demonstrated (Fig. 5.12). As Wilhelm Wadl stated, "the Slovenian component would be omnipresent in Carinthia's namescape without a single bilingual town sign" (2006: 182).

But linguistic assimilation toward German-speakers, the dominant group, proceeded. Social ascent was only possible by using the German language—very similar to assimilation toward Venetian-speakers (later Italian-speakers) in the Slavia Veneta (eastern parts of Friuli-Venezia Giulia), in Istria, and in other parts of the eastern Adriatic coast from the rise of Venice [Venezia] from the eleventh century onward. By the end of the Middle Ages a distinct language boundary within Carinthia had

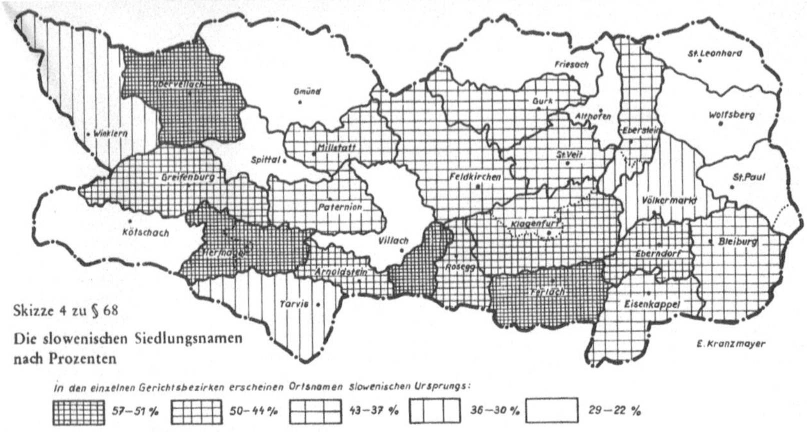

Fig. 5.12 Share of toponyms of Slavonic origin in pre-1919 Carinthia by court districts. (Source: Kranzmayer 1956, accompanying map folder)

developed, close to the ecclesiastical boundary between the ecclesiastical provinces of Salzburg and Aquileia along River Drau/Drava. This situation roughly persisted up to the middle of the nineteenth century as the famous ethnographic map of Karl Freiherr von Czoernig shows it (Fig. 5.13). The blue color represents a Slovene majority. A blue band encompasses the area with a Slovene minority.

Later, also on the Slovenian side of the boundary Slovenian consciousness eroded substantially. According to the 2001 census (by colloquial language, the last one in Austria[4]) 12,586 have declared to speak Slovene or 'Windisch',[5] that is, 2.4% of Carinthia's population (Statistik Austria 2020a, Fig. 5.14). This substantial decline cannot only be explained by social stratification resulting in assimilation toward the dominant group

[4] Although the census was asking for colloquial language, respondents understood this as a question for ethnic consciousness. The census may thus very well be regarded as an ethnic census.

[5] 'Windisch' is not another language spoken in Carinthia, but a vernacular term for the Carinthian Slavonic dialects. It figures in the population census 2001 and in earlier censuses as a choice for declaring one's colloquial language. People declaring to speak 'Windisch' are usually added to those declaring to speak Slovene, thus complementing the number of minority-language speakers in Carinthia, but have in the ethnic and cultural-political sense rather to be regarded as a separate group accepting their Slavonic roots and also the local Slavonic traditions and dialects, but distancing themselves from a Slovene identity as it has developed in the course of national awakening in the nineteenth century. A more detailed discussion of this item will follow later.

Fig. 5.13 Language boundary between German (red) and Slovene (blue) in 1851. (Source: Freiherr von Czoernig 1855, Private archive of Peter Jordan). (Color figure online)

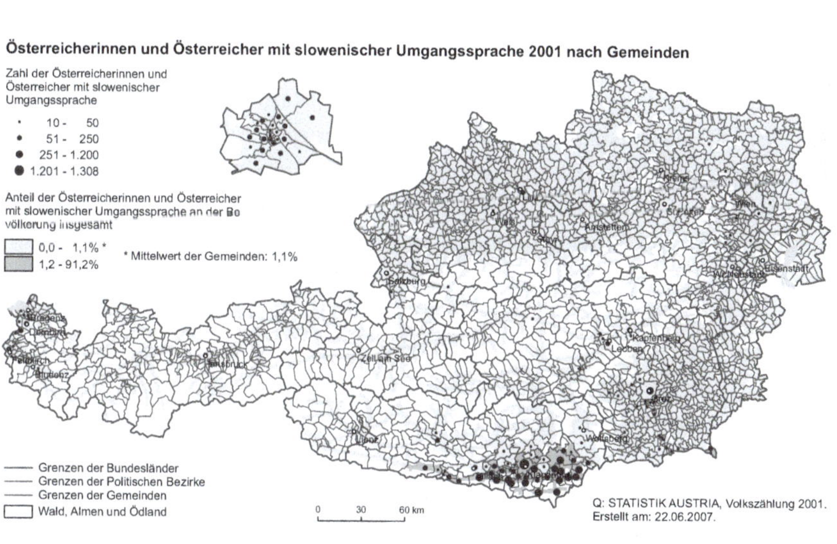

Fig. 5.14 Minority-language speakers in 2001. (Source: Population census 2001, Statistik Austria 2020b)

(as mentioned before), but is also due to societal change in general (conversion from rural to industrial and service-oriented societies) and the socio-economically peripheric situation of the lands, where the Slovenes even within Carinthia live (Carinthia in total is a periphery in socio-economic terms), connected with considerable out-migration.

In addition, political developments and events had their strong impact on the relations between majority and minority (see Vilfan 2001). The rise of the national idea and national awareness during the nineteenth century affected both sides and transformed a prevailingly common regional, Carinthian identity into a bipolar, partly antagonistic situation. German-Carinthians began to conceive themselves as affiliated to the German cultural nation boasting a relative majority in the Austrian Empire, later in the Austro-Hungarian Monarchy (1910: 23.4%, Rumpler and Seger 2010: 62) and being backed also by the larger German-speaking cultural sphere with Prussia, later (after 1871) the German Reich as an additional political power. Slovenian national consciousness manifested itself already most impressively by Peter Kozler's map of the "Slovenian lands and regions" as of 1853 (Kozler 1853, Fig. 5.15) comprising all of Carinthia and outlining the Slovenian-speaking or bilingual area. It faced the problem of fragmentation among several political entities (Carniola [Kranjska], Austrian Coastland [Österreichisches Küstenland/Littorale/Primorje/Primorska], Carinthia, Styria, Hungary proper) and a minority position in all of them except Carniola as well as a lack of societal elites and political leadership due to the fact that in "Slovenian lands" either Germans or Venetians (later Italians) had occupied these positions (Vilfan 2001). They were, however, replaced by Slovene clergy of the Roman-Catholic Church, to which the largest part of Slovenes had re-converted after having been adherents of Lutheranism, to whom they owe also a first linguistic standard thanks to the translation of the Holy Bible into Slovenian by the Lutheran reformer Primož Trubar (1508–1586). This clerical leadership of the Slovenian national idea established an intimate relation between Slovenes and the Roman-Catholic Church, which persisted in Carinthia (much more than in Slovenia) up to the present day. This has most recently (February 2020) been confirmed by the installment of a Carinthian Slovene, Josef Marketz, as the new bishop of the Carinthian Roman-Catholic diocese

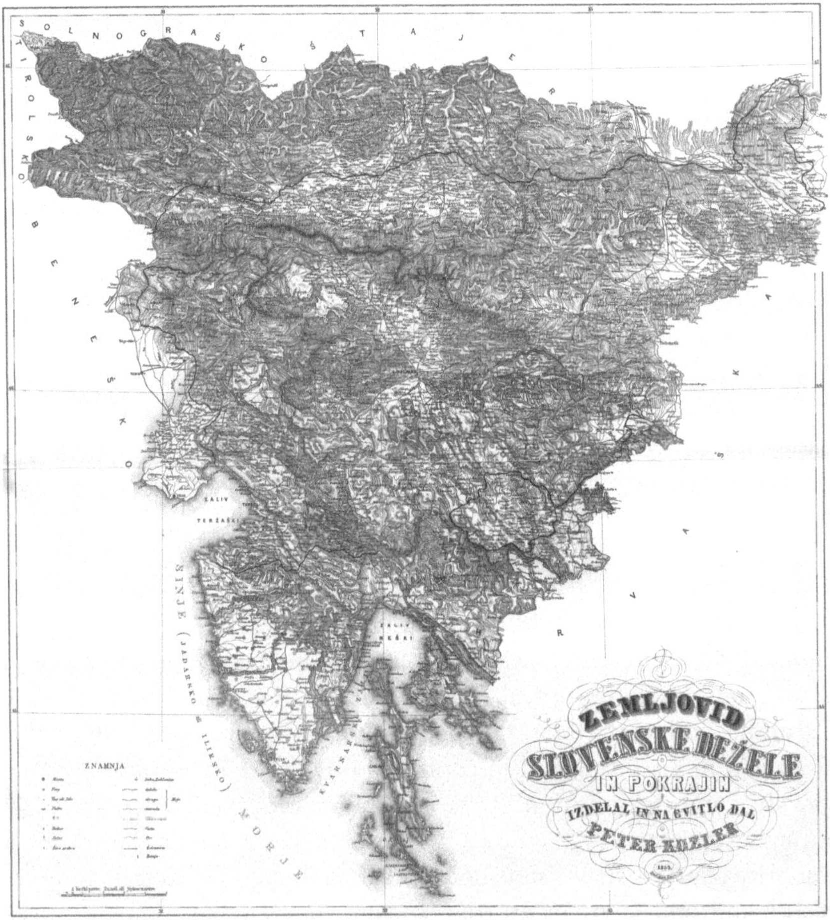

Fig. 5.15 "Slovenian lands and regions" by Peter Kozler 1853. (Source: Wikipedia)

Gurk-Klagenfurt coinciding with the federal province Carinthia. It is also documented by the activities of the Roman-Catholic educational center Tainach/Tinje.

The rise of national ideas in the nineteenth century had in Carinthia also the effect that the dominant German-Carinthians began to conceive themselves as the defenders of the pre-national land's unity for partly very practical economic reasons, while the Slovenian minority could now refer their identity to the Slovenian nation extending across several historical lands aiming at unification and being regarded—justified or not—by the German-Carinthian majority as potential threat to the land's unity (see Moritsch 2000). It is this situation that offers a key to understanding the growing antagonism between majority and minority that reached its climax, when the political power behind the minority—not necessarily invited or even welcomed by it—was strong, as it was between 1918 and 1991 (SHS State, Yugoslavia). It offers also a key to understanding the seemingly irrational attitudes of a majority of more than 90% toward a demographically small and economically weak minority. The rise of minority conflict in Carinthia (not only between majority and minority, but also inside the minority) in the later nineteenth century and culminating during World War II can be outlined by the simple formula of a clash between the regional and the ethnic principle in political community building.

In a book comparing a Czech and an Austrian minority situation it deserves in this context to be noted that the clash between the regional and the ethnic principle affected in a quite similar way also the Bohemian Lands (Bohemia [Čechy], Moravia [Morava], and Silesia [Slezsko, Śląsk]) with traditions as historical-cultural entities not much shorter than Carinthia and with their German minority—much larger than the Slovenian in Carinthia and in contrast to the latter in a politically and socially dominant position. Thus, the Paris peace conference in 1919, strongly supported by the new Czechoslovak state, denied the German minority contrary to Wilson's doctrine of national self-determination to secede and join German-Austria, a.o. on the argument that historical regional entities should not be divided. The French ambassador Jules Laroche as the head of the Commission on Czechoslovak Affairs, for example, argued with reference to the Germans in Bohemia that "the inhabitants of these regions were accustomed to live in close connexion [sic!] with the rest of Bohemia, and did not desire separation…" (Suppan 2019: 54). While the first part of this sentence reflects the real situation as it can also be found in Carinthia, the second does not stand verification: The German delegates of the Bohemian Lands insisted to remain in the Austrian parliament and left it after the Peace Treaty of Saint Germain only under solemn protest.

A second Europe-wide development with a significant impact on the minority situation in Carinthia was the attitude of national homogenization after World War I. After the dissolution of multinational empires like the Habsburg monarchy, but also the Ottoman and the Russian empires, practically all European countries conceived themselves as nation states in the sense of being the possession of the dominant nation. Thus, interwar Austria regarded itself explicitly as a second German nation state, which meant of course a serious challenge for all ethnic minorities. The State of the Serbs, Croats, and Slovenes [Država Srba, Hrvata i Slovenaca/Država Srbov, Hrvatov in Slovencev, SHS], called Yugoslavia [Jugoslavija] from 1929, saw itself as the state of the Yugoslav nation composed of three 'tribes', Serbs, Croats, and Slovenes.[6] Not only in the two of them it was conceived as legitimate to grant the state nation a privileged position and to regard non-dominant groups just as tolerated, not deserving to enjoy the same, if any collective rights. The Republic of Austria and the SHS State did not really differ in this respect.

Interwar Austria, titling herself *German-Austria* up to the Peace Treaty of Saint Germain (1919), on the grounds of her German national identity and a self-estimation of economic weakness, developed even a strive for unification with Germany—especially vigorous up to the Peace Treaty of Saint Germain, when it was forbidden, but persisting throughout the interwar period and rising again when Hitler had come to power in Germany in 1933. This strive promoted by all major political forces with the later exception of the Christian Socials was, however, significantly stronger in other parts of Austria (like Salzburg and Tyrol) than in Carinthia, where all German-Carinthian parties did in principle not object, but also not explicitly support it—well aware that this would mean the land's division (Rumpler 2005: 19, 28). Remarkably enough, the SHS State and later Yugoslavia supported Austria's unification with

[6] The idea of a "Yugoslav" nation was officially abandoned in the second, Communist Yugoslavia, but remained in the background as a 'silent hope' for a final convergence of the various national identities. Up to the last Yugoslavian population census of 1991 "Yugoslavian" was offered as a choice for ethnic affiliation and at the peak of Yugoslavia's socio-economic development in 1981 chosen by 8%—mainly the younger, urban, better educated, and in ethnically mixed areas. This category was, however, not conceived by its originators as a national, but supranational identity in the sense of refraining from any national or ethnic affiliation but declaring just to be a Yugoslavian citizen (see Jordan 2016a).

Germany—very likely considering that this would be a good excuse to claim southern or eventually all of Carinthia (Rumpler 2005: 62f). Representatives of Carinthian Slovenes also recommended—very likely for the same reasons—voting for unification with Germany at the referendum of 10 April 1938 (Rumpler 2005: 64).

The idea of national homogenization on the Austrian side was additionally promoted by the fact that Carinthia and Styria had lost after World War I (by the Peace Treaty of Saint Germain 1919) some territories settled or conceived as settled predominantly by non-German speakers to the first Yugoslavian state as well as to Italy (Fig. 5.16). Losses of Carinthia were the commune of Seeland [Jezersko] as well as the Mießtal [Mežiška dolina] and Unterdrauburg [Dravograd] with Slovene-speaking majorities awarded to the SHS State and the Kanal Valley [Val Canale/ Val Cjanâl/Kanalska dolina/Kanaltal] awarded to Italy, although before World War I (1910) the Kanal Valley had only a 'felt' Slovene majority, while by official declaration Slovene-speakers amounted not more than to a quarter. This resulted in the attitude to regard the remaining parts as 'ours' where it was justified that Germanization proceeds.[7]

Specific for the situation in Carinthia were the attempts of the first and second Yugoslavia (after World War I and II) to occupy larger parts of Carinthia on the pretext that they were settled by Slovenes (Yugoslavs). They confirmed the 'primeval fear' [Urangst] among German-Carinthians that acknowledging the existence of a Slovene minority in Carinthia, granting them minority rights and—even more—recognizing and delimitating a bilingual territory would mean risking the partition of Carinthia, this 'entity by nature', if not its entire incorporation into a Yugoslavian state.

Yugoslavian occupation after the end of World War I was met by armed resistance ('Carinthian fight of defense' [Kärntner Abwehrkampf]) based on regional initiatives[8] and not supported by the federal authorities in Vienna. After a U.S. fact-finding mission under Sherman Miles, it led to the decision of the peace conference in Paris that a plebiscite had to be organized to decide about political affiliation to Austria or the SHS State.

[7] Very instrumental in this respect was the cultural association "Deutscher Schulverein Südmark" founded in 1925 (Rumpler 2005: 41).

[8] The Carinthian parliament decided on 5 December 1918 to resist Yugoslavian occupation by military forces (Rumpler 2005: 16).

The plebiscite was to be held first in Zone A (Zone 1 in Fig. 5.16) with a 70% majority of Slovene-speakers (according to the last population census of 1910), and after a decision in favor of the SHS State also in Zone B (Zone 2 in Fig. 5.16) including Klagenfurt and areas north of it with a much lower share of Slovene-speakers. The voting in Zone A took place on 10 October 1920, still in presence of SHS occupation forces,[9] and resulted—against most expectations—in a 59% vote for remaining with Austria. A voting in Zone B was consequently dropped. 10 October 1920 is ever since a symbolic date in Carinthia celebrated annually by a regional

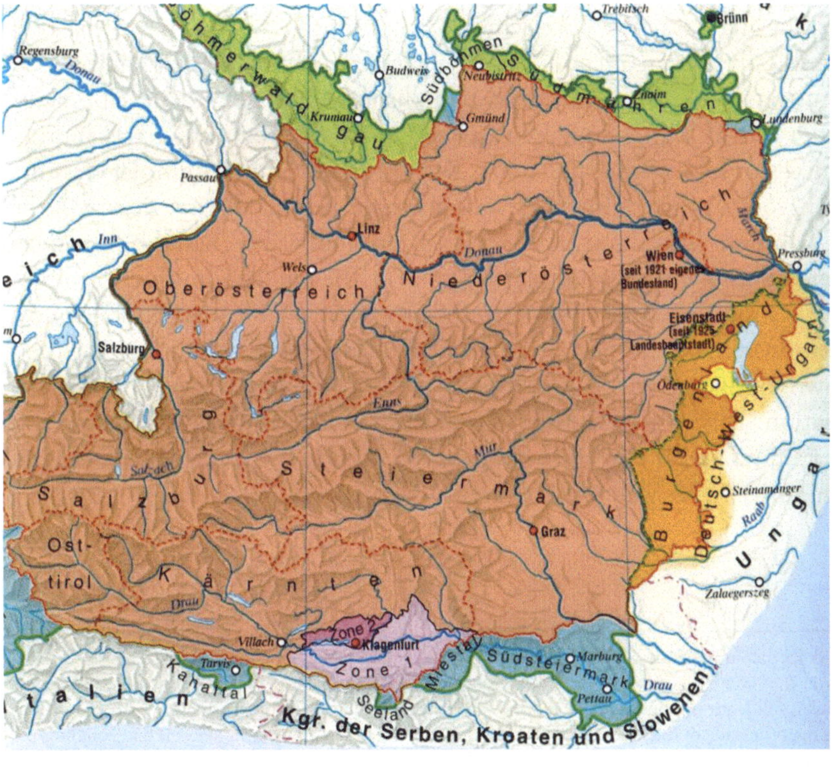

Fig. 5.16 Austria's territorial gains and losses 1918–1921 and zones of the Carinthian Plebiscite 1920. (Source: Hölzel 2004: 64)

[9] Occupation by SHS forces may, however, have even had a negative impact on voting for SHS due to the, in many cases, inadequate behavior of soldiers toward civilians (Suppan 2019: 116).

holiday and commemorated by street names in almost every city and town. The 100th anniversary in 2020 is an occasion for major festivities.

The rather surprising result of the referendum and the fact that also a large share of the Slovenian community (about 40%[10]) voted for Austria can mainly be explained by the feeling also of Carinthian Slovenes that in spite of all national awareness-raising since the nineteenth century they were prevailingly integrated into a Carinthian community and society—if only in socio-economic terms. As a predominantly rural farming population Carinthian Slovenes had their markets in the nearby Carinthian cities and towns and not at the other side of the mountain range. The voting may also have been affected by the view on Austria as the economically better-off and better organized country, although interwar Austria's economy was rather demolished compared to developments after World War II and also to contemporary Czechoslovakia as another successor state of the Monarchy. The historian Helmut Rumpler classifies this vote, however, much less a decision for Austria than a "victory of the Carinthian idea", that is, the idea of an 'indivisible' natural, historical, and cultural entity (Rumpler 2005: 17). The referendum result can perhaps also be taken as a proof for the fact that even in situations of utmost national charging, Slovenes in Carinthia feel as a community of their own, not necessarily as a part of the wider Slovenian or even Yugoslavian nation. This positive Slovenian attitude toward Carinthia and Austria was not, however, at that time rewarded by Austria.[11] Expectations for better minority rights were not fulfilled.

[10] A forthcoming study of Guido Tiemann to be published in "Historical Social Research" postulates that even a majority of Carinthian Slovene-speakers (51%) had voted for Austria. He derives this result from the finding that about a quarter of German-Carinthians (mainly with possessions at the other side of the border and people sympathizing with a monarchic system) had voted for the SHS State (science.ORF.at 12 October 2020).

[11] With the limited exception of the not really serious offer of a 'cultural autonomy' (see Chap. 4, Sect. 4.1.6). According to the historian Claudia Fräss-Ehrfeld also ideas of splitting Carinthia into two provinces and of elevating political districts [Politische Bezirke] to self-governing administrative units, in some of which the minority would have a better standing, were raised (Fräss-Ehrfeld 2005: 105, 110f).

The National-Socialist regime ruling Austria as a part of the German Reich from March 1938 to May 1945 conducted Germanization[12] to the extreme not only in Carinthia, but also in Carniola [Kranjska] and Lower Styria [Štajerska] prompting fierce and violent resistance. Toward the end of the Nazi regime southern Carinthia became a theater of heavy fighting between partisans and German forces, of all kinds of war atrocities like expulsion and execution provoking again Yugoslavian occupation of Carinthia's southern parts after the capitulation of the German Reich connected with Tito's aspirations of including all of Carinthia into the second, Communist Yugoslavia (see Anonymous s.a. 1945). These aspirations had to be abandoned after the rupture between Tito and Stalin in 1948, when Tito lost Soviet support. Yugoslavia remained, however, up to its dissolution in 1991/1992 a signatory power of the Treaty of Vienna in 1955 re-establishing Austria as a sovereign state and thus also a kind of an official protector of the Slovene minority in Austria.

The role of Carinthian-Slovenian partisans is divergently reflected in German-Carinthian and Slovenian-Carinthian historiography. While the Slovenian-Carinthian side highlights their reactive fight against the Nazi regime and their merits related to the Moscow Declaration of 1943 proclaiming the restoration of an independent Austria on the condition of domestic resistance (in which Carinthian-Slovenian partisans had a major share), the German-Carinthian side focuses on their intention of incorporating southern Carinthia into Communist Yugoslavia. It cannot be denied that both perspectives are justified.

All these events contributed to animosities between majority and minority and aggravated a relationship that had already earlier been burdened by social stratification. On the one hand they may have promoted the erosion of the minority community, but on the other have contributed to a stronger coherence of the remaining.

What is evident, however, is that they prompted a split of the minority into two factions (see Moritsch 1995; Wörsdörfer 2002), one conceiving themselves as 'Carinthian Slovenes' in the sense of an ethnic group of the (multilingual) Austrian cultural nation, but associated also with Slovenes

[12] Slovenes were expelled from southern Carinthia and replaced by German 'optants' from South Tyrol [Südtirol/Alto Adige] and the Kanal Valley [Val Canale/Val Cjanâl/Kanalska dolina/Kanaltal].

in Slovenia and recognizing the Slovene standard language as a tool of community building. The other, calling themselves 'Windische', recognize their Alpine-Slavonic descent and cultivate the Alpine-Slavonic dialects spoken in Carinthia as well as other Alpine-Slavonic traditions, but also emphasize that they were a community different from Slovenes in Slovenia and do not accept Slovene as their standard language. In political terms they side with German-Carinthians, distance themselves from the minority politics conducted by Carinthian Slovenes, and do not demand any minority rights or make use of the existing.

When ethnic or national affiliation is considered as essentially subjective, it is the image of self that exclusively counts (Gellner 1983; Symmons-Symonolewicz 1985). People conceiving themselves as a cultural/ethnic group by their own criteria are to be accepted and respected as such. From the Carinthian-Slovenian side they are frequently disqualified as renegades. Hard-core German-Carinthians usually refer to them as the co-operative and loyal part of the minority. When confronted with the reproach of being renegades, the 'Windische' would respond to have on their part and in contrast to the core group of Carinthian Slovenes rejected political Catholicism as well as denied to join a Balkan state in 1920 and a Communist Yugoslavia in 1945.

The 'Windische' are an almost invisible community and it is difficult to estimate their size. Their core group may correspond to the small number of people declaring to speak 'Windisch' as their colloquial language up to the population census of 2001 (568 in 2001, Statistik Austria 2002). However, it seems that they are not just a relic or curiosity (Wörsdörfer 2002: 156), the end of which is inevitable after the fundamental change in political relations, especially after Slovenia joined the EU in 2004. In 2012, the Association of the Carinthian Windische [Verein der Kärntner Windischen] was founded comprising members from all parts of the bilingual area (see Verein der Kärntner Windischen 2020).

At this point at the latest it needs to be mentioned that *Windische* or *windisch* is a multifaceted term and can easily be misunderstood in Carinthia. In principle, it is the older German word for all kinds of Slavonic communities. In this sense it can be found in several place names also outside Carinthia like in *Windischgarsten* in Upper Austria or (former)

Windisch Matrei (later *Matrei in Osttirol*) in Tyrol and corresponds to the other German word *Wenden*. When it is applied in Carinthia, it is usually used as a—nearer to the vernacular language—synonym for the Carinthian Slovenes or as a designation of the/an area inhabited by them. It has in principle no pejorative meaning. It is, however, also the name of this particular Carinthian group of Alpine-Slavonic descent described before, that is, a cultural-political faction. Finally—and this makes its use delicate—it can also have a pejorative and even offensive meaning, when it is used by German-Carinthians to disqualify Carinthian Slovenes in their strive for adequate minority rights and for getting their contribution to a common Carinthian culture recognized. This latter use of the word can be attributed to the fact that at least up to the most recent past German-Carinthians felt a kind of superiority over Slovenes resulting from a paternalistic attitude: They regarded them in a rather jovial and friendly manner (with the subtext "Let us nevertheless remain good friends!") as less conforming to standards and benchmarks set (of course) by the German-Carinthian side. Slovenes reflected this attitude in a way by behaving subordinate and regarding, for example, their language not as equally important as German and of lower prestige.

What also is evident is that 'Windisch' is not a linguistic category (Pohl 1982), although Austrian population censuses up to 2001 were offering it as a possible choice for defining one's colloquial language. That it is a standard language is also not claimed by the group defining themselves as 'Windische', but they regard 'Windisch' either as an umbrella term for the Carinthian Slavonic dialects or as a colloquial language with a lot of German elements without paying too much attention to linguistic classifications. They confirm to speak a local Alpine-Slavonic dialect and refer to German as their standard language, not to Slovene. With this attitude they are no significant exception when we look at other minorities in Europe: The plenty of German groups in Hungary speak their (mostly Alemannic) dialects while using Hungarian as their standard language. The Alemannic minority in French Alsace refers to French as their standard language. Friulian-speakers have no problem to accept Italian as their standard language, although Friulian is also a standard language in the sense of an idiom disposing of a codified grammar and vocabulary. And Venetian, earlier the prestigious language of a major European power and in linguistic terms having developed in parallel to Italian from vulgar Latin, thus actually a 'sister' of Italian, is today conceived to be a dialect of the Italian standard language.

It is (of course) always the prestigious language of a dominant power that is in all these cases chosen as the standard language. And the choice may also be regarded as an indication of the choice of the cultural sphere in which a certain group wishes to participate. A case in point is Austria, when it decided after World War II not to separate from the wider German-speaking cultural sphere by abstaining from the construction of an 'Austrian' different from German, besides other reasons for the sake of remaining part of this larger cultural sphere (see also Chap. 4, Sect. 4.1.4).

More recent events regarding the Slovene minority need not to be discussed anymore in this wider context, but with a focus on the minority area—that is, under item 5.1.2. What should, however, still be mentioned here are the changes in international relations after World War II with an effect on Carinthia and its minority.

These are first of all the dissolution of Yugoslavia in 1991/1992 and the independence of Slovenia in 1991. They meant for German-Carinthians the fading away of a potential threat taking into account that Yugoslavia had been a major political and military power in the region and had occupied Carinthia two times. Slovenia as its northernmost successor state has not at all the capacities of former Yugoslavia to exert political pressure. "Inviting Yugoslavian territorial claims by outlining the bilingual area in Carinthia through Slovenian place names in public space or on maps" was not a valid argument anymore. Slovenia is except Liechtenstein the smallest country in its neighborhood and sometimes ironically called "Austria's 10th federal state". It feels nevertheless some responsibility for the Slovenian minority in Carinthia and engages itself as a mother country of all Slovenes outside Slovenia.

Austria's and Slovenia's EU accession in 1995 and 2004, respectively, and their subsequent common membership of the EU internal market, the currency union, and the Schengen zone resulting in completely open borders rather reduced the minority's exclusiveness of relations to Slovenia than providing it with an additional bridge function. Minorities—due to their command of both languages and familiarity with both cultures—have always a certain advantage in bilateral cross-border relations.

However, already Communist Yugoslavia, after it had developed into a frequent tourist destination with many visitors from Austria in the 1960s, 1970s, and 1980s, was not anymore 'the strange other' for Austrians and especially for Carinthians. Already the Working Community Alps-Adriatic from 1978 and even more so European integration drove this development further through economic, administrative (e.g., partnerships of communes), cultural and scientific cooperation and exchange, and after 2013 also by labor migration (mainly in the direction from Slovenia to Austria). Carinthia in total, not only the Slovenian minority, was now faced with the challenge of being Austria's bridge to the wider Adriatic space.

5.1.2 The Minority Area

In this section, the minority area will be equaled with the so-called bilingual area in the sense of the part of Carinthia with a considerable minority-language speaking resident population and as outlined by Fig. 5.17.

5.1.2.1 Nature as a Place of Human Activities

This section is not to present a comprehensive physical-geographical assessment of the minority area, but just to give an imagination of under which natural conditions the minority lives—also compared to other parts of Carinthia and Austria.

Carinthia in total is an Alpine region with a predominantly sub-Alpine climate, that is, harsh winters with a lot of snow and warm, but not yet Mediterranean summers. The Mediterranean (Adriatic) influence manifests itself most significantly with precipitation, when it results in the highest amounts (more than 2000 mm) in the southern mountain ranges that function as a second (behind the Julian Alps [Julijske Alpe/Alpi Giulie]) barrier for the moist air flowing north from the Adriatic Sea and force it ascending and condensing as well as in a precipitation maximum in autumn (and not—as in most other parts of Austria—in spring). While thus far Carinthia's southern mountain ranges used to have traditionally sufficient snow for skiing from December to April supporting a

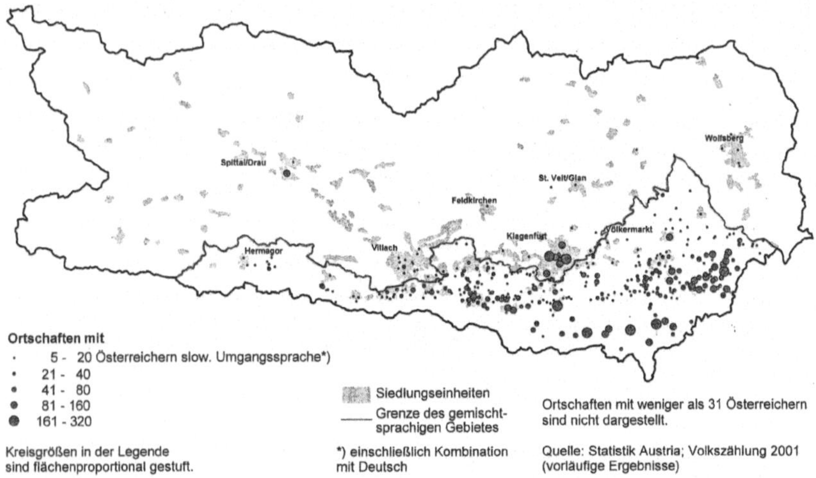

Fig. 5.17 Minority-language speakers (circular symbols) by population census 2001 and 'bilingual area' (outlined in red). (Source: Wonka 2002: 112). (Color figure online)

rise of winter(-sports) tourism after World War II, climate warming caused precipitation occurring rather as rain than as snow making artificial snow production necessary. Also summer temperatures are higher today than they used to be and promote bathing tourism at the plenty of lakes, for which Carinthia in general is well-known and that reach water temperatures up to 28 centigrades in July and August.

Geomorphology in the southern part of Carinthia is characterized by a barrier-like sequence of mountain ranges along the borders with Slovenia and Italy in the West-East direction with a sequence of wide glacial valleys at their northern bottom. Mountains rise up to 2237 m in the eastern range, the Karawanken/Karavanke forming the border with Slovenia, and to 2780 m in the western range, the Carnic Alps [Karnische Alpen/Alpi Carniche] along the border with Italy, with the most impressive rock faces right toward the North (Fig. 5.18; see also Fig. 5.4). Especially the Karawanken/Karavanke convey the impression of a real barrier. Mountain passes used by roads portion the ranges into sections: from East to West Seeberg/Jezerski vrh (1215 m), Loibl/Ljubelj (1369 m), and Wurzen/

Fig. 5.18 North-bound rock faces of the Koschuta/Košuta in the central part of the Karawanken/Karavanke. (Photo by Peter Jordan 2017)

Korensko sedlo (1071 m) in the eastern range connecting the Drau/Drava and the lower Gail/Zilja valleys in Carinthia with the Ljubljana Basin [Ljubljanska kotlina] and the Sava Valley [Savska dolina] in Slovenia; Nassfeldpass/Passo Pramollo (1562 m) and Plöckenpass/Passo di Monte Croce Carnico (1360 m) in the western range connecting the Gail Valley [Gailtal/Ziljska dolina] with the Kanal Valley [Val Canale/Val Cjanâl/Kanalska dolina/Kanaltal] and the Carnia region, respectively, in Friuli (Italy). The two ranges are separated by the Gailitz/Ziljica [Slizza] Valley near Tarvisio providing for a convenient gateway for all kinds of land transportation, that is, road, motorway, high-speed railroad between Carinthia and Italy with Villach as the Carinthian and Udine as the Italian hubs.[13]

[13]With rail cargo, Fürnitz near Villach and Cervignano south of Udine function as the respective hubs.

In the longitudinal direction, the two mountain ranges are dissected by smaller valleys that offer only exceptionally opportunities for permanent settlement and are, together with areas above the steep valley slopes, rather used for Alpine pastures, for which especially the Carnic Alps are famous (Fig. 5.19). A major exception is the larger valley of River Vellach/Bela leading to Seeberg/Jezerski vrh with some basins accommodating, for example, the town Eisenkappel/Železna Kapla. While the Karawanken/Karavanke are constructed mainly of Triassic limestone, the Carnic Alps consist mainly of a Paleozoic crystalline basic layer with dolomite rocks with a landmark function placed on them.

The glacial valleys of the Gail/Zilja and the Drau/Drava at the bottom of the mountain slopes were subjected to frequent flooding up to the regulation of their rivers starting in the later nineteenth century and the construction of hydropower stations after World War II. Until then they

Fig. 5.19 Alpine pasture Egger Alm/Brška planina in the Carnic Alps near Hermagor at an altitude of 1422 m with a 'village' inhabited by farmers only in summer, when cattle and horses are grazing there. (Photo by Peter Jordan 2012)

offered no place for permanent settlement. This had to be confined to gravel fans and river terraces at the valley fringes. Most traditional villages and towns are therefore not located at the valley bottom, but at least in an elevated location. Nevertheless, flooding—most frequently as a result of sudden snow smelting in spring or heavy rain in autumn—caused disasters almost periodically. While the valley bottoms could thus earlier be used only as wetlands, for example, for horse raising, they are today prevailingly fertile agricultural soil, most frequently used for cultivating maize.

The sequence of glacial, but tectonically preformed valleys is just a part of a longer sequence of Alpine valleys in the West-East direction, the so-called Peri-Adriatic Seam, the borderline between the African and the European tectonic plates. This means to be exposed to earthquakes that indeed occurred several times. The most severe were a pre-historical earthquake causing the coverage of larger parts of the Lower Gail Valley [Unteres Gailtal/Zilja] by rocks from Mount Dobratsch/Dobrač, an earthquake in 1348 causing a second rockslide at the same location and the earthquake of 1976 with its epicenter in Friuli, but affecting randomly also southern Carinthia, especially the Gail/Zilja Valley.

The northern slopes of the Gail Valley [Gailtal/Ziljska dolina] are formed by the Gailtal Alps [Gailtaler Alpen/Ziljske Alpe] with Dobratsch/Dobrač or Villacher Alpe as their highest (2166 m) and easternmost peak. They are—similar to the Carnic Alps and the Karawanken/Karavanke—covered by extensive montane mixed forests of beech (Fagus silvatica) and conifers (Abies alba and Picea abies) enabling a prosperous forestry. They, too, have wide Alpine pastures favoring animal husbandry, but also skiing. Like the Karawanken/Karavanke they are built mostly of Triassic limestone. Another similarity is accompanying layers of Greywacke containing deposits of lead and zinc. Indeed, Bleiberg-Kreuth in the Gailtal Alps as well as Cave del Predil/Raibl/Rabelj (today in Italy) to the West and Mežica (today in Slovenia) to the East of the Karawanken/Karavanke were important mining sites for lead and zinc and the basis of a manufacturing industry up to the postwar period before they were abandoned.

North of the southern Carinthian section of the Drau/Drava valley[14] to be discussed here up to Lake Wörth [Wörthersee/Vrbsko jezero] extends a mild hilly range of altitudes between 800 and 900 m composed of Tertiary conglomerates called Sattnitz/Gure (Fig. 5.20). It is covered by oak-hornbeam forests and oligotrophic oak and oak-pine forests as well as agricultural areas that, however, suffer from a lack of water due to a water-permeable underground. A chain of small lakes with Lake Keutschach [Keutschacher See/Hodiško jezero] as the largest is—like the much larger Lake Wörth in the North—of glacial origin and an asset for summer tourism.

Fig. 5.20 Central section of the Drau/Drava valley with river terraces at the left and the Sattnitz/Gure at the right. (Photo by Peter Jordan 2017)

[14] *Drautal/Dravska dolina* is nowhere in Austria a proper name and the generic *valley* is therefore written with a small initial letter. The valley in Austria is—very likely due to its geomorphological divergence—composed of sections with individual proper names, that is, Pustertal, Lienzer Becken, Oberes Drautal, Lurnfeld, Unteres Drautal, Rosental/Rož, and Jauntal/Podjuna.

West and east of Sattnitz/Gure, the vicinity of the Drau/Drava valley is characterized by glacial moraines and, again—in between the hills—by several lakes, that is Lake Faak [Faaker See/Baško jezero] in the West and Lake Klopein [Klopeiner See/Klopinjsko jezero] in the East offering all possibilities for summer tourism.

Klagenfurt am Wörthersee, Carinthia's capital and not anymore part of the bilingual area, occupies the center of the Klagenfurt Basin [Klagenfurter Becken] at an altitude of less than 500 m and in a former wetland east of Lake Wörth that had to be drained to be ready for settlement. A wide plain east of it is today used for intensive agriculture such as gardening and orchards. The North of the Basin in the narrower sense is marked by Paleozoic hills rising up to about 1000 m and surrounding a former wetland, later drained, called Zollfeld/Gosposvetsko polje. Its fringes are the location of most of Carinthia's symbolic historical sites.

East to the Klagenfurt Basin in the narrower sense and where the bilingual area reaches its northernmost extension lies the Crystalline mountain range of the Saualpe/Svinška planina rising up to 2079 m and characterized by Alpine pastures up to the highest elevations. With its soft slopes in the South it offers places for permanent settlement up to higher altitudes. In contrast to the mountain ranges mentioned earlier it extends in South-North direction and is already part of the Central Alpine mountain system, while Gailtal Alps, Carnic Alps/Alpi Carniche as well as Karawanken/Karavanke constitute together with the Julian Alps [Julijske Alpe/Alpi Giulie] in Slovenia and Italy as well as the Dolomites [Dolomiten/Dolomiti] in Italy the Southern Limestone Alps.

Thus, by nature and up to recent decades southern Carinthia with its bilingual area offered—even when only compared to other parts of Carinthia—not very favorable possibilities for human settlement and economy. In respect to agriculture it is certainly second to the areas north of the Zollfeld/Gosposvetsko polje and the Lavant Valley [Lavanttal] in the East, and safe settlement areas were confined to the fringes of the large valleys, to less steep mountain slopes, and to selected places in narrow mountain valleys.

Comparative advantages were just the abundance of forests that could be used for forestry and the mountains and lakes that later—from the late nineteenth century onward—attracted tourism, as well as mineral deposits whose exploitation, however, ended after World War II. River regulation and drainage of wetlands from the later nineteenth century onward have improved possibilities for land use in the large valleys, but even this could not support southern Carinthia escaping the position of an area not favored by nature and less attractive for most kinds of human activities. The fact that the minority has from the Late Middle Ages up to the present withdrawn or been pushed back to this area must therefore be regarded as a retreat typical for non-dominant groups.

5.1.2.2 History with a Focus on the Minority

This section just continues the presentation of history in a wider context of Sect. 5.1.1 with developments after World War II and a focus on the development of minority rights and the overall relations between majority and minority in Carinthia.

World War II and especially the events toward the end of the war in southern Carinthia meant a complete disruption of relations between majority and minority. Being occupied until 1955 by the four Allies (in Carinthia by British forces) and aiming at re-establishing its sovereignty Austria undertook in this early postwar period some efforts to reconcile the minority in Carinthia and to implement adequate minority rights. But these efforts met already a very small and further declining Slovenian group (Table 5.1; see also Jordan 1988, 1992; Land Kärnten 1990; Veiter 1980), 42,095 (8.9% of the Carinthian population) according to the population census of 1951. Driving forces were the 'international community' (as we would call it today) and the Austrian federal government against the (at times) fierce opposition of the majority population in Carinthia (at least their political representatives).

Since after 2001 population censuses did not document anymore colloquial language or any other criterion that would allow to estimate ethnic affiliation, enrollment in bilingual schools may serve as a substitute indicator for the development ever since. In elementary schools enrollment declined from the school year 2000/2001 to the school year 2018/2019 from 5989 to 4704, while it increased in secondary schools [Hauptschulen, Neue Mittelschulen] from 271 to 322 and in high schools from 684 to 970 (Bildungsdirektion Kärnten 2019a). When the

Table 5.1 Numerical development of the minority in Carinthia as reflected by population censuses 1880–2001

Year of the population census	Absolute number of minority-language speakers on the territory of modern Carinthia	Share of minority-language speakers in Carinthia's total population
1880	85,051	26.2
1890	84,667	25.1
1900	75,136	21.9
1910	66,463	17.9
1923	37,292	10.0
1934	26,796	6.6
1939	43,179	10.4
1951	42,095	8.9
1961	25,470	5.1
1971	17,011	3.3
1981	14,204	2.7
1991	13,962	2.6
2001	12,554	2.4

Sources: Jordan (1988: 22), Population censuses 1991, 2001: Statistik Austria (2020b)

Notes:

(1) All data refer to the current territory of Carinthia. Parts of Carinthia awarded to other countries after 1918 are not taken into account also with censuses 1880–1910

(2) Population total taken for reference: up to 1910: domestic population present; 1923: population present; 1934, 1951, 1961, 1981: resident population; 1939: citizens of the German Reich; 1971, 1991, 2001: Austrian citizens

(3) Linguistic/ethnic categories conceived to form the 'minority': 1880–1910: colloquial language Slovene; 1923: mental language [Denksprache] Slovene; 1934: Slovene cultural sphere; 1939: mother tongue Slovene, Slovene-German, Slovene-'Windisch', 'Windisch', 'Windisch'-German; 1951, 1971: colloquial language Slovene, Slovene-German, Slovene-'Windisch', 'Windisch', 'Windisch'-German; 1961: colloquial language Slovene, Slovene-German, Slovene-'Windisch', 'Windisch', 'Windisch'-German "in selected areas of Carinthia, in which in 1951 a considerable number of people with non-German colloquial language was registered"; 1981: colloquial language Slovene and German, Slovene, 'Windisch', German-'Windisch'; 1991, 2001: colloquial language Slovene, 'Windisch'

proportion between pupils and the overall number of the minority are assumed the same in 2000/2001 and 2018/2019, this would result in a minority number of 10,840 in 2018/2019 suggesting that the faint decline of the minority after 1971 continues.

Enrollment between 2000/2001 and 2018/2019, however, oscillated, very much in line with relevance of the minority question in the public debate: Enrollment was higher in phases of inter-ethnic conflict and a controversial public debate as between 2001 and 2006, when the judgment of the Constitutional Court of Justice as of 2001 had aroused it; and it faded down in phases of more relaxed intercultural relations as already some years before and ever since the place-name compromise of 2011. This phenomenon may be explained by a rise of ethnic 'confession' and solidarity in times of conflict and a prevailing of practical considerations when things go the usual way. Certainly, also random factors like the attractivity of teachers play some role.

When the widespread network of 58 elementary schools of the school year 2018/2019 is examined school by school, it can be noticed that oscillations are more significant at the peripheries of the bilingual area than in its core (see Bildungsdirektion Kärnten 2019a).

As already briefly addressed in Chap. 4, Sect. 4.1.6, on 3 October 1945 a new school ordinance was adopted, which provided bilingual education for all children in the traditional settlement area of the Carinthian Slovenes (regardless of their ethnic group affiliation). Bilingual education should take place in the first three grades, after which Slovene was a compulsory subject. After the signing of the Austrian State Treaty in 1955 and the implicit resolution of the hitherto unresolved question of the course of the Austro-Yugoslav border, protests against this model began, which culminated in school strikes in 1958. As a result of this development, in September of the same year, the Carinthian Governor Ferdinand Wedenig (Socialist Party) issued a decree allowing parents to deregister their children from bilingual education. In March 1959, the teaching system was changed again so that from that point on pupils had to register for bilingual education. Due to the consequent obligation to confess minority affiliation (conceived as implicit with registering for bilingual education), the number of pupils in the bilingual system dropped considerably. In 1958, only 20.9%,[15] in the 1970s, only 13.9% of bilingual pupils were registered for German-Slovene teaching (Cillia et al. 1998).

[15] According to Fräss-Ehrfeld, it was 17% (Fräss-Ehrfeld 2005: 131).

The Austrian State Treaty (officially 'Treaty of Vienna' or 'State Treaty regarding the restoration of an independent and democratic Austria'), its meaning for minority legislation as well as subsequent legislation elevating minority toponyms to official status have already been discussed in Chap. 4, Sect. 4.2.1.5, in the context of Austrian minority legislation and need not to be repeated.

Apart from minority legislation, specific for Carinthia and milestones for the Slovene minority were the foundation of a bilingual high school in Klagenfurt in 1957 and its extension in 1975 as well as the foundation of a University for Educational Sciences in Klagenfurt in 1970, in 1993 renamed and extended to the University of Klagenfurt. The bilingual high school started to provide Carinthian Slovenes with an educational elite and to function also as a focal point for their cultural activities. It was founded by initiative of the federal government and can be regarded as a rather generous offer to the minority taking into account that Carinthia had just a handful of high schools at that time and a minority high school was much more than proportional related to population shares. In the meantime, in an environment of a much higher density of high schools since the later 1960s, three more bilingual high schools have been established in Carinthia (Bildungsdirektion Kärnten 2019b).

The university, the first in the federal state of Carinthia, offered not only higher education in Slovene language and culture, but its staff represented also a (prevailingly external) intellectual layer in Carinthia that behaved at least neutrally in minority affairs and acted at times even as a spearhead for effectuating minority rights (see, e.g., Österreichische Rektorenkonferenz 1988).

As regards political representation and organization, Carinthian Slovenes were never able to converge to a political collecting movement or ethno-regional party as it exists in South Tyrol [Südtirol/Alto Adige] for the German-speakers substantiated by the South Tyrolean People's Party [Südtiroler Volkspartei]. They were rather split into ideological directions and corresponding political organizations with the Central Association of Slovenian Organizations in Carinthia [Zentralverband slowenischer Organisationen in Kärnten/Zveza slovenskih organizacij na Koroškem] at the left and the Council of Carinthian Slovenes [Rat der Kärntner Slowenen/Narodni svet koroških Slovencev] on the right side of the political spectrum. In 2003, a faction with a more flexible political approach titled Community of Carinthian Slovenes [Gemeinschaft der Kärntner Slowenen und Sloweninnen/Skupnost koroških slovencev in slovenk] even split off from the Council of Carinthian Slovenes. Efforts of creating a unitary representation in communal councils and the provincial parliament resulted just in small factions—like the Carinthian Unitary List [Kärntner Einheitsliste/Koroška enotna lista, KEL] founded in 1973 and later evolving to the Unitary List [Einheitsliste/Enotna lista, EL] as the most prominent examples.

At the provincial and federal levels Carinthian Slovenes usually felt to be best represented earlier by the Austrian People's Party [Österreichische Vollspartei, ÖVP], but were also voting for the Socialist, later Social Democratic Party of Austria [Sozialistische, Sozialdemokratische Partei Österreichs, SPÖ], and later by the Green [Die Grünen], who offered Carinthian Slovenes even leading positions and mandates in the Austrian national parliament like Karel Smolle. Generally speaking, the two largest political parties in Carinthia, the Freedom Party [Freiheitliche Partei, FPÖ] and the Socialist, later Social Democratic Party [Sozialistische, Sozialdemokratische Partei, SPÖ], have a tradition of rather representing German-Carinthians—the Freedom Party more than the Socialists or Social Democrats—while the Austrian People's Party, along with the Roman-Catholic Church, tended to support the minority. Carinthian governments led by the Freedom Party and its ephemeric successor, the Alliance Future Austria [Bündnis Zukunft Österreich, BZÖ] in 1989–1991 and 2000–2013, therefore rather aggravated the political climate in Carinthia between majority and minority and instrumentalized especially the place-name question as a playground for political maneuvering.

5.1.2.3 Subregions and Places

Southern Carinthia and the so-called bilingual area are in terms of space-related identity all but a homogenous region. They are not only by scientific literature, but also by the local population subdivided into four main subregions (Fig. 5.21): from West to East Gail Valley [Gailtal/Ziljska dolina], Rosental/Rož, Vellach Valley [Vellachtal/Dolina Bele], Jauntal/Podjuna.

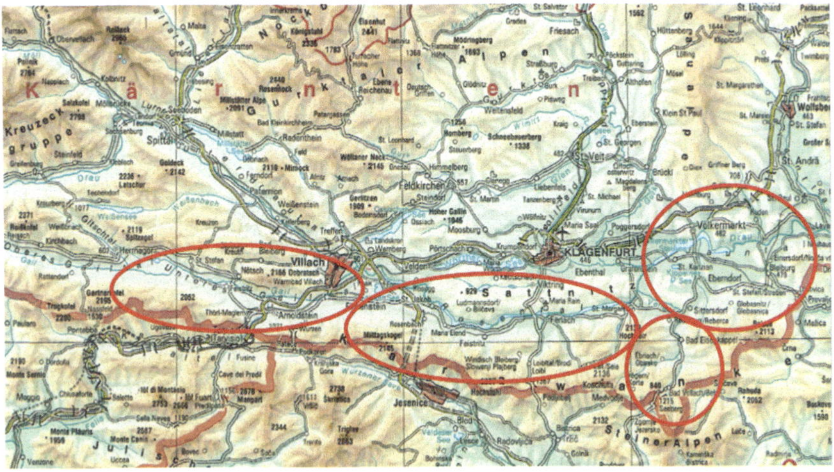

Fig. 5.21 Subregions in southern Carinthia. (Base map: Freytag and Berndt 2003)

They correspond to Slavonic dialect regions (Fig. 5.22) indicating that they were historically areas of internal closer communication. The dialect spoken in the Vellach Valley is titled Ebriach dialect [obirsko narečje].

Only the lower section of the Gail Valley [Gailtal/Ziljska dolina] called Lower Gail Valley [Unteres Gailtal/Zilja] is part of the bilingual area, and the Slovene name *Zilja* refers only to this part.[16] The other subregions mentioned above are, however, in their entirety part of the bilingual area. Up to 1918 also the Kanal Valley [Val Canale/Val Cjanâl/Kanalska dolina/Kanaltal] in modern Italy speaking the Gail Valley dialect and the Mieß Valley [Mežiška dolina] in modern Slovenia speaking a dialect of its own but related to the Carinthian Slavonic dialects were parts of the southern Carinthian bilingual area.

Since all of the southern Carinthian towns were mainly developed by Bavarian settlers and were focal points of the German-speaking population, they have not a prominent function for the identity of Carinthian Slovenes. This is rather true for some villages, from where prominent Slovenes (artists, scientists, politicians) descend.

[16] Information gratefully received from Martina Piko-Rustia, Slovene Ethnographic Institute Urban Jarnik, Klagenfurt/Celovec.

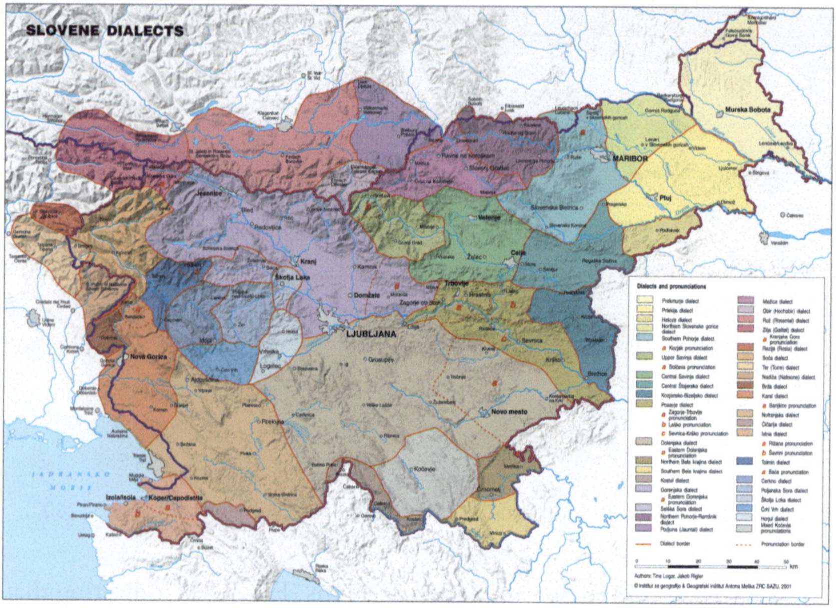

Fig. 5.22 'Slovene' dialects in and outside modern Slovenia. The reddish colors in the North indicate the Carinthian dialect group. (Source: Logar and Rigler 2001). (Color figure online)

Klagenfurt, however, although closely outside the bilingual area, has the function of a central place and also possesses symbolic value for Carinthian Slovenes. It is not only the capital of their wider home, but also the location of the most important Slovenian cultural institutions like the Hermagoras Publishers [Hermagoras Verlag/Mohorjeva založba] founded in the middle of the nineteenth century and a first disseminator of books in Slovene language, the bilingual high school or the Slovene Ethnographic Institute Urban Jarnik [Slowenisches Volkskundeinstitut Urban Jarnik/Slovenski narodopisni inštitut Urban Jarnik].

The most symbolic places for Slovenes in Carinthia and beyond are, however, situated north of Klagenfurt and thus even further outside the bilingual area: Karnburg (in Slovene *Krnski grad*), the seat of the rulers of Carantania, the Alpine-Slavonic principality in the Early Middle Ages, and Maria Saal (in Slovene *Gospa Sveta*), the place of pilgrimage and Salzburg-founded base of Alpine Slavonic Christianization. In the second line we find places like the Catholic education center in Tainach/Tinje, the church of Hemmaberg/Gora svete Heme, and two places of pilgrimage outside modern Carinthia or the bilingual area, that is, Monte Lussari in the Kanal Valley and Maria Luggau in the Carinthian Lesachtal, where small remnants of a Slovene minority still live or where Alpine-Slavonic settlement persisted up to the Late Middle Ages, respectively (for Lesachtal, see Neumann 1977).

5.1.2.4 Settlement Structure, Central-Place, and Administrative System

The minority area is in its entirety a rural area in the sense of an area with small and dispersed settlement and a still considerable role of agriculture and forestry. While in the valleys, mainly on gravel fans, in river terraces, and in accompanying hilly lands like the Sattnitz/Gure larger and smaller villages dominate besides some towns, the mountain area of the Karawanken/Karavanke is—with the exception of the town Bad Eisenkappel/Železna Kapla in the Vellach Valley—characterized by dispersed hamlets. When the social milieu of villages and hamlets is clearly rural in the sense of intimate neighborhood relations and a dense network of social responsibilities including membership in voluntary associations like fire brigades and associations of age groups, living in towns has already some aspects of an urban milieu such as a more distinct social stratification and a larger variety of lifestyles. Especially in the vicinity of the cities Klagenfurt and Villach suburbanization processes after World War II introduced urban lifestyles also in rural areas. Tourism development already before World War I, but even more intensively after World War II, had a similar effect on the lake areas, to a minor extent also on the bilingual area at large. An even more recent, but less intensive development affecting also remote parts of the rural area is amenity migration of urban dwellers, also from abroad, looking for an environment convenient for recreation.

A certain exception is Arnoldstein and its surroundings in the Lower Gail Valley [Unteres Gailtal/Zilja] due to its former heavy industry (lead smelting based on lead mining in nearby Bleiberg-Kreuth and Cave del

Predil/Raibl/Rabelj). Although this industry has been shut down, related settlement and social structures still remain.

A second exception, yet of a minor extent, is Ferlach in the Rosental/Rož with its old tradition of weapon manufacturing that later developed into industrial production.

Both exceptions are reflected already on a map of ethnic/linguistic structures in the interwar period (1923) by tearing gaps into the otherwise quite continuous share of Slovene-speakers (Wutte et al. 1925: map 46).

The highest-ranking central place for the minority area is, although located outside the bilingual area, Klagenfurt am Wörthersee, Carinthia's capital. It is, although with its population of 101,000 (2019), only a city of medium size by international standards, in the scope of Austrian centers a center of the second level and a federal state capital with a very good or good equipment with central facilities comparable to Graz, Linz, Salzburg, and Innsbruck (Seger 2019). It competes, however, at this level with Graz, the capital of Styria, with its larger, usually also more traditional and attractive facilities. This competition will very likely be enforced by the completion of a high-speed railroad track (Koralmbahn) in a few years reducing traveling time between the two cities to less than one hour. Klagenfurt's catchment area related to the minority area comprises at the federal state capital level all of it, at the lower level of a quarter capital[17] only its eastern part, that is, the political districts Klagenfurt-Land and Völkermarkt.

The competitor of Klagenfurt at this quarter-capital level is Villach, the second pole of the bipolar Carinthian central-place system, also located outside the bilingual area. Villach with its 62,200 inhabitants (2019) is significantly smaller than Klagenfurt and has the typical equipment of an Austrian quarter center like Leoben in Styria [Steiermark] or Wiener Neustadt in Lower Austria [Niederösterreich]. Traditionally it has the image of a trade center and traffic node due to its location at River Drau/Drava and at the crossroads of important railroad tracks and motorways in contrast to Klagenfurt that is rather known as an

[17] Quarter capitals are by Austrian central-place terminology centers of larger parts of a province, the name being derived from the cases of Upper Austria [Oberösterreich] and Lower Austria [Niederösterreich], where the province is indeed composed of four parts (quarters).

administrative and educational center. At the quarter-center level, the catchment of Villach comprises the western part of the minority area, that is, the southern parts of the Political District Villach-Land and the eastern section of the Political District Hermagor.

At the lower regional level two more centers exert central-place functions for the minority area: Hermagor (in Slovene *Šmohor*), located in the Gail Valley [Gailtal/Ziljska dolina] at the western edge of the minority area, and Völkermarkt (in Slovene *Velikovec*) in the Jauntal/Podjuna in the eastern part of the minority area, with a population number of less than 10,000 each. Both are seats of political districts, but with a weaker equipment with central-place facilities than other centers of political districts in Carinthia, that is, Spittal an der Drau, Sankt Veit an der Glan, and Wolfsberg (Seger 2010). They offer, however, in addition to high schools and the services of district and related offices, of a district court, and of a hospital, considerable shopping facilities. Hermagor's impact comprises in these respects western parts of the minority area in the Lower Gail Valley. Völkermarkt's catchment coincides more or less with the territory of its political district that is in total part of the minority area (Seger 2010).

At a still lower level, mainly in the commercial and cultural spheres, three more small towns can be attributed the role of central places: Ferlach (in Slovene *Borovlje*) for the central part of the Rosental/Rož, but very much in the shadow of Klagenfurt; Bleiburg/Pliberk for the eastern part of the Jauntal/Podjuna and Eisenkappel/Železna Kapla for the Vellach Valley. Ferlach has become prominent for the production of hunting guns, Bleiburg/Pliberk as a market place and more recently for the gallery of the painter Werner Berg, Eisenkappel/Železna Kapla is a historical traffic halt and market place at the bottom of Seeberg/Jezerski vrh along the earlier very important Iron Road from the Adriatic coast to historical iron mines in Carinthia and Styria.

The administrative subdivision of the minority area coincides to a high extent with the central-place system and is—in accordance with the rest of Austria—characterized by two levels below the federal state level: political districts and communes (Fig. 5.1).

In Austria, political districts are territorial-administrative units exerting deconcentrated state administration. Their seats function as regional 'branch offices' of central authorities of the Federation and the federal state with the aim to provide for a better accessibility of these otherwise centralized services. They are headed by a civil officer appointed by the federal state government and have no self-government through elected bodies. The minority area has a share in the eastern part of the Political District Hermagor, covers the South of the Political District Villach-Land, the largest part of the Political District Klagenfurt-Land, and the Political District Völkermarkt in total.

Political districts in Austria can be traced back to 1850, when they were established to complement the juridical or court districts in the administrative sphere. They displayed ever since a remarkable territorial continuity resulting in being deeply rooted in public consciousness. They are, however, rather conceived as technical units and have thus much less if any meaning for space-related identity, much in contrast to the regions and places mentioned before, from which they partly deviate—with the strongest discrepancies occurring in the Gail Valley [Gailtal/Ziljska dolina] and in Rosental/Rož. Carinthia, however, shows an exception concerning continuity of political districts: In 1982, the Political District Klagenfurt-Land has been divided into a northern part (Political District Feldkirchen) and a southern part (Political District Klagenfurt-Land). Thus Klagenfurt-Land coincides now almost completely with the minority area of the former district of the same name.

In Austria, the local territorial-administrative level is represented by self-governing communes with an elected communal council and in Carinthia (along with most other Austrian federal states) also an elected major. For the average citizen, communes are very likely the most relevant territorial-administrative category, since they are responsible for most basic civil services including real estate and infrastructure. The communal council also decides about the names of all populated places in the commune except dispersed hamlets, which can be named by the owner, as well as about all hodonyms, that is names of streets, squares, and so on.

In 2019, after a comprehensive state-wide administrative reform in 1973 merging rather small communes to much larger ones, Carinthia had 132 communes with an average number of 4250 inhabitants (including the larger cities Klagenfurt and Villach). Thirty-five of them have a share in the bilingual area. According to the population census of 2001,

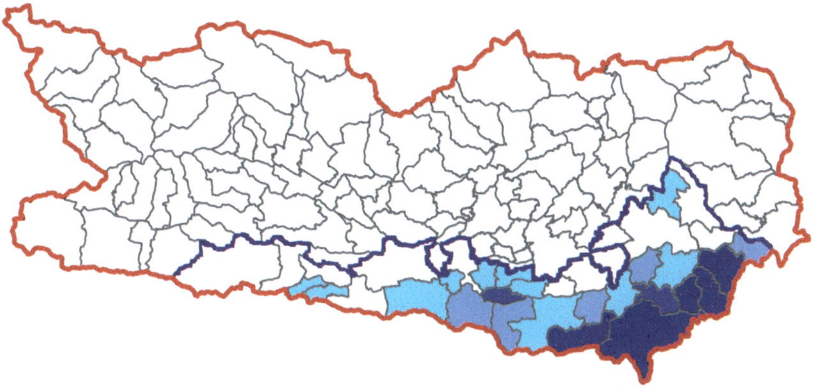

Fig. 5.23 Communes with a share of more than 5% of Slovene-speaking population. Legend: dark blue = >30%, blue = 20–30%, paler blue = 10–20%, pale blue = 5–10%; blue line = boundary of the bilingual area. (Source: Wikipedia). (Color figure online)

the only commune with a minority language-speaking majority was the tiny (691 inhabitants) Commune Zell (in Slovene *Sele*) high up in the Karawanken/Karavanke mountain range with a share of 90% (see also Table 5.2 and Fig. 5.23).

5.1.3 Basics on Austrian and Carinthian Toponymy

Place names for features of all categories in Austria are standardized in the sense of being approved by the legally responsible authorities. This means that these names are preferable to other names that may exist for the same feature.[18] Standardized names are documented by the "Digital Landscape Model (DLM)—Range Names" conducted by the Federal Office of Metrology and Surveying [Bundesamt für Eich- und Vermessungswesen],

[18] The Glossary of the United Nations Group of Experts on Geographical Names (UNGEGN) defines a standardized name as a "name sanctioned by a names authority as the preferred name from among a number of allonyms for a given feature. However, a single feature may have more than one standardized name. Example: Kaapstad and Cape Town (but not Capetown)" (Kadmon 2002: 19).

Table 5.2 Carinthian communes with more than 1% minority-language speakers (Slovene plus 'Windisch') in 2001

Commune with >1% minority-language speakers in 2001	Absolute number of minority-language speakers	Share of minority-language speakers in the overall population
Political district Hermagor		
Hermagor-Pressegger See	125	1.8
Sankt Stefan im Gailtal	27	1.6
Political district Klagenfurt-Land		
Ebenthal in Kärnten	306	4.2
Feistritz im Rosental	356	14.2
Ferlach	613	8.6
Keutschach am See	124	5.6
Köttmannsdorf	187	6.9
Ludmannsdorf	508	28.2
Magdalensberg	38	1.3
Maria Rain	75	3.9
Maria Wörth	13	1.1
Sankt Margarethen im Rosental	134	12.2
Zell	630	90.0
Political district Villach-Land		
Arnoldstein	107	1.7
Feistritz an der Gail	50	8.3
Finkenstein am Faaker See	438	5.7
Hohenthurn	63	7.9
Rosegg	107	6.3
Sankt Jakob im Rosental	752	17.5
Velden am Wörthersee	229	2.9
Political district Völkermarkt		
Bleiburg	1292	33.1
Diex	62	6.9
Eberndorf	560	9.8
Eisenkappel-Vellach	1038	39.0
Feistritz on Bleiburg	692	32.9
Gallizien	160	8.9
Globasnitz	718	38.6
Griffen	48	1.4
Neuhaus	183	15.2
Ruden	61	3.7
Sankt Kanzian am Klopeiner See	541	13.5
Sittersdorf	432	20.6
Völkermarkt	284	2.6

Source: Population census 2001, Statistik Austria (2020b)

the supreme and official mapping agency in Austria. A selection of standardized names—selected in accordance with the map scale—can be found on the official topographical Austrian Map in the scales 1:50,000, 1:250,000, and 1:500,000 that are revised in intervals of seven years and edited by this same agency. A selection of standardized names adequate to the map scale 1:500,000 supplemented by indications of pronunciation, feature category, location, spatial range (in cases of areal features), source (in the case of rivers), administrative affiliation, and location on a sheet of the Austrian Map 1:50,000 has first been published by Josef Breu in 1975 in print (Breu 1975) and in a revised and extended version in 2000 on CD-ROM (AKO 2000).

Up to the population census of 2001, names of populated places had been published after every population census, that is, in intervals of a decade, in the Gazetteer of Austria as well as in gazetteers of the individual federal provinces by Statistik Austria (earlier Central Statistical Office [Statistisches Zentralamt]), continuing a tradition that can be traced back to the Habsburg empire, when gazetteers were published in a similar mode for all crownlands.

In contrast to some other countries with more centralized structures, in Austria place-name standardization, colloquially speaking the definition of the 'correct' or official name, follows the bottom-up principle. It starts with the owner of a feature—a house or building, a farmstead, a piece of land, or a technical or infrastructure feature—who is entitled to name it as he/she likes. He/she, it can also be a juridical person like an enterprise or company, has to submit the name to the commune, but the proposal can only be denied, if it is 'against good morals'. So it is, for example, possible to name a farmstead, alpine pasture, or alpine hut in a minority language—as it is in some cases practiced in Carinthia—or to name an office building, hotel, shopping center, or airport in a foreign language like English, French, or Italian—as it is also frequently practiced. This certainly opens also the doors for commercial naming.

The right of naming one's own feature applies also to natural features like mountains, mountain peaks, or water bodies, if they are exclusively located on the plot of one owner—as it can easily happen with the Austrian Alpine Club, the Federal Forests, the Roman-Catholic Church, monasteries, or estates of noble families owning entire mountain ranges, forests, or lakes, but also with larger farmers. So far, their naming practice did not get into conflict with traditional names, but this cannot be excluded. So, there is some field of legal uncertainty.

Austrian Railroads [Österreichische Bundesbahnen, ÖBB], the enterprise conducting the vast majority of the Austrian railroad network and railroad stations, the Austrian Motorway and Expressway Financing Joint-stock Company [Autobahnen- und Schnellstraßen-Finanzierungs-Aktiengesellschaft, ASFINAG] as well as the road administrations of the federal provinces, legally based on the argument that the owner has the right of naming, name railroad tracks and stations, motorways, highways, and other roads with their signs according to their own principles, but in accordance with good practice in Austria and on the international scene. There are, however, significant discrepancies in road traffic signage between Central European countries (see Jordan 2009).

Names of populated places with more than one owner as well as names of streets and other urban and traffic areas like squares or parks are the prerogative of the commune except in cases of minority legislation related to populated places, where competences rest also with the Federation as it was practiced most recently with Federal Act No. 46/2011 (BKA 2020). Naming streets is not mandatory in Austria and up to the decision of the commune. It was earlier practiced only in cities and larger towns. In recent decades, however, it has become popular also in smaller towns and even villages, including Carinthia. Because commemorative naming in the sense of reminding by urban names of persons, institutions, or events has mushroomed resulting in local disputes (Who deserves a name?) and frequent name changes, in 2017 the Austrian Board on Geographical Names in accordance with UN resolutions passed a recommendation (AKO 2017) to avoid commemorative naming as much as possible in favor of descriptive naming—also in the sense of reminding of earlier situations or of field names. In a minority situation like in southern Carinthia, street naming would also mean to reserve a proportionate share of names to issues (descriptive or commemorative) important for the minority. This, in turn, can cause disputes. Since official street naming is promoted especially in southern Carinthia in the last few years, unofficial names reminding of local traditional features, very often with a Slovenian name, are frequently replaced by new 'neutral' names not typical for the place and disliked by the minority. In contrast to Burgenland, street and other urban naming in Carinthia has until recently not resulted in bilingual names. Only in 2020, the Commune Sankt Jakob im Rosental started a first initiative of this kind.

The name of the commune can be defined by the commune but has to be confirmed by the federal province. Names of political districts are defined by the federal province; names of federal provinces can be defined by the provinces but have to be confirmed by the Federation.

Names of natural features with the exception of features located exclusively on the territory of one owner (a physical or juridical person) are defined by the Federal Office of Metrology and Surveying after having explored name use by the local population. To explore it, with every revision of a map sheet the Office's topographers ask reliable local persons for their name use. Thus, the endonym principle rules, the name used by the local community counts.

There are, however, natural features named differently by the people residing in their surroundings. So it is with a mountain between the Lower Gail Valley [Unteres Gailtal/Zilja] and the Lower Drau Valley [Unteres Drautal] in Carinthia. It is called *Dobratsch* from the Lower Gail Valley and *Villacher Alpe* from the Lower Drau Valley, both in German language. This occurs on the grounds that the Lower Gail Valley has a longer Slovene-speaking tradition and is still partly minority-language speaking and its inhabitants thus prefer the orthographically Germanized version of the Slovenian name *Dobrač*, while the Lower Drau Valley is German-speaking for much longer. A second reason for this dual naming may be that the mountain looks completely different from both sides: While it has soft slopes and is covered with mild pastures on the Drau Valley side, it shows impressive steep rock faces on the other. In cases like this, both German endonyms have been standardized. Current Austrian official maps, however, mark in contrast to older ones by *Dobratsch* the peak and by *Villacher Alpe* the mountain massif based on an obvious misunderstanding.

Some natural features like mountain ranges extend over larger distances. Some of their names have only developed through cartographic surveys and geographical structuring of space from the eighteenth century onward and not been given by the local population. They are standardized in this 'artificial' form by the Federal Office of Metrology and Surveying, if they do not contradict local use.

The Federal Office of Metrology and Surveying, but also federal provinces and communes are consulted with their naming activities by the Austrian Board on Geographical Names [Arbeitsgemeinschaft für Kartographische Ortsnamenkunde, AKO], an expert body composed of scientists from all disciplines concerned with place names (linguists, geographers, cartographers, geodesists, historians, cultural anthropologists, lawyers) as well as of federal and provincial authorities (Federal Office of Metrology and Surveying, Statistik Austria, archives, libraries) and enterprises (publishers) providing place-name relevant products (see Bergmann and Jordan 2010). The Austrian Board cooperates on the international level with the Permanent Committee on Geographical Names [Ständiger Ausschuss für Geographische Namen, StAGN] that has a coordinative function for all German-speaking countries, and with the United Nations Group of Experts on Geographical Names (UNGEGN) as the global umbrella authority in this field.

For most Austrian federal provinces (Burgenland, Lower Austria [Niederösterreich], Salzburg, Styria [Steiermark], Tyrol [Tirol], Upper Austria [Oberösterreich], Vienna [Wien]) exist also provincial boards, in structure similar to the Austrian Board and first approached, if a question in a certain province arises. So far, a board in Carinthia could not be installed—not the least due to the place-name conflict that dominated the scene until recently. All boards, however, have only consultative function. They are not entitled to finally decide on any place name.

In Austria place names are in principle standardized according to the standard language and not in dialect except for place names given by the owner, where this cannot be excluded but is also not usual. This is today also true for place names in minority languages, although the practices of the minority place-name Act of 1972 and the Decree of Topography for Carinthia [Topographieverordnung für Kärnten] as of 1977 (BGBl. 306/1977, BKA 2020) were not in line with this principle but have been based on local use and historical records (Wadl 2006). It has, however, also to be noted that place names are more resistant to orthographic and other reforms and changes of a language and that for this very reason they frequently do not conform to the current language standard. This is especially true for names of populated places that are fossilized by their very official character.

While the maps of the Austrian Alpine Club tend to show also place names in their dialect version, official Austrian topographical maps as well as data files prefer the standard language principle based on the argument that dialects are primarily spoken languages and rarely written and that even dialect speakers would perceive it as strange and even satirizing to find place names written in dialect in official documents—although this is not always so and in some cases rather the other way round as the examples given further below show. Another argument is that the standard language is the common language of the country and standard language thus makes place names better manageable also for other parts of the country.

Nevertheless, many place names have been standardized in their dialect version, if they have a long tradition of being written in this version and are deeply grounded in literature—as in the cases of *Kirchbichl*, a populated place in Tyrol that would be called *Kirchbühel* ('hill with a church') in standard language, or *Koralpe*, a mountain range between Carinthia and Styria, that would have to be *Karalpe* ('Alpine pasture with cirques') in standard language. Standardization of fossilized dialect names, notably names of populated places, is also no major exception throughout Carinthia as the following examples quoted by Wilhelm Wadl (2006: 188) demonstrate: *Lambichl* ('loam hill') would be *Lehmhügel* in standard German, *Annabichl* ('Ann's hill') would be *Annahügel* in standard German, *Krumpendorf* ('curved village') would be *Krummes Dorf* in standard German, *Kerschdorf* ('cherry village') would be *Kirschendorf* in standard German.

The following is essentially based on the Toponymic Guidelines for Map and Other Editors for International Use edited by AKO (AKO 2012: 3–9) to which in the course of their seven editions the geographer Josef Breu and the linguists Otto Back, Hubert Bergmann, and Isolde Hausner have significantly contributed.

5.1.3.1 Place Names in German Language

German as the only nationwide official language and language of tuition in Austria is—like English—a polycentric language with three varieties, that is the German, the Swiss, and the Austrian. They differ slightly by vocabulary and orthography of the written form, but not by grammar. The spoken language, however, sounds quite different due to divergent melody and intonation, and colloquial language differs also by grammar. The German language is written in Roman script.

The German alphabet[19] consists of 26 letters (which are the same as the basic modern Latin alphabet). In addition, there are the so-called umlaut letters (diacritic letters) *Ä ä, Ö ö, Ü ü*, and the letter *ß* ('scharfes S' or 'Eszett'). In capitalized words, however, the letter *ß* is usually substituted by *SS*. In alphabetical indexes and gazetteers one can find three ways of sorting the umlaut letters: (1) treated like their base characters, (2) treated as decomposed letters: vowel plus *e* (*Ae ae, Oe oe, Ue ue*), and (3) treated like extra letters and placed after their base letters (after *Az-, Oz-, Uz-*). The letter *ß* is treated like *ss*.

The rules given below have been officially used by the Federal Office of Metrology and Surveying since 1969 and are generally in concordance with those in current use in the other German-speaking countries. Place names are normally not subject to the general spelling rules. They are, rather, spelled in the officially approved or customary form. However, in the following respects the German place names comply with the general spelling rules except for individual cases where another officially approved version exists.

Capitalization:

(a) Adjectives and participles forming part of geographical names are capitalized. This also holds true for the *-isch* derivatives. Examples: *Großer Priel; Hohe Tauern; Großes Walsertal; Hängender Stein; Mährische Thaya; Steirische Kalkspitze.*

[19] A a, B b, C c, D d, E e, F f, G g, H h, I i, J j, K k, L l, M m, N n, O o, P p, Q q, R r, S s, T t, U u, V v, W w, X x, Y y, Z z.

(b) Word forms with the suffix *-er* as derivations from place names are capitalized: Examples: *Gurktaler Alpen; Wiener Becken; Millstätter See.*

Use of one word, two words, or hyphen:

(a) Place and other names as determinative elements:

(aa) In general compounds consisting of a single or compound geographical name or a personal name plus a generic element are written as one word. Examples: *Wettersteingebirge, Glocknergruppe, Eberhardsbach, Leopoldsberg.*
(bb) A hyphen is often used in order to improve legibility, when the place name, which forms the determinative element, is followed by a compound generic element. Example: *Großglockner-Hochalpenstraße.* Where legibility is not impaired, writing in one word is preferred. Example: *Alpenvorland.*
(cc) Hyphens are used when the determinative element consists of several place names. Example: *Donau-Oder-Kanal.* Hyphens are, moreover, used when the determinative element consists of several words (e.g., title, first name, last name) including abbreviations. Example: *Otto-Ludwig-Haus, Dr.-Ignaz-Seipel-Platz.*

(b) Derivatives with the suffix *-er*:
(aa) Two words are retained when the derivatives with the suffix *-er* of place names designate the location. Examples: *Zillertaler Alpen, Neumarkter Sattel, Tuxer Joch, Ziersdorfer Bach, Lunzer See.*
(bb) There are place names with final *-er* which are not derivatives in the above sense. These names are written solid in accordance with place and other names as determinative elements. Examples: *Glocknergruppe, Brennerpass.* When the derivatives with the suffix *-er* of place names designate persons, writing in one word is preferred. Example: *Jägerkapelle, Lunzerkreuz* (after a peasant by the name *Lunzer); Steinerscharte* (after an alpinist by the name *Steiner*).
(cc) If the suffix *-er* is attached to a place name ending in *-ee*, only two *e*'s are written. Examples: *Altausseer See, Egelseer Bach.*

(c) Adjectives as determinative elements:

(aa) Spelling in one word is generally used for compounds with a non-inflected adjective such as *groß, klein, alt, neu, mittel, ober, unter, hoch, nieder* or points of the compass plus geographical names. Examples: *Mitteleuropa, Osttirol, Obersteiermark, Niederösterreich, Großglockner, Hochkar, Großnondorf.* On the other hand, official usage prescribes: *Klein-Ulrichschlag, Alt-Urfahr, Alt-Nagelberg.*

(bb) The hyphen is used with compounds consisting of uninflected adjectives ending in *-isch* which are derived from the names of places, countries, and peoples plus place names. Example: *Steirisch-Tauchen, Windisch-Grutschen.* On the other hand, official usage prescribes: *Kroatisch Minihof, Windisch Bleiberg.*

(d) Compounds consisting of place names: The hyphen is used when a place name is composed of two place names. Examples: *Rudolfsheim-Fünfhaus, Wünschendorf-Pirching.*

In 1996, all German-speaking countries adopted 'new rules for German orthography'. The implementation began in the academic year 1998/1999 and was brought to an end in 2005. In Austria, the new spelling rules are obligatory for institutions under federal control with respect to orthography, namely schools at all levels and administrative bodies since 2008. With respect to toponomy, the following categories of names for which there are no legally binding rules are to be considered: names of mountains, fields, rivers and lakes, glaciers, Alpine pastures, mountain huts, paths, ruins, passes.

With respect to place names the following aspects of the new orthography affect the spelling: *ss/ß*-spelling: the German letter *ß* only occurs after a long vowel or diphthong. Examples: simple and compound names with *Fass-, Fluss-/-fluss, Nuss-, Pass/-pass, Nass-, Ross-, Schloss*, and so on. There are exceptions in cases where non-settlement names are used as elements of place names: Example: *Schoberpass; but: Wald am Schoberpaß*; accumulation of consonants or vowels: in compounds triple consonants preceding a vowel are no longer reduced. Examples: *Schifffahrt, Irrriegel.* A hyphen is, however, recommended in some cases, for example, in compound words with three successive identical vowels. Example: *See-Eck.* Reference to the root word: Certain words are to be spelled according to their root words. Example: *Gämsalpe*. Variants are allowed in special cases. Example: *Waldschenke or Waldschänke*. On the other hand, official usage prescribes *Gemstel- (Gemstelpaß, Gemstelboden)*. Final silent *h*: The adjective *rau* drops the silent *h* to conform to other words. Example: *Raukopf*. Place names keep the spelling with *ß* according to official usage.

German spelling is not always clearly indicative of pronunciation. In many cases it does not say anything about stress or vowel length; with some letters or groups of letters it is not always certain which sound they represent (e.g., the letters *v* and *y*). All this is true of place names to an even higher degree than of general vocabulary, since most of these names are naturally less well known and since names often show conventional spellings and sometimes dialectal spellings, both deviating from the present norm. It is, therefore, not possible to give a simple pronunciation key covering the pronunciation of all place names.

5.1.3.2 Place Names in Slovene

Slovene, the official language of Slovenia, has the status of a regional minority language in Austria. It is written in the Roman script. The Slovenian alphabet has 25 letters.[20] In contrast to other Slavonic languages like Croatian, a *v* inside or at the end of a word is pronounced like *u* such as in *Ravne* or *Triglav*.

Slovenian names of geographical features other than populated places generally are represented on official Austrian maps in a form corresponding to German spelling rules, although translations from Slovenian into German are forbidden. The names must be written as they are spoken by the local population, but the sounds are represented in writing in accordance with German, not Slovenian, writing and reading conventions. Examples: *Gorelza, Hrewelnik*. An exception is made in the comparatively rare cases where a place name consists of more than one Slovenian word. Such syntactical units (mostly noun plus adjective) are written according to Slovenian spelling rules, using the diacritical marks of this language. Examples: *Tolsti vrh. Velka ravna*.

[20] A a, B b, C c, Č č, D d, E e, F f, G g, H h, I i, J j, K k, L l, M m, N n, O o, P p, R r, S s, Š š, T t, U u, V v, Z z, Ž ž

It also has to be noted that with a Slovenian place name composed of two words like in *Tolsti vrh* ('bold mountain'), *Velka ravna* ('large plain'), or *Suhi dol* ('dry valley') the generic is initiated by a lower-case letter. Exceptions are cases, where the second word is just seemingly a generic, but actually a proprial part of the name like in *Kranjska Gora*, where *gora* ('mountain') could be a generic but has a transferred meaning in this case. (It is actually the name of a populated place.)

5.2 Těšín/Cieszyn Region (Přemysl Mácha)

5.2.1 The Wider Geographical and Historical Setting

The historic Těšín/Cieszyn region lies in current northeastern Czechia and southern Poland (see Fig. 5.24). It is not a natural region that would be differentiated from its neighbors by distinct hydrological or

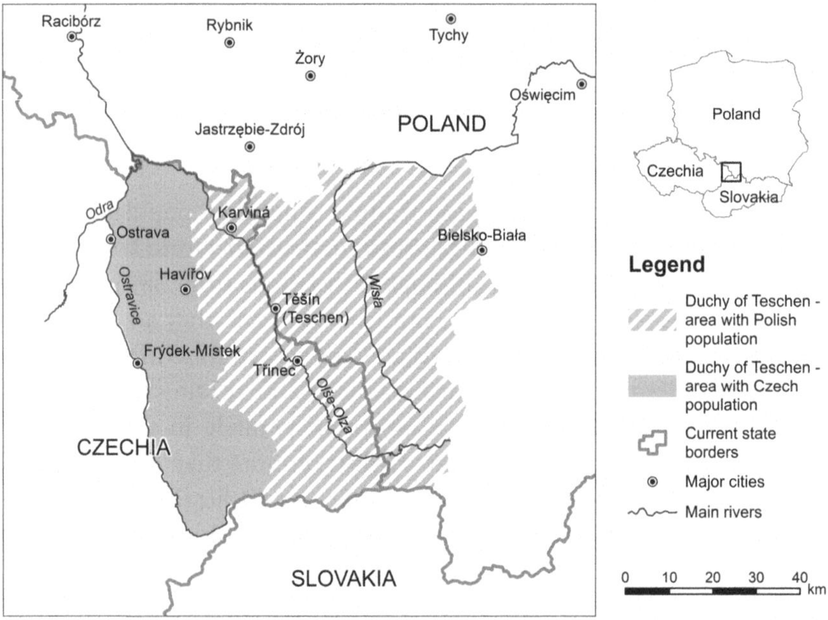

Fig. 5.24 Location of the Těšín/Cieszyn region. (Cartography: Luděk Krtička)

geomorphological boundaries. Rather, it gradually emerged as a political entity during the complex processes of state formation in Central Europe in the Early Middle Ages. The current Těšín/Cieszyn region of Czechia is a fragment of a larger historical administrative unit with semiautonomous standing called the Duchy of Teschen. This duchy was one of the Silesian duchies forming Silesia [Slezsko, Śląsk] which was integrated as a whole into the Bohemian Kingdom in 1335. The Silesian duchies were specific in the context of the Kingdom for their significant Germanization and, after the Thirty-Years War (1618–1648), for the strong presence of Protestantism which differentiated Silesia sharply from the overwhelmingly Catholic Bohemia [Čechy] and Moravia [Morava]. Unlike the majority of other Silesian duchies, however, the Duchy of Teschen remained Slavic. Still, the Lutheran faith was equally widespread there and continues to be even today.

From the late 1400s the region began to feel the influence of Wallachian colonization which spread mountain pastoralism along the Carpathian range, impacting southeastern Silesia and eastern Moravia. This cultural adaptation opened the previously uninhabited mountainous parts of the region to settlement and exploitation. It also significantly enriched local toponymy with Wallachian names such as *Kyčera, Gigula, Grúň, Grapa*, and others. Even more importantly, it contributed to the rise to the "Gorol" identity so important for the region today (see below). As a consequence, local folklore, traditional costumes, crafts, architecture, and music share much of their aspects with other Carpathian traditions while differing sharply from most other Moravian and Silesian regions.

In terms of the larger context, it is also important to note that in 1654 the Duchy of Teschen was acquired by the Habsburgs who owned and administered it until 1918. While this direct connection to the ruling family in the Austrian monarchy did not have any specific effects on the local population in terms of their rights and duties, the local branch of the Habsburg family (Saxon-Teschen) had a significant influence on local economy, industry, and politics as well as on forestry and agriculture. The availability of capital made it possible for the Habsburgs to invest heavily into coal mining and steel production, setting up many operations in the region and transforming it into the industrial heartland of the Austrian Empire. The need for wood initially for steel production and later for the reinforcement of mining shafts and the construction boom associated with the massive labor migration led to the reforestation of the Beskids [Beskydy, Beskidy] and the suppression of mountain pastoralism. The impact of the Habsburgs on local toponymy was mediated by the industrial operations (names of factories and coal mines) and numerous working-class residential neighborhoods on the outskirts of cities. In the late 1800s, the Habsburgs were also recognized in street names of Teschen/Těšín/Cieszyn. Memories of the Habsburgs are preserved in folklore and regional literature.

The majority of Silesia was ceded to Prussia after the Treaty of Berlin in 1742 and Austria was able to keep only the Duchies of Troppau/Opava, Jägerndorf/Krnov, and Teschen/Těšín/Cieszyn together with a fragment of the Duchy of Neisse (in German *Herzogtum Neisse*, in Czech *Niské knížectví*, in Polish *Księstwo Nyskie*). Austrian Silesia, nevertheless, kept its semiautonomous status as one of the lands of the Bohemian Kingdom with its seat in Troppau/Opava, a German city at that time. The Duchy of Teschen maintained extensive administrative functions and Teschen/Těšín/Cieszyn was a lively regional political and commercial center until the late 1800s.

In 1772, after the first partition of the Polish-Lithuanian Commonwealth, Austria obtained Galicia [Galicja, Galyčyna] (Kingdom of Galicia and Lodomeria) made up by the southeastern part of modern Poland and the western part of modern Ukraine. This had significant consequences not only for the Austrian Monarchy internally and internationally, but also for the Těšín/Cieszyn region. The latter became both an important trading crossroads and a destination for labor migration after coal mining and steel production began to develop in the early 1800s (Małysz and Kaszper 2009; Gawrecki 2017). Because the city of Teschen/Těšín/Cieszyn itself did not have either coal mines or steel mills, this migration impacted other areas of the Duchy more. By the late 1800s, Ostrau/Ostrava, where the concentration of mining and industry was the highest, had been transformed from an insignificant small town into the most important city at the heart of Austria's most industrialized region. Ostrau/Ostrava thus overshadowed both Teschen/Těšín/Cieszyn and Troppau/Opava and gradually took over their historic political, economic, and administrative functions.

The definite demise of the Duchy of Teschen was brought about by its division between Czechoslovakia and Poland in 1920. Following World War I, both Czechoslovakia and Poland claimed this area as their own. Poland claimed it on ethnic grounds, pointing out the prevailingly Polish character of the region based on the last Austrian population census (see the section on ethnic composition of the region below). Czechoslovakia, by contrast, argued that the region had been part of the Bohemian Kingdom since the 1300s. Both countries were interested in the coal deposits in the area, but the situation was more urgent for Czechoslovakia. The only railroad connecting the Czech and Slovak parts of the Republic, the Košice-Starý Bohumín Railroad (*Kaschau-Oderberg-Bahn*) was running through the train station of Teschen/Těšín/Cieszyn and its re-routing would come at an immense cost for the new state, should the entire Duchy be incorporated into Poland.

In the first days after the end of the War, both Poles and Czechs living in the area established 'national councils' which started discussing the postwar division of the region. Poles created the National Council of the Duchy of Teschen [Rada Narodowa Księstwa Cieszyńskiego] while Czechs founded the National Land Council for Silesia [Zemský národní výbor pro Slezsko] representing the Czechoslovak side. The two bodies quickly agreed on a temporary border with no bearing on the eventual political decision regarding the final solution. However, in preparation for the first parliamentary elections in Poland in January 1919, the pre-election campaign started also in the provisionally Czechoslovak part of the region and it was generally assumed by the Polish representatives that the elections were going to take place there as well (Gąsior 2009). This, however, was seen as a worrying step by Czech politicians.

The rising tension between the two countries culminated in the Seven-day War in January 1919 ending in a stalemate. In order to resolve the conflict, it was initially agreed that the local population would have a chance to decide its fate in a plebiscite organized by the allied forces (as was the case around the same time in Schleswig, Upper Silesia, and East Prussia); however, the ongoing tensions and disagreements between the Czechoslovak and Polish side made that impossible and the voting was canceled (Baran 2010). The division of the region was eventually agreed on at the Spa conference in July 1920. The majority of the Těšín/Cieszyn region was given to Czechoslovakia, which received 1270 km^2 of the area inhabited at the time by approximately 290,000 people. Poland received 1012 km^2 inhabited by only 39,000 people. At the end, then, Czechoslovakia obtained the part it was most interested in, which included the railroad access as well as coal mines and steel mills. Together with them, however, it also gained a sizable Polish minority inhabiting the area (Gąsior 2009). This meant that the local population in the Těšín/Cieszyn region that had been living there for centuries and using its own dialect suddenly became a minority population in a state of another nation.

As a result of international politics, the Duchy ceased to exist. Both Czechoslovakia and Poland moved away from feudal administration and land organization. Czechoslovakia outlawed aristocratic titles and expropriated all Habsburg property, which impacted the Těšín/Cieszyn region extensively because the Habsburgs were, by far, the largest landowners in the region. While the land system continued to exist, districts were reorganized and created to accommodate the needs of the localities. The newly created Český Těšín/Czeski Cieszyn became the seat of one small district, permanently losing its previous control over the territory of the Duchy. During the twentieth century, the industrialization of the region continued and Ostrava and other cities such as Třinec, Frýdek-Místek, Karviná, Orlová, and Havířov grew to prominence thanks to migration from different parts of Czechoslovakia. This definitely relegated Český Těšín/Czeski Cieszyn to the position of a small regional town. The historic part of Těšín/Cieszyn remained on the Polish side where it also played a diminished role as the seat of a district. The historic glory of Těšín/Cieszyn thus ended, never to return again.

A peculiar movement grew to prominence in the last decades of the Austrian period with a strong support across the Duchy. After the creation of Czechoslovakia, it became a dominant political force. This movement elevated the Silesian identity as a unique identity different from Czech, Polish, or German. Politically, it sided with local Germans and during the conflict over the region between Czechoslovakia and Poland, it sided with Czechoslovakia. *Ślonzáci* ('Silesians' in the dialect) organized in the Silesian People's Party [Slezská lidová strana] which became the strongest party in the region while Josef Koždoň, the party's leader, served as the mayor of Český Těšín/Czeski Cieszyn during most of the interwar period. Due to their political influence, the "Šlonzak" category was added to the population census but only in combination with other categories—Silesian-Czech, Silesian-German, and Silesian-Polish. Altogether, in the 1921 census, some 47,000 people (i.e., 42% of all inhabitants) declared Silesian identity in combination with Czech (51.4%), Polish (45.7%), and German (2.9%) in the political districts of Fryštát/Frysztat and Český Těšín/Czeski Cieszyn which formed the core of the Polish minority area. For political reasons, these census results were never published and people declaring Silesian identity were counted as Czechs, Poles, and Germans (Kadlec et al. 2016: 102–103). Problematic as the census was, it did, however, reveal that national identification of local inhabitants was not yet clearly consolidated at this time. As we show below, the process of consolidation of national identities was slow and the Silesian identity was to resurface again during World War II.

In 1938, Poland annexed a large section of the Czech portion of the former Duchy. The two cities were officially re-united and large-scale renaming took place, purging all public spaces of Czechoslovak names. Czech schools were closed, and Czech officials were replaced by Poles (Zahradnik 1997). The Polish annexation, however, lasted only shortly. In 1939, the entire area was occupied by Nazi Germany which controlled it until 1945. Most of the region became part of the Third Reich, only the southwestern flank remained in the Protectorate of Bohemia and Moravia. Both events left a lasting mark in the memory of local inhabitants because they involved forced population movements, property confiscations, imprisonments, executions, and repressions. After the Polish annexation, thousands of Czech families had to leave the area for the interior of the country. After the German occupation, the Poles became a particularly sought-out target for the Nazi oppressors, suffering severe persecutions in turn. When the war ended, returning to the pre-war and pre-annexation situation proved impossible and the historical grievances associated with this period have remained alive in the memory of local families up to this day.

One issue that had long-lasting effects on the minorities in the Těšín/Cieszyn region of that time was the so-called Volksliste. The categorization system introduced by the German administration separated people into four categories depending on how likely they were to be 'Germanized'.

Each category had its benefits as well as obligations, which affected how one was treated by the state apparatus (Siwek et al. 2001). Especially in the Těšín/Cieszyn region, many members of the Polish minority signed on to this list to avoid further persecution, claiming that they were members of the Silesian nation. Those who were added on the list were conscripted into the German army and, after World War II, were often seen as traitors who had helped the Nazis (Borák 2009).

After the renewal of Czechoslovakia and Poland, tensions between the two countries over the Těšín/Cieszyn region continued for some time, as discussed in Chap. 4, Sect. 4.2. This conflict impacted the region in many ways. It placed suspicion on the Polish minority for not demonstrating sufficient loyalty toward the Czechoslovak state. A relatively strict border regime also made it especially difficult to maintain transborder family and business relations. Communist governments emerged on both sides of the border and the actual border policies changed over the course of the next four decades with periods of relative openness succeeded by total border closures. Although Polish cultural and educational activities were supported by the Polish government and many members of the Polish minority went to universities in Poland, the border regime remained prevailingly restrictive during the entire period. While the agreement of 1947 between Poland and Czechoslovakia prohibited any attempts of the Czechoslovak state to force assimilation of the minority, it was generally assumed that the process of assimilation is irreversible and will eventually lead to the minority effectively dissolving in the majority society (Zahradnik 1997; Friedl 2009).

After the attempted political reforms in 1968, the situation of the Polish minority did not fundamentally change. The early 1980s saw a complicated situation in the Těšín/Cieszyn region as Poland proclaimed martial law and its borders were closed overnight. This had a lasting impact on the Polish minority in the Těšín/Cieszyn region, which became suddenly cut off from its kin state and support networks. While border crossing during Communism was never easy and involved a complicated system of permits, it was possible to navigate. After 1981, the border became first almost hermetically sealed, and only gradually it became possible to cross it again, under a much stricter regime (Josiek 1997). Members of the Polish minority were again treated with suspicion as the

Solidarity movement gained momentum in Poland and Czechoslovakia feared the spread of the reform ideas (Nowak 2009).

During the period of the Communist regime, the social and cultural life of the minority was supported by the state, but the political expressions of minority requests were discouraged and ultimately punished (Zahradnik 1997). The Czechoslovak government continued to support coal mining, iron smelting, and machinery industries in the region. Older factories and mines were extended, and new ones were built. This required a large number of workers who came from the interior and from Slovakia (both Slovaks and Roma). The newcomers often had a very limited understanding of the history of the Polish minority and were thus surprised to find Polish-language schools and signs. Extensive residential neighborhoods were built, including a new model Socialist city (Havířov) in place of a predominantly Polish village. The changing demographic make-up of the region further diluted the share of the Polish minority progressively weakened by growing assimilation. In addition, coal mining had a negative spatial impact on many communities in the Karviná region, often predominantly Polish, which were demolished to give space to coal slag heaps and cleaning ponds or to prevent their spontaneous collapse due to undermining. Especially the northern part of the Těšín/Cieszyn region was dramatically transformed both physically as well as demographically during the Communist regime, leaving the Polish minority on the margins (Rusková 1997; Zahradnik 1997).

The end of the Communist era brought about significant and mostly positive changes both in the international as well as national and regional contexts. The border gradually opened, making it possible for people from both sides of the border to (re-)establish social and business ties. The admission of Poland and Czechia into the European Union (EU) and the subsequent entrance into the Schengen area effectively erased the border entirely, allowing for historical healing and at least a partial practical re-unification of the region. The Polish minority was free to organize without state intervention, and new organizations and activities began to appear (see below). The minority also once again began receiving financial support from Poland, which allowed for a richer cultural, educational, and social life.

Not all developments were entirely positive, however. Bilingual signs on state-owned enterprises (shops, restaurants) disappeared as these were privatized. The question of the return of properties confiscated by the Nazis and subsequently nationalized by the Communist regime was also not resolved (Borák 1998). Furthermore, the population censuses of 1991, 2001, and 2011 confirmed the long-term trend of assimilation of the Polish minority into the mainstream society. Paradoxically, the flowering of Polish minority life promoted by new opportunities, freedoms, and national and international support was undermined by unfavorable demographic changes.

The political representation of minorities at the national level was never strong, and after the dissolution of Czechoslovakia minorities struggled to be heard. To be sure, minorities had their representatives in the first post-Communist parliament. A political party called *Coexistentia* was formed by representatives of the Polish, Hungarian, and Ruthenian minorities. This party, however, relied on the size of the Hungarian minority in Czechoslovakia in order to promote a minority-focused agenda. Once the country split into Czechia and Slovakia in 1993, the Hungarian minority found itself in Slovakia (Kopeček 2003). While Coexistentia has been more successful in regional and local elections than on the national scene, a number of politicians coming from the Polish minority made their way to national politics through mainstream political parties. Faced with a steady decrease in population numbers, the Polish minority has tried to use the political changes best to its benefit. It was active in setting up several organizations, promoting cross-border cooperation, and putting pressure on regional and national governments to support minority education, cultural activities, and other important issues. While at present the Polish minority is very well organized and highly visible and influential in the region, the changing demographics may eventually set back many of the gains made in the last three decades. The future of the Polish minority in the Těšín/Cieszyn region thus remains open.

5.2.2 The Minority Area

It is rarely easy to delimit a minority area without risking a heated political debate. This is doubly true for our case. Since the late 1800s, the delimitation of the Polish minority area has been a controversial issue. It played a central role in the conflict between Czechoslovakia and Poland over the Těšín/Cieszyn region, and as some examples of recent events given in Chap. 6 indicate, it is yet to be resolved. It is particularly challenging to attempt to delimit the area during the Austrian period since we would be forced to anachronistically project current identities into the past and thus misrepresent the results of the Austrian censuses on which such delimitations are usually based. In spite of the aforementioned, however, we could argue that the Austrian and Czechoslovak census data indicate a stable core Polish area flanked by transitional areas to the West and Northwest.

As mentioned in Chap. 4, the Austrian census used colloquial language (*Umgangssprache*) as the criterion. Since the Těšín/Cieszyn dialect was classified as a dialect of the Polish language, all dialect speakers were recorded as Polish speakers. In reality, however, we have little knowledge about their national identity, and we can only speculate about it from later population censuses which used different criteria. Furthermore, even today the majority of the local population speaks the dialect but only a minority identifies as Polish. There is no reason to believe that the situation was different during the Austrian period.

One of the first indications of the ethnic composition of the Těšín/Cieszyn region is Czoernig's ethnographic map of the Austrian monarchy published in 1855 (Czoernig 1855) (see Fig. 5.25). The vast majority of the region is identified with "Wasser Polaken" with a large strip of Czechs along the west and pockets of Germans in Teschen/Těšín/Cieszyn and around Bielitz/Bílsko/Bielsko. What data were used in the preparation of the map is not clear; however, the boundary between Czechs and "Wasser Polaken" resembles closely the one between Czechs and Poles which emerged from the first Austrian census in 1880 and which remained

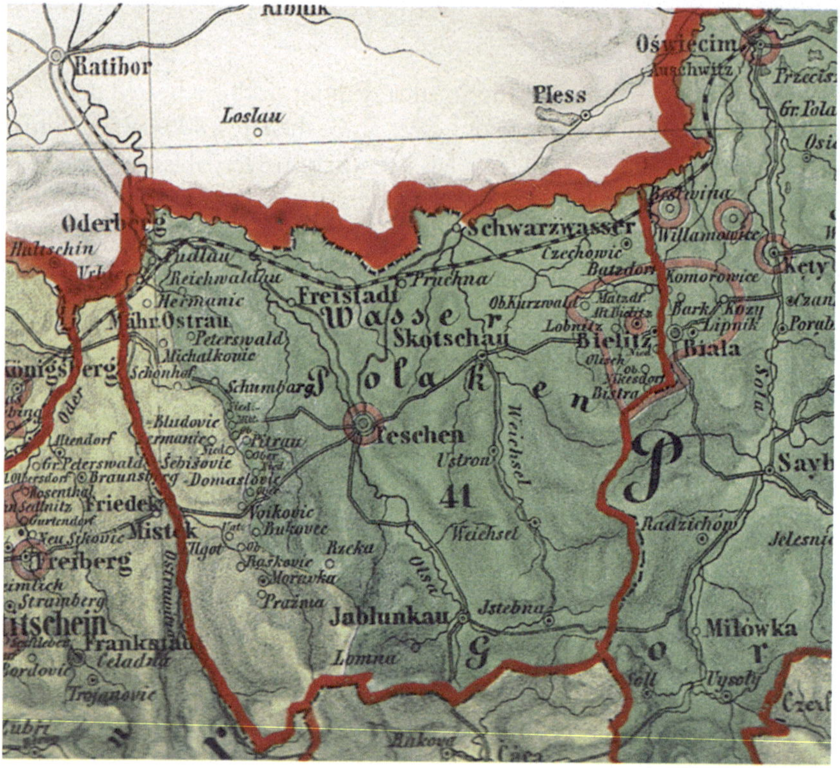

Fig. 5.25 The Duchy of Teschen on the ethnographic map of the Austrian Empire by Czoernig (1855). (Source: Private archive of Peter Jordan)

fairly stable in the following censuses. The term "Wasser Polak" itself refers to inhabitants of Silesia speaking mixed Czech-German-Polish dialects, mostly of Slavic ancestry. They differentiated them from Czechs, Poles, and Germans. In other words, it indicated a unique identity although we know little about the salience of this identity in everyday life in Silesia. With time, however, it gained a pejorative meaning for someone of unclear national origin, disappearing from general use by the mid-1900s (Hannan 1996; Holub 2011).

Figure 5.26 shows the share of people identified as Polish-speakers in the Austrian census of 1900 for that section of the Těšín/Cieszyn region which later became part of Czechoslovakia. As can be seen, the percentage of

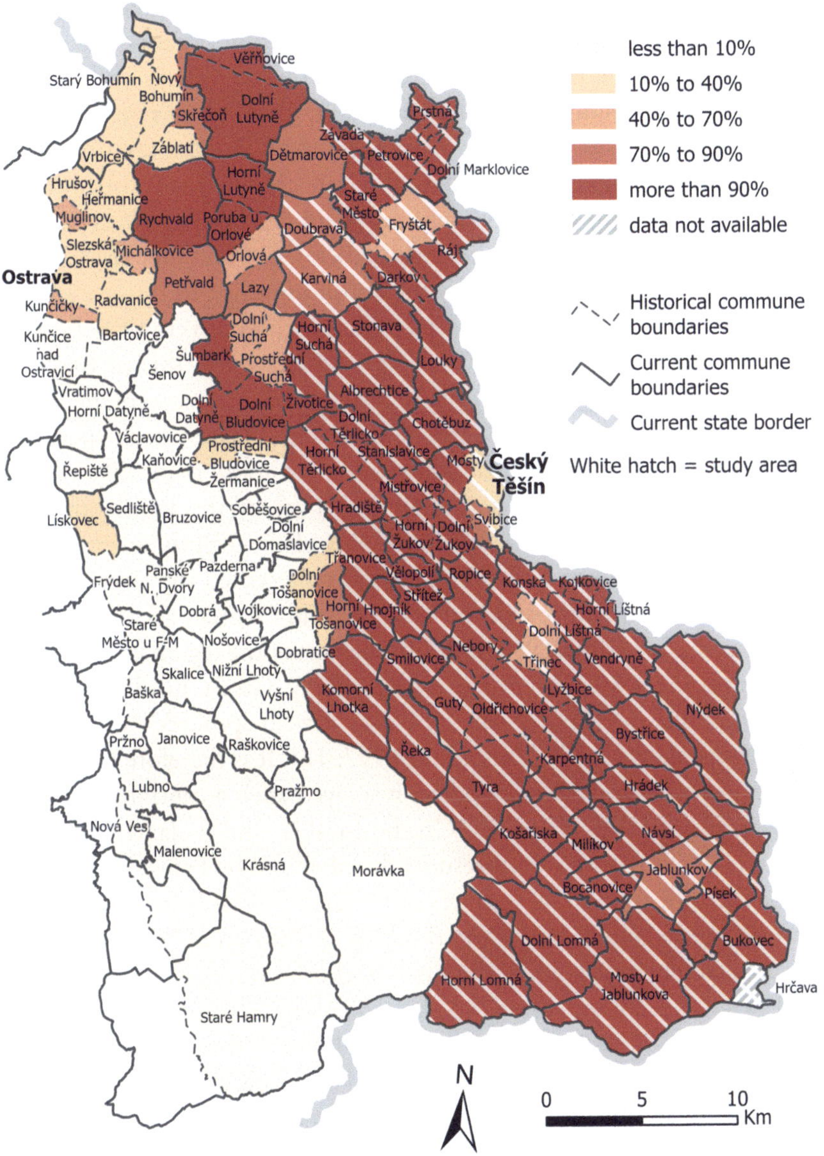

Fig. 5.26 Share of Polish-speakers according to the 1900 census. (Source: K.k. statistische Zentralcommission 1906)

Polish-speakers dropped rapidly in the western part of the region but dominated heavily in the eastern section, often to the point of monolingualism. It is this part of the region that we chose as our study area and that we analyze in more detail by the following maps. Our area of research, however, is slightly narrower in the Northwest than the one dominated by speakers of the Těšín/Cieszyn dialect who were classified as Polish-speakers by the 1900 census. This is because the share of Polish-speakers in areas outside of our research area decreased well below 5% and they ceased to be considered Polish-speakers even by the Polish minority. In this, we mostly follow Siwek (1996) and his research on the Czech-Polish ethnic boundary.

The Austrian census of 1910 brought no change in this picture (see Fig. 5.27). In total, 83% of people in our area of interest were identified as Polish-speaking, 11% as German-speaking, 4% as Czech-speaking, and 2% as other. In fact, most communes had more than 90% of presumed Polish-speakers. Even though the share of the Polish-speaking population in the future Czechoslovak part of the Těšín/Cieszyn region was significantly lower—48.5%, they still represented the largest group in the region with Czech-speakers (39.5%) and German-speakers (12%) well behind (Zahradnik 1992: 249). In the part of the region later claimed by Poland (i.e., without the overwhelmingly Czech juridical district Frýdek), Polish-speakers formed 69.2%, Czech-speakers 18.3%, and German-speakers 12.5% (Zahradnik 1997: 12). It is therefore easy to see why this census, and maps based on it, played such a central role in the postwar negotiations between Czechoslovakia and Poland in which Poland claimed the territory presumably inhabited by Poles.

The identification of all speakers of the dialect with Polish nationality, however, turned out to be wishful thinking on the part of Polish nationalists, as is clearly indicated by the first Czechoslovak census conducted in 1921. Figure 5.28 and Table 5.3 show a sharp decrease in the share of Polish speakers/Poles in our research area although there were no significant population movements. The share of Polish speakers/Poles decreased from 83% in 1910 to 46% in 1921 in relative numbers and from 89,000 to 54,000 in absolute numbers. By contrast, the number of Czech-speakers/Czechs increased from 4000 to 34,000, and their share grew from 4% to 30%. In the region as a whole, the proportion of Polish speakers/Poles decreased from 48.5% to 25.4%, while the proportion of Czech-speakers/Czechs increased from 39.5% to 65.1% (Zahradnik

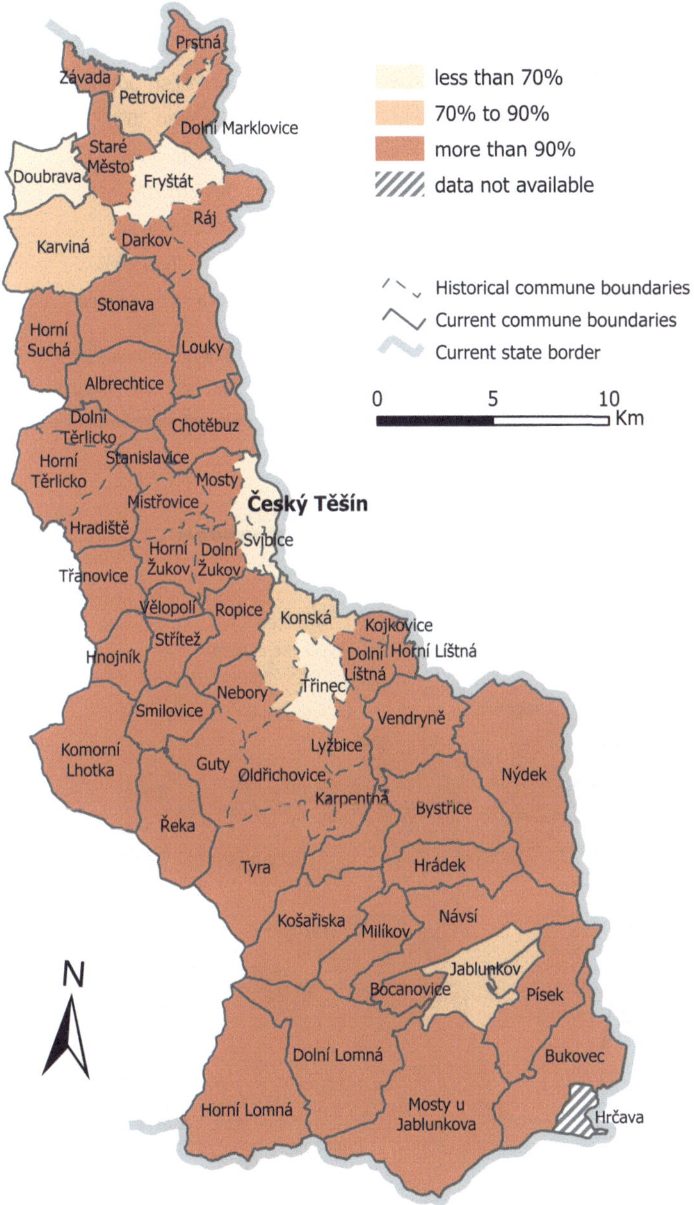

Fig. 5.27 Share of Polish-speakers by commune in 1910. (Source: Bevölkerung gegliedert nach den Nationalitäten laut Volkszählungen vom 31. Dezember 1900, 1910 und 15. Februar 1921. NAD 349, č. pom. 557, Zemský archiv Opava)

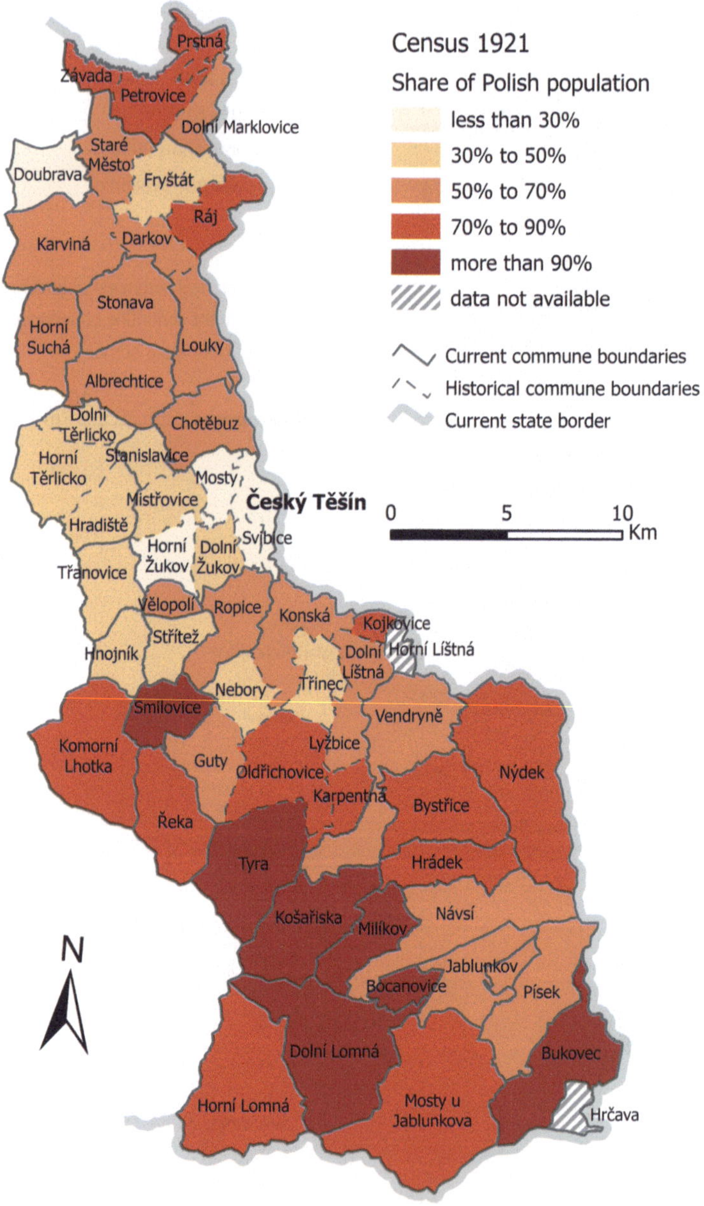

Fig. 5.28 Share of Polish population by commune in 1921. (Source: Bevölkerung gegliedert nach den Nationalitäten laut Volkszählungen vom 31. Dezember 1900, 1910 und 15. Februar 1921. NAD 349, č. pom. 557, Zemský archiv Opava)

Table 5.3 Share (in %) of colloquial languages (1910) and nationalities (1921–2011) in the research area 1910–2011

	1910	1921	1939	1948	1980	2011
Czech	3.8	29.9	9. 7	66. 3	69.8	60.7
Polish	83.1	46.4	24. 0	33. 0	21.1	13.1
German	10.8	8.5	14. 1		0.1	0.1
Jewish		1.1	0. 7			
Silesian			50. 2			1.3
Moravian						1.3
Slovak					7.8	3.3
Other	2.3	14.1	1. 4	0.7	1.1	20.3
Total	100.0	100.0	100.0	100.0	100.0	100.0

Sources: 1910 and 1921—Bevölkerung gegliedert nach den Nationalitäten laut Volkszählungen vom 31. Dezember 1900, 1910 und 15. Februar 1921. NAD 349, č. pom. 557, Zemský archiv Opava; 1939—Myška (1964), Zahradnik (1992, 2001), CZSO (2020)

1992). The failure of the Czechoslovak census to count Poles from former Galicia without Czechoslovak citizenship (Zahradnik 2014) can explain the sharp drop in the number of Poles compared to the Austrian 1910 census only partially. In other words, when people had the opportunity to declare their nationality freely, they often opted for a nationality different from the one associated with the colloquial language in the Austrian census. The census results were sharply criticized by representatives of the Polish minority (Kadlec et al. 2016: 95–114). As we shall still see, this was not the last time when a change in the methodology of the census led to a change in the ethnic composition of the region.

Even more dramatic changes came with the German census of 1939 when Silesian nationality was added as an option. Figures 5.29 and 5.30 show yet another drop in the share of Poles and an associated high number of Silesians. Which is more, we also observe a similar drop in the number of Czechs which cannot be explained simply by their forced expulsions after the Polish annexation in 1938. In other words, some people previously declaring Czech nationality opted for the Silesian one when given the choice. Finally, we also observe a significant increase in the number of Germans which cannot be explained solely by the presence of the German army. To sum up, the census results indicate how political circumstances and the methodology of the census influence declared identities and how problematic these metrics are in any appraisal

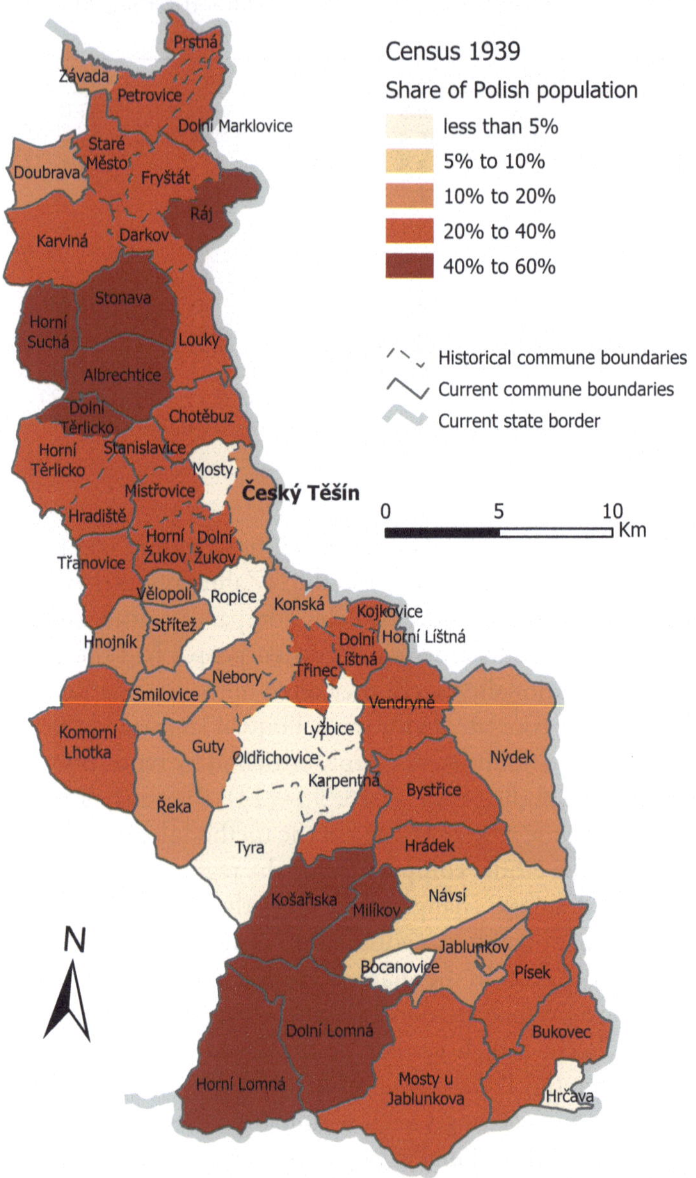

Fig. 5.29 Share of Polish population by commune in 1939. (Source: Myška 1964)

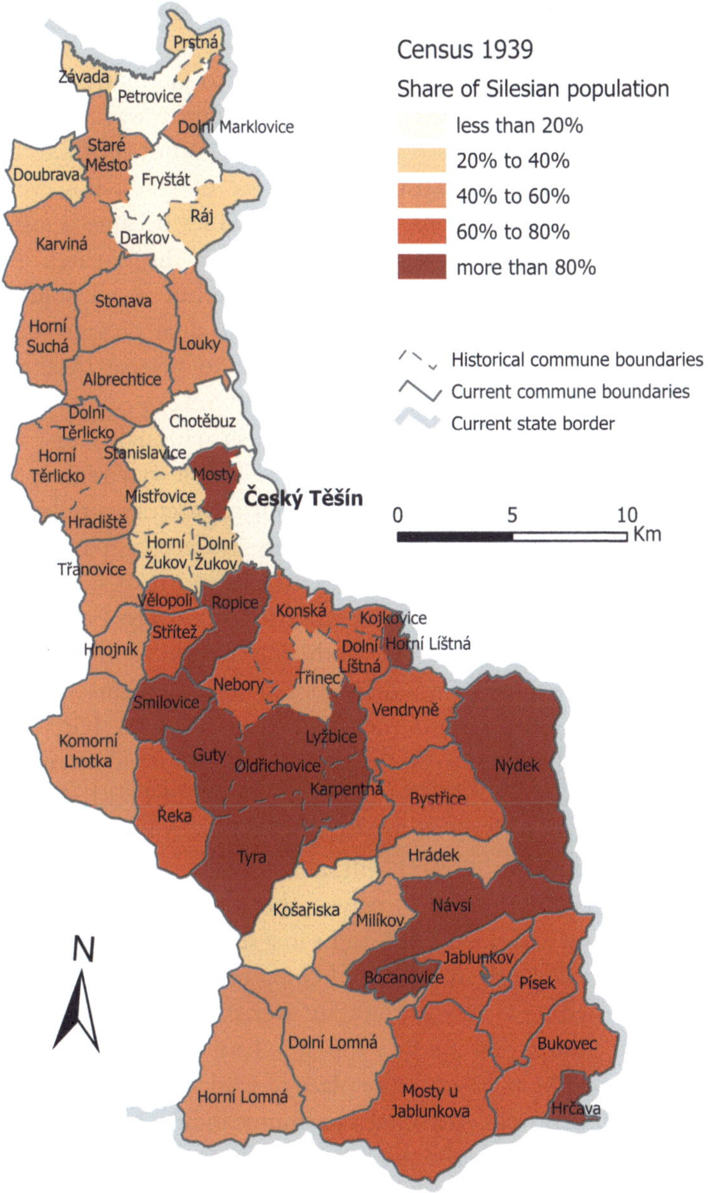

Fig. 5.30 Share of people declaring Silesian nationality by commune in 1939. (Source: Myška 1964)

of the ethnic composition of a given territory. However, for a lack of better sources, we are forced to rely on them.

After World War II, Germans were expelled to Germany and practically disappeared from statistics. The Silesian identity was no longer an option in the census, and we see an increase in the number of Czechs and Poles relative to the 1939 census. In the case of Poles, however, this increase did not match the pre-war level and Poles became a minority, representing only one-third of the population in the research area in 1948. This overall decrease was not only relative but also absolute, and the subsequent censuses only confirmed it. The dual pressure of labor immigration and assimilation was impossible to reverse (Małysz and Kaszper 2009; Zahradnik 2001) (Figs. 5.31 and 5.32).

In 2011, only 39,000 people declared Polish nationality in Czechia. The vast majority of them did so in the Těšín/Cieszyn region (Siwek 2017) and almost 23,000 (59% of all) in our research area only. It thus remains the stronghold of the Polish minority in Czechia, particularly in the southern half of the research area to the south of Český Těšín/Czeski Cieszyn. The size of the Polish minority has always been a subject of discussion and scrutiny from both the Czech (and before that Czechoslovak) state and the minority representatives themselves. As we look at the census numbers from the past century, we see a sharp decrease in numbers, and what used to be the majority turns into a minority. This trend is probably irreversible despite strong efforts of Polish representatives and organizations to promote Polish language, culture, and identity in the region. The census of 2021 will be the test.

Within the region, Hrčava/Hrczawa, the easternmost commune of Czechia, stands out as a pocket of Czechs in a sea of Poles. Originally part of Jaworzynka (in Poland), it chose to belong to Czechoslovakia in 1920 and has manifested its loyalty to the state ever since. The reasons for this decision are not entirely clear but they appear to have been mainly practical—the accessibility of the railroad to the place of work in the Čadca valley in Slovakia (Tadeusz Siwek—personal communication). It is yet another illustration of the fluid and arbitrary nature of national identities, especially in border regions with shifting national belonging (Figs. 5.33 and 5.34).

The overall change in the share of the Polish-speaking/Polish population in our research area is summarized in Table 5.4. Many historical communes lost their autonomy and were joined with larger communes

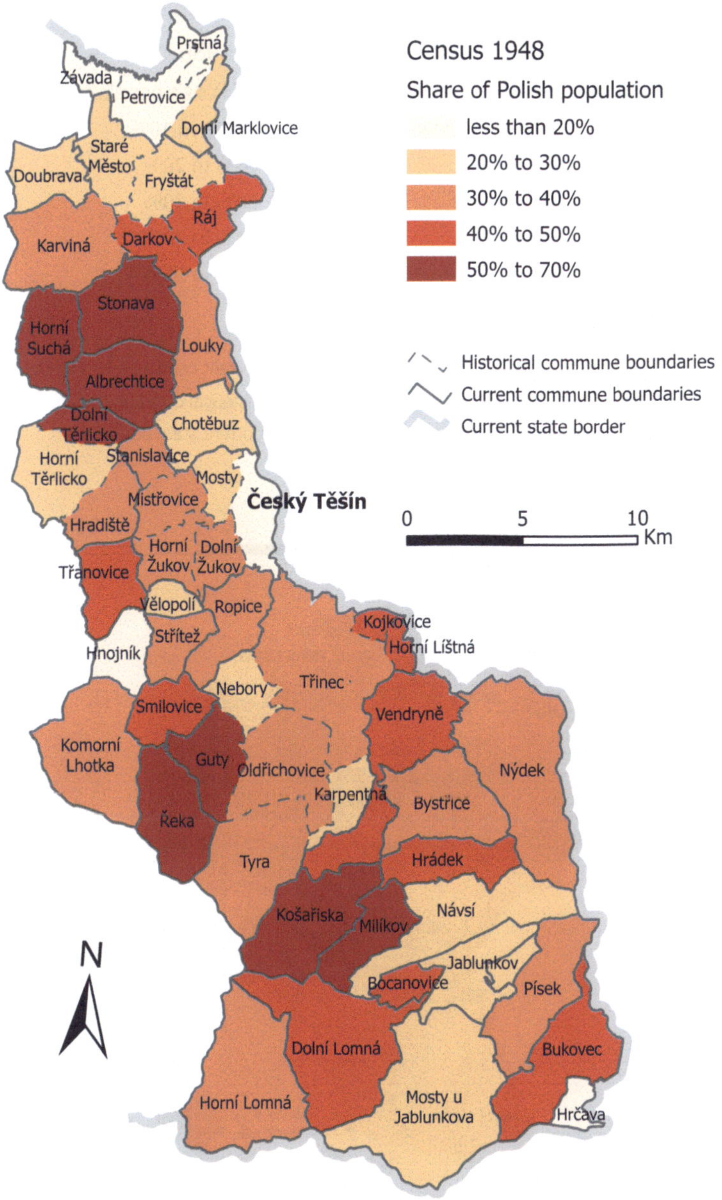

Fig. 5.31 Share of Polish population by commune in 1948. (Source: Zahradnik 1992)

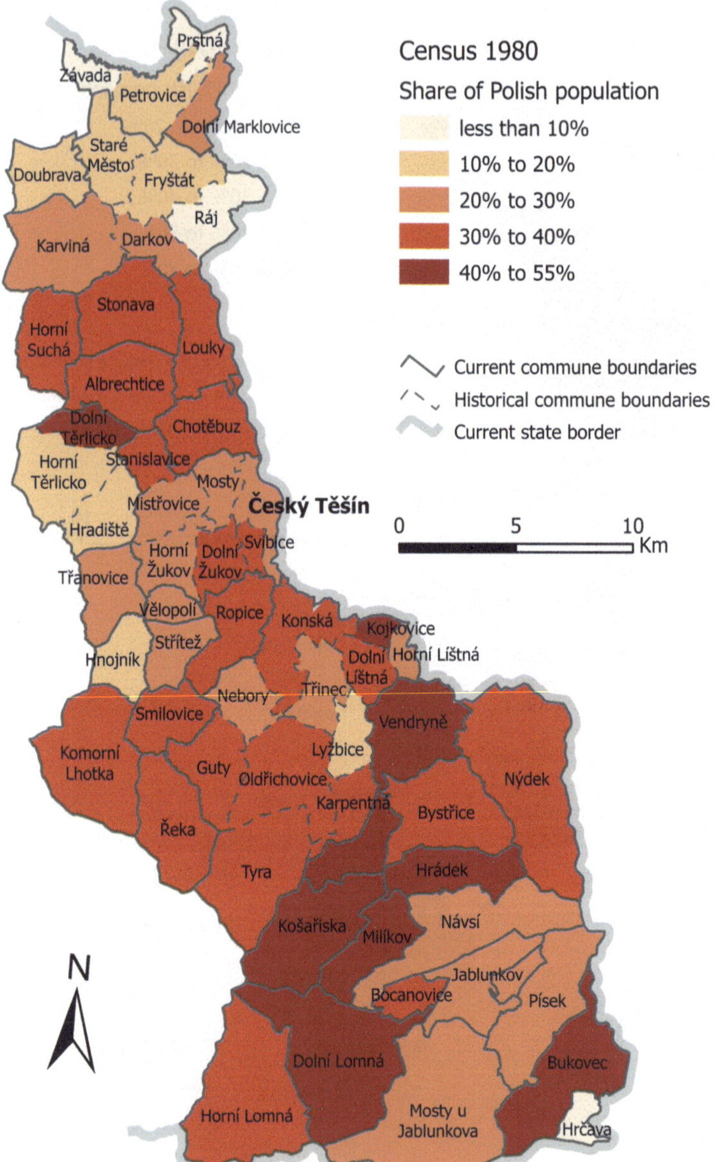

Fig. 5.32 Share of Polish population by commune in 1980. (Source: CZSO 2020)

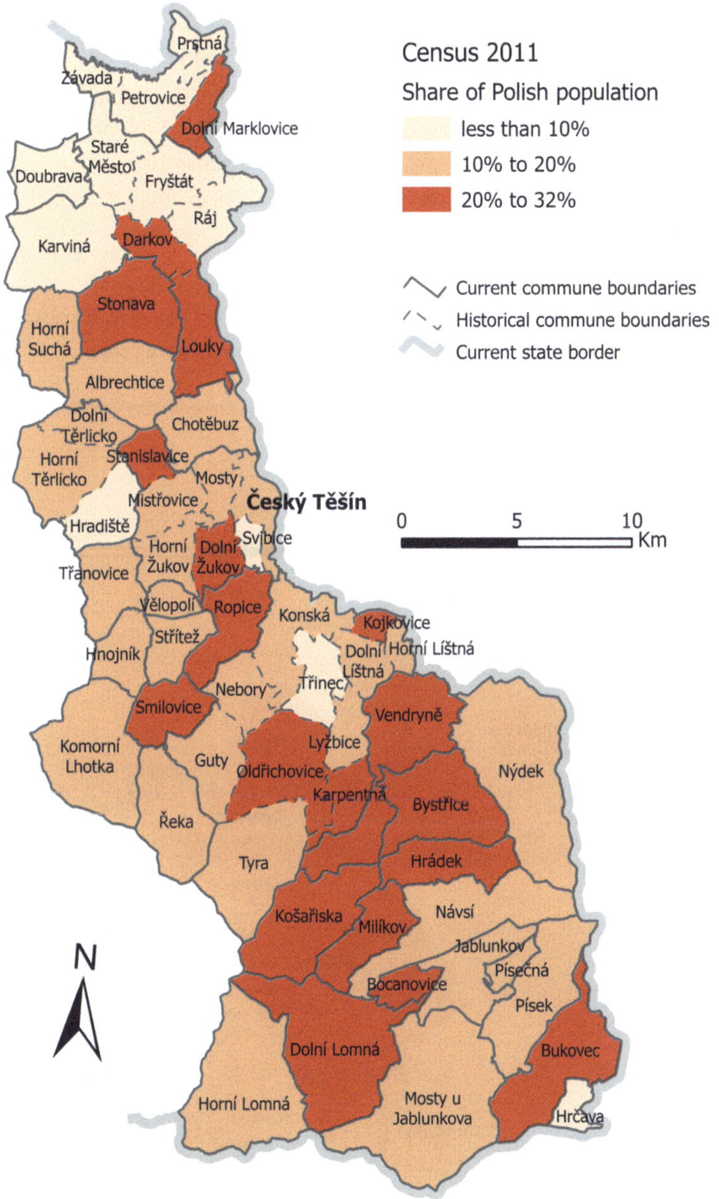

Fig. 5.33 Share of Polish population by historical communes in 2011. (Source: CZSO 2020)

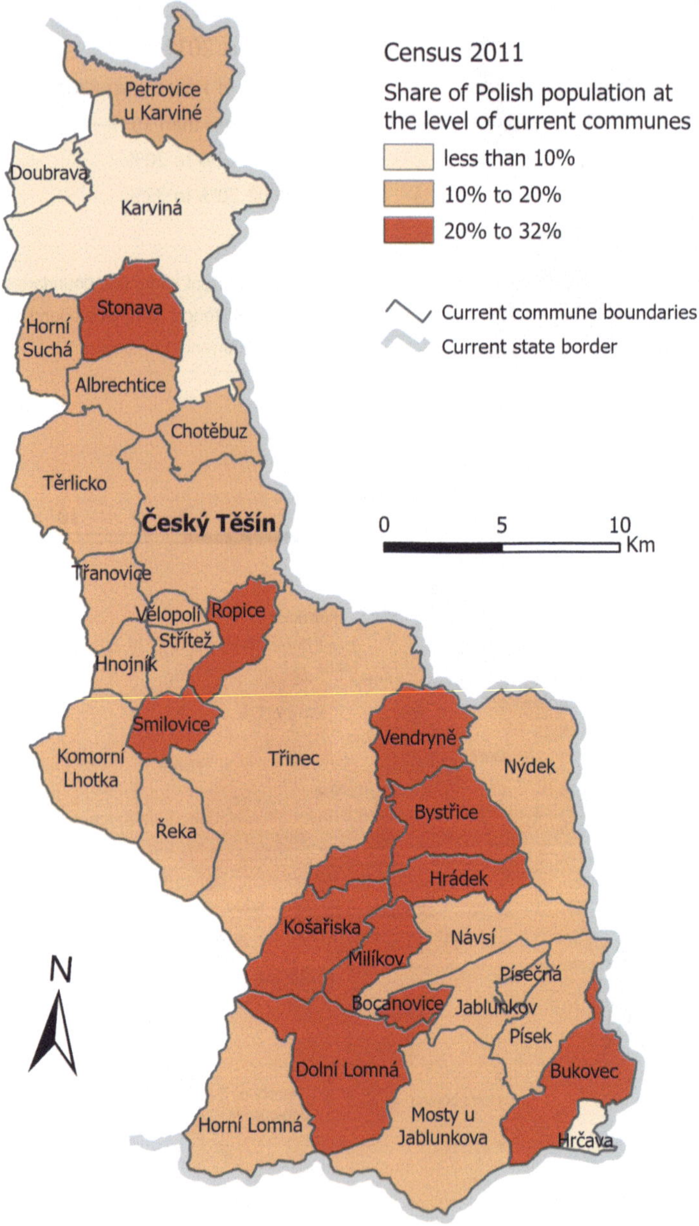

Fig. 5.34 Share of Polish population by current communes in 2011. (Source: CZSO 2020)

Table 5.4 Share (in %) of Polish-speakers (1910) and Polish population (1921–2011) in the research area by censuses 1910–2011

Historical commune	Current commune	1910	1921	1939	1948	1980	2011
Albrechtice	Albrechtice	99.5	62.4	45.5	61.3	34.2	18.6
Bocanovice	Bocanovice	99.1	96.8	0.3	43.5	39.9	28.1
Bukovec	Bukovec	99.8	94.9	30.5	41.4	41.0	25.9
Bystřice	Bystřice	98.2	75.4	23.3	38.7	38.2	24.9
Český Těšín	Český Těšín	33.7	7.5	15.1	16.2	20.1	15.6
Darkov	Karviná	96.5	61.9	30.6	41.8	27.9	23.5
Dolní Líštná	Třinec	97.3	69.6	28.7	49.6	31.2	13.6
Dolní Lomná	Dolní Lomná	95.1	94.3	50.6	24.3	42.8	23.2
Dolní Marklovice	Petrovice u Karviné	99.6	66.0	35.9	52.8	23.3	24.0
Dolní Těrlicko	Těrlicko	99.0	34.5	45.1	36.1	41.8	13.0
Dolní Žukov	Český Těšín	98.0	41.5	23.5	20.4	39.3	22.1
Doubrava	Doubrava	62.8	16.3	19.8	27.5	12.7	3.8
Fryštát	Karviná	59.5	32.2	29.0	53.7	11.0	5.5
Guty	Třinec	99.0	53.7	11.9	19.0	37.3	17.4
Hnojník	Hnojník	90.5	30.3	11.7	47.9	13.0	11.1
Horní Líštná	Třinec	99.1		14.8	35.5	27.7	13.9
Horní Lomná	Horní Lomná	97.9	79.7	40.4	52.6	32.7	19.9
Horní Suchá	Horní Suchá	96.0	67.7	43.2	27.5	37.5	17.6
Horní Těrlicko	Těrlicko	98.6	43.1	34.5	31.2	19.0	10.0
Horní Žukov	Český Těšín	98.8	21.8	22.8	50.0	25.7	15.2
Hrádek	Hrádek	98.7	78.7	39.5	33.4	50.4	32.6
Hradiště	Těrlicko	97.2	48.6	21.0	0.0	17.2	8.4
Hrčava[a]	Hrčava			0.4	27.2	1.6	3.7
Chotěbuz	Chotěbuz	95.2	56.2	31.6	24.2	31.0	16.1
Jablunkov	Jablunkov	84.4	63.2	12.1	27.3	26.5	16.3
Karpentná	Třinec	100.0	78.8	3.7	32.7	35.0	20.8
Karviná	Karviná	82.7	56.4	25.6	49.3	26.7	9.1
Kojkovice	Třinec	100.0	76.7	27.7	39.7	47.2	24.3
Komorní Lhotka	Komorní Lhotka	94.5	76.4	22.4	71.1	31.0	15.4
Konská	Třinec	85.0	59.8	16.5	35.0	34.1	19.8
Košařiska	Košařiska	98.9	97.5	58.5	63.7	55.4	28.2
Louky	Karviná	97.9	67.9	30.3	36.8	33.5	20.4
Lyžbice	Třinec	99.3	62.8	2.5	21.9	18.0	11.9
Milíkov	Milíkov	100.0	91.9	43.4	27.7	53.7	31.0
Mistřovice	Český Těšín	98.0	35.6	36.8	24.9	27.1	17.0
Mosty	Český Těšín	95.1	28.8	3.4	24.7	23.0	16.7
Mosty u Jablunkova	Mosty u Jablunkova	98.5	85.5	28.1	34.6	23.3	15.0
Návsí	Návsí	96.6	54.6	7.7	38.5	29.1	19.8

(*continued*)

Table 5.4 (continued)

Historical commune	Current commune	1910	1921	1939	1948	1980	2011
Nebory	Třinec	99.8	46.8	15.0	18.0	26.5	15.7
Nýdek	Nýdek	99.0	88.9	14.4	38.6	33.1	19.8
Oldřichovice	Třinec	96.1	79.8	4.7	17.0	33.5	20.3
Petrovice	Petrovice u Karviné	89.3	70.3	33.0	48.4	16.0	7.5
Písečná[b]	Písečná						19.2
Písek	Písek	97.5	65.1	26.2	38.9	29.8	17.0
Prstná	Petrovice u Karviné	97.3	71.5	23.1	50.2	8.4	7.1
Ráj	Karviná	97.3	78.3	43.2	46.6	9.8	6.0
Ropice	Ropice	95.5	67.9	3.7	32.4	37.5	24.8
Řeka	Řeka	97.7	82.2	19.8	21.1	30.0	14.2
Smilovice	Smilovice	96.0	92.7	13.1	52.8	37.0	22.4
Stanislavice	Český Těšín	99.0	31.4	39.9	33.6	33.4	20.4
Staré Město	Karviná	93.5	51.5	23.3	40.5	17.6	7.0
Stonava	Stonava	99.0	57.0	46.6	32.2	39.4	20.2
Střítež	Střítež	96.4	33.8	16.1	35.5	24.0	13.3
Svibice	Český Těšín	67.3	24.0	24.9	29.7	31.8	8.1
Třanovice	Třanovice	97.9	39.2	22.2	48.7	29.4	17.1
Třinec	Třinec	69.0	44.3	4.7	16.7	20.5	9.9
Tyra	Třinec	94.1	98.0	13.6		39.0	15.3
Vělopolí	Vělopolí	100.0	58.8	24.0		27.5	14.6
Vendryně	Vendryně	97.3	65.3	20.0		44.0	27.0
Závada	Petrovice u Karviné	95.2	80.1			9.4	4.9

Sources: 1910 and 1921—Bevölkerung gegliedert nach den Nationalitäten laut Volkszählungen vom 31. Dezember 1900, 1910 und 15. Februar 1921. NAD 349, č. pom. 557, Zemský archiv Opava; 1939—Myška (1964), Zahradnik (1992, 2001), CZSO (2020)

[a]Hrčava/Hrczawa belonged to the commune of Jaworzynka until 1924, even though it was physically in Czechoslovakia after the division of the Těšín/Cieszyn region between Poland and Czechoslovakia in 1920

[b]Písečná/Pioseczna separated from Jablunkov/Jabłonków in 2000

(mostly cities). We therefore list the names of the historical as well as current communes. The census results in 1980 and 2011 show the shares of Polish population in the historical communes, in order to preserve a maximum amount of statistical detail. However, it should be emphasized that for legal purposes, the 10% threshold is evaluated only at the current communal level. This means that even if the threshold is met at the level of a formerly independent commune (e.g., Darkov/Darków), when it is not met at the level of the current commune of which the historical commune is now a part (in the case of Darkov/Darków it is Karviná/Karwina), the minority is not entitled to bilingual signs and other rights anywhere in the commune. In other instances, however, the formerly independent communes with a large share of Polish population push the communal center with few Poles over the threshold line and make it possible for the Polish minority to demand bilingual signs in the entire commune. This is the case, for example, in Třinec, where the urban core did not meet the threshold but where all of the formerly independent, rural communes now joined to Třinec (Oldřichovice/Oldrzychowice, Karpentná/Karpętna, Tyra/Tyra, etc.) well exceeded it. This may be one of the principal reasons why the introduction of bilingual signs in Třinec/Trzyniec accompanied such a protracted conflict as the inhabitants of the urban core felt pressured to accept something that did not really apply to them. For a list of current communes in which the Polish minority exceeds the 10% threshold, see Table 5.5.

5.2.2.1 Nature as a Place of Human Activities

As we mentioned above, the Těšín/Cieszyn region (see Fig. 5.35) is not a natural region delimited by clear natural boundaries. The river Ostravice formed the southwestern boundary of the Duchy of Teschen and some sections of the remaining boundary followed other streams and rivers but there was no overall natural pattern. The region gradually slopes down from the Beskids [Beskydy, Beskidy] in the Southeast to the lowland areas in the Northwest where it is open to the Odra river plain. The Polish minority area is only a fraction of the former Duchy and its present extent is, above all else, determined by socio-economic factors. The natural character of the region, however, influenced to a significant degree its settlement structure, colonization history as well as economic exploitation. Let us, therefore, briefly describe it, especially as it is reflected in local and regional identities.

The Czech part of the Těšín/Cieszyn region inhabited by the Polish minority is referred to as *Zaolzie* by the Poles. This name literally means "area beyond

Table 5.5 Communes with a share of Poles above 10% in 2011 in descending order

Commune	Share of Poles (%)	Polish population	Total population
Hrádek	31.3	554	1768
Milíkov	30.4	387	1275
Bocanovice	27.9	119	426
Košařiska	27.1	96	354
Vendryně	26.1	1078	4137
Bukovec	25.4	340	1339
Ropice	24.2	359	1486
Bystřice	24.1	1254	5197
Dolní Lomná	22.4	189	843
Smilovice	21.7	155	714
Stonava	19.4	335	1728
Horní Lomná	19.3	68	352
Návsí	19.1	714	3743
Nýdek	19.0	378	1990
Písečná	18.6	167	899
Albrechtice	17.6	669	3795
Horní Suchá	17.1	741	4333
Třanovice	16.7	177	1057
Písek	16.4	289	1757
Chotěbuz	15.9	178	1122
Jablunkov	15.7	902	5732
Komorní Lhotka	14.9	186	1250
Mosty u Jablunkova	14.6	559	3830
Řeka	13.9	69	495
Český Těšín	13.7	3333	24,394
Třinec	13.4	4868	36,263
Vělopolí	13.2	37	281
Střítež	13.0	128	981
Petrovice u Karviné	12.8	695	5446
Hnojník	10.9	161	1479

Source: CZSO (2020)

the river Olza" as viewed from Poland. This is not geographically correct, though, since Olza begins in Poland, flows through the Czech interior, and becomes a border river only in its last section. Be it as it may, the name carries a strong emotional charge, positive for the Poles, and mostly negative for Czechs, reminding them of Polish irredentist claims. All, however, feel a connection to the river which weaves the region together. The historic and dialectal name of the river is *Olza*, etymologically associated with *Volga* and other Slavic names referring to streams with plentiful water. In Czech, however, it is

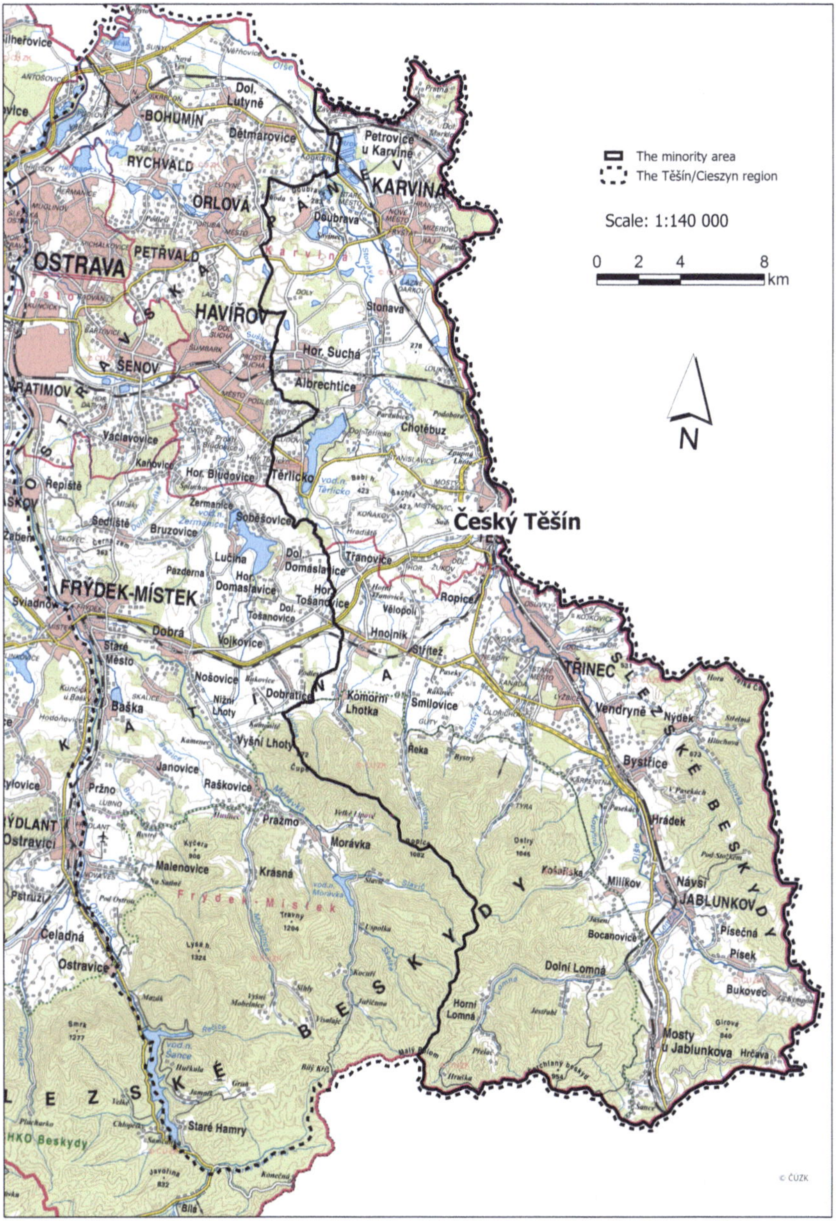

Fig. 5.35 Location of the minority area in the Těšín/Cieszyn region. (Thematic content: Přemysl Mácha; base map: Základní mapa ČR, ČÚZK 2019)

called *Olše* (after the alder tree *olše*), a name that was coined in the early 1900s to establish a Czech claim to the region because *Olza* was seen as Polish.[21] Olza/Olše flows through the entire region from East to West and connects as well as divides. Its symbolic value is expressed also in the anthem of the Polish minority called *Płyniesz Olzo po dolinie* ("Olza, you flow down the valley").

Olza begins in the Silesian Beskids [Slezské Beskydy, Beskid Śląski] which together with the Moravian-Silesian Beskids [Moravskoslezské Beskydy] form the southeastern part of the minority region. The highest peak is Ropice (1084 meters above the sea level) and there are many other peaks exceeding 900 meters. Some of them are associated with witches (Čantoryje), others with lost treasures (Godula). Traditionally, the mountains were exploited for pastoralism and later for logging. Towns and villages located in their proximity take a great pride in the pastoral tradition and see it as the defining feature of the Gorol ('mountain dweller') identity. Gorole ('Gorols') see themselves as culturally and linguistically unique, materially self-sufficient, and sharing many of their traits with other groups living in the Carpathians, while looking down upon Dolani ('valley dwellers'), that is, people living in the lower lying areas of the region. These, nevertheless, repay the Gorols with the same, seeing them in turn as backward, uneducated, and poor. In spite of this, the annual festival Gorol Fair [Gorolske Święto] in Jablunkov/Jabłonków is, by far, the most important cultural event attracting crowds from the entire region and beyond. Reenactments of pastoral traditions, folk songs, and Gorol dances are featured in the event, accompanied by speeches and performances in Polish and the dialect, rooting thus the Polish minority firmly in the cultural life of the region while producing a complex system of overlapping territorial, cultural, and national identities (Grygar 2006). Overall, mountains figure prominently in people's link to the region. In our interviews and surveys, mountain names were the most frequently mentioned types of names for places which held a high emotional value for our respondents (see Chap. 6) even though for most people, they are only places of recreational activities today.

[21] Ironically, in the same period, Polish nationalists in Těšín/Cieszyn attempted to rename the river to *Olsza* as well, because to them *Olza* sounded too German. They did not succeed and the river is still called *Olza* in Poland.

Compared to the mountainous part of the region boasting dramatic views, the lowland between Třinec/Trzyniec and Bohumín/Bogumin presents a very different type of landscape. It is formed by gentle hillocks and open valleys lying between 200 and 350 meters above sea level (see Fig. 5.36). The lowest point is at the confluence of Olza/Olše and Odra 195 meters above sea level. The lowland receives significantly less precipitation and has higher average temperatures than the mountains, even though these lie to the South. Unlike the mountains where traditional agriculture and pastoralism still survive, the lowland has been entirely transformed by intensive urbanization and industrialization resulting in an extremely high population and settlement density. As was mentioned above, coal deposits attracted investors from the early 1800s. The ironworks in Třinec/Trzyniec founded by the Habsburgs in 1836 still remains one of the largest employers in the region, alongside other large coal mining and iron-smelting operations (see Figs. 5.37 and 5.38). Although the region is now trying to diversify its economic base and move away from heavy

Fig. 5.36 Where mountains meet the lowland in the vicinity of Třinec/Trzyniec. (Photo by Přemysl Mácha 2020)

Fig. 5.37 *Werk*, the steel mill in Třinec/Trzyniec founded by the Habsburgs. (Photo by Přemysl Mácha 2020)

industry (not least because of the associated air pollution), the coal-mining and iron-smelting identity remains very strong and is associated nationally with the entire region while the Gorol identity remains only regionally known and tied to the southern part of the area. The coal miner/steel-worker identity also significantly influences electoral preferences with the traditional left-wing parties and the new populist movements gaining substantially greater support than in other parts of the country.

5.2.2.2 Settlement Structure, Central-Place, and Administrative System

The pivot of the minority area today remains in Český Těšín/Czeski Cieszyn (see Fig. 5.39). It only has approximately 24,000 inhabitants, but because of its central location and historical significance it continues to play a key role in minority life. All principal minority organizations

Fig. 5.38 One of the last functioning coalmines in Darkov/Darków, in place of demolished settlements. (Photo by Přemysl Mácha 2020)

and newspapers are based there, as is the only remaining Polish high school. The city's symbolic importance is also underlined by its location on the Czech-Polish border with the river Olše/Olza dividing Český Těšín/Czeski Cieszyn and (Polish) Cieszyn. Český Těšín/Czeski Czieszyn is thus also seen as a metaphorical and physical bridge to Poland. Thanks to the Schengen area and the absence of the border, many people begin to see Český Těšín and Cieszyn as one city again and use them for culture, recreation, social gatherings, employment, and study regardless of citizenship.

Nearby Třinec with a population of 35,000 plays a crucial economic role due to its steel mill and as a central place for the mountainous areas to the south. These areas consist of rural communes with the exception of Jablunkov/Jabłonków, a small town with a population of 5500 (see Fig. 5.40). Jablunkov/Jabłonków is often considered the capital of

Fig. 5.39 The town hall on the main square in Český Těšín/Czeski Cieszyn. (Photo by Přemysl Mácha 2020)

Goralia. This part of the region belongs to the Frýdek-Místek District. Even though districts [okresy] as administrative units were abolished in 2002, they still host many of the district institutions (court, archive, police headquarters, district hospital, etc.) and serve for electoral and statistical purposes.

The northern part of the minority area including Český Těšín/Czeski Cieszyn belongs to the Karviná District with its center in Karviná, a city with a population of 53,000. Its surroundings have been extensively altered by coal mining, and many settlements including the original town of Karviná were demolished in the 1950s and 1960s. New Karviná took the administrative role of the previous district center Fryštát which continued to exist as one of Karviná's parts. At its height in the late 1980s, Karviná boasted 90,000 inhabitants, its current population, however, illustrates the larger trend characterizing the entire region—residential

Fig. 5.40 Jablunkov/Jabłonków, the 'capital' of Goralia. (Photo by Přemysl Mácha 2020)

suburbanization of the middle-aged and emigration of the young to Prague [Praha], Brno, Olomouc, and other Czech cities due to high unemployment associated with the gradual dismantling of the coal-mining and iron-smelting industries in the last two decades.

Just outside the minority region, other major cities can be found. Ostrava with a population of almost 290,000 is the largest. It is the seat of the Moravian-Silesian Region [Moravskoslezský kraj] and a center of commerce and industry. It also hosts several universities. From the perspective of the minority, Ostrava is also important because the Polish radio and TV headquarters are based there. Other cities include Havířov (72,000), Frýdek-Místek (56,000), Orlová (29,000), and Bohumín (21,000), to name the most important ones. The settlement structure is of the dispersed Silesian type, meaning that in addition to town and communal centers individual houses and hamlets are scattered everywhere.

This differentiates the region from the rest of Czechia where compact settlements separated by fields and forest prevail. Because of this settlement structure, the lowland part of the Těšín/Cieszyn region has the highest population density in the country.

Currently, there are 30 communes in the Těšín/Cieszyn region in which the minority population exceeded the 10% threshold in the population censuses of 2001 and 2011 (see Table 5.5 and Fig. 5.34). The only large towns in the Region where the threshold is met are Český Těšín/Czeski Cieszyn (13.7%, i.e., 3333 Poles) and Třinec/Trzyniec (13.4%, i.e., 4868). In absolute numbers, there is also a large minority population in the towns of Karviná/Karwina (5.7%, 3226) and Havířov/Hawierzów (2.4%, i.e., 1873), but it does not reach the percentage threshold as these are bigger cities. In one commune (Těrlicko/Cierlicko), the percentage of Poles fell just below the threshold to 9.7% in 2011 which immediately led to the disappearance of bilingual signs. The percentage of Poles in all other communes in the region with the exception of Karviná/Karwina is well below 5%. In five cases, the percentage is between 3 and 4, in six cases it is between 1 and 3, and the rest are under 1. All but six of the communes where the share of Polish population exceeds 10% are located in the former Frýdek-Místek District, that is, in the southern, mostly mountainous half of the region. Communes to the West and North of Český Těšín/Czeski Cieszyn in the District of Karviná have mostly lost their Polish populations, both relatively due to immigration as well as absolutely due to assimilation.

The 30 communes in which the share of Poles exceeded the 10% threshold formed the basis of our research area. However, as explained in Chap. 4, exceeding the 10% threshold does not automatically lead to the introduction of bilingual signs. Meeting the threshold only makes it possible to apply for their introduction. Although the process itself is no longer complicated, in many communes the Polish minority chose not to apply for them. The actual number of communes with bilingual signs is only 17. Chapter 6 explains the reasons for such development.

5.2.3 Minority Life

The Polish minority in the Těšín/Cieszyn region has a rich cultural life and effective political representation unparalleled by any other minority in Czechia. It is very well organized and because of its spatial concentration, everyday social contacts make it possible to maintain a certain level of permanent national mobilization manifested in various individual and collective activities. Furthermore, the activities of the Polish minority reach a region-wide audience drawing in Czechs as well. As a result, visibility and impact of Polish cultural life in the region significantly surpass their share of the population and make the minority a crucial actor in regional culture.

5.2.3.1 Education

Education has always been highly valued in the Těšín/Cieszyn region. The Protestant tradition (see below) played a key role in the spread of literacy to the remotest parts of the region (Zahradnik 1997: 37). The Polish minority historically put great emphasis particularly on the importance of education in the minority language. An extensive network of Polish schools of all levels was maintained during the First Republic and continued to exist throughout the Communist period (Malá 1989). Although the number of schools has been steadily decreasing in the last decades due to the decrease of the Polish minority, Polish kindergarten and elementary schools can still be found in most communes with Polish population. Members of the Polish minority can still go through the whole educational system in their mother tongue. In 2015, there were 33 kindergartens with 852 pupils, 25 primary schools with 1781 pupils, and one secondary school and one Polish class in the business academy in Český Těšín/Czeski Cieszyn with 383 students offering education in Polish (Szymeczek 2017). The University of Ostrava [Ostravská univerzita] offers degrees in Polish studies which are widely used by minority students. Many, however, pursue their degrees directly in Poland. The Silesian University [Uniwersytet Śląski] based in Katowice has a campus in Cieszyn which makes it very easy for Polish students from Czechia to study a wide array of disciplines just across the border.

The number of students in schools is a constant source of concern for the minority. For many years, the numbers were steadily decreasing as people were having fewer children and assimilation processes were orienting many children toward Czech schools. The sustained pressure of the minority succeeded in changing state regulations whereby minority schools were allowed lower quotas for the minimal number of pupils in attendance. Even with this, the threat of school closures is a constant topic in minority media and an object of scrutiny every September when the new academic year starts. The recent years, however, have seen a slight reversal of this trend. As borders with Poland effectively disappeared, it became possible for children from the other side of the border to attend minority schools in Czechia. This phenomenon is seen as a double-edged sword for many schools who fear that their state funding (from the Czech state) might be curbed if the number of essentially 'foreign' students is too high.

Nowadays, we can also observe a small but steadily increasing number of Czech parents sending their children to minority schools. This can have several reasons ranging from smaller class sizes to learning another language (Ćmiel 2016; Szymeczek 2017). The availability of education in the minority language, nevertheless, remains one of the most important aspects of minority life in the region and also one of the most sensitive political topics because of its direct and substantial impact on communal budgets.

5.2.3.2 Community Life and Organizations

The Polish minority in the Těšín/Cieszyn region has always taken pride in its community life, strong networks, and cultural continuity. There are several organizations attending to every aspect of community life. The most important are the Polish Cultural and Educational Association [Polski Związek Kulturalno-Oświatowy, PZKO] and the Congress of Poles in the Czech Republic [Kongres Polaków v Republice Czeskiej, KP]. PZKO was established in 1947 and was the only allowed minority organization during the Communist regime (Hernová 1989). It played a crucial role in fostering Polish national consciousness providing a creative and supportive space for Polish intelligentsia and maintaining a high degree of national engagement of the wider minority population. Its structure consists of the main office located in Český Těšín/Czeski Cieszyn and 81 local chapters [miejscowe kola PZKO]. There is at least one local chapter in the vast majority of towns and villages in the region while bigger or more active communes often have two or more chapters. PZKO also has a network of 40 properties, so-called Houses of PZKO [Domy PZKO] which serve both the minority as well as the majority as venues and centers of social life. There is a PZKO house in the majority of towns and villages in the region. PZKO, as the name suggests, is predominantly focused on cultural and educational activities. It also publishes a monthly journal called *Zwrot*. The local chapters organize many events throughout the year, from Christmas parties in winter and carnivals and Easter celebrations in spring to fairs in summer and harvest festivals in autumn. PZKO has currently 12,000 paying members and it is the largest organization of the Polish minority in Czechia and one of the largest non-governmental organizations in the country.

The Congress of Poles was set up in 1990 as a response to wider democratic changes in society. It is also based in Český Těšín/Czeski Cieszyn. Like PZKO, it supports cultural activities and publishes a semi-daily newspaper titled *Głos Ludu*. However, unlike PZKO, it does not have local chapters, and its orientation is primarily regional and national in terms of scale and political rather than cultural in terms of content. The Congress represents the Polish minority in the governmental Council on National Minorities and intensively lobbies for minority rights.

During Communism, the only allowed minority organization was PZKO and those who did not want to belong to it had nowhere else to go. This historical PZKO monopoly is still resented by many Poles and continues to be a source of internal conflicts within the Polish community. Once this monopoly was broken, several new specialized organizations were established (e.g., Zaolzie Photographic Society, Tourist Association Beskid Śląski, Polish Teachers' Society). All these minority organizations were then united under the umbrella of the Congress of Poles. While individuals can be members of PZKO, only organizations can be members of the Congress. It is, however, important to note that since the Polish minority in the Těšín/Cieszyn region is a fairly compact group, a large number of people are members of both PZKO and several other organizations.

For many members of the minority the cultural and social life facilitated by close friendship networks and long-standing participation in community events and organizations is the most important aspect of belonging to the Polish minority. The importance of participation in community events and their preparation cannot be overstated. Many of these events are highlighted in the annual calendar of the whole community of a town or village and are attended by the Czech majority as well. The minority organizations are also open to the Czech majority, although Czechs belonging to these organizations are generally connected to the minority by marriage.

5.2.3.3 Cultural Life

As we already mentioned, the cultural life of the Polish minority in the Těšín/Cieszyn region is vibrant. The only Polish theater scene outside of Poland is located in Český Těšín/Czeski Cieszyn and the performances are very popular among both minority and majority population. There are also several smaller amateur theater groups organized by minority members, who stage plays in both Polish and the local dialect. Minority members are also active in a vast array of choirs, music groups, and dance troupes which perform traditional regional dances. The dance troupes play an important role in introducing the regional culture to children who can enroll from an early age, as well as to the majority society as they often perform in public events. It is common that membership in these groups is passed from generation to generation and is thus seen as crucial in supporting the local cultural practices. These troupes are also popular among the youth as they offer opportunities to socialize and build networks useful later in life.

5.2.3.4 Minority Language

The Těšín/Cieszyn region has a unique linguistic landscape where Czech and Polish intermingle with the local linguistic code, po naszymu (also *gwara* in Polish). Polish linguists classify it as a Polish dialect while Czech linguists see it as a transitional Czech-Polish dialect. Both Czechs and Poles who are not used to it have difficulties in understanding it. Over the course of the last century, the traditional dialect has moved both grammatically and lexically toward colloquial Czech (and colloquial Polish in Poland), and today, po naszymu (literally, 'in our language') is seen as a hybrid code combining the features of the dialect as well as standard national languages from both sides of the border (Bogoczowa and Bortliczek 2015). In spite of progressive homogenization, dialectal differences remain between the mountainous and lowland parts of the region. The dialect still lacks a codified form. There is a rich oral tradition of songs, stories, and other narratives recorded in the dialect as well as a considerable body of regional literature. However, few people ever write it or even know how to write it, and for most people it is a spoken language only.

The dialect has traditionally served as a regional lingua franca, widely used by both the Czech majority and the Polish minority. In recent years, however, its use among the majority population has seen a decline as our survey also showed. Nevertheless, for the majority of Poles po naszymu remains their mother tongue rather than standard Polish which is rarely spoken at home (Bogoczová 1997; Bogoczowa and Bortliczek 2015). This adds an important layer to many aspects of minority life including place naming. The conflict over bilingual signs is led over names written in languages that few people actually speak. Similarly, while minority schools teach in standard Polish, many pupils struggle to be proficient in Polish because they only speak the dialect (or Czech) with their families, friends, and sometimes even teachers.

While practically all members of the Polish minority in the Těšín/Cieszyn region are fluent in dialect and use it exclusively in their daily life, many claim the overt promotion of dialect use goes against the minority's best interest. After all, one of the distinct features of the Polish minority in the Těšín/Cieszyn region is its linguistic difference (based on the use of Polish). If this distinction is given up, they say, it will become even more difficult to maintain the Polish identity in light of wider social changes. This leads to a paradoxical situation when the promoted language (Polish) is rarely used in everyday life. At the same time, the local dialect—the communication code actually used—is often practically invisible in the linguistic landscape.

5.2.3.5 Religion

Religion has been a particularly strong factor in the Polish national consciousness in the Těšín/Cieszyn region. The beginnings of the Polish national movement are tied with the work of Catholic priests and Protestant pastors from both sides of the present border. Although the Polish national identity in Poland today is strongly associated with the Roman-Catholic Church, in the Těšín/Cieszyn region the Lutheran evangelical faith was particularly strong and for a long time nearly synonymous with the Polish minority. The Těšín/Cieszyn Poles thus differ from Poles in Poland not only in spoken language but also religiously. The twentieth century saw a diversification of church membership with many Czechs joining the Lutheran Church while Poles have always been members of the Roman-Catholic Church as well. Currently, the national composition of both churches is balanced, Poles forming approximately a half of all members. However, the level of religiosity among Poles is much higher than among Czechs (Szymeczek 2015). Services are regularly given in Czech and Polish (either simultaneously or subsequently). Many church leaders continue to play an active role in Polish cultural and political organizations including their highest representation.

5.2.3.6 Media

The Polish minority has its own media that provide a coverage of daily life of the minority as well as relevant issues from Czechia and Poland. *Głos Ludu* ('The Voice of the People') is a semi-daily newspaper founded in the 1940s and currently sponsored by the Congress of Poles. Furthermore, there are several magazines reaching adult and juvenile audiences. The monthly magazine *Zwrot* ('Turn of Events') is published by PZKO and reports on principal cultural events and issues. The Pedagogical Centre for Polish Minority Schools Český Těšín publishes two magazines, one for smaller children called *Jutrzenka* ('The Morning Star') which first came out already in 1911 as one of the activities of the local teachers' association and the second for teenagers called *Ogniwo* ('Connection') also with a long history. These are supported by funding from the Polish government. The minority is also provided with a daily broadcast (every working day) in the public radio station Czech Radio [Český rozhlas], which is produced in its regional office in Ostrava and is solely focused on minority issues and events. Similarly, a weekly newscast is provided by the public TV station Czech Television [Česká televize], also produced in Ostrava and centered around minority issues. The radio and TV broadcasting are funded by the Czech state. Although nowadays the access to Polish media via the internet is easy and direct, the minority media continue to be a key source of information about the life of the minority.

5.2.3.7 Current Issues

The Polish minority in the Těšín/Cieszyn region has undergone a number of changes in the past century. While it has maintained a visible presence in the region, it continues to struggle with several issues. The minority population is aging. While many younger people are active in various minority organizations and societies, membership numbers are still significantly lower than in previous generations. Another issue is the prevalence of mixed marriages, where one of the parents is Czech and the other Polish. This setup often leads to children going to Czech schools and not learning Polish, losing thus the touch with the minority community in general (Rusková 1997; Ćmiel 2016). This, combined with the fact that the minority population is following the general population trends in Czechia where couples tend to have fewer children than before, means that the number of Poles is steadily decreasing as evidenced by the census every ten years. There is also the issue of brain drain when young minority members tend to move away to Prague [Praha] or Warsaw [Warszawa] (or other metropolitan areas in either country) for jobs and better salaries. Gradual assimilation continues to be an issue although the support for minority identity and public discourse surrounding its distinctiveness are considerably better than in the past. Finally, the Polish minority continues to argue for the return of properties confiscated and nationalized by the previous regimes from minority organizations, but there seems to be little hope to move this topic forward as very little has changed since 1993.

One of the most repeated arguments we encountered when interviewing the members of the minority about the bilingual signs in the region was that they are an important reminder of the minority's presence in the area. Some went as far as to say that "they will be here when we are gone". This betrays a certain weariness and anxiety about the future survival of the minority. However, there have been several initiatives aimed at the future. There was a mass-media campaign ahead of the 2011 census promoting the idea of being Polish and declaring oneself as such in the census sheet. There is an ongoing project of the Congress of Poles called "Vision 2035" [Wizja 2035] outlining the possible paths for development for the minority population. The Polish minority is also active in raising funding from other entities than the state. A number of businesspeople from the minority have been successful in recent years and established a Fund for the Development of Zaolzie [Fundusz Rozwoju Zaolzia], which offers grants and funding for many community events and educational activities. Minority media and minority organizations make an active use of social media in attempts to engage the younger generation as well as to reflect wider cultural shifts. The cross-border cooperation with Poland has been flourishing, largely supported by EU funding. Despite many assertions to the contrary, the Polish minority in the Těšín/Cieszyn region remains a well-organized and vibrant community invested in maintaining its presence and visibility in the region well into the future.

5.2.4 Basics on Czech and Těšín/Cieszyn Region Toponymy

The seemingly homogenous toponymic landscape of contemporary Czechia is the result of dramatic historic events which marked its recent history as part of Czechoslovakia. Since the creation of the state in 1918, geographical names have been under direct supervision of the national government and its ministries. Unlike in Austria where the bottom-up principle is well established, in Czechoslovakia and later Czechia the naming process has been controlled from above, and in many instances the national government even directly intervened and imposed names on communes and geographical features against the wishes of the local populations. As we show in Chap. 6, this policy was clearly reflected in the rendering of names on state maps which preferred official Czech names even in minority areas. This is the result of the precarious position of the newly created state vis-à-vis its international enemies, perceived existential threats from its internal national minorities, and major political regime changes during the twentieth century. In the following lines we will only deal with the situation in the Czech lands, however, the situation in Slovakia and Ruthenia undoubtedly had an impact on the national toponymic and linguistic policies as well.

There were several important periods of renaming, partially driven by popular demand and always closely overseen by the national government. The first occurred directly after the creation of Czechoslovakia when names associated with the Habsburg monarchy were replaced by names associated with Czechoslovak statehood. In addition, communes with minority names were given parallel Czech names in order to emphasize the territorial integrity of the newly established state (e.g., *Reichenberg* became *Liberec*, *Brüx* became *Most*, *Troppau* became *Opava*, etc.). During the German occupation of the country during World War II, many Czech names were replaced by German ones. After the War and the near-to-total expulsion of the German-speaking population, there was as a strong pressure for toponymic "degermanization", which was often demanded by local populations and always strictly controlled by the national government. This wave went so far as to erase any reference to Germans ("Němci") even when there was no direct demographic connection (e.g., *Německý Brod* became *Havlíčkův Brod*). More importantly, tens of thousands of German names entirely disappeared from the toponymic landscape in the areas from which the German population was expelled. Only a fraction of those names survived in a Czechisized form or popular memory. The rise of the Communist regime in 1948 brought yet another wave of ideologically motivated renaming overseen by the national government, ranging from street names to whole cities and neighborhoods (e.g., *Zlín* became *Gottwaldov* after the Communist president Klement Gottwald, new residential quarters in Ostrava and Karviná were called *Stalingrad*, etc.). The Communist regime also heavily influenced field names through the operation of collective farms and the merging of small private fields into large farming plots which were given new names, erasing the old ones. After the Velvet Revolution in 1989, many Communist-motivated names were once again changed, not necessarily returning to the pre-Communist versions and adding thus yet another toponymic layer to the already complex toponymic situation.

The repeated toponymic overhauls of Czech geographical names have had several important consequences. First, minority names have been erased. Second, the state control of the naming process with a strong emphasis on name standardization has led to an almost complete erasure of dialectal diversity in names. The contemporary toponymic landscape is thus relatively unified, presenting an incorrect image of a linguistically and culturally homogenous nation. This is also the larger context which we have to take into account when interpreting the debates about bilingual signs in the Těšín/Cieszyn region today, the painful local history notwithstanding. And third, because of the aforementioned overhauls, multiple names for the same object originating in different periods are used, most commonly in cities but also in the countryside. Maps contain many errors and there is often much confusion about which name refers to which object and which form of a name is actually the correct one.

Currently, all geographical names used in the Czech Republic, including exonyms, are subject to standardization by the Czech Office of Surveying, Mapping and Cadastre [Český úřad zeměměřičský a katastrální, ČÚZK]. ČÚZK publishes toponymic guidelines which establish norms and procedures for naming. We refer the reader to the English version of the guidelines which contains a useful overview of Czech orthography, pronunciation, and toponymy (ČÚZK 2007). With the exception of a few Polish names in the Těšín/Cieszyn region, all other names in Czechia are strictly in Czech and even field names are predominantly rendered in the standard language. This often leads to the corruption of names and even changes in their meaning. Furthermore, only one name variant for a given feature is accepted, toponymic plurality is not allowed. This practice also elicits criticism, especially when maps contain errors and people are forced to use names nobody knows. While this policy makes sense from the perspective of governmentality, it leaves virtually no space for regional dialectal differentiation or minority field names while at the same time putting a great pressure on toponymic assimilation into the official toponymic system.

So far there is no comprehensive list of geographical names which would include field names. Names of populated places have been published (Profous 1947–1957; Profous et al. 1960; Hosák and Šrámek 1970) and there are studies of field names in some localities (e.g., David and Rous 2006; Mácha 2014; Lábus and Vrbík 2018). From the 1960s to the 1980s, a heroic attempt was made by a group of toponomists to collect all field names in Bohemia, Moravia, and Silesia, but the mostly successful results have not been made available to the public yet. The ČÚZK-run GeoNames database thus remains the only national database of geographical names covering the entire country. It contains, however, only a fraction of the Czech toponymic heritage.

The vast majority of Czech geographical names in the Těšín/Cieszyn region are of Slavic origin. In the case of settlements, they most commonly refer to the founder or the original owner of the settlement (e.g., *Těšín* from the personal name *Těcha* and the locative suffix *-ín,* or *Petrovice* from the personal name *Petr* and the common possessive-locative suffix *-ice*). In other instances, they may refer to natural features and qualities (e.g., *Bystřice* referring to a quick-flowing river or *Bukovec* referring to the beech tree *buk* followed by the locative suffix *-ec*). In a minority of cases, the settlement name may be of German origin (e.g., *Rychvald* from *Reichenwald* or *Šunychl* from *Schön Eichel*). Occasionally, German names may be used for other objects such as the Třinec steel mill generally referred to as *Werk*.

Field names are highly diverse, referring to ownership, occupancy, natural quality, agricultural and industrial exploitation, and other motives. Most of them exist only in dialect but some have been put on maps in standardized Czech. The transcription of dialectal names presents a particular challenge because the dialect contains phonemes for which the Czech alphabet does not have characters. When they are transcribed in Czech, such names are inevitably corrupted. Their transcription using the Polish alphabet, however, generates a different set of issues, and we comment on those in Chap. 6.

Polish names exist only for populated places and they have been secondarily derived from dialectal names. Like official Czech names, Polish names also appear in standardized Polish. As a consequence, all names of populated places exist in three forms—Czech, dialectal, and Polish—which may or may not be identical in writing and pronunciation. Sometimes dialectal and Polish names are identical and they differ from the Czech name (e.g., *Hrádek* in Czech, *Gródek* in dialect and Polish, *Hnojník* in Czech, and *Gnojnik* in dialect and Polish), in other cases all three names are distinct (e.g., *Vendryně* in Czech, *Wyindrynia* in dialect and *Węndrynia* in Polish or *Návsí* in Czech, *Nowsi* in dialect, and *Nawsie* in Polish), and in still other instances, all three forms are identical in both writing and pronunciation (e.g., *Tyra/Tyra/Tyra* or *Guty/Guty/Guty*). The last two possibilities elicit particular criticism, because in the first case, the official names are seen as corruptions of the true (dialectal) name, while in the second case, bilingual signs are seen as an absurd waste of public money. Finally, in a minority of cases, the form of the official Polish name may be identical with the dialectal name, but its meaning is very different. In such cases, the name also becomes a target for criticism for corrupting the meaning of the original name despite respecting its form (e.g., *Paseky* in Czech and *Pasieky* in dialect both mean 'forest clearings', the official Polish name *Pasieky*, however, means 'apiaries' in Polish). As one can imagine, the aforementioned intricacies of meaning, transcription, and pronunciation are predestined for conflicts, especially when names become politicized. And that is precisely what happened in the Těšín/Cieszyn region.

Sources and Literature

Anonymous (s.a. [post 1945]). *Maps relating to the ethnical structure of Slovenian Carinthia*. s.l.

Arbeitsgemeinschaft für Kartographische Ortsnamenkunde (AKO). (2000). *Geographische Namendatenbank Österreich/Geographical Names Database of Austria*. Wien: Verlag der Österreichischen Akademie der Wissenschaften.

Austrian Board on Geographical Names (AKO) (Ed.). (2012). *Toponymic guidelines for map and other editors for international use*. Retrieved April 25, 2020, from http://ortsnamen.at/wp-content/uploads/2017/12/Toponymic_Guidelines_AT_2012_07.pdf

Austrian Board on Geographical Names (AKO) (Ed.). (2017). Recommendation for the naming of urban traffic areas. *11th United Nations Conference on the Standardization of Geographical Names, New York, 8–17 August 2017*, E/CONF.105/21/CRP.21. Retrieved from https://unstats.un.org/unsd/geoinfo/UNGEGN/docs/11th-uncsgn-docs/E_Conf.105_21_CRP.21_9_Recommendations%20.pdf

Baran, I. (2010). Těšínsko v proměnách Československo-Polských vztahů 1918–1938. In R. Jež & D. Pindur (Eds.), *Těšínsko v proměnách staletí* (pp. 149–159). Český Těšín: Muzeum Těšínska.

Baum, W. (2002). *Klagenfurt. Geschichte einer Stadt am Schnittpunkt dreier Kulturen*. Klagenfurt, Wien: Kitab Verlag.

Bergmann, H., & Jordan, P. (Eds.). (2010). *Geographische Namen—Vielfalt und Norm. 40 Jahre institutionalisierte Ortsnamenforschung und -standardisierung in Österreich, 65. Geburtstag von Isolde Hausner*. Wien: Praesens Verlag.

Bildungsdirektion Kärnten (Ed.). (2019a). Das Minderheitenschulwesen 1958/59–2018/19. Retrieved June 15, 2020, from www.bildung-ktn.gv.at/Minderheitenschulwesen/Jahresberichte.html

Bildungsdirektion Kärnten (Ed.). (2019b). Standorte zweisprachiger Schulen. Retrieved May 1, 2020, from www.bildung-ktn.gv.at/Minderheitenschulwesen/Standorte.html

Bogoczová, I. (1997). Jazyk a jeho dynamika. In D. Kadłubiec (Ed.), *Polská národnostní menšina na Těšínsku v České Republice (1920–1995)* (pp. 85–106). Ostrava: Ostravská univerzita.

Bogoczowa, I., & Bortliczek, M. (2015). Gdy mowa twoja cię (nie) zdradza czyli o wielojęzyczności na Śląsku Cieszyńskim w Republice Czeskiej. *Socjolingwistyka*, XXIX, 39–62.

Borák, M. (1998). Problémy polské menšiny v ČR v 90. letech. In: O. Šrajerová, & G. Sokolová (Eds.), *Národnostní menšiny a majoritní společnost v České republice a v zemích střední Evropy v 90. letech XX. století* (pp. 230–241). Opava: Slezský ústav Slezského zemského muzea v Opavě.

Borák, M. (2009). Německá okupace Těšínského Slezska. In B. Małysz & R. Kaszper (Eds.), *Poláci na Těšínsku* (pp. 42–60). Český Těšín: Kongres Poláků v České republice.

Bottesch, M., Grieshofer, F., & Schabus, W. (Eds.). (2002). *Die siebenbürgischen Landler*. Wien: Böhlau.

Breu, J. (1975). *Geographisches Namenbuch Österreichs*. Wien: Verlag der Österreichischen Akademie der Wissenschaften.

Bundeskanzleramt der Republik Österreich (BKA). (2020). Retrieved March 17, 2020, from https://www.ris.bka.gv.at/.

Český úřad zeměměřický a katastrální (ČÚZK) (Ed.). (2007). Toponymic Guidelines of the Czech Republic. Praha: Czech Office for Surveying, Mapping and Cadastre. Retrieved July 10, 2020, from https://www.cuzk.cz/Zivotni-situace/Poradci-a-poradni-organy/Nazvoslovna-komise-CUZK/Geograficke-nazvoslovne-seznamy/Toponymic_guidelines_of_the_Czech_republic.aspx.

Český úřad zeměměřický a katastrální (ČÚZK) (Ed.). (2019). Retrieved May 26, 2020, from https://archivnimapy.cuzk.cz.

Cillia, R. d., Menz, F., Dressler, W. U., & Cech, P. (1998). Linguistic minorities in Austria. In C. Bratt Paulston & D. Peckham (Eds.), *Linguistic minorities in Central & Eastern Europe* (pp. 18–36). Clevedon, Philadelphia, Toronto, Sydney, Johannesburg: Multilingual Matters Ltd..

Ćmiel, M. (2016). *Tożsamość dwóch pokoleń Zaolzian*. Český Těšín: Sekcja Ludoznawcza PZKO.

Czech Statistical Office (CZSO) (Ed.). (2020). *Results of the 1980, 1991, 2001 and 2011 population census*. Retrieved July 10, 2020, from https://www.czso.cz/csu/sldb/historie_scitani

Czoernig, K. Freiherr von (1855). *Ethnographische Karte der Oesterreichischen Monarchie* [1:864,000]. Wien: Direction der administrativen Statistik.

David, J., & Rous, P. (2006). *Neviditelní svědkové minulosti. Místní a pomístní jména na Vysočině*. Praha: Academia.

Fräss-Ehrfeld, C. (2005). Zwischen Bundeskompetenz und Kärntner Realität. Die Kärntner Minderheitenproblematik in der Zweiten Republik 1945–1976. In C. Fräss-Ehrfeld & H. Rumpler (Eds.), *Kärnten und die nationale Frage*. Zwischen Staatsidee und Landesbewusstsein. Heyn: Kärnten und Wien.

Freytag & Berndt (Ed.). (2003). *Unterstufen-Atlas*. Wien: Freytag-Berndt und Artaria.

Friedl, J. (2009). Polská menšina v letech 1945–1956. In B. Małysz & R. Kaszper (Eds.), *Poláci na Těšínsku* (pp. 60–68). Český Těšín: Kongres Poláků v České republice/Kongres Polaków w Republice Czeskiej.

Gąsior, G. (2009). Rozdělení Těšínského Slezska v letech 1918–1920. In B. Małysz & R. Kaszper (Eds.), *Poláci na Těšínsku* (pp. 22–30). Český Těšín: Kongres Poláků v České republice/Kongres Polaków w Republice Czeskiej.

Gawrecki, D. (2017). *Jazyk a národnost ve sčítáních lidu na Těšínsku 1880–1930*. Český Těšín: Muzeum Těšínska.
Gellner, E. (1983). *Nations and nationalism*. Oxford: Blackwell.
Glaser, F. (2001). Die Römerzeit im Alpen-Adria-Gebiet. In A. Moritsch (Ed.), *Alpen-Adria. Zur Geschichte einer Region* (pp. 51–88). Klagenfurt/Celovec, Ljubljana/Laibach, Wien/Dunaj: Hermagoras Verlag/Mohorjeva založba.
Grygar, J. (2006). Borders and borderlands in the process of Europeanization: The case of Teschen Silesia. In L. Jesień (Ed.), *The borders and limits of European integration* (pp. 129–150). Krakow: Tischner European University.
Hannan, K. (1996). *Borders of language and identity in Teschen Silesia*. New York: Peter Lang.
Hernová, Š. (1989). K etnickému vědomí Poláků na Těšínsku. In A. Robek, D. Kadłubiec, & A. Sulitka (Eds.), *Etnické procesy v ČSSR. Polské etnikum* (pp. 155–164). Ostrava: Krajské kulturní středisko Ostrava.
Holub, Z. (2011). Šlonzáci, Vasrpoláci, Laši, Gorali, Moravci, Prajzáci a lidé Ponašymu. In I. Pospíšil & A. Zelenková (Eds.), *Literární historiografie a česko-slovenské vztahy* (pp. 211–227). Brno: Tribun EU.
Hölzel, E. (Ed.). (2004). *Hölzel-Universalatlas zu Geographie und Geschihcte*. Wien: Verlag Ed. Hölzel.
Hölzel, E. (Ed.). (2011). *Großer Kozenn-Atlas*. Wien: Verlag Ed. Hölzel.
Hosák, L., & Šrámek, R. (1970). *Místní jména na Moravě a ve Slezsku. 1.-2. díl*. Praha: Academia.
Jordan, P. (1988). *Möglichkeiten einer stärkeren Berücksichtigung slowenischer Ortsnamen in den heutigen amtlichen topgraphischen Karten Österreichs*. Wien: Österreichische Akademie der Wissenschaften, Institut für Kartographie.
Jordan, P. (1992). Slowenische Ortsnamen in den amtlichen topographischen Karten Österreichs. Heutiger Zustand und Vorschläge zu seiner Verbesserung. In H.-D. Pohl (Ed.), *Kärnten—Deutsche und slowenische Namen* (pp. 89–105). Klagenfurt: UBW Klagenfurt, Institut für Sprachwissenschaft.
Jordan, P. (2009). Use of place names on road signposts. Examples of practice in Central Europe. In: W. Ahrens, Sh. Embleton, & A. Lapierre (Eds.), *Names in Multi-Lingual, Multi-Cultural and Multi-Ethnic Contact. Proceedings of the 23rd International Congress of Onomastic Sciences, August 17–22, 2008, York University, Toronto, Canada* (pp. 551–565). Toronto: York University.
Jordan, P. (2016a). The ambivalent view on Tito in the successor states of Yugoslavia reflected by the preservation or replacement of place names commemorating him. In: P. Jordan & P. Woodman (Eds.), *Place-Name Changes. Proceedings of the Symposion in Rome, 17–18 November 2014* (= Name & Place, 5 (pp. 461–473). Hamburg: Verlag Dr. Kovač.

Jordan, P. (Ed.). (2016b). *Atlas des Habsburgerreiches*. Schleinbach: Winkler-Hermaden.

Josiek, W. (1997). Společenské organizace a instituce. In D. Kadłubiec (Ed.), *Polská národnostní menšina na Těšínsku v České Republice (1920–1995)* (pp. 149–165). Ostrava: Ostravská univerzita.

K.k. Statistische Zentralkommission (Ed.). (1906). *Gemeindelexikon von Schlesien*. Wien: K.k. Hof- und Staatsdruckerei.

Kadlec, P., et al. (2016). *Národnostní statistika v českých zemích 1880–1930. Mechanismy, problémy a důsledky národnostní klasifikace* (Vol. 2). Ostrava: Ostravská univerzita.

Kadmon, N. (Ed.). (2002). *Glossary of terms for the standardization of geographical names*. United Nations: New York.

Kopeček, L. (2003). Coexistentia-Soužití a politická reprezentace polské menšiny na Těšínsku. *Středoevropské politické studie, 5*(2–3). Retrieved July 9, 2020, from https://journals.muni.cz/cepsr/article/view/3941/5312

Kozler, P. (1853). *Zemljovid Slovenske dežele in pokrajin* [1:576,000]. S.l.

Krahwinkler, H., & Wolfram, H. (2001). Der Alpen-Adria-Raum im Frühmittelalter. In A. Moritsch (Ed.), *Alpen-Adria. Zur Geschichte einer Region* (pp. 89–122). Klagenfurt/Celovec, Ljubljana/Laibach, Wien/Dunaj: Hermagoras Verlag/Mohorjeva založba.

Kranzmayer, E. (1956). *Ortsnamenbuch von Kärnten. I. Teil: Die Siedlungsgeschichte Kärntens von der Urzeit bis zur Gegenwart im Spiegel der Namen*. Klagenfurt: Verlag des Geschichtsvereines für Kärnten.

Lábus, V., & Vrbík, D. (2018). *Toponyma v krajině a možnosti jejich výzkumu*. Liberec: Technická univerzita v Liberci.

Land Kärnten (Ed.). (1990). *Die Minderheiten im Alpen-Adria-Raum*. Klagenfurt: Arbeitsgemeinschaft Alpen-Adria.

Lausegger, J. (Ed.). (1992). *Arbeitsgemeinschaft Alpen-Adria. Organisation, Zeittafel, Schwerpunkte der Arbeit*. Klagenfurt: Evidenzstelle, Amt der Kärntner Landesregierung.

Lendl, E., Wagner, W., & Klein, R. (1972). *F.W. Putzger Historischer Weltatlas zur allgemeinen und österreichischen Geschichte*. Wien: Hölder-Pichler-Tempsky, Österreichischer Bundesverlag.

Logar, T., & Rigler, J. (2001). Slovene dialects. In J. Fridl et al. (Eds.), *National Atlas of Slovenia* (p. 119). Rokus: Ljubljana.

Lukan, W., & Moritsch, A. (Eds.). (1988). *Geschichte der Kärntner Slowenen*. Klagenfurt/Celovec, Wien/Dunaj: Hermagoras/Mohorjeva.

Mácha, P. (2014). *Tajemství v názvech ukrytá. Morávka, Krásná, Raškovice, Pražmo, Vyšní Lhoty*. Ostrava: Ostravská univerzita.

Malá, E. (1989). Postoje občanů polské národnosti k národnostnímu školství. In A. Robek, D. Kadłubiec, & A. Sulitka (Eds.), *Etnické procesy v ČSSR. Polské etnikum* (pp. 165–173). Ostrava: Krajské kulturní středisko Ostrava.

Małysz, B., & Kaszper, R. (Eds.). (2009). *Poláci na Těšínsku.* Český Těšín: Kongres Poláků v České republice/Kongres Polaków w Republice Czeskiej.

Moritsch, A. (1995). Das Windische—eine nationale Hilfsideologie. In A. Moritsch (Ed.), *Problemfelder der Geschichte und der Geschichtsschreibung der Kärntner Slovenen/Problemska polja zgodovinein zgodovinopisja koroških Slovencev* (pp. 15–29). Klagenfurt/Celovec, Ljubljana/Laibach, Wien/Dunaj: Hermagoras/Mohorjeva.

Moritsch, A. (2000). Nationale Ideologien in Kärnten. In A. Moritsch (Ed.), *Kärntner Slovenen/Koroški Slovenci 1900–2000* (pp. 9–28). Klagenfurt/ Celovec, Ljubljana, Wien: Verlag Hermagoras/Mohorjeva.

Moritsch, A. (2001). Geographische Voraussetzungen der Geschichte der Alpen-Adria-Region. In A. Moritsch (Ed.), *Alpen-Adria. Zur Geschichte einer Region* (pp. 11–36). Klagenfurt/Celovec, Ljubljana/Laibach, Wien/Dunaj: Hermagoras Verlag/Mohorjeva založba.

Moro, G. (1957). Zur Entstehung und Ausdehnung bambergischen Besitzes in Kärnten. *Carinthia I*, 147(1–2), 243–260.

Myška, M. (1964). *Z tajných zpráv NSDAP o Těšínsku.* Ostrava: Krajské nakladatelství.

Nećak, D. (2001). Die Alpen-Adria-Region 1945 bis 1991. In A. Moritsch (Ed.), *Alpen-Adria. Zur Geschichte einer Region* (pp. 485–516). Klagenfurt/ Celovec, Ljubljana/Laibach, Wien/Dunaj: Hermagoras Verlag/ Mohorjeva založba.

Neumann, D. (1977). *Das Kärntner Lesachtal. Werden und Wandlungen einer bergbäuerlichen Kultur- und Wirtschaftslandschaft.* Klagenfurt: Verlag des Kärntner Landesarchivs.

Nowak, K. (2009). Polská menšina v letech 1956–1989. In B. Małysz & R. Kaszper (Eds.), *Poláci na Těšínsku* (pp. 68–82). Český Těšín: Kongres Poláků v České republice/Kongres Polaków w Republice Czeskiej.

Österreichische Rektorenkonferenz (Ed.). (1988). Bericht der Arbeitsgruppe „Lage und Perspektiven der österreichischen Volksgruppen". Wien.

Österreichischer Rundfunk (ORF) (Ed.). (2020). Retrieved October 12, 2020, from science.ORF.at.

Pohl, H.-D. (1982). Linguistische Aspekte der Zweisprachigkeit in Kärnten. In F. Dotter (Ed.), *Kein einig Volk von Brüdern. Studien zum Mehrheiten-/ Minderheitenproblem am Beispiel Kärntens* (pp. 35–53). Klagenfurt: Verlag für Gesellschaftskritik.

Pohl, H.-D. (2008). Sekundäre Umformungen von (Kärntner) Ortsnamen im deutsch-slowenischen Sprachkontaktgebiet. In: P. Ernst (Ed.), *Namenarten in Österreich und Bayern. Vorträge der 4. Tagung des Arbeitskreises für bayerisch-österreichische Namenforschung am 28. und 29. September 2006 in Wien* (pp. 95–105). Wien: Praesens Verlag.

Pohl, H.-D. (2009a). Ortsnamen in einer historisch gewachsenen Kulturlandschaft unter besonderer Berücksichtigung Kärntens und Osttirols. *Europäisches Journal für Minderheitenfragen*, 2(2), 72–88.

Pohl, H.-D. (2009b). Sprachkontakt in Kärnten. In M. Elmentaler (Ed.), *Deutsch und seine Nachbarn* (pp. 117–132). Frankfurt a. M. et al.: Peter Lang.

Pohl, H.-D. (2010). *Unsere slowenischen Ortsnamen—Naša slovenska krajevna imena*. Klagenfurt: Hermagoras/Mohorjeva.

Profous, A. (1947–1957). *Místní jména v Čechách. Jejich vznik, původní význam a změny—1.-4. díl.* Praha: Československá akademie věd.

Profous, A., Šmilauer, V., & Svoboda, J. (1960). *Místní jména v Čechách. Jejich vznik, původní význam a změny—5. díl, Dodatky*. Praha: Československá akademie věd.

Rumpler, H. (2005). Die nationale Frage im Spannungsfeld von kärntnerischem Landespatriotismus, österreichischem Staatsbewusstsein und völkischem Nationalismus 1918–1938. In: C. Fräss-Ehrfeld, & H. Rumpler (Eds.), *Kärnten und Wien. Zwischen Staatsidee und Landesbewusstsein* (= Kärnten und die nationale Frage, 4) (pp. 9–82). Klagenfurt/Celovec, Ljubljana/Laibach, Wien/Dunaj: Hermagoras/Mohorjeva.

Rumpler, H., & Seger, M. (2010). *Die Gesellschaft der Habsburgermonarchie im Kartenbild. Die Habsburgermonarchie 1848–1910, Band IX/2: Soziale Strukturen. Verwaltungs- Sozial- und Infrastrukturen. Nach dem Zensus von 1910. Atlas thematischer Karten*. Wien: Verlag der Österreichischen Akademie der Wissenschaften.

Rusková, H. (1997). Společenské a kulturní procesy. In D. Kadłubiec (Ed.), *Polská národnostní menšina na Těšínsku v České Republice (1920–1995)* (pp. 67–85). Ostrava: Ostravská univerzita.

Schabus, W. (1996). *Die Landler. Sprach- und Kulturkontakt in einer alt-österreichischen Enklave in Siebenbürgen (Rumänien)*. Wien: Edition Praesens.

Seger, M. (2010). *Kärnten. Landschaftsräume—Lebensräume*. Klagenfurt: Naturwissenschaftlicher Verein des Landes Kärnten, Geschichtsverein des Landes Kärnten.

Seger, M. (2019). *Österreich. Raum und Gesellschaft*. Klagenfurt am Wörthersee: Naturwissenschaftlicher Verein für Kärnten.

Siwek, T. (1996). *Česko-polská etnická hranice*. Ostrava: Filozofická fakulta Ostravské univerzity.

Siwek, T. (2017). New tendencies in Czech Minority policy. In: Jordan, P. (Ed.). *10 Years of EU Eastern enlargement. The geographical balance of a courageous step. Proceedings of the Symposion in Vienna, 3–4 December 2014* (= ISR-Forschungsberichte, 42) (pp. 461–468). Wien: Verlag der Österreichischen Akademie der Wissenschaften.

Siwek, T., Zahradník, S., & Szymeczek, J. (2001). *Polská národnostní menšina v Československu 1945–1954*. Praha: Ústav pro soudobé dějiny AV ČR.

Statistik Austria (ed.). (2002). Volkszählungen 1971–2001. Umgangssprache Kärnten. Gemeinden und Ortschaften (Auswahl). Retrieved March 21, 2020, from http://www.statistik.at.

Statistik Austria (ed.). (2020a). Volkszählung 2001. Umgangssprache in Kärnten. Pressekonferenz mit Dr. Ewald Kutzenberger und Johann Ladstätter, April 16. Retrieved March 21, 2020, from http://www.statistik.at.

Statistik Austria (ed.). (2020b). Retrieved March 21, 2020, from http://www.statistik.at

Suppan, A. (2019). *Imperialist peace order in Central Europe. Saint-Germain and Trianon 1919–1920*. Wien: Austrian Academy of Sciences Press.

Symmons-Symonolewicz, K. (1985). The concept of nationhood: Toward a theoretical clarification. *Canadian Review of Studies in Nationalism*, XII(2), 215–222.

Szymeczek, J. (2015). Życie religijne. In: K. Nowak (Ed.), *Śląsk Cieszyński w latach 1945–2015* (pp. 353–368 and 611–618). Cieszyn: Starostwo Powiatowe v Cieszynie.

Szymeczek, J. (2017). Szkolnictwo polskie w Czechosłowacji (Republice Czeskiej) od czasów jej powstania do dzisiejszego dnia. *Rocznik SNPL*, 17, 254–274.

Veiter, Th. (1980). *Die Kärntner Ortstafelkommission* (= Das gemeinsame Kärnten/Skupna Koroška, 8). Klagenfurt: Hermagoras/Mohorjeva.

Verein der Kärntner Windischen (Ed.). (2020). Retrieved May 22, 2020, from https://www.kaerntnerwindische.com/.

Vilfan, S. (2001). Historische Stereotype in der Alpen-Adria-Region. In A. Moritsch (Ed.), *Alpen-Adria. Zur Geschichte einer Region* (pp. 37–49). Klagenfurt/Celovec, Ljubljana/Laibach, Wien/Dunaj: Hermagoras Verlag/Mohorjeva založba.

Wadl, W. (2006). Ortsnamen sind ein wertvolles Kulturgut. In: Amt der Kärntner Landesregierung, Abteilung 1—Landesamtsdirektion/Volksgruppenbüro, Dokumentationsstelle für Zeitgeschichte (Ed.). *Die Ortstafelfrage aus Expertensicht. Eine kritische Beleuchtung* (= Kärnten Dokumentation, Sonderband 1) (pp. 180–195). Klagenfurt: Verlag Land Kärnten.

Wonka, E. (2002). Die räumliche Verteilung der Österreicher mit slowenischer Umgangssprache in Kärnten auf der Grundlage einer detaillierteren Statistik als der auf Gemeindebasis. *Mitteilungen der Österreichischen Geographischen Gesellschaft*, 144, 101–124.

Wörsdörfer, R. (2002). Ethnisch-nationale Differenzierung in den Ostalpen: ‚Deutsch-Windisch-Slowenisch' (1920–1991). In: Müller, M.G. & Petri, R. (Eds.). *Die Nationalisierung von Grenzen. Zur Konstruktion nationaler Identität in sprachlich gemischten Grenzregionen* (= Tagungen zur Ostmitteleuropa-Forschung, 16) (pp. 137–160). Marburg: Verlag Herder-Institut.

Wutte, M., Paschinger, V., & Lex, F. (1925). *Kärntner Heimatatlas*. Wien: Österreichischer Bundesverlag für Unterricht, Wissenschaft und Kunst.

Zahradnik, S. (1992). Národnostní struktur na Záolží podle výsledků sčítání lidu (1880–1991). In M. Borák & D. Gawrecki (Eds.), *Nástin dějin Těšínska* (pp. 161–253). Praha: Česká národní rada.

Zahradnik, S. (1997). Nástin historického vývoje. In D. Kadłubiec (Ed.), *Polská národnostní menšina na Těšínsku v České Republice (1920–1995)* (pp. 10–43). Ostrava: Ostravská univerzita.

Zahradnik, S. (2001). Od většiny k menšině—Poláci na Zaolzí. In I. Kufová (Ed.), *Minulost a současnost národnostních menšin na Těšínsku* (pp. 47–53). Český Těšín: Pedagogické centrum pro polské národní školství.

Zahradnik, S. (2014). *Polska społeczność Zaolziańska i jej prekursorzy*. Třinec: PZKO.

Zemský archiv Opava. (2020). Bevölkerung gegliedert nach den Nationalitäten laut Volkszählungen vom 31. Dezember 1900, 1910 und 15. Februar 1921. NAD 349, č. pom. 557 (hand-written document).

6

Research Results

In this chapter we present the principal findings from our research in southern Carinthia and the Těšín/Cieszyn region. Since we already provided the general methodological context in Chap. 3, we will now focus on the results. The three-year investigation yielded a wealth of data, and it unfortunately goes well beyond the scope of this chapter to report on all of them in detail. However, we will attempt to provide as much detail on as many findings as possible, so the readers can judge for themselves if our data is sound and our interpretation of them correct. The structure of this chapter follows that of Chap. 3; that is, we will go over the principal findings from archival research first through the linguistic landscape and media analyses to interviews and questionnaires. This will allow for a systematic overview of the findings for both regions independently in a transparent way. A complex synthesis of our research is then offered in Chaps. 7 and 8, where results from both regions are critically examined and compared (Chap. 7), and where general conclusions are offered (Chap. 8).

P. Jordan et al., *Place-Name Politics in Multilingual Areas*, Palgrave Studies in Minority Languages and Communities, https://doi.org/10.1007/978-3-030-69488-3_6

6.1 Carinthia

6.1.1 Diachronic Development of Minority Place-Name Use (with a Focus on Maps and Gazetteers) (Peter Jordan)

The following subchapter will focus on the use of minority place names on official Austrian topographical maps and related gazetteers as an expression not only of the legal status of the minority, but also of their general embedding into the wider political community and its recognition as a cultural and identity group. Private mapping will be left aside, since maps of private publishers pursue varying interests and purposes and apply in consequence various representation modes also related to place names, while official topography and mapping have in principle the same approach throughout political periods. It may therefore be regarded as a good indicator of how the very subject of this investigation, the use of minority place names as an expression of the legal status of the minority as well as their general embedding into the wider political community—disregarding other influences as much as possible—has developed over time and under varying political situations.

The treatise will start with the Austrian land surveys from 1764 to 1915 conducted first by the Royal Military Geographical Institute [Reale Istituto geografico militare] in Milan [Milano] and from 1839 onward by the Military Geographical Institute [Militärgeographisches Institut, MGI] in Vienna [Wien] and proceed with the official topographical maps of the Republic of Austria edited and published from 1923 onward by the Federal Office of Metrology and Surveying [Bundesamt für Eich- und Vermessungswesen, BEV] in Vienna. Findings as regards minority place-name use will be related to the contemporary political, cultural, societal, and demographic background in cases, where the background is necessary for an adequate understanding and had not yet been provided sufficiently by earlier chapters or where it is important to stress it.

After the First or Josephinian Land Survey [Erste oder Josephinische Landesaufnahme] (1764–1787) and the Second or Franciscan Land Survey [Zweite oder Franziszeische Landesaufnahme] (1806–1869), the topographic-cartographic rendering of the Habsburg empire was almost completed by the Third or Francisco-Josephinian Land Survey [Dritte oder Franzisko-Josephische Landesaufnahme] (1869–1887). All of them encompassed also Carinthia [Kärnten]. The Fourth or Precision Survey [Vierte oder Präzisionsaufnahme] (1896–1915) had to be canceled incomplete in World War I and only a few maps were published (Arnberger and Kretschmer 1975: 437ff).

The First or Josephinian Land Survey and partly also the Second or Franciscan Land Survey were conducted in periods, in which a Slovenian standard language and a Slovenian ethnic/national consciousness were not yet or sufficiently developed, and German was the Empire's only official language. It is true that a first standard of Slovene can be traced back to the Lutheran Reformer Primož Trubar (1508–1586) and to Jurij Dalmatin (1547–1589), who published religious texts in a dialect of Lower Carniola [Dolenjska]. But only due to the publication of comprehensive dictionaries by Blaž Kumerdej (1738–1805), Valentin Vodnik (1758–1819), and Jernej Kopitar (1780–1844)—all of them from Upper Carniola [Gorenjska] and therefore emphasizing the dialect of this region—Slovene received roughly its current shape at the end of the eighteenth and the beginning of the nineteenth centuries with Urban Jarnik (1784–1844) and Anton Janežič (1828–1869) contributing elements of the Carinthian-Slovene dialects. The works of the poets and writers France Prešeren (1800–1849), Fran Levstik (1831–1887), Ivan Cankar (1876–1918), and Oton Župančič (1878–1949) elevated Slovene finally to the heights of a language of literature (Grdina and Stabej 2002; Lukan 1978; Ramovš 1936; Svane 1958).

Based on Enlightenment during the eighteenth century raising for the first time the question of ethnic/national affiliation in the modern sense (replacing dynastic, religious, and regional loyalties and affiliations that had dominated earlier), the French Revolution at the end of the same century, and after a consciousness of a South-Slavonic cultural communality had developed in the Napoleonic Illyrian Provinces (1809–1813) including Carniola [Kranjska] and western parts of Carinthia [Kärnten], a Slovenian ethnic/national consciousness manifested itself for the first time in the 'year of revolutions' 1848 and most impressively in Peter Kozler's map of the "Slovenian lands and regions" of 1853 (Kozler 1853,

Fig. 46). With its national-political intention, it applied of course all Slovenian place names available, also for features outside the Slovene-speaking area in the sense of exonyms that may have been Alpine-Slavonic endonyms in former times like *Belak* [Villach], *Sovodenj* [Gmünd], *Gornje Belani* [Obervellach], or *Vispriani* [Weißbriach], or in the sense of Slovenian exonyms that had never been endonyms like *Solnograško* for Salzburg or *Tirolsko* for Tyrol [Tirol] (Fig. 6.1).

Until 1848, however, German was the only official language and language of tuition in Carinthia, and it was not before 1850 that Slovenian place names were standardized (Zdovc 1974: 289); 1848, the year of revolutions, however, brought also a turn-around: Between 1848 and 1869 the language of tuition was by law the mother tongue of the majority of pupils in the entire Slovene-speaking area of Carinthia. In fact, however, it was the church language of the parish, which meant that also

Fig. 6.1 Section Upper Carinthia [Oberkärnten] of Peter Kozler's map of the "Slovenian lands and regions" as of 1853. (Source: Wikipedia)

in areas with a majority of German-speakers Slovene was the language of tuition (Hugelmann 1934: 502).

On the background of this linguistic and cultural-political situation it is understandable that the First or Josephinian Land Survey (1764–1787) in the scale of 1:28,800 shows in cases of multiple places names mostly the German name or the Slovenian name in German orthography. Only exceptionally a name version nearer to the Slovenian name is preferred. Recognizing minority names was all the more not appropriate as this map work was conceived to be used exclusively by the Army, whose language of command was—except for the Hungarian homeland defenders (Honvéd)—German up to the end of the Habsburg empire. The German orthography of names is near to local pronunciation but looks inconsequent from a modern point of view as the examples from Section 107 of the sequence of sections presenting Austria Interior [Innerösterreich] demonstrate (Fig. 6.2). They stand for three villages in the Lower Gail Valley [Unteres Gailtal/Zilja] in close vicinity having names with the same Slovenian suffix *-ic* that is today in German orthography spelled as *-itz*. The First Survey, however, renders them *Feistricz* [Feistritz an der Gail], *Achomitz* [Achomitz/Zahomec], and *Drasisch* [Draschitz]—the *-icz* conforming to Hungarian orthography as it should have been used for all Slovenian names according to official instructions (Rajšp 1998: XXXII). Section 137 shows by *Blasnicza* [Blasnitzen/Plasnica] a name that is exceptionally rendered closer to Slovenian than to German, however, again with a Hungarian *cz* for Slovenian *c* (Fig. 6.3). In contrast, the same Section has *Weißenbach* as a full German translation of Slovenian *Bela* [Vellach] in an area that had at that time a significant majority of Slovene-speakers and is up to the present a core region of the minority. *Capel* in the South of Fig. 6.3 stands for Bad Eisenkappel/Železna Kapla, the urban center of the Vellach Valley [Vellachtal/Dolina Bele].

The First Survey was started already under the rule of Mary Theresa [Maria Theresia], but continued and completed under her son Emperor Joseph II. It was—after the Carte géométrique de la France 1:86,400 by César-François Cassini de Thury—the second comprehensive topographic survey of an entire country, but even in a larger scale than the French. It portrayed the entire contemporary Habsburg empire by 3589 (later 4096) sections (Arnberger and Kretschmer 1975: 435) except Tyrol

Fig. 6.2 Part of Section 107 of the First Land Survey in Austria Interior [Innerösterreich] showing some villages in the Lower Gail Valley [Unteres Gailtal/Zilja] and the lower part of the Gailitz/Ziljica [Slizza] Valley. (Source: Austrian State Archive)

Fig. 6.3 Part of Section 137 of the First Land Survey in Austria Interior [Innerösterreich] showing the lower Vellach Valley [Vellachtal/Dolina Bele]. (Source: Austrian State Archive)

[Tirol], that had already been mapped earlier and of which a later attempt of mapping it was interrupted by the Napoleonic wars and never completed (Beimrohr 2012). Head of the team of topographers mapping Carinthia [Kärnten] and Carniola [Kranjska] was Colonel Jeney, a Hungarian, in 1785 also temporarily chief of the general staff of the Austrian Army (Rajšp 1998: XXXVI). As it had been conceived just for the head army commanders, it was produced only in two hand-drawn and hand-colored copies with four to five colors, today kept by the Austrian National Library [Österreichische Nationalbibliothek] and the Austrian State Archive [Österreichisches Staatsarchiv], both in Vienna [Wien], and digitally accessible via Mapire. Cartographically the map work excels by long curved mountain hachures, partly also cross ruling and hill shading (Arnberger and Kretschmer 1975: 436) thus emphasizing relief representation so important for military purposes.

The Second or Franciscan Land Survey (1806–1869) was initiated to equalize the cartographic inequalities between the sections of the First Survey and to base all map sheets on a consistent triangulation providing for a much higher geometric accurateness. It was accompanied by the first nationwide cadastral survey, the so-called Franciscan Cadaster [Franziszeischer Kataster] between 1817 and 1861 constituting the basis for a complete land register so important for real estate management and the levying of taxes, and the Survey could be based on it. The original plan of 2515 survey sections in the scale 1:28,800 was later expanded to 3333, but the survey was never completed. Relief representation was performed by Lehmann-type slope hachuring, vegetation types were differentiated by colors. While the survey sections in the 1:28,800 scale were never published, derived map sheets in the scale 1:144,000 ("Special Map" [Spezialkarte]) and 1:288,000 ("General Map" [Generalkarte]) appeared after 1810 and 1812, respectively. In 1822 also a "General Map of the Austrian Empire" [Generalkarte des Österreichischen Kaisertums] in the scale 1:864,000 and in 1856 a "General Map of the Austrian Empire" [Generalkarte des Österreichischen Kaiserstaates] in the scale 1:576,000 followed (Arnberger and Kretschmer 1975: 436f).

Regarding place names in the bilingual area of Carinthia, the Second Survey is characterized by a more consistent rendering of German names compared to the First Survey avoiding, for example, inconsistencies such

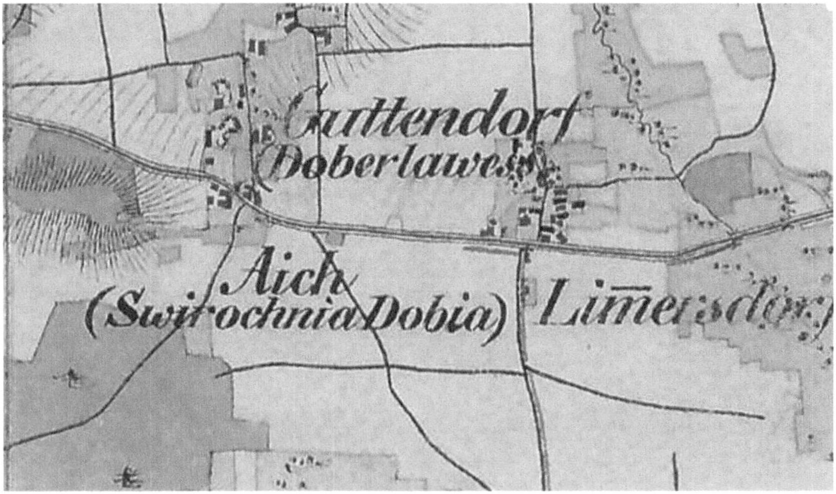

Fig. 6.4 Section of the Second Land Survey showing villages east of Klagenfurt am Wörthersee. (Source: Austrian State Archive)

as *Feistricz* [Feistritz an der Gail], *Achomitz* [Achomitz/Zahomec], and *Drasisch* [Draschitz], and practicing in principle naming in German orthography also in the core minority regions without adding a Slovenian name. Surprisingly enough dual naming occurs in areas without major minority shares, that is, east of Klagenfurt up to Tainach. Thus, Fig. 6.4 shows *Guttendorf (Doberlawes)* [Gutendorf] and *Aich (Swirochnia Dobia)* [Aich an der Straße]—both villages with their Slovenian name in brackets but written in German orthography. Figure 6.5 presents dual naming with four more villages—the Slovenian name written in German orthography and dialectal Slovenian *ves* ('village') written with *ss* not conforming to pronunciation[1]: *Zinnsdorf (Zinsewesa)* [Zinsdorf], *Pubersdorf (Purasche)* [Pubersdorf], *Hörtendorf (Drassnavess)* [Hörtendorf], *St. Johann (Tschanschawess)* [Sankt Johann]. Figure 6.6 finally has *Teinach (Tinie)* [Tainach] and *Pakein (Hum)* [Hum] as villages with bilingual names, while in the entire rest of the minority area—although at that

[1] The spelling with *ss* can be explained by the directive to topographers to apply a Hungarian orthography to Slovenian names. While a Hungarian *ss* is pronounced like *s* in German or Slovene, a Hungarian *s* would be pronounced like *sch* in German and *š* in Slovene.

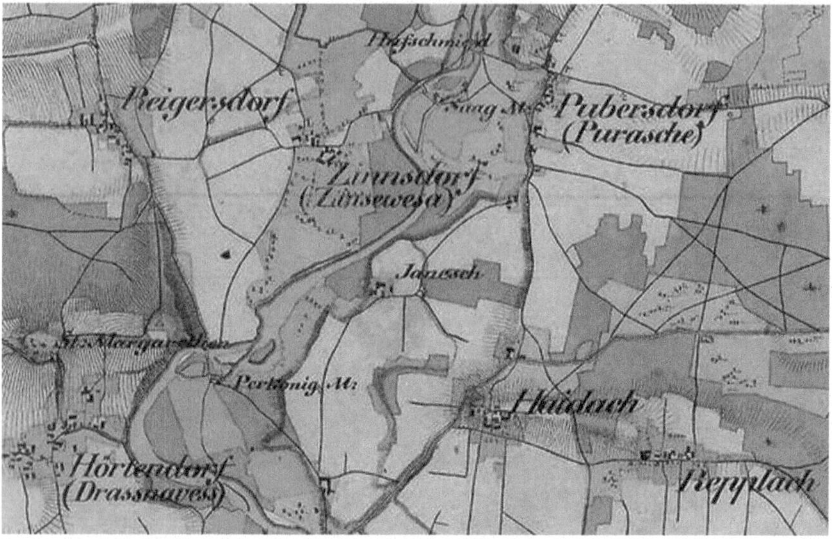

Fig. 6.5 Section of the Second Land Survey showing villages east of Klagenfurt am Wörthersee along river Gurk. (Source: Austrian State Archive)

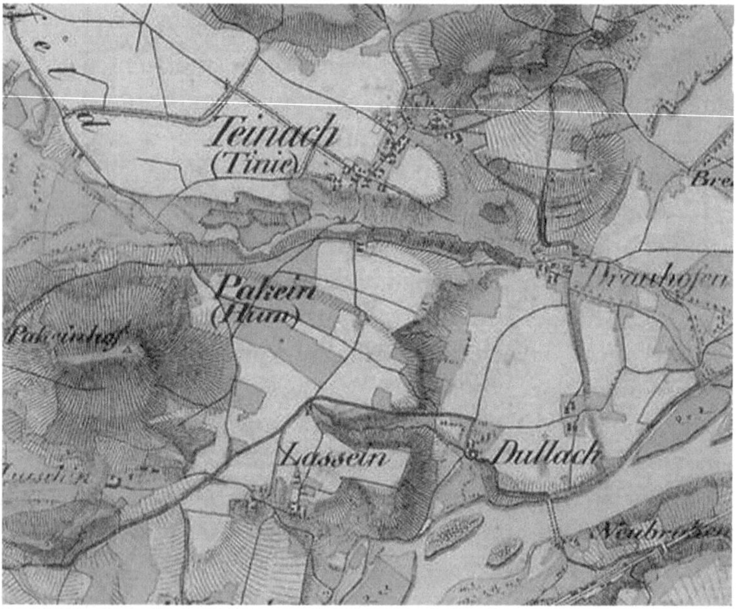

Fig. 6.6 Section of the Second Land Survey with Tainach on River Drau/Drava. (Source: Austrian State Archive)

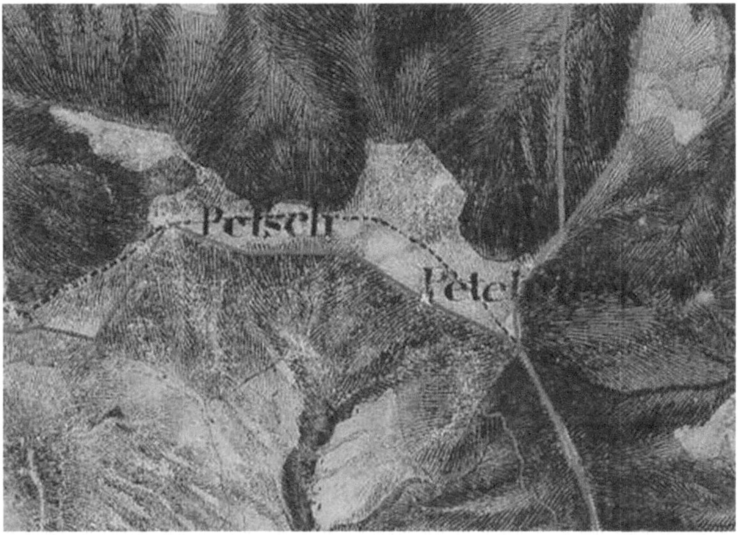

Fig. 6.7 Section of the Second Land Survey showing Petsch, today *Dreiländereck/Peč/Monte Forno*. (Source: Austrian State Archive)

time still populated by a high share of Slovene-speakers—no dual naming occurs at all. Also the Lower Gail Valley [Unteres Gailtal/Zilja] is completely free of dual naming except the mountain dominating it, Dobratsch, where the peak is named *Dobracz*, while the massif bears the name *Villacher Alpe*. While *Dobracz* seems to be a last remnant of the Hungarian-style rendering of Slovenian names as it was practiced in most of the First Survey, naming the peak differently from the massif is a misunderstanding continued in modern Austrian official maps. It is also interesting to see that the modern Dreiländereck/Peč/Monte Forno at the border between Slovenia, Italy, and Austria bearing earlier the German name *Ofen* (instead of the rather touristic modern naming *Dreiländereck*) is named *Petsch* in the Second Survey (Fig. 6.7), that is, by the Slovenian name in German orthography.

The Third or Francisco-Josephinian Land Survey was based on an inter-ministerial agreement and was for the first time intended not only for military but also for civil purposes. The contents of the maps have been extended accordingly. From 2780 basic recordings in the scale

1:25,000 (in urban areas partly 1: 12,500, in Transylvania [Ardeal] 1:28,800) maps in the scales of 1:75,000, 1:200,000, and 1:750,000 were derived (Arnberger and Kretschmer 1975: 437f). On the recordings of the scale 25,000, the terrain was represented by Lehmann's slope hachures as well as by elevation lines in an equidistance of 100 m, in flat terrain of 50 m. The recordings remained unpublished and were distributed as black photographic platinum prints only after the turn of the century.

The sheets of the Special Map [Spezialkarte] in the scale of 1:75,000 consisted of four recording sections each. The first series of 752 sheets was designed and colored in the same way as the recording sections and issued for 16 years (1873–1889). Later, the sheet number of the Special Map grew to 1079, due to the addition of maps of foreign territories. Between 1888 and 1914 the Special Map underwent two revisions. Some sheets received green forest overprint. In addition, sheets with brown slope hachures, brown elevation lines (100 m equidistance), green forest overprint, and blue or black waters or sheets with brown elevation lines in 20 m equidistance (in flat terrain 10 m) instead of slope hachures, with blue waters and green forest overprint appeared.

Derived from the Special Map and having the same sheet line (eight pages of the Special Map constitute a General Map) the General Map of Central Europe [Generalkarte von Mitteleuropa] in the scale of 1:200,000 appeared from 1889 onward in at first 265, later, especially after comprehensive mappings during World War I, 282 sheets (Dörflinger 1986). It also captured much more than just the lands of the Austrian-Hungarian Monarchy and even went beyond the area of the Special Map. The General Map was usually printed in four colors with brown mountain hachures and brown rock representation, blue water network, black settlement, traffic network representation, and green forest areas. However, there were also maps with red fillings of important roads and church symbols as well as red, blue, and yellow bands accompanying boundaries. Overall, however, the graphic design of this map work is more uniform than that of the Special Map.

While maps in the 1:28,800 scale of the Josephinian Survey show in cases of multiple names only the majority name and maps in the same

Fig. 6.8 Section of Special Map 1:75,000, Sheet No. 53 52 Villach, edited 1881. In bilingually German-Slovenian speaking southern Carinthia, most towns and villages and many hamlets bear besides their German also their Slovenian name. It is positioned in smaller letters and in brackets below the German name. (Source: Austrian State Archive)

scale of the Franciscan Survey add minority names only in exceptional cases, the Special Map in the 1:75,000 scale of the Francisco-Josephinian Survey displays in Carinthia a much more minority-friendly attitude.

In bilingually German-Slovenian speaking southern Carinthia, where the map sheets were recorded between 1876 and 1888, most towns and villages and many hamlets bear besides their German also their Slovenian name. It is positioned in smaller letters and in brackets below the German name (Fig. 6.8). A larger number of hamlets and scattered farmsteads remain without a Slovenian name added. Their frequently Slovenian-sounding name is rendered in German orthography (e.g., *Arneutz, Wutz*). This may have the reason that it was (and still is) the right of the house owner to define the name of his house and its orthography and that also among Slovene-speakers a larger share existed regarding a German ren-

Fig. 6.9 Section of Special Map 1:75,000, Sheet No. 53 51 Bleiberg and Tarvis, edited 1881. Potschach (Poče) [Potschach/Potoče] is the westernmost bilingual populated place in Carinthia, while Hermagor (Sv. Mahor) [Hermagor] and Möderndorf (Modrinjaves) [Möderndorf] are already located outside the bilingual area. (Source: Austrian State Archive)

dering of their name as the more appropriate—a group corresponding to the 'Windische' as described earlier.[2]

But even outside the bilingual area some larger populated places are named bilingually, for example, *Hermagor (Sv. Mahor)* and *Möderndorf (Modrinjaves)* right outside the western Slovenian language boundary in the Gail Valley [Gailtal/Ziljska dolina] (Fig. 6.9), *Afritz (Koberca)* and *Winklern (Vogliče)* northwest of Villach, or *Tigring (Tigerče)* between Klagenfurt and Feldkirchen. None of these places had Slovene-speakers according to the population census of 1880, so their Slovenian names

[2]When modern historiography lets sometimes the impression arise that the 'Windische' as a cultural-political fraction of the Carinthian minority did only emerge after World War I, attitudes attributed to this group existed certainly also before and were perhaps rather common in times before the 'national awakening'.

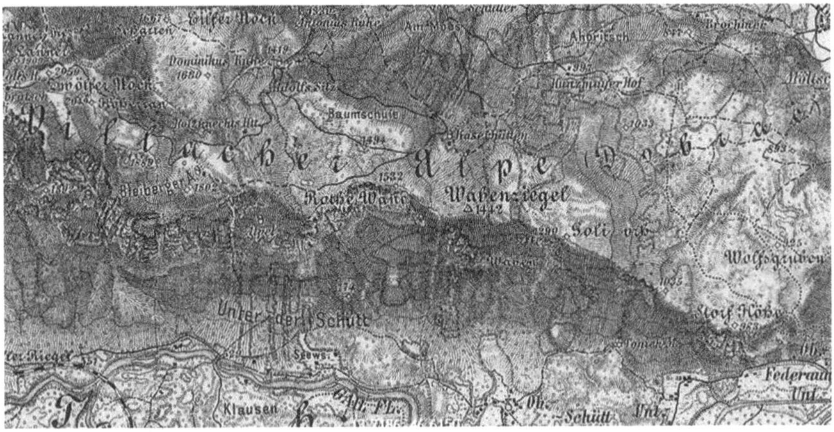

Fig. 6.10 Section of Special Map 1:75,000, Sheet No. 53 51 Bleiberg and Tarvis, edited 1881. *Villacher Alpe (Dobrač)* stand in the mode of a dual majority/minority naming correctly (and in contrast to the Second Survey and current practice of the Austrian Map) for the entire mountain massif. (Source: Austrian State Archive)

were actually exonyms, that is names from without, not used by the local community. The Slovenian naming of places relatively far off the bilingual area and even not very prominent like *Afritz* or *Winklern* may be explained by the fact that Slovenian farmers used to bring their cattle to the ample Alpine pastures in the North of Carinthia, while the mountains in the bilingual area east of River Gailitz/Ziljica are mainly built of limestone offering less convenient Alpine pasture lands. Thus, among Carinthian Slovenes places along the cattle trade routes had well-known Slovenian names (Kronsteiner 1974: 11)—apart from the fact that they had anyway Slavic inhabitants and Slavic names earlier in history. Earlier endonyms have thus been converted into exonyms.

Names of mountains, fields, and water bodies are on the Special Map predominantly rendered in German only, exceptionally also bilingually with the Slovenian name in smaller letters and in brackets. Examples for the latter case are *Villacher Alpe (Dobrač)* (Fig. 6.10) and *Gailitz Bach (Zilica)*. Exceptionally, the Slovenian name is in the first position, like with *Peč (Ofen)*. When with mountains, fields, and water bodies only one name is given, many more names are rendered in Slovenian orthography than in the current Austrian official topographical maps. Thus, on sheet

53 51 (Bleiberg and Tarvis) portraying the Lower Gail Valley [Unteres Gailtal/Zilja] the following cases occur: *Dobrava* (word for 'wood' in standard Slovenian), *Travnik* (standard Slovenian for 'meadow'), *Nučova laz* (*laz*: standard Slovenian for 'forest clearing'), or *Kališca* (word for a swampy location in the local Slovenian dialect). To most Slovenian names of mountains, fields, and water bodies, however, German generics were added, for example, with *Gračenica Graben, Ardešica Graben*, or they were adapted to German by a phonetic transcription of the Slovenian name, mostly of its dialect version, into German orthography, for example with *Pridou* (transcribed from the Slovenian dialect word *predol* for 'mountain pass'). Even more frequent is the rendering of Slovenian names in German orthography with German generics added, for example with *Bistritz Graben* (from the Slovenian *Bistrica* meaning 'torrent').

It also happens that a name has—very likely not intentionally, but by mistake—a mixed Slovenian and German spelling, like in *Gorčach*, where a consequent German spelling would result in *Gortschach* and a consequent Slovenian spelling would lead to *Gorčah*.

Title, legend, and technical remarks on the map frame are, however, only given in German. Nevertheless, the first version of the Carinthian sheets of the Special Map represents the all-time climax of minority names rendering in Austrian official topographic mapping. Never before and never later minority language and minority names received so much recognition.

As already shown by Jordan (1988), the contemporary official place-name gazetteers go even beyond the treatment of Slovenian place names on the Special Map. The Special Place-Name Gazetteer of Carinthia [Special-Orts-Repertorium von Kärnten], for example, edited by the Central Statistical Commission [K.k. statistische Central-Commission] in Vienna [Wien] in 1883 (K.k. statistische Central-Commission 1883), has besides its German title a Slovenian in the same font size and type and can be opened up by explanations and registers in Slovenian as good as in German. Slovenian names—in the same font size and type as the German—are not only given for communes and populated places in the bilingual area, but also for communes and populated places without Slovene-speakers and partly far-off areas with Slovene-speakers. Thus 21 of the 108 names of the Juridical District Kötschach, 82 of the 302 names of the Juridical District Feldkirchen, and 41 of the 118 names of the Juridical District Wolfsberg are rendered bilingually. According to the population census 1880 the Juridical District Feldkirchen had no Slovene-speakers at all, the Juridical District Kötschach just a small number in five populated places.

Not any juridical district in Carinthia remains in the Gazetteer of 1883 without a Slovenian name, even if Slovene had faded away as a spoken language from its territory already centuries ago. This exceptionally benevolent treatment of minority names is confirmed especially by the following examples:

Political District Villach-Land:	
Juridical District Paternion:	*Mooswald—Močirje*
	Kreuzen—Krajcen
	Kamering—Kamerče
Political District Spittal an der Drau:	
Juridical District Winklern:	*Winklern—Kot*
	Heiligenblut—Sv. Kri
Juridical District Obervellach:	*Obervellach—Zgornja Bela*
	Kolbnitz—Holmic
Juridical District Spittal:	*Spittal—Spital*
	Lind—Lipa
	Lieserhofen—Jezerhofen
Juridical District Millstatt:	*Klein-Kirchheim—Mala Cirkica*
Juridical District Greifenburg:	*Berg—Gora*
Political District Hermagor:	
Juridical District Kötschach:	*Kötschach—Kotje*
	Mauthen—Muta

These names are only exceptionally Slovenian exonyms that had earlier been endonyms in the sense of names used and accepted by an earlier local Slavonic population. According to Eberhard Kranzmayer, many of them are constructed replacing a truly traditional name that had fallen into oblivion. Thus, the secondary Slovenian *Kotje* for *Kötschach* replaces a primary Slovenian *Hotešev(o)* as it can be deducted from Middle-Latin and Friulian derivates in upper-Carnian documents (Kranzmayer 1958: 125).

The Slovenian names for the contemporary Slovenian settlement area are partly close to dialect—as exemplified by dialectal *ves* instead of *vas* ('village'), the latter corresponding to the standard language already at that time.

The next gazetteer of 1894 (K.k. statistische Central-Commission 1894), restricts the use of minority names, while the gazetteer of 1905, edited in separate German and Slovenian versions, shows again a somehow more generous treatment, but less than the gazetteer of 1883. Due to the war, the Slovenian version of the gazetteer of 1918 (Statistische Zentralkommission 1918), was not published anymore.

In parallel to the gazetteers also the revised editions of the Carinthian Special Map sheets published up to 1899 display a reduced number of Slovenian names. Thus, in the Political District Völkermarkt with 67% Slovenian-speaking population according to the population census of 1900 only the German names of some larger populated places are accompanied by the Slovenian name. And even these names are written in smaller letters than in the edition before. Places with a distinct Slovenian majority according to the population census of 1900 like *Eberndorf* [Eberndorf/Dobrla vas], *Sankt Kanzian* [Sankt Kanzian am Klopeiner See], *Globasnitz* [Globasnitz/Globasnica], or *Sittersdorf* [Sittersdorf/Žitara vas] remain without a Slovenian name. With smaller villages like *Grafenstein (Grabštenj)* [Grafenstein], *Stein (Kamen)* [Stein im Jauntal], *Ebenthal (Žrelc)* [Ebenthal in Kärnten], *Stemeritsch (Smeriče)* [Stemeritsch], or *Tuzach (Tuče)* [Tutzach/Tuce], however, Slovenian names still appear. The selection looks inconsequent and does not correspond to the spatial distribution of Slovene-speakers. With names of farmsteads and hamlets the practice of the first edition was continued. Also with names of mountains, fields, and water bodies almost nothing has changed.

All later editions of the Special Map up to World War I present themselves in a similar way. A comparison between Sheet Bleiberg and Tarvis of the Special Map edited in 1881 and the same sheet with the same areal section edited in the course of the Fourth or Precision Survey (1896–1915) published in 1914[3] results in a reduction of Slovenian names of populated

[3] Special Map sheets of the Fourth Survey have of the parts of Carinthia populated by Slovene-speakers only been published on the Gail Valley due to the fact that this valley had to assume the role of a war front against Italy.

places in brackets from 83 to 9. The nine remaining populated places with their share of Slovene-speaking population according to the population census of 1910 are:

Förolach (Borlje) [Förolach]	87%
Görtschach (Goriče) [Görtschach]	90%
Vorderberg (Blače) [Vorderberg]	96%
Kerschdorf (Črešnje) [Kerschdorf]	62%
Emersdorf (Smerče) [Emmersdorf]	52%
Feistritz an der Gail (Bistrica na Zili) [Feistritz an der Gail]	82%
Dreulach (Drevlje) [Dreulach]	95%
Hohenthurn (Straja ves) [Hohenthurn]	89%
Seltschach (Sovče) [Seltschach]	59%

Again, Slovenian names in brackets are not indicative for the spatial distribution of Slovene-speakers at this point in time. Populated places with similar or even larger shares of Slovene-speakers like Achomitz [Achomitz/Zahomec] (100%), Sankt Paul an der Gail (99%), Draschitz (91%), Nampolach (90%), or Radendorf (84%) remain named only in German.

On the sheets of the General Map of Central Europe [Generalkarte von Mitteleuropa] in the scale 1:200,000 only one populated place is named bilingually in Carinthia: *Ebenthal (Žrelc)* [Ebenthal in Kärnten] east of Klagenfurt (Fig. 6.11). It is a small village and has no prominence also in the Slovenian context. Its Slovenian name has very likely been shown just by mistake.

In the Republic of Austria after World War I the responsibility for the publication of official topographical maps shifted after a certain transition period to the Federal Office of Metrology and Surveying [Bundesamt für Eich- und Vermessungswesen]. The scale 1:50,000 became its basic map series (Austrian Map [Österreichische Karte] 50) with derived series in the scales 1: 200,000 (Austrian Map [Österreichische Karte] 200) and 1:500,000 (Austrian Map [Österreichische Karte] 500). In this context, it would mean going too far to explain the development of the map work and of its cartographic design. It may be sufficient to state that relief representation underwent—compared to the Special Map—a fundamental change from slope hachuring to elevation lines and hill shading enabling a better readability of topographical content (populated places, traffic

Fig. 6.11 Section of the General Map of Central Europe [Generalkarte von Mitteleuropa] in the scale 1:200,000, Sheet No. 32 47 Klagenfurt, edited 1894. *Ebenthal (Žrelc)* is the only bilingual name in all of Carinthia. (Source: Austrian State Archive)

networks), also of place names in mountainous areas like Carinthia. Brown rock representation, a blue water network, black settlement, and traffic network representation as well as green forest areas complement the cartographic appearance.

As regards minority place names, representation principles did not significantly change between the first sheets in the interwar period and the editions in the late 1980s. They can be described as follows:

1. Bilingual naming of populated places is excluded. Hamlets, villages, and towns are (with the exception of *Dražja vas*) only labeled by their German name.
2. Some names of dispersed farmsteads and other individual buildings are rendered in Slovene orthography like *Breznik, Bukovnik, Kališnik, Meležnik, Pristovnik, Ravnjak*. This is based on the right of the owner

to name his/her property that can only be denied, when the name 'is against good morals'. The number of such names grew in the 1970s and 1980s.

3. With names of natural features of Slavonic origin Slovenian phonemes are usually rendered according to German orthography: *Gračenica* is converted to *Gratschenitza*, *Borovnica* to *Worounitza*, *Grintavec* to *Grintoutz*—with one major (see item 4) and some minor (see items 5 and 6) exceptions.
4. The major exception from item (3) are syntactical units (mostly noun plus adjective) in Slovene like *Tolsti vrh* ('bold mountain') *Velka ravna* ('large plain'). They remain written according to Slovenian spelling rules using the diacritical marks of this language. The explanation provided by Josef Breu, who proposed to do this, is "that these are not words fossilized as proper names, the meaning of which remains frequently opaque also for speakers of Slovene, but transparent syntactical configurations of the Slovenian language" (Breu 1970: 160). Following Josef Breu's proposal, who made it in his capacity of the contemporary chair of the Austrian Board on Geographical Names (AKO) and Austrian UNGEGN delegate, the number of such names grew, also by purifying mixed Slovenian/German spellings like replacing *Velka Rauna* by *Velka ravna*.
5. The number of single (not composed) names of natural features in purely Slovenian orthography, however, decreased up to the late 1980s from edition to edition down to the number of 27 due to a more consequent application of German orthography (see item 3) or the complete omission of names. Thus, the 27 are just the poor relics of a once rich toponymy in Slovenian orthography. They remained, since there is neither a German counterpart nor a version in German orthography in local use.
6. Exceptionally natural features are subject to dual naming, the second name in brackets being the Slovenian name in Slovenian or German orthography. This is practiced to present the autochthonous, still popular, and locally used Slovenian name mainly of mountains and mountain peaks in addition to the German name that was usually created in the course of a touristic opening up in the later nineteenth century by Alpine clubs. In some cases, the Slovenian name establishes

a relation to the name of another feature in the surrounding—as it is with *Kosiak*, the Slovenian bracket name of a mountain in the Commune Feistritz im Rosental with the German name *Geißberg* in the first position and with a cirque called *Kosiakkar*. Also mountains at the border with Slovenia (at that time Yugoslavia) are only named bilingually (the German name first and the Slovenian name in brackets), if one of the reasons mentioned applies (the Slovenian name is the locally popular name; reference to another Slovenian name). Examples are *Steinberg (Kamnik)* or *Koschuta (Košuta)* as the name of a mountain massif. The latter requires the bracket name in Slovenian orthography to refer to the peaks *Mala Košuta* and *Tolsta Košuta* that—according to the principle of rendering syntactical units (item 3)—had to be spelled in Slovenian orthography (Fig. 6.12).

The map series in the 1:200,000 scale (Austrian Map [Österreichische Karte] 200) shows more Slovenian names of natural features than one would expect having studied the map in the 1:50,000 scale. This larger

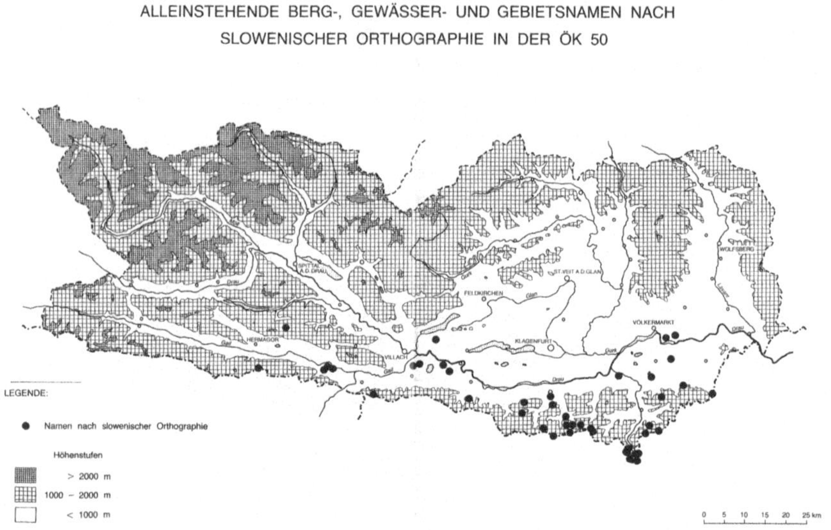

Fig. 6.12 Distribution of syntactical units (item 4) plus single names (item 5) in Slovenian orthography on the Austrian Map 1:50,000 of the late 1980s. (Source: Jordan 1988)

number is, however, only due to a more frequent dual naming of mountains and mountain passes at the border to Slovenia as it is usual with map works in smaller scales showing larger parts of neighboring countries. In contrast, due to generalization no more purely Slovenian names occur inside Austria.

The map series in the 1:500,000 scale (Austrian Map [Österreichische Karte] 500), however, quite unexpectedly, does not continue the trend toward bilingual names at the border with Slovenia, but has no such names at all.

As already explained in Sect. 4.1.6, the Decree of Topography for Carinthia [Topographieverordnung für Kärnten] of 1977 (BGBl. 306/1977, BKA 2020) ruled that 91 populated places [Ortschaften] in Carinthia had to have officially bilingual names. This Decree was, however, only gradually effectuated in physical reality (by town signs) and not at all on official Austrian topographical maps. Only a directive of the relevant federal ministry to the Federal Office of Metrology and Surveying resulting from a political process based on a scientific publication (Jordan 1988) initiated a gradual, but never complete adaptation of the maps.

This situation remained unchanged until Federal Act No. 46/2011 (BKA 2020) ruling that 164 populated places [Ortschaften] in Carinthia with a share of minority-language speakers of at least 17.5% have officially bilingual names. These names were completely and immediately incorporated into the Austrian Map 1:50,000. Both names figure in the same font type and size, but the Slovenian name is in brackets (Fig. 6.13). It would better conform to the concept of equally ranking official names, if both names were divided just by a slash.

An odd exception occurs in the Commune Zell, where a hamlet bears a single Slovenian name in Slovenian orthography: *Dražja vas* (Fig. 6.14). This would only be covered by federal and provincial law, if the hamlet consisted of only one inhabited building and was the property of the same owner. The name is, however, not new in the Austrian Map, but occurred already in editions long before 2011.

In the two smaller map scales, bilingually named populated places decrease in number, since they are to their major part small and subject to generalization. Just 19 of them remain:

Fig. 6.13 Bilingual naming of populated places in the Sattnitz/Gure on the Austrian Map 1:50,000 according to Federal Act No. 46/2011. (Source: BEV 2020)

Fig. 6.14 A monolingual Slovenian name of a hamlet in the Commune Zell: *Dražja vas*. (Source: BEV 2020)

Aich/Dob
Bad Eisenkappel/Železna Kapla
Bad Vellach/Bela
Bleiburg/Pliberk
Eberndorf/Dobrla vas

Einersdorf/Nonča vas
Globasnitz/Globasnica
Gösselsdorf/Goslena vas
Loibach/Libuče
Ludmannsdorf/Bilčovs
Maria Elend/Podgorje
Rinkenberg/Vogrče
Schwabegg/Žvabek
Sittersdorf/Žitara vas
St. Michael ob Bleiburg/Šmihel pri Pliberku
St. Primus/Šentprimož
Waidisch/Bajdiše
Windisch Bleiberg/Slovenji Plajberk
Zell-Pfarre/Sele-Cerkev

Two more officially bilingual populated places figure on the map in the 1:500,000 scale but are not named bilingually—obviously due to their length and a lack of space: *Feistritz ob Bleiburg/Bistrica pri Pliberk*u and *St. Jakob im Rosental/Šentjakob v Rožu.*

Names of mountains and mountain passes along the border are not consequently, but usually rendered bilingually, in one case also trilingually, either by adding the non-German name(s) in brackets as in the case of the *Dreiländereck (M. Forno, Peč)* or—much more frequently—by placing the other name in a different font type, but equal in size and without brackets at the opposite side of the border—as in the case of *Mittagskogel Kepa.*

Concluding this section on the diachronic development of minority place-name use (with a focus on maps and gazetteers) it can be stated that official topographic maps precisely reflect the development of minority legislation and the overall political, cultural, and societal climate between majority and minority. It has, however, to be borne in mind that state-official mapping emerges from mapping for military purposes and that minority names on maps in general and even more so on official maps have a very symbolic meaning. It can therefore be no surprise that the earlier Austrian surveys, top secret and devoted exclusively to military purposes, reflect minority toponymy insufficiently. It is rather a surprise that official topographical maps deviate from legislation—as it was with the first edition of the Third Austrian Land Survey in the positive direction and with the Austrian Map between 1977 and 2011 in the negative. Such deviations may be taken as indications of extraordinary efforts to reconcile ethnic/national animosities in the first case and of serious ruptures between state and minority in the second.

Indicators of these facts and developments are not only the relative share of minority names and their spelling according to the orthography of the minority or majority language, but also topographical consistency (Does a certain mode of minority-name rendering only affect certain feature categories or all of them?) and consistency across map scales (Do minority names only or predominantly occur in the largest scale or are they—in line with generalization—also present in smaller ones?). Another indicator is the visual representation of minority in relation to majority names: Are both represented in the same font size and type? Are they separated by a slash or is the minority name set in brackets?

Regarding these criteria the representation of minority names in Carinthia is—with the limited exception of the Third Austrian Land Survey's first edition of the Special Map 1:75,000—insufficient on Austrian official topographical maps. This is also true in comparison to other European countries as it can be stated based on Ferjan Ormeling's seminal work (Ormeling 1983) that investigated minority place-name rendering in the at that time political West of Europe including Austria. In 1988, the author of this subchapter attempted to attribute minority place-name rendering on official topographical maps in West-European minority situations based on Ormeling to seven quality levels (Jordan 1988: 11):

1. If features have minority names, minority names are represented not only in the minority area, but all over the country besides majority names or even without them. (This is an exceptional situation occurring only in Ireland, where the language of a small, but traditional minority has been elevated to the rank of the second national language besides English as a main identity marker of the Irish nation.)
2. If a feature has a minority name, the minority name is represented in the minority area without a majority name.

3. If a feature has a minority name, the minority name is represented in the minority area in most cases without a majority name. More important features bear a majority name in addition.
4. If a feature has a minority name, the minority name is represented in the minority area besides the majority name.
5. In the minority area the minority name is not shown with all features having a minority name. If the minority name is shown, it is represented as the only name of the feature.
6. In the minority area the minority name is not shown with all features having a minority name. If the minority name is shown, it is represented either as the only name of the feature or besides the majority name.
7. The map shows no names in the orthography of the minority language. Minority names either appear in the orthography of the majority language or are replaced by majority names.

The situation in Carinthia was corresponding at that time (after the Decree of 1977 and before the improvements following 1988) to quality level 6 and is still corresponding to it also after the Act of 2011 and the placement of 164 minority names of populated places on the Austrian Map 1:50,000.

6.1.2 Linguistic Landscape (Peter Jordan and Marika Balode)

The analysis of the linguistic landscape is particularly interesting in the Carinthian bilingual area, because bilingual signs in public space were the core issue of conflict between majority and minority in Carinthia in the period between the Treaty of Vienna (1955) and the so-called town-sign compromise [Ortstafelkompromiss] of 2011. It culminated in the so-called town-sign storm [Ortstafelsturm] of 1972, when bilingual town signs were vandalized (see Sects. 4.1 and 5.1).

As already explained in Chap. 3, the Carinthian case has in respect of the linguistic landscape been approached and analyzed not in the same way as the case of the Těšín/Cieszyn region. In southern Carinthia, the linguistic landscape was documented only by a selective survey of signs in 16 communes (see Fig. 6.15) due to the region's large extension and ter-

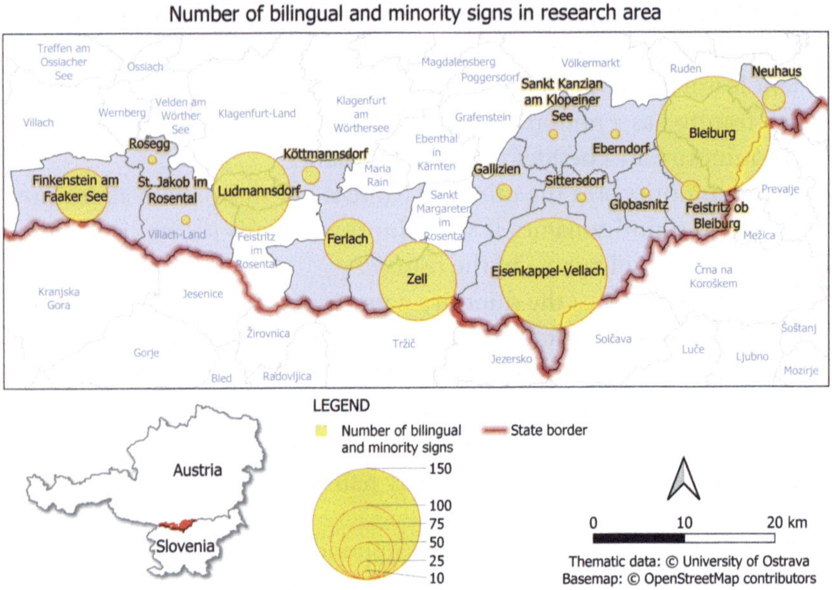

Fig. 6.15 Numerical structure of bilingual and minority language signs as (selectively) documented in southern Carinthia not including the Lower Gail Valley. (Cartography: Luděk Krtička)

ritorial shape, that is, forming a narrow fringe along Austria's southern border with little internal cohesion and composed of four historical-cultural landscapes (see Sect. 5.1.2).

Figure 6.15 shows the 16 communes surveyed. They do not cover the entire research area, that is, not the Lower Gail Valley [Unteres Gailtal/Zilja] forming the western end of the Carinthian bilingual area and composed of the communes Arnoldstein, Hohenthurn, Nötsch im Gailtal, Feistritz an der Gail, Sankt Stefan im Gailtal, and Hermagor-Presseggersee. In the area surveyed 2901 photos were taken. In contrast to the Těšín/Cieszyn region not all signs were covered. In contrast to the Czech case, the survey could therefore not be evaluated statistically-quantitatively. Evaluation and interpretation are just based on the photos taken as well as on the personal experience of the surveyors with the linguistic landscape of the region.

A more comprehensive survey was, however, conducted in two places: the town Ferlach (Borovlje) and the Commune Zell (Sele)—the first as an example for an urban center in the bilingual area with a distinct majority of German-Carinthians, the second as the only remaining commune with a numerical majority of the minority. In Ferlach near to all signs were documented (1401), in the Commune Zell at least a significant number (148).

The subchapter will at first refer to the situation in southern Carinthia in general followed by the two cases closer investigated and end with a conclusion referring to the hypotheses formulated in Sect. 3.2.

6.1.2.1 The Situation in Southern Carinthia in General

The following treatise will proceed by sign and linguistic landscape categories starting with mandatory bilingual signs in the sense of signs for which bilingualism is defined by law, continuing with voluntary bilingual signs and signage placed by the public sector (public authorities, NGOs, public institutions including religious communities, associations) followed by signs and other elements of the linguistic landscape placed by the private sector.

In the cases of mandatory bilingual signs, it will examine, whether the law has consistently been effectuated or whether it has been effectuated in a rather liberal or restrictive sense. In the cases of voluntary bilingual or monolingual Slovenian signs it will describe their relative share in the overall signage of a place in the relevant sign category, their visibility, their social or prestige level, regional differences in signage as well as spatial distribution related to the center-periphery and urban-rural dichotomies, try to explain the motives behind them and elucidate historical, political, cultural, and other reasons and backgrounds. It will also hint at cases, in which a bilingual sign or a bilingual linguistic landscape could have been expected according to standards, but is missing, as well as highlight exceptional cases in both directions.

Slovenian Place Names Are Mandatory

Slovenian place names in combination with German names are mandatory in Carinthia when this is explicitly defined by Federal Act 46/2011 and its Annex 1 (see Sect. 4.1.6, BKA 2020).

Town Signs

Town signs are certainly the element of the linguistic landscape with the highest symbolic value due to the facts that they are official, define (by their language) the identity of the populated place (see Chap. 2), and stand out also in terms of visibility. It is therefore no surprise that they were in the focus of political negotiations and disputes, while, for example, street names or names of communes were much less a concern.

Act 46/2011 and its Annex 1 define that 164 town signs in Carinthia must be bilingual (see Appendix A) and that both place names have to be reflected in the same font size and type. A memorandum of the same year opened the door for additional town signs, but on a less sound juridical basis. Indeed, in the meantime (in 2020) some additional town signs have been implemented in two communes (Sankt Jakob im Rosental, Sittersdorf). Town signs are placed along roads where populated places begin and end and appear in two versions: Town signs in larger populated places surrounded by a blue frame have at the same time the function of traffic signs ruling a speed limit of 50 km/h for drivers. Town signs in smaller villages, mostly just hamlets, are smaller in size, appear without a blue frame, and lack the traffic sign function.

Act 46/2011 and its Annex 1 have consequently been effectuated as regards town signs of both versions. They correctly show both official place names in the same font size and type, the German name in the first position. The examples in Fig. 6.16 are from the Jauntal/Podjuna.

Fig. 6.16 Mandatory bilingual town signs of larger (left) and smaller (right) populated places. (Photos by Marika Balode 2016)

Road Signs

According to Federal Act 46/2011/6/2, populated places defined by the Act as officially bilingual also have to be named bilingually on road signs hinting at these places, if these road signs are located on the territory of the respective populated place (BKA 2020). A verbal interpretation of the Act[4] would suggest that only place names on road signs located in the cadastral commune of the officially bilingual populated place are mandatorily bilingual, while a bilingual naming of populated places on road signs in larger distance from the destination (outside the respective cadastral commune)—common at least with larger, more important populated places—was not mandatory. Since cadastral communes in southern Carinthia (and elsewhere in Austria) are rather small and road signs hinting at a certain place are of course only necessary outside the same place, the Act sounds rather restrictive.

It is, however, practiced by communes with Slovene as the second official language according to Act 46/2011/8/1 that also outside the

[4] Federal Act 46/2011/6/2: "The obligation according to Paragraph 1 [to install bilingual signs] applies to the signs 'town sign' and 'end of the populated place', **but also to other signs** [road signs with names of officially bilingual places] **in the scope of the areas specified in Annex 1** [defining the 164 officially bilingual place names], **which refer to areas covered by Annex 1**" (for Annex 1, see Appendix A) [Die Verpflichtung gemäß Abs. 1 gilt für die Hinweiszeichen "Ortstafel" und "Ortsende", aber auch für sonstige Hinweisschilder im Bereich der in der Anlage 1 bezeichneten Gebietsteile, mit denen auf von der Anlage 1 erfasste Gebietsteile hingewiesen wird] (Bundeskanzleramt 2020).

respective cadastral commune an officially bilingual destination is indicated bilingually. The officially bilingual Market Commune Feistritz ob Bleiburg is a case in point. A road sign on the territory of this commune gives directions to two officially bilingual populated places, although the road sign is not located in the cadastral commune of either of these places (Fig. 6.17). But the Act is not always interpreted in this liberal sense—not even in the core area of the minority, for example, in the Market Commune Eisenkappel-Vellach (Fig. 6.18). What strikes there is that even the officially bilingual name of an important town like Bad Eisenkappel/Železna Kapla on a road sign along a major road is not represented bilingually (Fig. 6.18).

The naming of a mountain pass at the border between Austria and Slovenia (Seeberg/Jezerski vrh), that is of a transborder feature, on this same road sign is still another case. Naming practice of this kind of features is not defined by law, but it would just be logical to name it in

Fig. 6.17 Road sign in the officially bilingual Market Commune Feistritz ob Bleiburg directing to officially bilingual populated places with both of their names. (Photo by Marika Balode 2016)

Fig. 6.18 Road signs in the officially bilingual Market Commune Eisenkappel-Vellach pointing at the officially bilingual town Bad Eisenkappel/Železna Kapla as well as at the mountain pass Seeberg/Jezerski vrh at the border between Austria and Slovenia only in German. (Photo by Marika Balode 2016)

the languages of both countries—not only when the road sign is located in the bilingual area. In fact, it is just indicated by its German name (Fig. 6.18). But there are also proofs of the contrary, for example, a road sign hinting at Paulitschsattel/Pavličevo sedlo in the same commune (Fig. 6.19).

Slovene Is an Official Language

The use of the Slovenian language in combination with German is mandatory with the next three categories of the linguistic landscape insofar as Federal Act 46/2011/8/1 and its Annex 2 define that Slovene is a second official language in the commune. This does not mean that the name of the commune is officially bilingual. The Act can be interpreted as suggesting that all official texts at the façade of and inside public buildings of

Fig. 6.19 Road sign in the officially bilingual Market Commune Eisenkappel-Vellach pointing at a mountain pass at the border between Austria and Slovenia in both languages. The Slovene name is misspelled on the sign and should be *Pavličevo sedlo*. (Photo by Marika Balode 2016)

the commune to be bilingual. While the name of the commune is not explicitly defined as officially bilingual, it can be derived that in the context of official Slovenian signage and texts the Act implicitly allows the rendering of the commune name also in Slovene, since names are parts of a language.

Signage of Communal Offices

All communal offices of officially bilingual communes according to Federal Act 46/2011/8/1 and its Annex 2 are signed bilingually with the name of the commune also in both languages—not only at the façade (Fig. 6.20) and inside the building, but also as regards orientation signs hinting at it.

With communal institutions and activities like, for example, waste collection points or information boards, it can, however, be different

Fig. 6.20 Bilingual signage at the office entrance of the officially bilingual Market Commune Feistritz ob Bleiburg. The non-official Slovenian name of the commune is also shown (Photo by Marika Balode 2016)

(Fig. 6.21), sometimes just for very practical reasons, for example, because stickers are only available in German. But usually they are bilingual like the communal information board of a toll road in the Market Commune Eisenkappel-Vellach (Fig. 6.22).

Signage of Other Public Institutions

Three district courts in Carinthia (Ferlach, Eisenkappel, Bleiburg) are officially bilingual according to Federal Act 46/2011 and its Annex 2, and all are indeed signed bilingually, but not including the place name as shown by the example of the District Court Eisenkappel (Fig. 6.23).

Fig. 6.21 Monolingual German signage heading the information board of the officially bilingual Commune Neuhaus. (Photo by Marika Balode 2016)

Fig. 6.22 Bilingual communal information board of a toll road in the officially bilingual Market Commune Eisenkappel-Vellach. (Photo by Marika Balode 2016)

Fig. 6.23 The officially bilingual District Court Eisenkappel in Bad Eisenkappel/Železna Kapla. (Photo by Marika Balode 2016)

Signage of Schools and Kindergartens

In southern Carinthia there are 60 bilingual elementary schools, 17 bilingual secondary schools, and 3 bilingual high schools (Bildungsdirektion Kärnten 2019). It could be expected that all of them display a bilingual visual appearance. But this is not always the case. If a kindergarten or school is located in an officially bilingual village or town and in an officially bilingual commune, it is indeed without any exception signed bilingually, for example, in Ludmannsdorf/Bilčovs (Fig. 6.24).

Sometimes also schools in officially bilingual villages or towns and not officially bilingual communes or vice versa are named bilingually. Bilingual kindergartens and schools in places without any official bilingual status, but located in the bilingual area, are also usually named and signed bilingually—as the bilingual elementary school in Köttmannsdorf in the Sattnitz/Gure shows (Fig. 6.25). The building of the fire brigade in the same village is only signed in German.

Fig. 6.24 Bilingual kindergarten in the officially bilingual commune and village Ludmannsdorf/Bilčovs in the Sattnitz/Gure. (Photo by Marika Balode 2016)

Fig. 6.25 Bilingual kindergarten Köttmannsdorf in the Sattnitz/Gure, where neither the village nor the commune is officially bilingual. (Photo by Marika Balode 2016)

Fig. 6.26 Bilingual elementary school Egg in the Gail Valley. (Photo by Peter Jordan 2020)

Outside offcially bilingual communes, however, bilingual kindergartens and schools show only German names, even if their village/town has a significant minority tradition, as it is with the bilingual elementary school in Egg in the Gail Valley [Gailtal/Ziljska dolina], where the façade is only marked by German *Volksschule* ('elementary school') (Fig. 6.26).

Voluntary

This section includes all sign and linguistic landscape categories, whose language is not defined by law. They are classified according to the owner or the juridical or physical person responsible for signage into signs defined by the public or the private sector of society. Both appear and are accessible in public space.

It is this type of signs that best reflects actual practice and functionality of the minority language and its place names in the public and thus the situation 'on the ground', although in a democratic society like modern Austria also legal regulations cannot be classified as just 'top-down'. They, too, are the result of public opinion-building and it would not have been possible to implement them against the will of the majority at the federal, the provincial, and the communal level—as the unsuccessful attempt of 1972 vividly demonstrated it.

The private sphere as opposed to the public sphere is still another realm, and when the minority language is not presented by private owners in the public sphere very much, this would not exclude its more intensive use in the family.

Texts in/on sacral buildings and sacral monuments and even more so on graves may be regarded as an interface of public and private sphere. Being certainly parts of public space, they touch at the same time the most personal feelings of believers. Graves are not just public monuments, but also represent invested feelings of individuals and families.

Public Sector (Public Authorities, NGOs, Public Institutions Including Religious Communities, Associations)

Street Signs

It could be argued that in an officially bilingual populated place of an officially bilingual commune it is indirectly mandatory to name streets bilingually, even more so as street naming is a prerogative of the commune. In recent years, street signs have been introduced in many small towns and villages in Carinthia. However, until 2020 all street signs were, in contrast to Burgenland, the other Austrian province with autochthonous ethnic minorities, always only in German. In 2020 (and not anymore documentable by a photo in this book) the Commune Sankt Jakob im Rosental with a minority share of 17.5% in 2001 made the first advance. Until then, no bilingual street signage existed even in core places of the minority like Ludmannsdorf/Bilčovs with a minority share of 28.2% in 2001, where not only the village but also the commune is

Fig. 6.27 The usual kind of monolingual German street names in the officially bilingual village and commune Ludmannsdorf/Bilčovs. (Photo by Marika Balode 2016)

officially bilingual (Fig. 6.27). Conscious Carinthian Slovenes overrule this fact by signing their house numbers also with the officially bilingual village name.

The striking difference to Burgenland can be explained by the historical burdens influencing the Carinthian situation (see Chaps. 4 and 5)—in addition by the fact that today the Slovenian ethnic group is in a minority position in all Carinthian communes except in the small mountain commune Zell, that is, it does not have the power to decide about street names. It requires the good will of the majority or at least of their political representatives in the commune, in practice mainly the mayor's, to get bilingual street names granted. The Commune Zell, composed of dispersed hamlets without 'streets' in the classical sense, has not introduced street naming so far.

Welcome Signs

Welcome signs of communes are located, where a road crosses the commune's boundary—so usually not in a settled area, but in open space—and contrary to town signs they are not uniform, but very individual, even artistic, alluding sometimes to the character or identity markers of the commune. Even welcome signs of officially bilingual communes rarely have bilingual texts. This may be explained by their rather touristic than official character. A telling example is the welcome sign of the Commune Neuhaus in the Jauntal/Podjuna—a by Federal Act 46/2011 and its Annex 2 officially bilingual commune with an officially monolingual (German) name (Fig. 6.28). The sign shows (according to the Act) only the German name of the commune, but the texts are quadrilingual (German/Slovenian/Italian/English). Is it to please international tourists or to avoid bilingualism? The Slovenian version is at least placed in the second position. In the same commune a plate in front of a bridge with a warning presents a text in German, Italian, and Slovene—the Slovene version this time the last in sequence.

Also Sittersdorf in Jauntal/Podjuna is by law an officially bilingual commune, but its welcome signs are only German (Fig. 6.29). A contrasting example is Ludmannsdorf/Bilčovs with the same status in minority language terms as Sittersdorf, but with a partly bilingual welcome sign (Fig. 6.30). However, not the descriptive text (Erholungsgemeinde 'commune for recreation'), but the name of the commune, which is not officially bilingual, is presented bilingually.

Fig. 6.28 Quadrilingual welcome sign in the bilingual commune Neuhaus. (Photo by Marika Balode 2016)

Orientation Signs

These are signs hinting in the streets of populated places at features of public interest like the communal office, the police station, employment exchange, school, kindergarten, church, medical doctors, and so on. They are mostly placed by the commune, but sometimes also by institutions and private persons in accordance with the commune. They are usually small and have a very practical meaning and no representative function at all. Even in core bilingual areas like Bad Eisenkappel/Železna Kapla they are only occasionally bilingual (Fig. 6.31).

Tourist Signs

Tourism is economically important all over the bilingual area with concentrations on the lakes (bathing and summer recreation) and the

Fig. 6.29 Monolingual German welcome sign in the officially bilingual commune Sittersdorf. (Photo by Marika Balode 2016)

Fig. 6.30 Welcome sign in the Commune Ludmannsdorf, where Slovene is the second official language, but the name of the commune is not officially bilingual. (Photo by Marika Balode 2016)

Fig. 6.31 Orientation signs in Bad Eisenkappel/Železna Kapla. Two (*Arzt* 'medical doctor', *Polizei* 'police') are only in German, one is linguistically seemingly neutral (*Extrem Café*, although *extrem* would be *ekstremno* in Slovene) and one (*Gemeindeamt/občinski urad* 'communal office') is bilingual. (Photo by Marika Balode 2016)

Karawanken/Karavanke mountain range (hiking, mountaineering, skiing). But also places in between have their tourist attractions like historical town centers, places of pilgrimage, castles, or cultural events (see also Sect. 5.1). Generating markets are mainly the rest of Carinthia and Austria as well as Germany suggesting that a monolingual German signage was sufficient. A stronger internationalization of tourism started only after the fall of Communism in Eastern Europe but did not affect the bilingual area too much. Slovenia as a generating market plays only some role in winter(sport) tourism in the way of day trips. Since even the minority is not regarding a multicultural situation as a tourism asset (see interview results, Sect. 6.1.4), also this seems to be no argument for bilingual tourist signs.

Tourism advanced to an economically remarkable extent by the completion of the Carinthian Railroad [Kärntner Bahn] as a branch of the Southern Railroad's main route between Vienna [Wien] and Trieste, opened up to Klagenfurt and Villach in 1863 and 1864, respectively. The resulting influx of summer guests primarily from Vienna had been a major factor in pushing back the language boundary and interrupting the former continuous bilingual region in the lake areas. This is obvious in the linguistic landscape up to the present day.

Nevertheless, at least outside these tourism hot spots and due to local initiatives, some tourist signs are bilingual (Fig. 6.32), but in their majority they are only German. This applies even to the minority's core area of the Karawanken/Karavanke mountain range like in Ebriach/Obirsko (Fig. 6.33). Characteristically enough, due to the Church's role for the minority, the only bilingual sign in this sign conglomerate is the one hinting at the Roman-Catholic church.

Exceptional situations have emerged by Austrian-Slovenian touristic transborder cooperation projects supported by the European Union such as the climbing path Türkenkopf/Turško glavo in the Market Commune Eisenkappel-Vellach (Fig. 6.34) or the museum "Welt der Geologie/Svet geologije" in the same commune (Fig. 6.35).

Fig. 6.32 Bilingual tourist signs placed by a local initiative in the Slovenian majority commune Zell. (Photo by Nanti Olip 2010)

Fig. 6.33 Predominantly monolingual German tourist signs in Ebriach/Obirsko, an officially bilingual village in the Market Commune Eisenkappel-Vellach, where Slovene has the status of a second official language. (Photo by Marika Balode 2016)

Fig. 6.34 Bilingual sign hinting at the transborder project of a climbing path near Blasnitzen/Plasnica in the Market Commune Eisenkappel-Vellach. (Photo by Marika Balode 2016)

Fig. 6.35 Bilingual signage at the transborder project of a geological exhibition in Bad Eisenkappel/Železna Kapla. (Photo by Marika Balode 2016)

A remarkable innovation that would not have been imaginable two decades ago is also a board for tourist information in the officially bilingual village Globasnitz/Globasnica with a map of the wider area where all place names are presented in both languages (Fig. 6.36). Similar tourist and hiking maps, albeit with a focus on field and house names, have in recent years been published of and with the support of several (in majority German-Carinthian) communes (see Piko-Rustia 2016).

Inscriptions of Monuments

In general, monuments are very symbolic and representative. They are usually placed in the center of populated places (main square), in the vicinity of public buildings (communal office, school) or at churchyards. Monuments commemorating victims of the two world wars, the Carinthian Fight of Defense after World War I or the Carinthian Plebiscite of 1920 are very common in southern Carinthia. Monuments

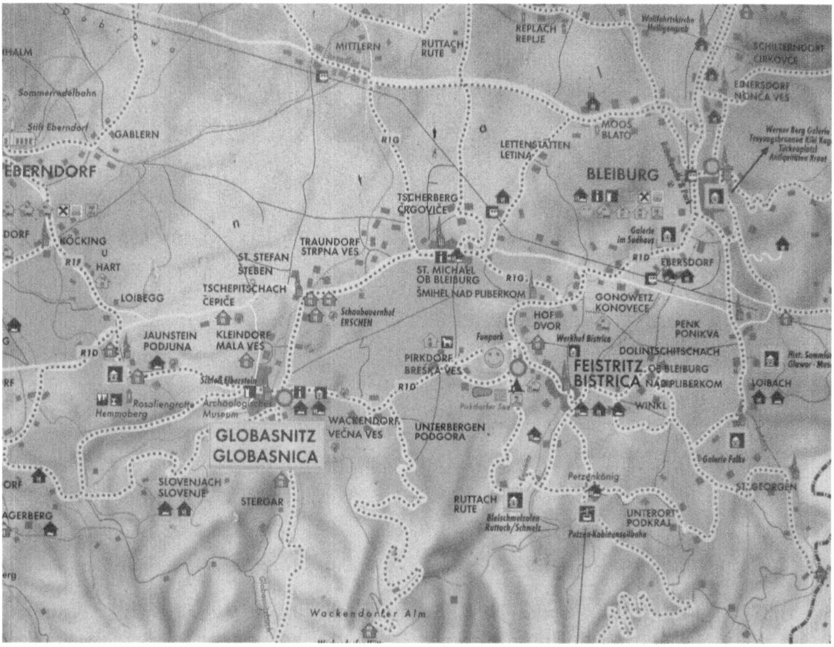

Fig. 6.36 Bilingual map on a tourist information board in the officially bilingual village Globasnitz/Globasnica, Jauntal/Podjuna. (Photo by Peter Jordan 2017)

commemorating partisan fights against the Nazi regime in World War II are comparatively rare and usually located at remote places. A specific and rather frequent variant of monuments is represented by plates reminding of prominent persons (artists, politicians, etc.) placed on walls of houses where they were born or had resided.

It depends on the institution or association behind the monument, whether it is only in German or bilingual. War monuments and monuments commemorating the Carinthian Fight of Defense after World War I as well as the Plebiscite of 1920 had usually been erected by veteran and homeland organizations and associations. They are usually monolingually German—as can be seen on a memorial of the Carinthian Fight of Defense combined with a monument reminding of the Carinthians deported to Yugoslavia at the end of World War II from the cemetery of Bad Eisenkappel/Železna Kapla (Fig. 6.37). However, the absence of a third monument—one to the Slovenian victims of the Nazi regime—is a telling indication of the continuing societal polarization in Carinthia.

Fig. 6.37 Monolingual German monuments reminding of the Carinthian Fight of Defense (left) and of Carinthians deported to Yugoslavia at the end of World War II (right) at the cemetery of Bad Eisenkappel/Železna Kapla. (Photo by Marika Balode 2016)

Monuments reminding of the partisan fights against the Nazi regime, usually implemented by Slovenian organizations, are bilingual. The example is taken from the village Abtei (Fig. 6.38) in the Commune Gallizien—both do not fall under Act 46/2011.

In recent years, plates commemorating prominent Carinthian Slovenes have appeared rather frequently and are always bilingual like the plate commemorating the priest, ethnographer, Slavist, and politician Matija Majar-Ziljski in Göriach in the Lower Gail Valley [Unteres Gailtal/Zilja] (Fig. 6.39).

Texts in/on Sacral Buildings, on Monuments, and on Graves

Besides language, Catholicism is certainly the most important identity marker of the Carinthian minority. The Roman-Catholic Church is its spiritual and emotional 'home' even after decades of secularization and

Fig. 6.38 Bilingual partisan monument in the village Abtei, Rosental/Rož. (Photo by Marika Balode 2016)

was at times—and to some extent still is—its organizational stronghold and backbone. Adherence to Catholicism marks to a significant degree the difference between Carinthian Slovenes and German-Carinthians, the latter having remained much more affected by Protestantism (see Sect. 5.1).

Not differing from all other parts of Austria, the cultural landscape of the bilingual area is 'impregnated' by Catholic monuments of all kinds

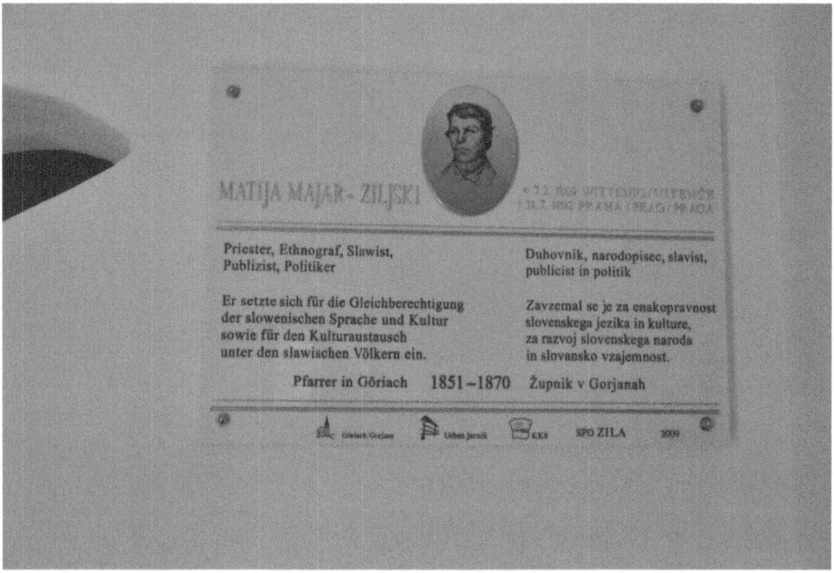

Fig. 6.39 Bilingual plate reminding of Matija Majar-Ziljski in Göriach, Gail Valley. (Photo by Milka Olip 2017)

ranging from the plenty of field crosses and wayside shrines to chapels, churches, and monasteries. They all contribute to the fact that in the linguistic landscape the Roman-Catholic Church is also a stronghold of the minority.

With churches it is not so much the façade of the building that mirrors Slovenian Catholicism, but the interior. Thus, ways of the cross paintings, painted glass windows, and altar cloths frequently have only texts in Slovene. A strong presence of Slovene is also typical for Roman-Catholic organizations like Caritas or cultural associations (Fig. 6.40). Even the usual announcements of church services along roads and at the entrance of populated places are sometimes bilingual (Fig. 6.41).

This applies, however, rather to the core area of the minority, that is, southern Carinthia east of Lake Faak [Faaker See/Baško jezero], not so much to places west of it. Thus, the Roman-Catholic church of Egg (Brdo) in the Gail Valley [Gailtal/Ziljska dolina] east of Hermagor, not an officially bilingual place by law, but the traditional center of the wider

Fig. 6.40 Monolingual Slovene sign of a Catholic cultural association in the officially bilingual town Bleiburg/Pliberk. (Photo by Marika Balode 2016)

Fig. 6.41 Bilingual announcement of a Catholic church service at the entrance of the officially bilingual town Feistritz ob Bleiburg/Bistrica pri Pliberku in the Jauntal/Podjuna. (Photo by Marika Balode 2016)

Fig. 6.42 Information board with (only) the place name in both languages at the entrance to the churchyard in Egg, Gail Valley. (Photo by Peter Jordan 2020)

bilingual area and a bilingual parish, presents itself rather ambivalently. While the information board at the entrance of the churchyard has a bilingual headline (Kath. Pfarrgemeinde Egg/Brda,[5] Fig. 6.42), the visitor would not deduce from the interior of the church that the parish is bilingual. Just the parish newsletter "farni list Brdo[6]/Pfarrbrief Egg" has a decorative bilingual appearance, while all the texts are only in German, and song books in German as well as Slovene are available. Thus, bilingualism even in this very traditional bilingual (and earlier Slovenian) parish is rather decorative than functional.

---

[5] *Brda* is the plural form of *Brdo* ('Hill') and differs from the form usually used for this place in Slovene.

[6] Here the singular form *Brdo* is preferred.

Fig. 6.43 Monolingual-Slovene grave at the cemetery of the bilingual town Bad Eisenkappel/Železna Kapla. (Photo by Marika Balode 2016)

Texts on graves in cemeteries (Fig. 6.43) are perhaps the closest approximation to the spatial distribution of the minority and its temporal development in the linguistic landscape, although with a time lag. Both can be derived from the texts accompanying the names of the faded away and their dates of birth and death. Even political impacts like the effects of World War II and the 'town-sign storm' of 1972 can be traced. What has to be considered, however, is (1) a certain bias in favor of the minority, since cemeteries are mostly part of the Roman-Catholic, that is, minority-friendly realm, (2) that they are substantially shaped by family traditions in the sense that a family grave with an old inscription is continued, although members of the same family having died later would not feel the same close tie to the minority anymore resulting in Slovenian inscriptions on a grave that would later have been German. Thus, churchyards may be regarded as reflecting minority shares of some decades ago prompting overestimation of current shares.

Private Sector

It can be assumed that language use by the private sector as compared to the public sector is less guided by considerations directed to the society at large and ethical principles but rather by very practical, personal, and commercial motives. If thus the part of the linguistic landscape contributed by the private sector reflects practical use and functionality of the language in the public sphere more than linguistic landscape elements contributed by the public sector, Slovene in southern Carinthia is in a very weak position.

This, however, allows neither to conclude to language use in the private sphere, that is, 'at home', in the family, nor to Carinthian-Slovenian group consciousness. This is confirmed by the examples of the Russian language in Ukraine or of English in Ireland, where the dominant use of Russian and English in the public sphere does not at all exclude a Ukrainian or an Irish national consciousness, respectively. Similar to Carinthia also in these cases a historical social stratification and the subsequent gradient in language prestige have caused a retreat of the submerged language into the private sphere, while an ethnic or national consciousness nevertheless developed.

Names of Bus Stops

Bus lines in the bilingual area are maintained by regional private enterprises operating also in other parts of Carinthia. Due to the fact that most bus lines had been operated by state-owned enterprises in a not too distant past, bus stops still have a rather 'official' image.

All over southern Carinthia, they have their standard stop signs in German and apply them usually without any modification. This is also true for the timetables to be found at these stations. The only exception discovered occurs in the 90% Slovenian-speaking Commune Zell (Fig. 6.44), while otherwise also core areas of the minority have no bilingual signs.

Fig. 6.44 Bilingual bus sign in Zell-Pfarre/Sele-Cerkev in the 90% Slovenian-speaking Commune Zell. (Photo by Nanti Olip 2010)

Names of Railroad Stations

Signage of all railroad stations in the bilingual area falls under the competence of the infrastructure branch of the Austrian Federal Railroads [Österreichische Bundesbahnen, ÖBB], a holding still state-owned and regularly subsidized by the federal budget, but de iure independent for about 20 years. Its image is nevertheless rather official, and so is that of the stations. They are practicing a uniform method of signage all over Austria. Thus, also the railroad station of Bleiburg/Pliberk is signed in German only, although town and commune are officially bilingual (Fig. 6.45).

This is much in contrast to practices in other European countries, where, for example, in the Scottish Highlands (United Kingdom) all railroad stations are also named in Scottish Gaelic, although hardly any active speaker of this language can be found there. But it contrasts also with Austrian traditions as it can be seen from the old station building of Bad Eisenkappel/Železna Kapla that was named bilingually. Figure 6.46

Fig. 6.45 Monolingual-German signage of the railroad station Bleiburg/Pliberk. (Photo by Marika Balode 2016)

Fig. 6.46 Restored former railroad station of Bad Eisenkappel/Železna Kapla. (Photo by Marika Balode 2016)

shows the building restored and used for other purposes by a private owner after the railroad station with finally only the German name on it had lost its function in 1966 due to the closing of the railroad track. From 1902, when railroad track and station were opened, up to at least 1930 it was named bilingually (Schmiedt et al. 2020: 49). The restored bilingual name corresponds by font size and color exactly to the historical example.

Signage of Production and Trade Facilities

Branches of enterprises operating nationwide or regionally (in Carinthia) present themselves usually in their standard outfit, that is, with signage only in German. This is also true for 'semi-public' companies like the Carinthian electricity, gas and heat supplier Kelag (Carinthian Electricity Corporation [Kärntner Elektrizitäts A.G.]) with labels of their branch offices, or warning plates at transmission lines and transformer stations. However, Trgovina ('shop') Krivograd, a larger trader with windows and doors of local origin, in the officially bilingual village Sankt Michael ob Bleiburg/Šmihel pri Pliberku of the officially bilingual Market Commune Feistritz ob Bleiburg, is signed in Slovene (Fig. 6.47), although even with this shop billboards are only in German.

Exceptions from the rule that enterprises operating nationwide present themselves without any modification also in the bilingual area are Raiffeisen (a bank) and Spar, a chain of supermarkets for foodstuff. Both cooperate in the bilingual area with enterprises of local origin and a significant Carinthian-Slovenian orientation or have integrated them: Raiffeisen with Posojilnica Bank and Spar with the company Zadruga. Thus, facilities of both companies in the bilingual area are signed by both names conveying a bilingual impression. By this cooperation and bilingual outfit, Raiffeisen underlines their intimate relation to local farmers with a strong Slovenian tradition. Spar emphasizes in this way their image of regionality, of selling regional products (Fig. 6.48).

But the cooperation between Raiffeisen and Posojilnica Bank is obviously in decline. While the abandoned facility of the bank in Feistritz an der Gail, where Slovene is not an official language but has strong tradi-

Fig. 6.47 The shop for windows and doors Trgovina Krivograd in Sankt Michael ob Bleiburg/Šmihel pri Pliberku. (Photo by Marika Balode 2016)

Fig. 6.48 Spar-Zadruga supermarket in the officially bilingual town Bleiburg/Pliberk. (Photo by Marika Balode 2016)

Fig. 6.49 Branch of Raiffeisen in Feistritz an der Gail with trilingual (but not Slovenian) signage. (Photo by Peter Jordan 2017)

tions, was signed by both labels, the new building of the bank bears a trilingual (German/English/Italian) signage (Fig. 6.49). This may be explained by the declining bilingualism in the Gail Valley [Gailtal/Ziljska dolina] and the vicinity of the Italian border.

The 'normal' appearance of small shops even in officially bilingual places and communes like Bleiburg/Pliberk, however, is purely German (Fig. 6.50).

Signage of Inns, Restaurants, and Hotels

As already explained, tourism is not a factor supporting bilingualism in Carinthia, rather the opposite. Thus, inns, restaurants, and hotels are only rarely named in both local languages, while fancy English, French, or Italian names can more easily be found, for example, in Egg am Faaker See (Fig. 6.51).

Fig. 6.50 Small shop for glass, porcelain, household and kitchen utensils in the officially bilingual town Bleiburg/Pliberk. (Photo by Marika Balode 2016)

An exception is Globasnitz/Globasnica with its Gostišče-Restaurant Juenna (Fig. 6.52), a commune that for a longer time (up to population census 1971) was the only Slovenian majority commune outside the Karawanken/Karavanke mountain area. A very telling example is the hand-written daily offer of dishes of an inn in the 90% Slovene-speaking commune Zell that is only in German (Fig. 6.53).

Texts on Billboards

Billboards have no symbolic value and are ephemeric. It has also to be considered that they are usually produced for a wider range of shops, not just for those in the bilingual area. They are almost always only in German—even with companies with a strong Carinthian-Slovenian tradition like Trgovina Krivograd in the bilingual village Sankt Michael ob Bleiburg/Šmihel pri Pliberku (Fig.6.54) or in the only commune with a Slovenian majority, Zell (Fig. 6.55).

Fig. 6.51 Orientation board on tourist accommodation facilities in Egg am Faaker See, a tourism hot spot in the bilingual area. (Photo by Marika Balode 2016)

Fig. 6.52 Bilingual signage of a restaurant in the bilingual village and commune Globasnitz/Globasnica, Jauntal/Podjuna. (Photo by Marika Balode 2016)

Fig. 6.53 Monolingual-German hand-written daily offer of dishes of an inn in the 90% Slovene-speaking commune Zell. (Photo by Marika Balode 2016)

Fig. 6.54 Monolingual-German billboards at the entrance of the otherwise bilingual shop Trgovina Krivograd in Sankt Michael ob Bleiburg/Šmihel pri Pliberku. (Photo by Marika Balode 2016)

Fig. 6.55 Monolingual-German billboards in the only Slovenian majority commune Zell. (Photo by Marika Balode 2016)

Texts on Posters

What has been said about billboards with regard to social rank and continuity may be applied to posters as well. In the bilingual area, language on posters varies depending on location and organizers of the events advertised. They can have a share of bilingual or even monolingual Slovenian texts in core areas of the minority, especially if the event has a folkloristic character. But even there we find German-only posters as the example of the soccer club of the officially bilingual village and commune Ludmannsdorf/Bilčovs demonstrates (Fig. 6.56).

6.1.2.2 The Two Cases Closer Investigated

In the two cases investigated closer we tried to document the complete linguistic landscape of a town area (Ferlach) and a commune (Zell) except texts on graves. We classified it by sign categories and ownership and

Fig. 6.56 Monolingual-German poster announcing a game of the local soccer club in the officially bilingual village and commune Ludmannsdorf/Bilčovs. (Photo by Marika Balode 2016)

mapped it with the aim of discovering regularities of its spatial distribution. The two cases are extremely contrasting by size, structure, and character as their following description will show.

Ferlach (Borovlje) is located in a section of the Drau/Drava valley named Rosental/Rož, at the bottom of the Loibl/Ljubelj mountain pass (1368 m) along an important road directly connecting Klagenfurt am Wörthersee, Carinthia's capital, and Ljubljana, the capital of Slovenia. Its location only 16 kilometers south of Klagenfurt am Wörthersee incorporates it closely into this almost exclusively German-speaking city's catchment in terms of labor and educational commuting as well as shopping. Ferlach's own catchment, however, extends well into the mountain area south of it with a significant share of Slovene-speakers. Like all other towns in the bilingual area, Ferlach was historically shaped by a German-speaking urban population, mainly crafts- and tradesmen. In the sixteenth century a production of weapons was initiated that in the later nineteenth century developed into an industry of hunting weapons, for which Ferlach became famous in Austria and beyond. It made the town also the location of a professional high school for this field that attracts students from all of Austria. Weapon production was later complemented by industrial wire and steel production that is up till today continued by the production of tools (Köstler 1986). These manufacturing and industrial traditions resulted in an almost monolingual German-speaking population and at times in particularly German-national attitudes as they were represented by the town's prominent son, the poet Josef Friedrich Perkonig (1890–1959), who—not at all denying and frequently alluding to his Slavonic roots—joined the extreme German-national side in the later interwar period (see Rumpler 2005: 72).

The Commune Zell (Sele) spreads in the Karawanken/Karavanke mountain range along the border to Slovenia at an altitude of 948 m of its communal center. It is characterized by dispersed settlement composed of several small villages, hamlets, and individual farmsteads with all public institutions concentrated in a centrally located small village (Zell-Pfarre/Sele-Cerkev). Its peripheric location in the mountains with forestry and agriculture (animal husbandry) as the main local activities (except commuting to external centers) may explain its position as the only predominantly (90%) Slovene-speaking commune in Carinthia according to the population census of 2001 (see Tables 6.1 and 6.2).

Table 6.1 Basic characteristics of the cases closer investigated

Case	Population (2019)	% of Slovene-speakers (2001)[b]	Number of signs	Communal area (ha)	Signs/ inhabitant	Signs/ ha
Ferlach	4478[a]	8.7%	1401	11,731	0.31	0.12
Zell	702	90.0%	187	7544	0.21	0.02

[a]Inhabitants of the town only, not of the entire commune
[b]Slovene- and 'Windisch'-speakers added. Figures refer to the whole commune

Table 6.2 Signs by language in Ferlach and Zell

Case	German only	Bilingual	Slovene only	Multilingual	Other	Total
Ferlach	1258	53	10	23	41	1401
	90%	4%	1%	2%	3%	100%
Zell	59	62	27	0	0	148
	40%	42%	18%	0%	0%	100%

Ferlach

Judging from the linguistic landscape, Ferlach is an almost exclusively German-speaking place. The share of bilingual German-Slovene and monolingual Slovene signage amounts to no more than 5%. Multilingual signs—mostly with English and Italian—and signs with other languages—mainly connected with Turkish migrant workers—make up for another 5%. Multilingual signs are a matter of private owners (Fig. 6.60), mainly of inns and restaurants (Fig. 6.61).

As the three maps show (Figs. 6.57, 6.58, and 6.59) bilingual and purely Slovenian signs are, with the limited exception of some private information boards, confined to the town center and to the cemetery northeast of it. Thus, the symbolic places around the main square, the town hall, and the Catholic parish church as well as schools and kindergartens are clearly the peak of bilingual signage, while monolingual-Slovenian signage is more or less confined to the Roman-Catholic sphere—that is, church and cemetery. These findings are in line with observations in Bleiburg/Pliberk and Bad Eisenkappel/Železna Kapla, the other two urban centers of the bilingual area except the district center Völkermarkt.

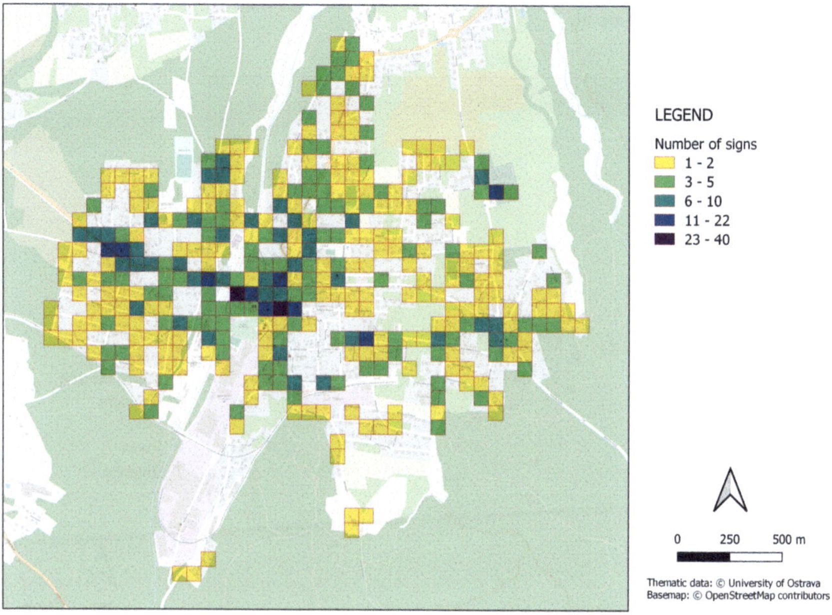

Fig. 6.57 Spatial distribution of signs in Ferlach in 2017. (Cartography: Luděk Krtička)

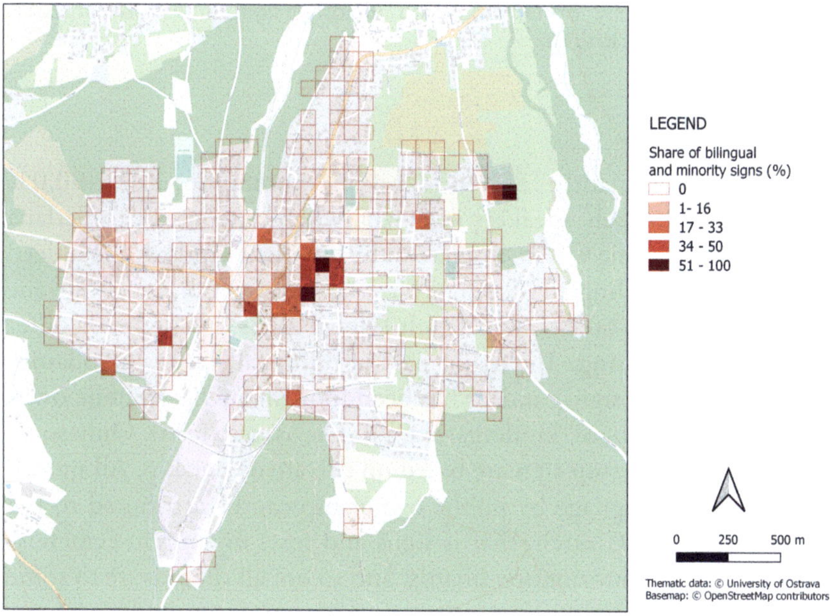

Fig. 6.58 Share of bilingual and minority signs in Ferlach in 2017. (Cartography: Luděk Krtička)

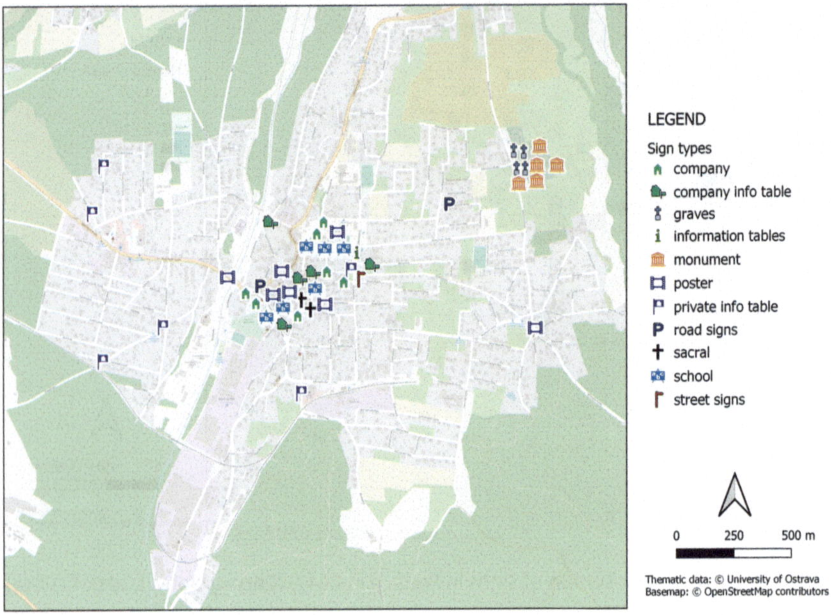

Fig. 6.59 Topography of bilingual and minority signs in Ferlach in 2017. (Cartography: Luděk Krtička)

Zell

In the predominantly Slovene-speaking commune Zell (90.0% Slovene + 'Windisch' speakers in 2001), where in contrast to Ferlach the entire commune was investigated, the linguistic landscape is indeed dominated by bilingual German-Slovene and monolingual-Slovene signs (42 + 18 = 60%), while multilingual signs and signs in other languages are missing. The major part of bilingual signs is mandatory (town signs, road signs) and/or has been implemented by public owners (commune, school, kindergarten, NGOs, associations). Only some posters and a bus-stop sign are based on private ownership. All monolingual-Slovene signage by public ownership can be attributed to the Roman-Catholic Church, that is signs and texts in church, cemetery, parish office, on information boards, and so on; all the private to some obviously very conscious Slovenian house owners naming their houses

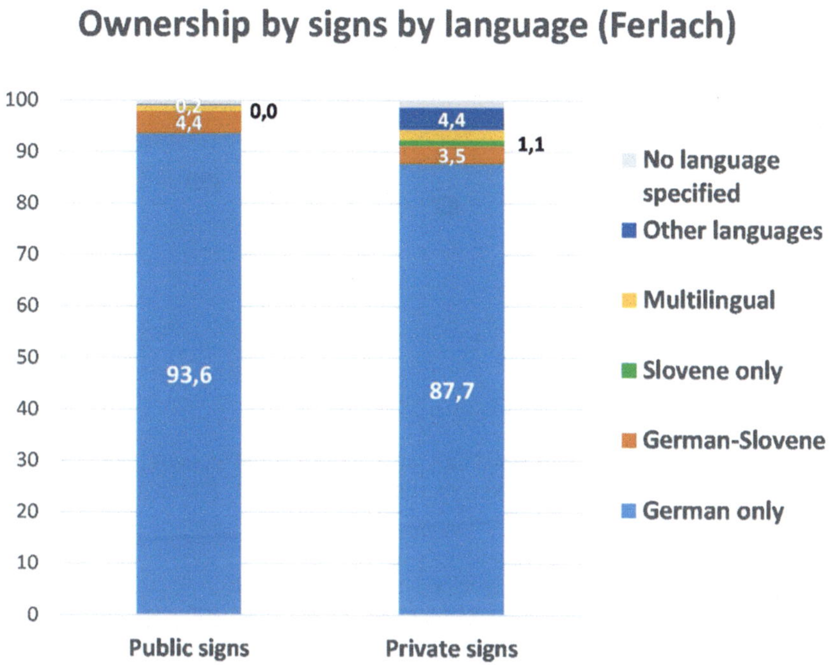

Fig. 6.60 Signs by language and ownership in Ferlach in 2017. (Graphics: Luděk Krtička)

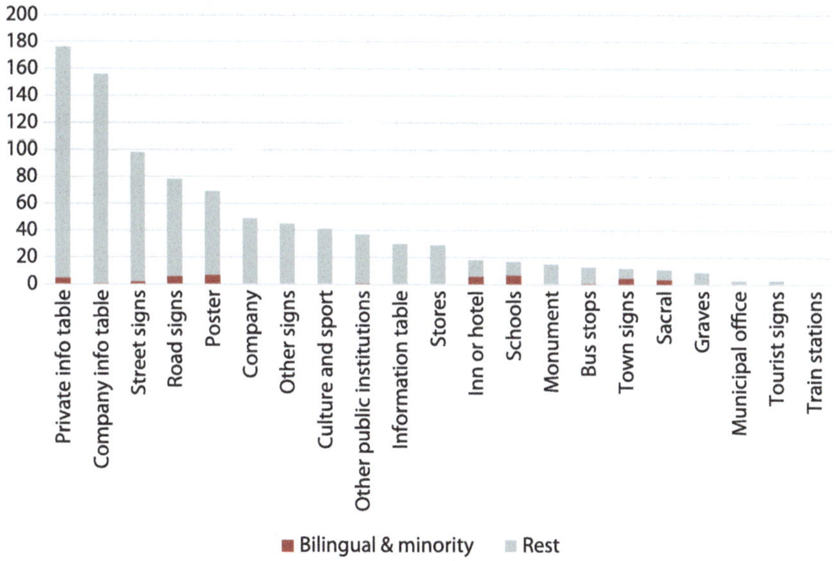

Fig. 6.61 Signs by type and language in Ferlach in 2017. (Graphics: Luděk Krtička)

Fig. 6.62 Bilingual German-Slovenian and monolingual-Slovene signs in the main village Zell-Pfarre/Sele-Cerkev of the commune Zell. (Cartography: Luděk Krtička)

in Slovene. Monolingual-German signage dominates with private ownership. The signage and texts of shops, inns, billboards, and posters are practically exclusively German. With signs of public owners, monolingual-German signs dominate with older tourist signage, while more recent signs tend to be bilingual.

Figure 6.62 shows only the main village Zell-Pfarre/Sele-Cerkev of the commune with its bilingual and Slovenian signs marked by symbols (for legend see Fig. 6.59). It is clear enough that minority signs also concentrate in the center and its symbolic places. However, if the map extended to the entire communal area, it would also show that especially bilingual signs are much more dispersed than in Ferlach, mainly due to the high number of tourist signs in the area frequented by mountain hikers and bikers, but also due to the mandatory bilingual town signs of all the hamlets spreading over the commune.

6.1.2.3 Conclusion

Let us finally reflect on the hypotheses mentioned in Sect. 3.2 on the background of our findings.

- Hypothesis: Collective identity is absent, at least in top-down flow, where power relations work against minorities.

A collective identity of the minority in Carinthia is not absent although power relations worked against the minority during long periods of history. As regards identity, the minority is composed of a core group or core groups and a fringe with a wide gradient. This is also reflected by the linguistic landscape, where the core group around the Catholic Church stands out by monolingual-Slovene and bilingual signs. The town-sign compromise of 2011 may be regarded as a turning point in power relations: For the first time since the later nineteenth century mandatory bilingual signs were placed (and not removed) where no other component of the linguistic landscape would suggest that there is a minority. Thus, the result of a "top-down initiative" (if it can be called so) results in a more minority-friendly face of the linguistic landscape than most bottom-up activities. This in turn can be explained by the low prestige of the minority language due to former power relations.

- Hypothesis: Where there is official policy promoting minority languages, the linguistic landscape associated with minorities may be more visible in top-down signs than the bottom-up flow.

  This hypothesis can be confirmed related to the period after 2011, that is, mainly due to the town-sign compromise, but also due to many other activities like bilingual tourist signage and bilingual information boards that started to flourish in a relaxed political climate.

- Hypothesis: Where multiculturalism is taken for granted, minority signs should be equally frequent in the top-down and bottom-up flows.

  This is certainly also a valid hypothesis and is confirmed by the Carinthian situation insofar as it can still not be regarded as a situation, where multiculturalism is taken for granted. Although politics and inter-ethnic relations are certainly more in favor of the minority than 15–20 years ago, the situation still suffers from historical burdens and animosities. In consequence, top-down and bottom-up flows are not at all equally frequent with minority signs.

- Hypothesis: The location of bilingual and minority signs is a simple function of the location of all signs (i.e., where there are more signs, there will be more bilingual and minority signs).

This hypothesis is falsified by the Carinthian situation. The Lower Gail Valley [Unteres Gailtal/Zilja], for example, has a certain share of minority population and quite a density of signs, but the minority is almost invisible in the linguistic landscape. The lake district around Lake Klopein [Klopeiner See/Klopinjsko jezero] has a higher minority share and due to its intensive tourism a very dense linguistic landscape, but also there the minority remains almost invisible.

- Hypothesis: The location of bilingual and minority signs is associated with symbolic points in the populated place (schools, communal office, settlement boundaries).

  The Carinthian findings seem to agree with this hypothesis. As systematically tested in the cases of Ferlach and Zell and by accidental evidence found also in many other situations, the center of a populated place with most of the symbolic sites is also the peak of minority signage. In addition, the very symbolic town signs along settlement boundaries are bilingual and in the Carinthian case also the result of fierce, finally successful fighting for minority rights.

- Hypothesis: The number of bilingual and minority signs correlates with the number of minority members in the populated place (ethnolinguistic vitality hypothesis of Landry and Bourhis 1997).

  This hypothesis could not systematically be tested in the Carinthian case, since no complete survey of the linguistic landscape was conducted. Accidental evidence, however, suggests that it is valid. Thus, the Karawanken/Karavanke mountain area with a higher share of the minority compared to the Gail Valley [Gailtal/Ziljska dolina] or the Rosental/Rož, or the village Ludmannsdorf/Bilčovs with a higher share compared to other villages of the Sattnitz/Gure region can be taken as proofs of it.

What could additionally be concluded from the Carinthian case is that also tourism and industrialization have a strong differentiating impact on the share of the minority in the linguistic landscape by substantially reducing the share of minority signs. This results in the fact that the linguistic landscapes of the tourist regions around Lake Faak [Faaker See/Baško jezero] and Lake Klopein [Klopeiner See/Klopinjsko jezero] are almost monolingual-German, although they have not the lowest minority shares. The same is true for areas with major industry and/or industrial traditions, especially Arnoldstein and Ferlach. The reasons have already been explained in this subchapter and need not to be repeated here.

A final observation relevant in this context is that some parts of the bilingual area have a lower share of minority signs not due to the objective reasons and regularities mentioned before, but due to specific local traditions and attitudes, for example, regarding Slovene—or rather Alpine-Slavonic dialects—as a private and not a public affair and affiliating themselves not to the Carinthian-Slovenian community in the cultural-political sense. This applies, for example, to the Lower Gail Valley [Unteres Gailtal/Zilja].

6.1.3 Media Discourse on Minority Place Names (Peter Jordan and Alexis Sancho Reinoso)

As already explained in Chap. 3, for southern Carinthia opinions published in newspapers, both in German and Slovene, were collected. For newspapers in German the main source was the Austrian Press Agency's [Austria Presse Agentur, APA] digital database (APA 2002), which documents all media in German in Austria and South Tyrol [Südtirol/Alto Adige] as well as selected media in other German-speaking countries (Germany and Switzerland). In practice, most of the analyzed items were press releases by APA. In addition, articles from the following newspapers were included in the sample: "Kleine Zeitung" (Klagenfurt), "Kärntner Tageszeitung" (Klagenfurt, closed in 2014), "Der Standard" (Vienna [Wien]), "Die Presse" (Vienna), "Kurier" (Vienna), "Wiener Zeitung" (Vienna), "Wirtschaftsblatt" (Vienna), "Vorarlberger Nachrichten" (Schwarzach, Vorarlberg), and "Frankfurter Allgemeine" (Frankfurt am Main). As regards newspapers in Slovene, both Slovenian newspapers issued in Carinthia ("Novice" and "Nedelja") were researched via a digital repository. No media in Slovenia were included in the analysis. Keywords used were "Ortstafel/mestni znak" ('town sign') and "Kärnten/Koroška" ('Carinthia'). The time span investigated extended from 2001 to 2017. In addition to the digital database also a private collection of newspaper articles about the town-sign affair in Carinthia [Ortstafelkonflikt] was evaluated. The sample amounted to 199 articles fairly balanced by language: 87 in German and 112 in Slovene, that is, 44% and 56% (see Fig. 6.63).

Events and issues resulting in the temporal distribution of newspaper articles on our topic as represented by Fig. 6.63 were the following:

- 2005: 50th anniversary of the Treaty of Vienna (1955) that laid by its Article 7 the foundations of minority legislation in postwar Austria;
- 2005–2006: Federal government (administration Wolfgang Schüssel) starts negotiations on bilingual town signs in Carinthia;
- 2009–2011: After Jörg Haider's, the Carinthian governor's, death (2008) and elections to the Carinthian provincial government (2009), the new Carinthian provincial government led by Governor Gerhard Dörfler and the federal government (administration Werner Faymann) start a next attempt to solve the question of bilingual town signs in Carinthia;
- 2011: Carinthian "town-sign compromise" settled;
- 2013: Elections to the Carinthian provincial parliament resulting in a fundamental change in majority from the successor of the Freedom Party to Social Democrats;

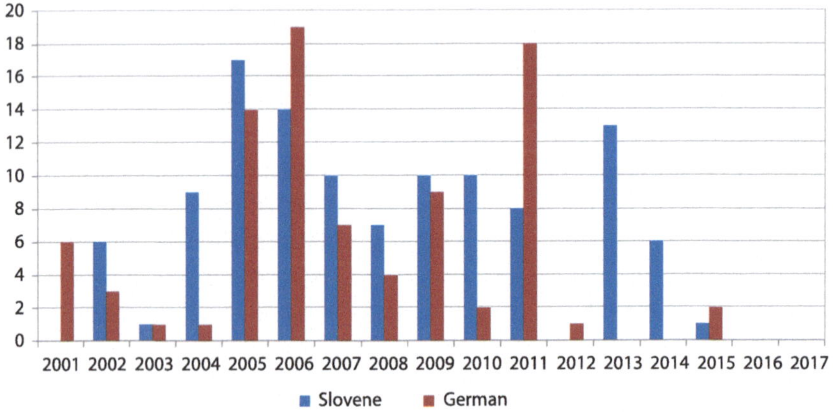

Fig. 6.63 Number of newspaper articles analyzed by year and language. (Graphics: Luděk Krtička)

- 2014: Elections to the European Parliament with a Carinthian Slovene (Angelika Mlinar) as the Austrian top candidate of the political party "Neos".

The following analysis presents arguments classified by character into "pragmatic" and "symbolic" as well as by principal political direction into "for" or "against" bilingual signs following in principle the scheme applied in the Těšín/Cieszyn region.

Before going into it, it deserves, however, to be noted that for most newspapers in German the town-sign question was not so much a topic discussed in its own right. It rather served as the smokescreen for addressing the political situation in Carinthia in general that presented itself in most of this period (up to 2013) rather exceptional. In contrast to all other Austrian federal states and also to the federal level, where the Freedom Party [Freiheitliche Partei Österreichs, FPÖ] and its successor, the Alliance Future Austria [Bündnis Zukunft Österreich, BZÖ], were just in the position of the junior coalition partner of the administrations Wolfgang Schüssel I (2000–2003) and Wolfgang Schüssel II (2003–2007)—Federal Chancellor Schüssel representing the conservative Austrian People's Party [Österreichische Volkspartei, ÖVP]—Carinthian governments were led by the Freedom Party and its successors and their representatives as governors of Carinthia, that is, the administrations Jörg Haider II (1999–2004), Jörg Haider III (2004–2008), Gerhard Dörfler I (2008–2009), and Gerhard Dörfler II (2009–2013).

This political exception has also historical reasons as explained in Sects. 4.1 and 5.1 and was as such a frequent topic of discussions in Austrian media, all the more as Governor Jörg Haider had remained the leader of the Freedom Party also when it was part of the Austrian government and liked to comment on and interfere into federal politics from his Carinthian governor position. Even after his party had left the Austrian federal government (from 2007 on) and up to his mortal accident in 2008 he continued to influence federal politics and to be a significant and controversial political figure in Austria and beyond. Demonized as an exponent of right-wing populism as well as admired as a Robin Hood fighting for the underprivileged (and many in peripheric Carinthia feel to be underprivileged) he attracted media interest nationwide and even internationally.

The town-sign question served thus just as the most tangible and popular feature for commenting on him and his politics in general—all the more as he liked to 'play' with it by at one moment opposing "inappropriate demands" of the Carinthian Slovenes and thus soliciting for sympathy from German-Carinthians and in the next moment extending 'generous offers' to the minority, for example, by 'granting' it an additional bilingual town sign (Bleiburg/Pliberk) and posing photogenically in front of it, although the Slovenian sign was much smaller and just attached to the sign with the German name. Figure 6.64 may be taken as satirizing Carinthia as Haider's playground marked by town signs.

Thus, Carinthia was certainly a theater of federal politics at least up to Haider's death in 2008. But it continued to attract nationwide interest also up to 2013 as long as offsprings of the Freedom Party were ruling in Carinthia and the Freedom Party was the strongest opposition on the federal level. It was interesting to watch this Party's performance in its Carinthian stronghold.

Table 6.3 presents a summary of arguments classified by character into "pragmatic" and "symbolic" and by political direction "in favor" or "against" minority place names in southern Carinthia. As it can be seen,

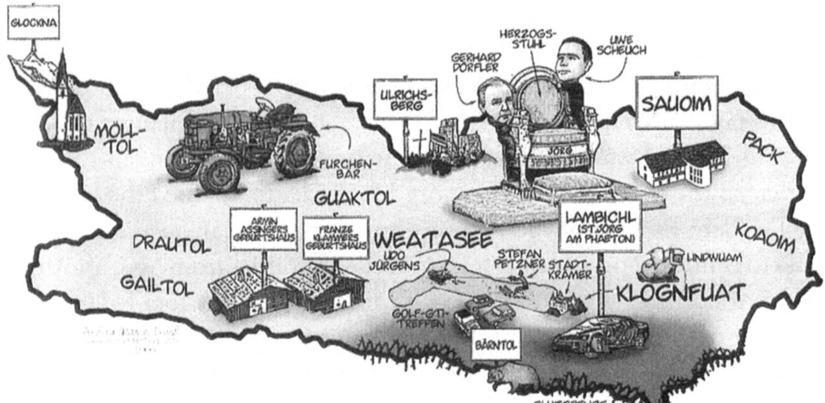

Fig. 6.64 Carinthia with "Haider's places" indicated a.o. by town signs with names in satirizing German-Carinthian dialect orthography. (Design: Andrea Maria Dusl)

Table 6.3 Structure of arguments on bilingual signs found in the sample of newspaper articles

	Arguments	For (%)	Against (%)	Total
German	Pragmatic	31 (54)	26 (46)	57
	Symbolic	18 (60)	12 (40)	30
Slovene	Pragmatic	64 (100)	0 (0)	64
	Symbolic	48 (100)	0 (0)	48

both with newspapers in German and Slovene pragmatic arguments prevail, while with newspapers in Slovene the share of symbolic arguments is higher. This can be explained by a more distanced relation of Austrian newspapers outside Carinthia to the Carinthian place-name question provoking a rather rational than emotional approach and regarding pragmatic arguments like "official bilingual place names conform to minority rights" and "their presence also on town signs is just consequent" as fully sufficient. They are neither affected by the German-Carinthian 'primeval fear' [Urangst] phenomenon addressed earlier nor by the feeling of Carinthian Slovenes not to be sufficiently acknowledged as a constitutive part of Carinthian culture and identity.

It needs to be mentioned in this context that the Carinthian place-name dispute has in Austria outside Carinthia rarely, also before the period investigated, been discussed more profoundly, that is, including its historical reasons and emanating animosities, but rather at the level of adequate or insufficient minority legislation disregarding the fact that conflicts on place names are usually just covering deeper societal cleavages. Not a few Austrian journalists and other intellectuals outside Carinthia tended even to feel ashamed of the Carinthian case on the international scene and classifying the Carinthian 'camps' as irrational.

This is also the main reason for the prevalence of arguments in favor of bilingual town signs—besides political correctness as a high value in published opinion. The exclusiveness of arguments in favor of bilingual town signs with newspapers in Slovene is no surprise, since they are all published by Carinthian-Slovenian organizations and express thus the opinion of the minority's political elite. Arguments against bilingual town signs in newspapers in German can practically only be found in newspapers published by the Carinthian homeland organizations, not by independent media whether they appear in Carinthia or elsewhere in Austria.

Table 6.4 Pragmatic arguments disaggregated (gray fields = no occurrences)

	For				Against			
	German		Slovene		German		Slovene	
Argument type	*N*	%	*N*	%	*N*	%	*N*	%
Language preservation			15	23				
Orientation function			1	2				
Costs					1	4		
Thresholds	11	35	7	11	13	50		
"External influence"	15	48	31	48	6	23		
Economic value	5	16			2	8		
Other			10	16	4	15		
Total	31	100	64	100	26	100	0	0

As regards **pragmatic arguments** (see Table 6.4), roughly half of the articles **in favor of bilingual signs** both in German and in Slovene refer to "external influence". Typical for this kind of arguments in Carinthian-Slovenian media is for instance: "We (the Carinthian Slovenes) have a legal right to the signs as part of what was established in the Treaty of Vienna 1955. Otherwise this would mean disobedience of the rule of law. It's a shame that we have to claim our rights in Vienna."

Another common topic is the numerical threshold for minority rights in general and bilingual town signs in particular including arguments like the following: "There are enough Slovenes to claim rights; no census is needed; all town signs in southern Carinthia could be bilingual or at least all of populated places above 10% Slovene-speakers; street plates should also be bilingual."

Major divergencies between articles in German and Slovene arise related to the meaning of bilingual signs for language preservation and their economic value. While newspapers in Slovene often argue that "bilingual signs help to keep the language alive and in daily use", this kind of arguments could not be found in media in German.

On the one hand, this has to do with the concern of Carinthian Slovenes that with the loss of their language they would lose their main identity marker. On the other hand, it is also associated with little up to no acquaintance of Austrians outside the minority with the minority's cultural disposition. Hardly any Austrian, especially outside Carinthia, is informed about the language(s) spoken by the Carinthian minority: Is this standard Slovene, a common dialect (perhaps 'Windisch'?) or do they speak several dialects? How are these linguistic variants developing? What do they mean for the minority's identity in relation to other identity markers like Catholicism? That arguments of this kind are absent in media in German language can thus be taken as an indication of the fact that (also intellectual) interest in the minority goes hardly beyond observing (at times) spectacular conflict over the place-name question.

The opposite happens with arguments related to the economic value of bilingual signs. They occur only in articles in German, not in Slovene, for example, arguing that "linguistic minorities in Europe help thriving European economy; bilingual town signs may become a touristic attraction for the region". Their authors may have thought of the successful examples of the Croats in Burgenland, of South Tyrolians and Ladins, of the so-called German language islands in Friuli-Venezia Giulia and Veneto like Sauris/Zahre or Sappada/Bladen, of Saxon cities and fortified churches in Romanian Transylvania [Ardeal] or even of historical minority situations without a current minority population but with attractive architectonical witnesses of the past like the Spiš region (in German: *Zips*) or Banská Štiavnica in Slovakia. The lack of any reference to this argument in newspapers in Slovene confirms conclusions from our interviews (see Sect. 6.1.4) that even core-members of the Slovenian minority underestimate the value of their language as well as the cultural wealth (and in consequence touristic attractivity) of a region shaped by two cultures. When even newspapers in Slovene, that is newspapers published by the minority's elites, make no attempts to raise this consciousness, there is little surprise when it rests at a modest level with the wider minority population.

Among arguments **against bilingual signs** almost a half refer to numbers and thresholds like the following: "The number of Carinthian Slovenes is so small; the minority is not entitled to impose its wishes on the majority; Carinthian Slovenes always want more and more and will never be satisfied; new signs can only be granted based on a census and a minimum threshold (of 25%)."

Among the rest of the arguments one stands out, that is so-called external influence with wordings like "The federal government in Vienna is telling us (Carinthians) what to do; we (the Carinthians) have the right to decide and not Viennese politicians or lawyers, since they don't understand the local situation and don't defend Carinthia against external attacks from Slovenia; the Treaty of Vienna was signed by Yugoslavia and not by Slovenia".

It cannot be denied that most of these arguments are based on three factual grounds: (1) the low level of acquaintance of Austrians outside Carinthia with the historical backgrounds of the Carinthian place-name conflict—as already mentioned; (2) the delegating of minority legislation from the local and regional to the federal level bearing the danger of a 'missionary' attitude at this supreme level as it materialized at least in 1972 resulting in a 'town-sign storm'; (3) the historical experience that 'Vienna' (in the sense of the federal Austrian government) disliked or even obstructed Carinthian homeland defense after World War I, without which Carinthia would not exist as a political unit anymore. These arguments express, however, also a denial of all external (also international) benchmarks and criteria in dealing with minorities and the claim that decisions on the minority are exclusively in the hands of the regional majority. The argument that Yugoslavia was a signatory state of the Vienna Treaty and not its successor Slovenia is valid from a juridical point of view, but reduces the granting of minority rights to a question of international obligations (and external pressure) instead of recognizing it as a value in its own right.

In the opinion of Carinthian newspapers in German and especially those published by the Carinthian homeland organizations, "external influence" is, however, not only exerted by the federal authorities in Vienna, by Yugoslavia or Slovenia, but also exerted by Austrian mainstream media and even by the University of Klagenfurt—as strange as the latter may sound.

Austrian mainstream media are reproached of indulging in "stylish multiculturalism", while not honoring the deeper reasons of the 'Carinthian problem' as there would be (and admittedly are): (1) national awakening and polarization starting in the nineteenth century; (2) the fear of the Carinthian majority of losing control over larger parts of their dominion; (3) the "martyrdom syndrome" of Slovenes in general (Vilfan 2001); (4) threat and repeated military occupation by South-Slavonic states; (5) sympathy and cooperation of not a few Carinthian Slovenes with the latter.

The University of Klagenfurt—acting indeed repeatedly as a spearhead for the improvement of the minority situation as, for example, in 1988 (Österreichische Rektorenkonferenz 1988)—is classified as an intellectual exclave in Carinthia with a bias for the minority. This reproach is based on the background that in contrast to the old Austrian universities of Vienna, Graz, and Innsbruck, the University of Klagenfurt is rather young (founded in 1970) and at least in the first decades of its existence could not draw on regional intellectual resources, but had to invite professors mainly from Vienna, after 1995 increasingly from Germany, thus developing as an 'intellectual island' and introducing external views on the Carinthian minority situation hardly compatible with that of exponents of regional historiography like the Carinthian Historical Association [Geschichtsverein für Kärnten] and the provincial archive [Kärntner Landesarchiv].

Further pragmatic arguments against bilingual signs—in Table 6.4 subsumed under "Other"—refer to orientation ("Bilingual signs would look overcrowded"), geographical factors ("There is often no compact settlement area to locate a sign"), or legal regulations ("Bilingual signs do not correspond to the road traffic law"). The first sounds very similar to the argument brought forward by the Federal Office of Metrology and Surveying in the late 1980s (see Sect. 4.1.6) that there would not be space enough for placing bilingual names on official topographical maps.

The argument of lacking reciprocity (no bilingual signs in German-speaking minority situations in Slovenia) occurring in some interviews was, however, not found in the newspapers. It would indeed be difficult defending it, since in contrast to Italians on Istria [Istra] and Hungarians in Prekmurje, who are officially recognized as ethnic minorities in Slovenia and enjoy substantial minority rights including town signs, German-speakers (sometimes called 'Old Austrians' [Altösterreicher]) are today—after resettlement (of the Germans in the Kočevje region, in German *Gottschee*) as well as liquidation and expulsion during and after World War II—a tiny and dispersed minority and do thus lack official minority status. If numbers and thresholds should, however, not be valid criteria for granting minority rights (as some argue in the Carinthian case), also an official recognition of this community would be justified.

As can be seen in Table 6.5 **symbolic arguments** are less concentrated on some categories than pragmatic. Only two categories were not found

Table 6.5 Symbolic arguments disaggregated (gray fields = no occurrences)

	German		Slovene		German		Slovene	
	For	%	For	%	Against	%	Against	%
Power	5	28	7	15	2	17		
Appreciation	1	6	19	40	2	17		
Inclusion	7	39	2	4	6	50		
Assimilation			1	2	1	8		
History	3	17	13	27	1	8		
Other	2	11	6	13				
Total	18	100	48	100	12	100	0	0

in newspapers in German, that is, "assimilation" among arguments in favor of bilingual signs and "other" among those not in favor. All categories of symbolic arguments in favor occur in newspapers in Slovene, none of them against.

The most frequent arguments **in favor of bilingual signs** in articles published in German can be categorized under the term "inclusion". Typical wordings are: "It is a way of including the minority into the mainstream society; it makes us feel more at home and flourishing." Second in frequency with newspapers in German are arguments related to "power" like "Politicians and the general public in Carinthia are unable to accept a multicultural society; Carinthians do not mind to be managed by populists, while politicians only care about their own interest".

In contrast, the most frequently mentioned arguments by newspapers in Slovene are related to "appreciation", for example, "Bilingual signs mean an acknowledgement of the minority's presence" and—secondarily—to "history", for example, "The minority has been living in this area for centuries, it is an autochthonous community; there were bilingual signs here in the past; Slovenia has no territorial claims on Carinthia; Carinthian-Slovenian partisans fought against Nazi Germany resulting in Austria declared to be victim of the Nazi aggression by the Moscow Declaration of 1943 and allowing Austria regarding itself as a victim after World War II".

These are all valid arguments, and at least one deserves to be even extended compared to its relatively modest wording here: The Slovenian minority has not only been residing in Carinthia for centuries, but it also is the older of the two communities in question, having contributed essentially to a common culture (see Sect. 5.1). "Appreciation" would therefore mean to acknowledge its contributions to this common culture and Carinthia's specific identity. Even if the minority is rather small today, it can be proud of its cultural heritage that has widely influenced also the culture of the German-Carinthian majority and without which Carinthia would—from a cultural-geographical perspective—not be the land it is.

"Inclusion", however, also stands out among symbolic arguments **against bilingual signs** with opinions like "No further signs are needed. The Slovenian minority in Carinthia is one of the best-treated minorities in Europe (in terms of subsidies and social acknowledgement)." While it is questionable whether this opinion can withstand an objective review, it lacks acknowledgment of personal and group identity as constituent factors of human existence and well-being.

In conclusion and focusing on the issue of place names it could be stated with some risk of simplification that

- newspapers in Slovene, both of them representing the opinion of Carinthian-Slovenian organizations, stress the high symbolic function of bilingual town signs and regard them as the symbolic surface of the wider and deeper 'Carinthian problem';
- newspapers in German published by Carinthian homeland organizations attribute bilingual town signs the same symbolic function on the same background, but regard them as threats to Carinthian unity;
- other newspapers in German published in Carinthia regard bilingual town signs very well in the context of the wider historical and cultural background without attributing them utmost importance;
- newspapers in German published outside Carinthia can hardly hide their amazement about styling the 'petitesse' of place names so important and a source of conflict, deal with them in a rather pragmatic way without more profoundly exploring historical and cultural backgrounds and use the place-name issue frequently just as an occasion to highlight Carinthian politics in general.

6.1.4 Interviews (Peter Jordan and Alexis Sancho Reinoso)

Interviews in southern Carinthia followed the methodology described in Sect. 3.4 and did also not deviate from the methodology applied in the Těšín/Cieszyn region. This also applies to the selection of the 132 respondents that followed the snowball principle—that is, by addressing first a key person, who recommended other contacts. Most interviews were conducted inside the bilingual area, in an environment convenient for the respondent, but not all respondents were permanent residents of this area. The interviews were mainly conducted in German, but partly also in Slovene, if this was preferred by the respondent. It has again to be emphasized that the two interviewers were by purpose not native Carinthians, not even Austrians and by pronunciation and language intonation certainly easily recognizable as 'externals'—a fact that may have avoided affiliating them to one of the local 'camps' and prompted more frank and less 'prefabricated' answers.

As Table 6.6 shows, the structure of respondents by gender and age does not differ so much from the average population, while respondents with tertiary education are highly overrepresented. Also, the share of 44% Slovene-speakers is certainly disproportional, although strong efforts were made to reach a balanced proportion. The reasons why many German-Carinthians refused to be interviewed can be manifold. They can range from not being affected by the place-name question as much as Slovene-speakers to wishing "not again to be bothered by this nasty place-name affair" up to knowing which answers academic interviewers expect and not willing to perform a 'politically correct' academic fingering exercise.

Table 6.6 Interview sample in southern Carinthia

Characteristic	Structure of respondents
Gender	50% male and 50% female
Age	<19: 13.1%; 19–39: 44.0%; 40–59: 28.6%; >60: 14.3%
Education	No certificate at all: 2.0%; elementary education: 12.3%; apprenticeship: 15.5%; secondary education: 29.7%; tertiary education: 40.5%
Mother tongue	German: 51.2%; Slovene: 44%; both: 4.8%

Others may be similar to that of some respondents not particularly interested in the issue and mentioned later, namely: unfortunate personal experiences with minority issues; dissensus on this issue in their current personal environment (family, workplace, neighborhood); the intention not to expose themselves when declaring their opinion.

Like the interviews conducted in the Těšín/Cieszyn region, the interviews in southern Carinthia circled around three main topics: (1) space-related identity, (2) language usage in daily life, (3) perception and usage of (bilingual) toponyms. Questions to be answered, although not necessarily in this order, are shown in Appendix C.

A caveat needs to be exclaimed related to categorization and classification of discourses into 'minority-friendly' and the opposite. Categorization and classification are indispensable scientific methods if we want to describe and understand complex reality, although they are in danger of disguising and oversimplifying it. Reality is always much more colorful and varied than our perception of it, and it is very much so also in this case.

In this way also Fig. 6.65 is to be understood: There are favorable opinions toward bilingualism and the rights of the Slovene-speaking minority

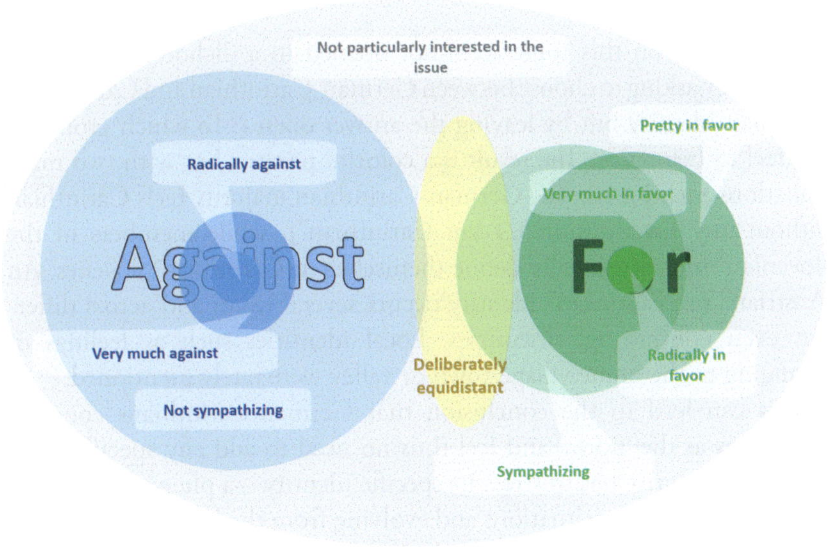

Fig. 6.65 A rough classification of discourses as regards the dichotomy 'minority-friendly' versus 'minority-unfriendly'. (Draft and graphics: Alexis Sancho Reinoso)

including language use and presence of their place names in public space, while there are also discourses against. Even when Fig. 6.65 classifies them into several grades, it cannot portray all the nuances that have been presented. Discourses identified in Fig. 6.65 as 'deliberately equidistant' can be understood as being positioned in the middle by purpose, but not necessarily neutral. In principle, they recognize and support minority language and minority rights including the right on the minority's own place names, but express at the same time their understanding for certain arguments of the German-Carinthian majority usually hinting at the place-name conflict and possible ways out.

What has been classified as 'not particularly interested in the issue' are not only answers of persons neglecting the issue of (bilingual) place names or insensitive to it, but also of people fully aware of the situation but distancing themselves from this topic for several reasons already mentioned.

In the following, the interview topics will be analyzed individually. The most representative opinions will be quoted, in most cases, however, not as verbal quotations but compiled of similar quotations.

6.1.4.1 Group Affiliation

The questions on this topic were not opened in a dichotomic way, for example, by asking to choose between German-Carinthian and Carinthian-Slovenian identity but by leaving the answer open ("To which group do you feel to belong?"). The result is a colorful picture, but with two main affiliations mentioned: The German-Carinthian majority feels Carinthian without any specification ("I am Carinthian"), while members of the Slovenian minority usually define themselves as Carinthian Slovenes. An "Austrian" or "European" identity occurs several times and across different, even contrasting, discourses. Local identities such as feelings of belonging to the home village/town or valley were rarely mentioned.

This can lead to the conclusion that German-Carinthians conceive themselves as the 'norm' and feel thus no need to add any specification, while the minority has to stress its specific identity—a phenomenon very common in minority situations and evolving from the fact that minorities are usually in the defensive position and have to demonstrate that they exist. The Slovenian historian Sergij Vilfan would explain this also by the fact that in the course of the 'national awakening' in the nineteenth century German-Carinthians had learned to conceive the land, where they constitute majority and social elites as 'theirs', while Carinthian Slovenes had developed rather an ethnic consciousness across pre-national cultural-historical territorial units like Carinthia (Vilfan 2001: 40). A Yugoslavian atlas of the immediate postwar period without an author or editor mentioned (Anonymous s.l.) speaks even of a "Slovenian ethnic territory" in Carinthia as if (1) an ethnic group could be defined by objective criteria, and (2) this group was permanently bound to a certain territory.

This result also shows that regional identity and, in the case of the minority, group-related identity are clearly outmatching Austrian (national) and Slovenian (national) identities. This is not a surprise as regards German-Carinthians related to Austrian identity, since—as we tried to show in Sect. 4.1.4—Austrian cultural-national identity is strong today, but it is overshadowed by land identities in most Austrian provinces. This is especially true for Carinthia with its long tradition as a political-cultural unit. It is not as evident with minority members. They could have also affiliated themselves with the Slovenian nation—as it is practiced 'from the outside' by the 'mother nation' on the other side of the border (see, e.g., Anton Melik Geographical Institute and Institute of Geography 2001). Instead they prefer to declare their closer group identity and would also certainly not refrain from declaring "Austrian" as their third-ranking identity (after "Carinthian"), when asked for it. This is possible, because the Austrian cultural-national identity—as developed after World War II—is not defined by language, but mainly by common history (see Sect. 4.1.4).

That a "European" identity is only rarely declared is no surprise, in comparison to either other parts of Austria or other European countries, since a European identity is only in its infancy everywhere.

The lack of mentions related to local and sub-regional identities can be explained by the strong positions of regional (mainly with German-Carinthians, but not much less with Carinthian Slovenes) and of ethnic group identity (with Carinthian Slovenes) in Carinthia. Much more in the shadow than, for example, quarter identities in Upper and Lower Austria or Gau identities in Salzburg, they are nevertheless strong and would certainly have been revealed during a longer discussion and if the 'conventional' space-related identities would not fall into one's mind first while others much later. The questionnaire survey (see Sect. 6.1.5) has confirmed this.

Very inspiring are also the answers related to the factors causing group affiliation or identity. Two of them—"roots" and "family"—were most frequently mentioned. They are in line with the constructed character of group- and space-related identities, which are never based on 'hard facts', but on subjective deliberations based on what was usual in the family or in a person's wider surrounding. What distinguishes minority-friendly discourses from discourses not in favor of the minority here is the mentioning of (the Slovenian) language by minority-friendly discourses. This points to the fact that minorities need an identity marker, be it language, denomination, or something else. Right in a situation, when the Carinthian-Slovenian dialects are fading away in the family and thus the traditional identity marker is losing ground, the rather new identity marker of the Slovenian standard language, actually imported from Carniola and not traditional in Carinthia, grows in importance, if group identity should be maintained. Language as a distinguishing and thus community-building factor is all the more necessary as German-Carinthians and Carinthian Slovenes have been a cultural community for centuries and thus certainly very close in many other cultural aspects.

When referring to the role of childhood in the identity-shaping process, most discourses for and against the minority argue that it played an important role, while equidistant discourses tend to attribute less importance to childhood in favor of a gradual process of conscious choices. This is in line with their deliberately chosen position.

In response to the question whether they were active in organizations or associations, the Carinthian Defenders' Federation [Kärntner Abwehrkämpferbund], a traditionally anti-Slovenian patriotic association, appeared several times in minority-unfriendly discourses, while a wide plethora of local and regional associations (mainly cultural and religious) was mentioned on the opposite side.

Beyond statements strictly related to interview questions, feelings of being victims like the following were brought forward by respondents from both poles of the spectrum: "Slovenes have never enough, although they are one of the best-subsidized minorities in Europe." "Slovenes are always the ones to draw back; the conflict has perpetuated in history, there hasn't been a revision of the past in Austria."

6.1.4.2 Language Use in Daily Life

The interviews show that the native and family language is predominantly German, which is by no means surprising when the history of the region is taken into account (see Sects. 4.1 and 5.1). The domain of German as the mother tongue extends even far into the realm of people sympathizing with and feeling to be part of the minority. Many of them mention that their ancestors used to speak 'Windisch', although this is not the term for a language according to the state of the art in linguistics (see Pohl 2004). Only respondents explicitly fighting for the cause of the minority state that their native language is Slovene. No mentions of 'Windisch' can be heard from them.

6.1.4.3 Importance of Learning Slovene in Carinthia

On the importance of learning Slovene as a second language in Carinthia a clear divide appears between discourses against and the rest including not only minority-friendly discourses, but also equidistant and even insensitive ones. While the former avoid answering the question explicitly and, instead, call for the free choice to learn a language, the latter agree on the importance of learning Slovene, albeit with manifold nuances of this statement. Minority-unfriendly discourses usually try to downgrade Slovene to an irrelevant language as it appears in the following: "Anyone should learn Slovene if he/she wishes. However, Slovene is disappearing anyway. There are more important languages, such as English and Italian."

In contrast, the discourses in favor of the minority tend to highlight the importance of preserving the bilingual character of the region, and of language as a matter of maintaining one's cultural roots and intimacy: "To learn Slovene is very important, because through this some people find their lost roots again, because multilingualism is important, because Carinthia was always bilingual. It's a reason to be involved in cultural life and politics. It's something intimate."

Discourses less sentimental, mainly those qualified as equidistant, use quite different arguments and are particularly relevant, because they help to identify real problems related to the minority language, such as the gap between local dialects spoken in the family and standard Slovene taught in schools: "There's a huge difference between learning Slovene at home or just in school. The latter usually doesn't work."

When asked about the importance of speaking Slovene in the professional context, a majority of respondents agree ("Yes, it is an advantage"), while nobody strongly against the minority shares this opinion. Among the variety of jobs, education-related jobs (e.g., teachers) are frequently mentioned as examples for the relevance of having a command of Slovene.

6.1.4.4 Slovene as a Tourism Asset of the Minority Region

Another aspect addressed in the context of language was the role of the minority language in the socio-economic development of the region. To the question "Does German/Slovene bilingualism make southern Carinthia a more attractive tourism destination?" respondents provided contrasting answers and mentioned a wide range of arguments. Answers denying this question, however, dominated and were part of all kinds of discourses. This can be explained both by (1) a misunderstanding and (2) by regarding Slovene and the bilingual character of the region as less important. The misunderstanding is that Slovene would only be an asset, if tourists from Slovenia would have a larger share in local tourism (which is—except some non-stationary winter sports and shopping tourism—not the case) or if by presenting and speaking Slovene in public space and in hotels, restaurants, and shops many more tourists from Slovenia could be attracted (which to expect is very optimistic).

The real aim of the question, however, was to explore, whether a multicultural situation due to its 'otherness', its difference, or augmented variety in cultural manifestations (not confined to language) compared to the 'monotony' of the mainstream society could be of special interest for (at least cultural) tourists—as we are frequently traveling to see the 'other', the 'unusual', the 'exiting', and as it is, for example, successfully practiced in Burgenland, where dancing and Tamburica music groups of the Croatian minority contribute a lot to the colorful tourism image of this province.

When it may be found less surprising that minority-unfriendly discourses regard bilingualism not as an asset for tourism (They regard bilingualism not as an asset at all.), the same opinion expressed also in minority-friendly discourses may be regarded as rather strange. This, however, has its very reason in the not only external, but also internalized and traditional self-estimation of the minority to speak a less important language. It coincides with the observation that the linguistic landscape of major destinations of international tourism in the bilingual area like Lake Faak [Faaker See/Baško jezero], the southern shores of Lake Wörth [Wörthersee/Vrbsko jezero], or Lake Klopein [Klopeiner See/Klopinjsko jezero] lacks practically any private sign or label in Slovene (see Sect. 6.1.2). Even owners of tourist accommodation and catering enterprises with a pronounced minority background and not hiding their minority affiliation in all other situations do not 'dare' to name their tourist facilities bilingually or in Slovene.

The following statement is exceptional insofar as the respondent has really understood the question very well. Minority-friendly and presented by a member of the minority it illustrates that the minority expects activities rather from public authorities than from its own ranks: "Bilingualism can make Carinthia a more attractive tourism destination. There's a big potential as South Tyrol shows. The Slavic touch for people from German-speaking countries can be exploited. However, public authorities don't consider bilingualism as an asset."

The latter part of this statement may also be regarded as coinciding with the opinion of the Slovenian historian Sergij Vilfan that the "politically colored Slovenian historiography" perpetuated up to the present day the concepts of (Slovenian) "martyrdom" and "servitude", that is, not to be the master of one's own fate due to external domination (Vilfan 2001: 46). It could be added that the martyrdom syndrome is shared by many minorities and (if only by self-estimation) non-dominant groups of all kinds.

6.1.4.5 Minority Toponyms on Official Town Signs

Questions on this topic so crucial for our study have been posed in two different ways, that is, by asking for the opinion about using bilingual names in southern Carinthia in general as well as by asking for the use of bilingual names in the respondent's own home. The dual questioning departed from the assumption that place names are identity markers (see Chap. 2) and that even somebody in favor of minority rights and bilingual signage in general may be hesitating when it comes to the decision, whether his/her place is defined in this or the other way. The latter is even more delicate as in Carinthia the strong symbolic power of place names became obvious through open conflict up to the recent past. The latter question can thus intrinsically also have the meaning: Do I wish to expose myself and my closer environment to such risks? What is more important for me: declaring my identity or remaining innocent? Do I want to be categorized and stigmatized? Although these options may have been addressed too drastically now, it is thus no surprise that respondents needed or dedicated longer time to answer.

Most of the respondents providing minority-unfriendly discourses mentioned to have no personal relation to official signs and identify them with conflict. They usually are against bilingual town signs using pragmatic arguments: "Bilingual town signs don't mean anything for me. They aren't useful, because everyone understands German."

Some minority-unfriendly respondents would accept the Slovene name on a town sign, if it was complemented by names in other languages: "If they add more languages, Slovene could be one of them." This argument, looking pragmatic at first sight, is in fact an argument against bilingual town signs, since installing multilingual official town signs is impossible in the bilingual area of Carinthia. No populated place there has other than German or Slovenian names—neither endonyms nor exonyms. Also for populated places in Carinthia outside the bilingual area only two names in other languages can be found: Villach (Slovenian *Beljak*) is addressed by Italians as *Villaco* and by Friulians as Vilac, and Hermagor (Slovenian *Šmohor*) is called *Emochar* in Friulian. But these are—and would in additional cases be—exonyms in the sense of names used not within the local community, but from the outside (see Chap. 2).

And it is not the intention and task of official town signs or official signs in general to reflect exonyms, but the names used locally, the endonyms.

Names in the status of exonyms—this applies of course also to Carinthia—are just in use for more important features like countries, larger cities with a long historical record, trans-boundary features, or traditional sub-national units like provinces. Carinthia with its long tradition as a political entity is certainly a case in point and 'honored' by a long list of exonyms in many languages, but also Styria [Steiermark]. Styria has—somehow in reaction to the Carinthian place-name conflict and ironizing it—in the early 2000s initiated an "art project" (as it was officially called in response to a question of one of the authors) presenting at main provincial road boundary crossings plates with all exonyms (24 by number) for Styria in foreign languages (Fig. 6.66). However, since exonyms have not the function of marking local identities (see Chap. 2)

Fig. 6.66 Plate at the provincial boundary between Carinthia and Styria nearby Dürnstein in der Steiermark displaying 24 exonyms for Styria. (Photo by Peter Jordan 2007)

these plates cannot be taken as a proof for Styria's more liberal attitude toward non-German names (as it was perhaps the secret intention to demonstrate taking into account that there is some rivalry between the two provinces). The real proof would be to install bilingual German-Slovenian town signs in the South of Styria (Urban Commune Bad Radkersburg), where a small Slovene-speaking community resides. But only town signs with the German name can be found there (and hardly anybody in Austria takes note of it).

Some minority-unfriendly discourses provided rather symbolic arguments when referring to bilingual town signs at home: "Bilingual town signs in my hometown would mean an undesired Slovenization."

However, statements of this kind are—by content, not by wording, certainly not using 'Slovenization'—not only part of minority-unfriendly discourses but they even correspond to the opinion of some minority members. Some of them just do not want to see their familiar Slovenian name to which they are used in intimate communication officially represented on the town sign. They conceive it as their private, intimate name for their place not to be exposed to a wider public. It is an attitude not unsimilar to the use of nicknames or love words within a family or between partners, which they would also not mention in conversation exceeding this intimate circle.

Others—as mentioned before—wish on the background of political conflict over bilingual town signs not to be 'stigmatized' and weigh carefully any statement on this matter. It is, thirdly, also the historically rooted feeling that Slovene is a language second to German with a lower prestige and a weaker claim on being represented officially that makes them behave in this way. This latter interpretation may sound strange to somebody acquainted with French or Hungarian minorities but it will be rather familiar to those acquainted with the situation of the Slovenian minority in northeastern Italy (see Stranj 1992) or of Catalan outside Catalonia [Cataluña/Catalunya] (Marí Mayans 2011).

A statement like this can, however, also originate from a respondent affiliated to the community of the 'Windische', who recognize their Alpine-Slavonic roots and cultivate the corresponding cultural heritage, but distance themselves from the Carinthian Slovenes in the sense of a political-cultural community. They would prefer no bilingual names at all or—if it comes nevertheless to a bilingual naming—they would prefer names in the local dialect and not in standard Slovene since they feel to be members of the very local ethnic group and neither part of the 'larger' community of Carinthian Slovenes nor of the Slovenian nation. For them names in standard Slovene would indeed mean 'Slovenization'.

In other discourses close to the Slovene-speaking minority we find both pragmatic and symbolic arguments. They often appear mixed, like in the following statement, where the respondent claims the personal right on using the Slovenian place name and connects it with her identity: "I use more the name *Loče* than *Latschach*, and I'm there at home, and I think that where I'm at home, I should be able to decide what happens. I also decide about what's going on in Vienna, and I'd love to be listened as active citizen" (Woman, 20–30 years old, Latschach, in Slovene: *Loče*).

When referring to personal attitudes toward bilingual signs, most of the discourses in favor of the minority highlight issues like 'normality' meaning that bilingual signs express acceptance of minority members and visibility of their presence as it is usual with all other citizens.

Again, particularly interesting arguments were brought forward in equidistant discourses. They stress their positive attitude and express their tolerance toward the minority, but also recognize and, in some cases, stress the need of dealing carefully with this issue: "Bilingual town signs in my hometown would indicate that bilingual people live there and that they appreciate their language. But this is actually not (anymore) the case in my town. Bilingual signs would bring agitation. Many people are against them and would associate negative clichés." The statement hints at the historical burden lasting—in contrast to Burgenland—on the Carinthian situation and can be taken as a proof for the symbolic power of place names and their function as identity markers. It also shows that the minority is—also by 'equidistant' respondents and people regarding minority rights as justified—not appreciated as a cultural wealth of Carinthia, as having contributed to a common Carinthian culture and a constitutive factor of Carinthian identity. If the latter was the case, current language use would not be classified as relevant for minority place names in public space and any argument would be good enough to demonstrate the bilingual character of the province and its identity stemming from Bavarian as well as Slavonic roots. A good example for the styling of such an identity based on minimal factual grounds is the neighboring Italian Autonomous Region Friuli-Venezia Giulia and especially its Province Udine, where bilingual Italian-Friulian town signs spread down to the coast of the Adriatic Sea, although Friulian is actively spoken only in its mountainous part with a focus on the region Carnia.

As also confirmed by our questionnaires, the question which percentage of Slovene-speakers in the overall population number should be sufficient for bilingual town signs is also intensively discussed. While discourses against the minority complain about the low percentage of Slovene-speakers considered as sufficient for bilingual town signs (according to the Federal Act of 2011: 17.5%), discourses in favor of the minority argue that the threshold was too high and should be reduced to 10% (as it had been proposed by the Constitutional Court of Justice in 2001). This 'percentage' discussion, however, disguises just a deeper divergence of opinions on the treatment of minorities in a democracy. While discourses against bilingual names appeal to the majority principle ("Majority decides"), discourses in favor of bilingual names appeal to minority rights. ("Minorities should be respected irrespective of their number and percentage.") Also the political discussion was at times conducted with these arguments.

Beyond the percentage discussion, a few minority-friendly discourses plead for extending official bilingual names, for example, to the whole southern fringe of Carinthia regardless of the actual presence of speakers, just because Slovene was present or dominant earlier. They implicitly support the argument of the minority as a constituent factor of Carinthian culture and identity as presented earlier. The example of the Sorabian minority in Germany[7] was mentioned repeatedly to demonstrate that this is possible, and also the example of Friuli could have been mentioned here. This opinion, however, is rare compared to being in favor of bilingual signs just where bilingualism is actually practiced—even without referring to any percentage.

[7] The Sorbs are a Slavic-speaking group in the eastern part of Germany. Although the Sorabian language is hardly spoken in families anymore, bilingual signposts can be found across the area and in every village (see a.o. Bundesministerium des Inneren 1999; Eberhardt 2015; Jordan et al. 2006).

Some responses alluded explicitly to the historical background in Carinthia and may explain, why even conscious Carinthian Slovenes or at least Carinthians conscious of Slavonic roots are not necessarily in favor of bilingual town signs in their home. They are quoted literally. The first was provided by a young man from a Slovene-speaking family actively engaged in the cause of the minority: "I wouldn't find it particularly useful [to have bilingual signs in my village] … because we are the last Slovene-speaking family in the village. (…) For me it isn't particularly important, because in my family no one was deported."

The second comes from a man from the same commune (yet from a different village) engaged in a cultural association that tries to preserve local dialects classifying them, however, not as Slovenian but as 'Windisch' dialects, that is, an active member of this group of Carinthians recognizing and appreciating their Alpine-Slavonic roots and cultural traditions, but wishing not to be called 'Slovenes' and distancing themselves from the Carinthian-Slovenian minority in the sense of political-cultural community: "No one would probably die [if a bilingual town sign would be put in the village], no one … But if they do as they did in the past, then here we go again, because then certain organizations will come and say: 'This is a huge mess' and 'we are going to be stamped as Yugoslavs'" (Man, 70–80 years old, Hohenthurn).

6.1.4.6 Minority Place Names on Maps, Bilingual Street Names

Since this topic is rather specific and people hardly have an opinion on it, it is difficult to recognize a clear trend in the interviews. Much less than about official bilingual town signs they are informed about legal basis, responsibility and (with maps) purpose, contents, and publishers. There is also no public discussion on it and no prefabricated answers exist.

Out of the few answers received not surprisingly minority-unfriendly discourses vote against minority place names on both official and non-official maps (e.g., tourist, hiking maps) as well as on street plates, which are not covered by federal minority law, fall into the responsibility of communes and were up to 2020 nowhere in Carinthia—in contrast to

Burgenland—bilingual.[8] Minority-friendly discourses tend to be rather in favor of including them. Opinions against usually expose pragmatic arguments ("The official language in Austria is German"; "Bilingual names would generate confusion among visitors"). Also minority-friendly discourses prefer pragmatic arguments like "It's a matter of rule of law", although right in this field not very much is regulated by law: It is up to map publishers (including the Federal Office of Metrology and Surveying as the editor and publisher of official topographical maps) to respect minority names, especially as regards names of natural features, and it is up to communes to decide about street naming.

The arguments for and against bilingual street plates are particularly interesting, because they escape the typical polarization that characterizes other issues and are not so much affected by public discourses. Some (typically minority-unfriendly) respondents argued that there is no tradition of street plates in small villages, so no discussion is needed. However, in recent years, street names have been introduced on the initiative of the State and the national postal services, so that new names come up and thousands of postal addresses have to be changed with the new street name replacing the name of the village plus house number. These situations have been mentioned by several (typically minority-friendly) respondents. They explained that by these new street names (in German) in several cases former postal addresses just composed of the bilingual village name and the house number are replaced—hiding in fact the bilingual character of the place. Some raised the opinion that the new names were frequently just invented and no relation between them and the place can be found. Many argue that traditional field names (in the bilingual area frequently of Slavonic origin) should be used to strengthen people's ties with their place.[9] The following statement illustrates this idea

[8] As already mentioned (see Sects. 4.1, 5.1, and 6.1.2) the Commune Sankt Jakob im Rosental was in 2020 the first to introduce bilingual street names in Carinthia.

[9] In 2017, the Austrian Board on Geographical Names [Arbeitsgemeinschaft für Kartographische Ortsnamenkunde, AKO] passed a recommendation to use, a.o., field names when it comes to the naming of streets (AKO 2017). This recommendation was passed right with a look at rural villages and towns, where field names constitute a respectable part of the toponymic corpus, but are in danger to fall into oblivion, since fields are merged and settlement structures spread into fields. Applying their names on streets and other urban features would preserve them and refer the street to a traditional location at the same place. Since in the bilingual area of Carinthia (and other

very well: "In Sankt Jakob [im Rosental], (…) from now on, the old field names are erased from the postal addresses, and instead invented names such as 'ancient sight path old' or 'clove path' have become the official ones. Thereby, the old Slovene designations disappear. Now, if I treat this issue from a neutral point of view, something that has centuries of lifetime, which is the result of a natural development, is being substituted by something fanciless, mundane, because these names exist all over the world. It really surprises me that [local] German-speaking people don't see this and don't say 'This isn't correct and we support the old names'" (Man, 50–60 years old, Klagenfurt a.W.).

6.1.4.7 Relations Between Minority and Majority

To the question whether they were satisfied with the way the political conflict has been settled by the compromise adopted in 2011 ruling that 164 populated places must have bilingual names (see Sect. 4.1.6) one could have expected low satisfaction at the two 'poles' (for and against minority names) and more satisfaction with equidistant respondents. This came true only to a certain extent, since also some of the latter are skeptic. A frequent answer is that the conflict might have been settled "at the top" (in the sense of the political level), but "at the bottom" it still exists and will probably last: "Relationships between the different groups have become better, but only in politics and among well-educated strata, not among the ordinary people, who still have huge difficulties to gain a better understanding of the conflict."

This doesn't mean that people don't see the atmospheric convergence in society. Many respondents, particularly the minority-friendly, stated that the relations were better now, but that the political compromise in 2011 was a consequence of this improvement, rather than its trigger. Some respondents mentioned the dissolution of Yugoslavia and Slovenia's EU accession as important milestones toward an improvement of the atmosphere in Carinthia.

minority areas in Austria) field names originate frequently from the minority language, this recommendation means also a support of minority names.

To the question what would still have to be done to strengthen political and social relations between majority and minority in Carinthia, by the minority skeptics the idea of sticking to the 'right of majority' as a 'democratic principle' was mentioned repeatedly, and very often they demanded to count the minority as a precondition for granting minority rights resuming an idea promoted by the homeland-defender organizations in the 1970s and heavily opposed by the political organizations of the minority.[10] Among the minority-friendly, tolerance toward the minority is demanded. The strategy should be to keep on expanding their rights, which includes more public representation of their place names. An ethnic or linguistic census would not help to solve the situation.

The last interview extract shows how complex this issue is and how important place names and their public representation can be. The respondent is a man with Slovenian roots but having been raised only in German language. He uses his social position in his home village and among his farmer peers to initiate constructive discourses. The extract informs about his talk with the former (and highly controversial) governor of Carinthia, Jörg Haider: "As a representative of farmers, I met once Jörg Haider (…) He called me and invited me to a private conversation (…). We began to address each other by our first name and suddenly he asked me about the problem of the town signs. I said: 'I would be happy if we make steps forward and find a solution'. Then I asked him: 'How about you? Why do you always act in a provocative way?' And he said: 'Look, it's very easy: I've met with the [Carinthian] Slovenes and I told them: You can have everything you want. But please leave the town signs alone, a couple of governors have stumbled on this topic in the past, and I wouldn't allow myself [to do the same], because … This by no means, but I'm generous in everything else'" (Man, 40–50 years old, Eberndorf/Dobrla vas).

[10] Although the official Austrian population censuses conducted up to 2001 asked for the 'objective' criterion of the colloquial language and not for ethnic/national affiliation and representatives of the Slovenian minority liked to claim that the minority comprises a much larger number, it can rather be assumed that respondents to the census did not weigh their language use when answering this question, but answered following their ethnic consciousness. Thus, these censuses are very likely to portray the ethnic corpus of the minority quite sufficiently and a special census of the minority was never necessary.

6.1.4.8 Conclusion

In summing up the experiences with and the results of interviews it has first to be stated that it was both an advantage and a disadvantage to have engaged interviewers not from the region and not even from Austria. Their (relative) absence of ties to the region made it possible to approach the research field without any pre-existing local ethnic, social, or political group links which could have influenced respondents' answers. Respondents thus may have felt less need to strategize in their responses. Even the interviewers' certainly lower extent of familiarity with the situation may have not only been a disadvantage, but prompted respondents to explain their arguments more profoundly, to reflect their answers thoroughly instead of just offering prefabricated standard responses. It has, however, also to be admitted that the interviewers' lower acquaintance with local situations and toponymy did obviously not generate such colorful stories and associations about and with respondents' most familiar places and their etymologies as presented in the Czech case (see Sect. 6.2.4).

Interviews on the topic of bilingual place names are in Carinthia anyway in danger of being strongly influenced by the political discussion conducted for decades and widely reflected by the media with the consequence that just answers prepared and constantly repeated by political parties and the two 'camps' are offered. Our interviews were in this respect in general no exception. Many, if not most discourses reflected just the two main lines of arguments. Nevertheless, there were also several widening the spectrum. On the German-Carinthian side they ranged from very minority-friendly and place-name sensitive to the opposite, the latter frequently disguised by reasonably sounding pragmatic arguments against bilingual town signs. The low number of respondents, however, does not allow a projection on their share in the total population. On the minority side with its larger number of respondents and its higher extent of being personally affected by the place-name question an even wider variety of opinions could be revealed that goes well beyond the usual statements and positions of Carinthian-Slovenian organizations and political representatives. They may have even been surprising at least for readers not so acquainted with the Carinthian situation and especially its historical burdens and also hardly digestible for the political representatives of the minority. It may be worthwhile to highlight them briefly once again.

Thus, even some 'confessing' members of the minority regard their language as second to German or not as adequate for official communication. This makes them not strongly supporting officially bilingual place names at least in their home place. Another reason that makes them not wishing to see their place names on official town signs is the intimacy they feel that their names have and that they should therefore not necessarily be exposed to the public. A third reason mentioned is the discrepancy between the names in local dialect and in standard Slovene that is to be applied with official names alluding to the difference between the local ethnic group and the Slovenian nation. A last reason—only understandable on the background of history and certainly going very far—is the fear of stirring up conflict (with neighbors) and being stigmatized. Opinions like these are certainly exceptional in their rigidity and frankness but shed some light on the overall constitution of the minority and minority-majority relations in Carinthia in general. Very indicative in this context is also that it was hardly understood by any respondent (even from the ranks of the minority core group), why German-Slovene bilingualism should be a tourism asset.

This illustrates the delicate position of the minority's political representation with the task of defending and augmenting minority rights. In pursuing these aims, they can certainly count on the support of most minority members but act in a situation that does not exclude feelings and attitudes like those expressed before. It can be expected that a certain 'naturalness' of using improved minority rights may gain ground with all minority members in times to come. At present, however, this is not yet the case and the shadows of the past seem to be long.

Concerns related to bilingual naming of one's own home place demonstrate the symbolic power of place names most impressively. While it seems to be not as problematic to affiliate oneself openly to the minority, to be active in minority organizations, to learn and speak Slovene, attitudes become much more hesitant and careful, when it comes to accepting that one's own place is named in Slovene—up to the level of fearing to be stigmatized.

6.1.5 Questionnaires (Peter Jordan and Alexis Sancho Reinoso)

As already explained in Chap. 3, the questionnaire that was used in southern Carinthia consisted of 11 questions plus 7 more devoted to personal data. The sample is essentially smaller than the one obtained in the Těšín/Cieszyn region: 589 (102 in Slovene, 487 in German) of the questionnaires could be evaluated (compared to 1804 in the Těšín/Cieszyn region). While gender proportion of respondents coincided quite well with the population average, the proportion of age groups had a bias in favor of the younger, and the educational structure is even more significantly tilted in favor of tertiary education (see Table 6.7).

The questionnaire asked about all project topics, that is, language use, place names, and identity, with the main intention of testing the validity of the interviews. In the following the responses will be analyzed by major topic groups.

6.1.5.1 Language Use in Daily Life

A question on this topic referred to language use in seven different situations (within the family, with other relatives, with friends speaking the same language, in the neighborhood, in shops/restaurants, at the communal office, in school/at work) and three levels of language use (dialect, colloquial language, standard language) were offered for choice for each language and two (dialect and colloquial language) for mixed language use (German and Slovene). Thus, respondents had eight options for defining their language use in seven different situations.

Table 6.7 Questionnaire sample in southern Carinthia

Characteristic	Structure of respondents
Gender	48.5% male and 51.5% female
Age	<19: 19%, 19–39: 25%, 40–59: 30%, >60: 26%
Urban-rural dichotomy	34% town residents, 66% village residents
Education	22% elementary, 30% secondary, 40% tertiary, 8% other (mainly college education)

Unfortunately, this question was answered by many respondents not in the expected way, that is, instead of ticking only one they were choosing more than one option for each situation. Their answers could be evaluated only by simplifying the data. In addition, it had to be taken into account that only bi- or multilingual respondents had the full choice between all options. In contrast to the Těšín/Cieszyn region, where practically every minority member has a command of the local dialect, in southern Carinthia not even every person feeling affiliated to the minority speaks or understands Slovene. Thus, as regards this particular question (not the rest of the questions) only respondents stating to use Slovene in at least one of the situations were included in the analysis, that is, 60%.

Figure 6.67 shows that they use the minority language mainly in private situations (in the family, with relatives, friends). In the public sphere mixed language also prevails in informal situations (shopping, restaurants, with neighbors) while the share of German grows with formality. In contact with public authorities such as at the communal office German is by far prevailing, while a mixed language use occurs in school and at work, where bilingual Carinthians usually attend lessons in Slovene and meet or interact with other Slovene-speakers, respectively.

The overall result indicates that Slovene is the language of the private and intimate sphere. It is of course present in school, because standard Slovene is taught there and serves as an identity marker. But Slovene can

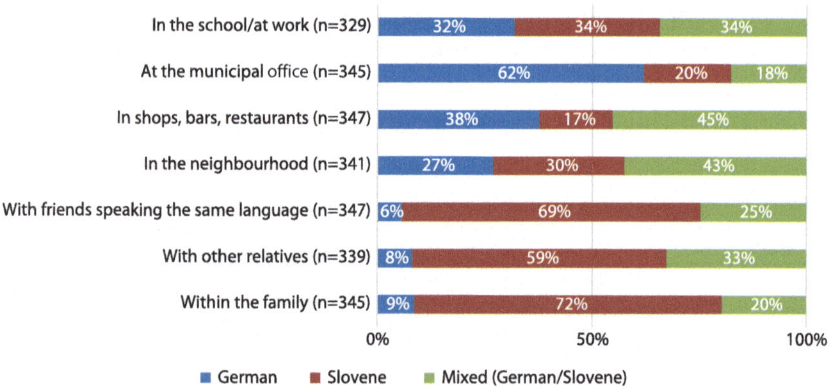

Fig. 6.67 Prevailing language use in different situations by bilingual respondents. (Graphics: Luděk Krtička)

hardly be heard in the public—contrary to Hungarian in all countries neighboring Hungary, or French, where it is the minority language in a Swiss canton (Berne [Bern/Berne]).

Had it been possible to statistically evaluate the answers by language levels (dialect, colloquial language, standard language; see above), this would certainly have revealed that dialect is prevailing in the private sphere, while the language in school is standard Slovene (certainly mixed with colloquial elements). It can, however, be assumed that (1) a colloquial Slovene with dialectal elements is gradually replacing the traditional Carinthian-Slovene dialects, (2) even the local German-Carinthian dialects are partly going to replace any Slovenian idiom in the private sphere, and (3) standard Slovene (and its colloquial variants) are just learned as an identity marker and used only in very official minority-centered communication (festivities of the minority, Roman-Catholic church services, meetings of Slovenian organizations, etc.).

It was interesting to see how language use in daily life influences perception of bilingual signs. For this purpose, the results of three questions (for the language spoken at home, opinion on bilingual signs in southern Carinthia, opinion on bilingual signs in one's home) were compared—this time, however, also including monolinguals, that is German-only speakers. The results of this comparison (see Fig. 6.68) show a clear difference between German-only speakers and all others (speakers of Slovene or German and Slovene): While practically all respondents speaking

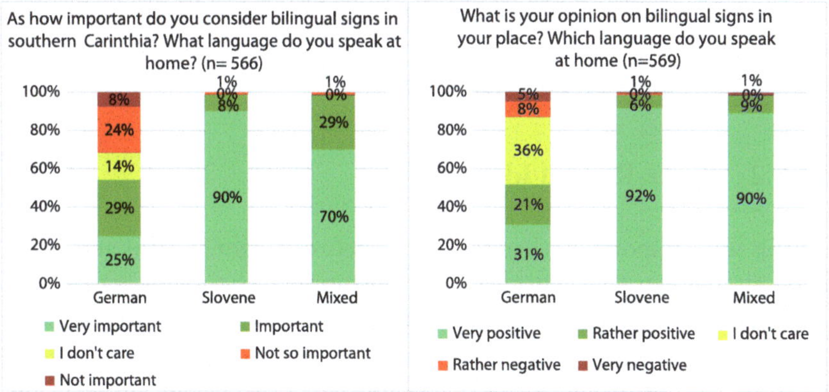

Fig. 6.68 Perception of bilingual signs by language spoken at home. (Graphics: Luděk Krtička)

Slovene or mixed German-Slovene at home ("within the family") perceive bilingual signs in southern Carinthia and at home as "very important" or "important", for respondents speaking only German at home they are significantly less important. Nevertheless, 25 and 31% German-only speakers perceive bilingual signs as "very important" and 29 and 21% as "important" accumulating to a quite surprising 54 and 52%.

The only major difference between the two correlations (Fig. 6.68, diagrams to the left and to the right) is the much larger percentage of German-speaking respondents answering "I don't care/I wouldn't care", when asked for their opinion on bilingual signs at home versus in southern Carinthia (36% compared to 14%). This may at first sight be surprising, since our hypothesis was that people tend to be more sensitive about the issue of bilingual town signs if their own village or town is concerned. It deviates also from responses received in the interviews. This result may, however, also be taken as confirming this hypothesis, if one takes into account that filling in a questionnaire and delivering it (to in some cases very official institution like a high school directorate) has a much more formal and official character than responding to interview questions in an informal talk—even more so, when the interviewer was obviously not an Austrian and thus not involved in local affairs. The higher extent of formality may have caused respondents to be careful and prefer a 'neutral' answer right on issues affecting them most and conceived to be delicate. Another explanation could be that some of the German-only speakers were from the Upper Gail Valley [Oberes Gailtal] already outside the bilingual area and with no places that could be named bilingually. For them the answer "I don't care/I wouldn't care" may just have been a substitute for "This question does not apply to me".

6.1.5.2 Space-Related Identity

On this topic respondents had the choice between eight answers and space for an alternative. They had also the opportunity to choose more than one answer and to rank identities. Aggregated results show that almost half of them (48%) selected "Carinthian Slovene". The most frequent answer, however, was "Austrian" (73%), followed by "Carinthian" (65%) and "European" (64%). Only 2% selected the option "German- Carinthian", while about two-thirds of respondents who selected "Carinthian Slovene" also selected "Carinthian". This contradicts a finding of the interviews that Carinthian identity was a characteristic only of German-speakers in Carinthia and supports the assumption that the region-related identity in Carinthia is so significant and strong that it bridges and overshadows ethnic group identities—even in the ranks of the minority, whose members tend to have a stronger group affiliation and are thus rather inclined specifying their identity. This may lead to the general conclusion—applicable also to minority situations in other countries—that strengthening regional identities, for example, by trying to achieve a coincidence between historical-cultural and administrative regions with major competences of self-government can mitigate antagonistic situations in ethnic or national terms (see Jordan 2005). Another (less favorable) interpretation could be that German-speaking Carinthians conceive themselves as the 'norm' and thus not in need of any specifying adjective.

The high percentages of "Austrian" (73%) and "European" (64%) identities must be attributed to the fact that the questionnaire offered choices starting with "global citizen", followed by "European" and "Austrian"—"Carinthian" and sub-regional identities following suite. This was very likely a mistake, since it suggested identities, which most respondents would not have claimed out of themselves. It can be assumed that the interview results are in this respect much more reliable, when they made clear that—at least in Carinthia—Austrian (state) identity ranks significantly behind Carinthian (land) identity and that a European identity is just in its infancy.

6.1.5.3 Perception of Bilingual Signs

When the responses to space-related identity were compared with responses to the question on the importance of bilingual signs in Carinthia and their very own home place, the difference between respondents conceiving themselves as "Carinthian Slovenes" and the rest is striking (see Figs. 6.69 and 6.70). Almost all respondents having ranked "Carinthian Slovene" as their main identity consider bilingual signs as very important in southern Carinthia and have a very positive opinion on them in their home place. The fact that among the rest of the respondents the gap

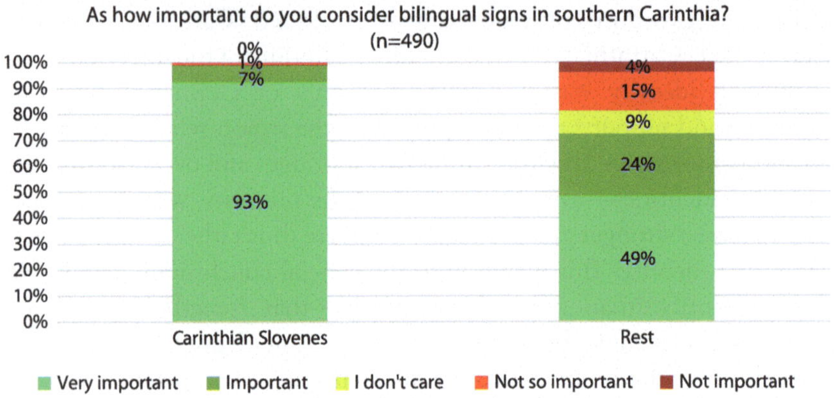

Fig. 6.69 Opinion on the importance of bilingual signs in Carinthia with "Carinthian Slovenes" versus other space-related identities. (Graphics: Luděk Krtička)

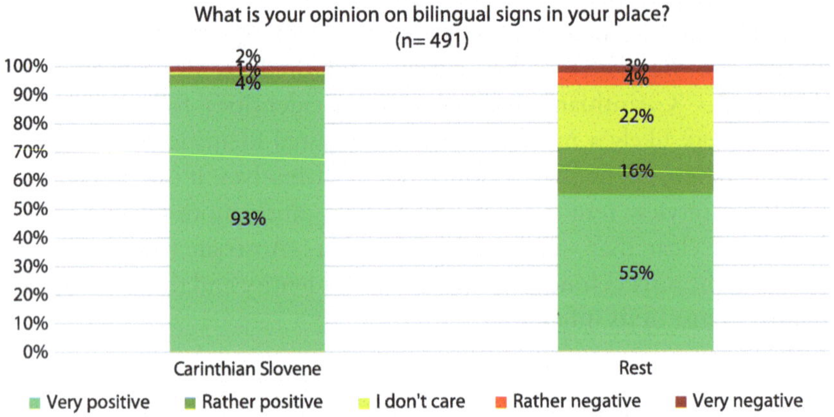

Fig. 6.70 Opinion on the importance of bilingual signs at home with "Carinthian Slovenes" versus other space-related identities. (Graphics: Luděk Krtička)

between 'neutral' answers to the question related to southern Carinthia and their home place is again high (9 vs. 22%) may again be explained by the attitude of avoiding negative statements in rather official situations on delicate issues. It can be assumed that in a less formal situation the majority of respondents having answered "I don't care" in Fig. 6.70 would have found bilingual town signs as "not as important" or "not important".

Another way of investigating into differences in the perception of bilingual signs is to ask for the relationship with the bilingual area. To this question eight possible answers plus space for alternatives were offered. Most respondents selected at least one of the following three answers: "I live there", "I was born there/I grew up there", or "I have relatives there", that is, answers indicating personal closeness to the area.

When correlating the opinions of both respondents living in and having no relationship with the bilingual area with their opinions on bilingual signs in southern Carinthia and at home, the results were again distinct (see Fig. 6.71). For respondents living in the area, bilingual signs are "very important" or "important", both in southern Carinthia and in their home place. For respondents having no relationship to this area, bilingual signs seem to be an unimportant issue, and a vast majority seems not to care about bilingual signs in their homeplace.

The gap in 'neutral' responses related to bilingual signs in the wider region and at home with respondents living in the area (5 vs. 12%) has already been explained before. The even larger difference of 18 versus 58% with respondents not having any relationship with the area can, however, indeed be attributed to the fact that people living outside the area are not personally affected by the question related to their own place. They have just an outsider's opinion on the issue and the response "I don't

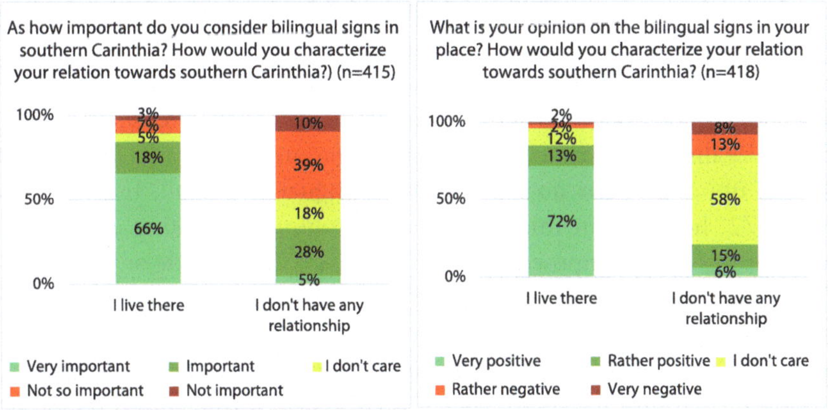

Fig. 6.71 Perception of bilingual signs by personal relation toward the bilingual area. (Graphics: Luděk Krtička)

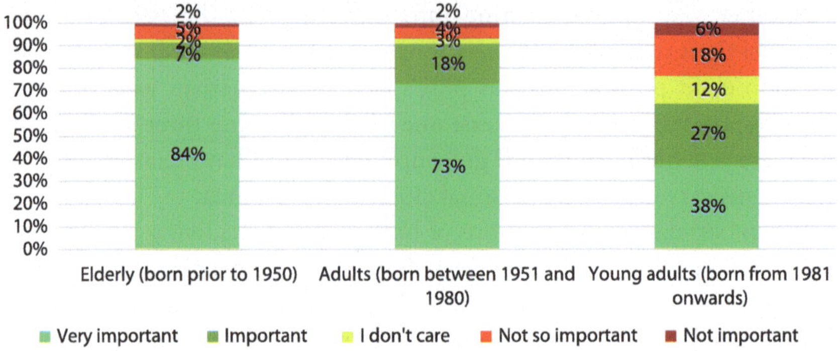

Fig. 6.72 Perception of bilingual signs by age groups. (Graphics: Luděk Krtička)

care" should perhaps be taken rather as a substitute for "This question is irrelevant in my case, does not apply to me".

It would not be justified concluding from this high percentage of 58 that Carinthians residing outside the bilingual area were indifferent toward the bilingual sign issue or that the younger had this attitude. (Many respondents with no personal relation to the bilingual area were high-school students from the Upper Gail Valley.) What, however, can be concluded is that among younger people living outside the area positive and negative opinions on bilingual signs are rather balanced.

No relevant differences between men and women regarding their perception of importance of bilingual signs could be found while it seems to decrease with age (see Fig. 6.72). This finding, however, may again have been distorted by the fact that a larger share of young respondents were 16- to 18-year-old high-school pre-graduates from the Gail Valley [Gailtal/Ziljska dolina], where bilingual signs are scarce in its lower part and its upper part has no share in the bilingual area at all. If one reads Fig. 6.72, however, in this way that the column for young adults has more or less been composed of respondents residing outside or just at the fringe of the bilingual area and that to the columns for adults and the elderly have contributed more or less only residents of the bilingual core area, a share of in total 65% regarding bilingual signs important or very important is quite high and does not contradict earlier interpretations that the younger outside the bilingual area display a more positive attitude toward bilingual signs. What has also to be considered with the interpretation of these diagrams is that the corpus of respondents had a very strong bias in favor of Carinthian Slovenes.

As regards education, perception of bilingual signs is clearly different between respondents with tertiary education and the rest (secondary education at the maximum). The former consider bilingual signs as more important than the latter (see Fig. 6.73). It has again to be taken into account that the corpus of respondents shows a dominance of Carinthian Slovenes and that the results of a survey completely outside the minority area would most likely look different. It is, however, also true that younger academics all over Carinthia have a more liberal and minority-friendly attitude also due to the University of Klagenfurt, founded in 1970 and known for its liberal and minority-friendly climate from the very beginning and at times acting as a kind of spearhead in minority affairs. But also other universities attracting students from Carinthia (mainly Graz and Vienna [Wien]) work into the same direction.

Respondents living in the bilingual area were also classified according to their places of residence into rural and urban, whereby the category 'urban' is in our case just represented by rather small towns functioning as urban centers of rural regions. Slightly more than half of the respondents (55%) live according to this definition in rural places and only 27% in urban. The rest of 14% has not been included into the evaluation, since they live in places outside the bilingual area.

In general, acceptance and appreciation of bilingual signs in rural places are slightly higher than in urban (see Fig. 6.74). This contradicts the usual cliché attributed to urban dwellers to be more tolerant toward

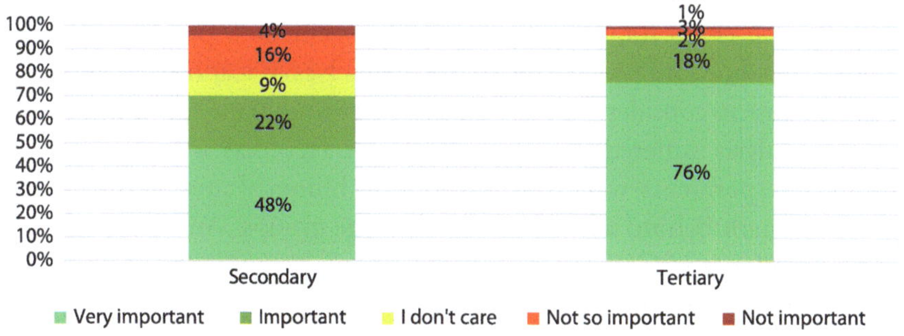

Fig. 6.73 Perception of bilingual signs by education levels. (Graphics: Luděk Krtička)

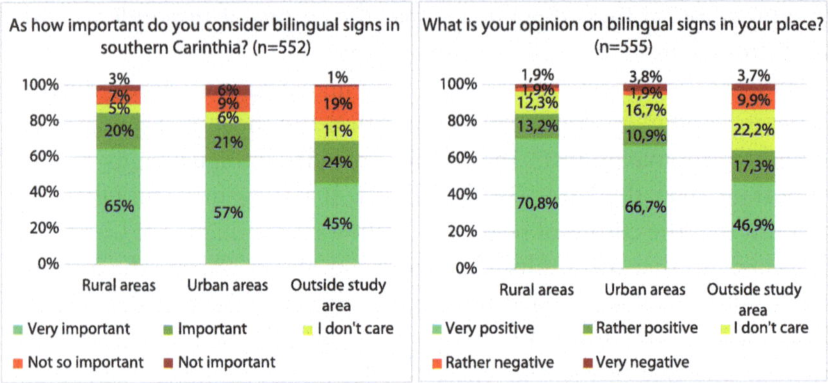

Fig. 6.74 Perception of bilingual signs by place of residence. (Graphics: Luděk Krtička)

diversity. It needs, however, to be reminded of the historical fact that the Bavarian settlers were the founders of towns and remained dominant there as craftsmen and traders, while the countryside was and has remained (in the bilingual area) the realm of the Slavonic, later Slovenian population (see Sects. 4.1 and 5.1).

6.1.5.4 Conclusion

Summing up the questionnaire analysis it can be said that for younger people, town dwellers, persons with no tertiary education, people speaking German at home and persons not conceiving themselves as Carinthian Slovenes the presence of bilingual signs in the bilingual area is less important than for the elder, rural dwellers, university graduates, persons speaking Slovene at home and people conceiving themselves as Carinthian Slovenes. The latter use to consider bilingual signs as "very important" or "important".

The analysis suffered, however, from the fact that a high share of younger respondents was from outside the bilingual area setting a big question mark behind all issues related to age groups. Another problem is the prevalence of Carinthian Slovenes among respondents. Thus, it has to be conceded that a corpus of respondents corresponding to the average proportion between minority and majority in the bilingual area would have surfaced rather different results.

The results of the questionnaire survey must thus for the reasons mentioned and due to an (after the work was done becoming obvious) unfortunate research design be regarded as not fulfilling the original requirement of testing the interview results. They have, however, provided some complementary information.

6.1.6 Conclusion (Peter Jordan)

The minority situation in Carinthia and more specifically the issue of place names is characterized by essentially a common culture all over the province nourished by Germanic and Slavonic ingredients from the Early Middle Ages and expressing itself most impressively by a mixed and closely interrelated namescape. It is, however, also characterized by a socio-economic stratification from the Early Middle Ages; the emerging of ethnic or national identities that questioned the former common regional identity; the later rise of Slovene than German as a standard and literary language combined with a gradient in demographic and political power and a gap in prestige; the fact that modern standard Slovene is not based on local Carinthian Slavonic dialects, but 'imported'; the repeated and very realistic threat to an old cultural-political entity to be divided, whereby the majority would have lost control over a larger part of its dominion; terrible atrocities during World War II that had been started by the German side; the feeling of the minority not to be adequately recognized as a contributor to common Carinthian culture; the feeling of majority and minority (but mainly of the majority) not to be adequately understood by the 'outside world'. All this diversified and even polarized the minority and let a core group (or core groups) as well as a wide transition zone between minority and majority arise.

The Slovenian geographer Jernej Zupančič reflects in his portrait of the Carinthian Slovenes this diversification quoting Marija Jurič (1987), who categorizes the minority into the following groups (Zupančič 1999: 60f)[11]:

[11] The original wording (Zupančič 1999: 60f):

1. Zaveden Slovenec, politično aktiven:

- vedno govori slovensko, četudi lahko sledijo posledice,
- označuje se kot (koroški) Slovenec,

(1) Conscious Slovene, politically active:

- always speaks Slovene, even if consequences may follow,
- identifies himself as (Carinthian) Slovene,
- is critical of politics,
- criticizes anti-Slovenian (ethnocentric, nationalist) policies,
- is engaged (politically and culturally) for the interests of the Slovenian national community,
- advocates the preservation and achievement or construction of the rights of the Slovenian minority,

---

- je kritičen do politike,
- kritizira protislovensko (etnocentrično, nacionalistično) politiko,
- je angažiran (politično in kulturno) za interese slovenske narodne skupnosti,
- zavzema se za ohranitev in doseganje oziroma graditev pravic slovenske manjšine,
- poroča o posebej številnih izkušnjah, s katerimi utemeljuje trenutni položaj.

2. Zaveden Slovenec:
- govori slovensko, kolikor je največ mogoče,
- izreka se za (koroškega) Slovenca,
- kritizira protislovensko (etnocentrično, nacionalistično) politiko,
- zavzema se za ohranitev in doseganje oziroma graditev pravic slovenske manjšine,
- poroča o številnih izkušnjah, s katerimi utemeljuje trenutni položaj.

3. Tisti, ki omahujejo med kulturama (nem. "Kulturpendlerji"):
- govori slovensko, če je slovensko ogovorjen,
- pri samopredstavitivi poudarja afiniteto do slovenske in nemške kulture,
- kritizira nemško nacionalne in slovenskonacionalne kroge ("skrajneže")

4. Asimiliranec:
- le redko govori slovensko, in sicer le, če je slovensko odgovorjen,
- označuje se za Vindišarja ali je negotov pri samoopredelitvi,
- vztraja, da ne zna slovensko, temveč vindiš; to narečje naj bi bilo bliže nemščini kot slovenščini,
- zmerja "skrajneže" in na splošno politiko,
- obstoječe pravice manjšine ocenjuje za zadovoljive, vsako nadaljnjo zahtevo pojmuje kot opeharjenje in osornost do večine.

5. Radikalni asimiliranec:
- ne govori slovensko, morda še pasivno razume,
- vztraja, da ne zna slovensko, temveč vindiš; to narečje naj bi bilo bliže nemščini kot slovenščini,
- označuje se za pripadnika večine,
- na splošno kritizira Slovence,
- ne obiskuje prireditev manjšine, temveč prej prireditve Slovencem sovražnih krogov,
- obstoječe pravice manjšine ocenjuje za zadovoljive, vsako nadaljnjo zahtevo pojmuje kot opeharjene in osornost do večine in
- boji se ekspanzije večine.

- reports on a particularly large number of experiences to justify the current situation.

(2) Conscious Slovene:

- speaks Slovene as much as possible,
- identifies himself as (Carinthian) Slovene,
- criticizes anti-Slovenian (ethnocentric, nationalist) policies,
- advocates the preservation and achievement or construction of the rights of the Slovenian minority,
- reports on a number of experiences to justify the current situation.

(3) Those who oscillate between cultures (in German: "Kulturpendler"):

- speak Slovene, if they are addressed in Slovene,
- in self-presentation emphasize the affinity to Slovenian as well as German culture,
- criticize German national and Slovenian national circles (as "extremists").

(4) Assimilant:

- speaks Slovene only if he is addressed in this language,
- identifies himself as 'Windisch' or is uncertain in self-determination,
- insists that he does not know Slovene, but 'Windisch' as a dialect that is said to be closer to German than to Slovene,
- insults "extremists" and politics in general,
- assesses the existing rights of the minority as satisfactory, perceives any further request as defrauded and rude to the majority.

(5) Radical assimilant:

- does not speak Slovene, perhaps still passively understands,
- insists that he does not know Slovene, but 'Windisch' as a dialect that is said to be closer to German than to Slovene,
- identifies himself as a member of the majority,
- criticizes Slovenes in general,
- does not attend minority events, but rather events of circles hostile

to Slovenes,
- assesses the existing rights of the minority as satisfactory, considers any further request as defrauded and rude to the majority, and.
- fears the expansion of the majority.

In analogy and based on our research also the Carinthian majority[12] could be categorized into the following groups:

(1) Conscious German-Carinthian, politically active:

- identifies him/herself as German-Carinthian,
- is a Carinthian regionalist,
- is an active member of a Carinthian 'homeland association',
- regards Carinthia as the land of the majority and the minority as "guests",
- criticizes "external influence" on the minority and minority politics in Carinthia,
- criticizes current minority rights as exaggerated and a danger for the land's unity,
- regards learning Slovene as neither necessary nor useful.

(2) Conscious German-Carinthian:

- identifies him/herself as German-Carinthian,
- is a Carinthian regionalist,
- is member of a Carinthian 'homeland association',
- likes to allude to the traditional dominance of the majority in Carinthia, the importance of and the threats to Carinthian unity due to the existence of the minority (although not necessarily caused by it),
- criticizes "external influence" on the minority and minority politics in Carinthia,
- regards current minority rights as "more than adequate",
- would not prefer learning Slovene to Italian.

[12] According to the principle of self-determination of ethnic (and other) social groups also Category 5 of the minority could be counted as a part of the majority.

(3) The in cultural-political affairs less- or insensitive:

- identifies him/herself as Carinthian,
- regards the land's unity as important and has a strong regional consciousness,
- recognizes the minority as autochthonous,
- regards current minority rights as adequate,
- avoids core groups of both sides,
- is not interested in minority questions,
- accepts Slovene as the land's second language but would not learn it.

(4) Multiculturalist:

- identifies himself as global citizen, European, Carinthian, or inhabitant of a certain place rather than as Austrian and certainly not as German-Carinthian,
- is skeptic opposite national feelings in general,
- regards the minority as a cultural wealth and would appreciate in general a more multicultural milieu in Carinthia,
- favors an extension of minority rights,
- is interested in minority questions,
- is in favor of learning languages in general and "of course" also Slovene,
- tends to enroll his/her children in bilingual kindergartens and schools.

(5) Advocate of a common Carinthian culture:

- identifies himself as Carinthian,
- looks at the minority as an essential ingredient of Carinthian culture and as making it specific,
- favors an extension of minority rights,
- is interested in all minority affairs,
- would learn Slovene, because it is the other local language,
- enrolls his/her children in bilingual kindergartens and schools.

As our research has shown, almost all these categories have their specific attitudes toward minority place names, especially toward minority place names on bilingual town signs as the most symbolic and most debated application of minority place names.

Conscious, politically active Slovenes regard the visibility of minority place names in public space as symbols of the minority's presence for centuries and of its share in Carinthia's culture and identity. They engage themselves actively and without any mental reservations for an extensive use throughout the bilingual area in all fields of application, but with a focus on town signs, since these are the most prominent and visible for everybody, while, for example, minority place names on maps have just a restricted audience, that is, the community of map users. The ups and downs in the numbers of official bilingual town signs have thus become the main indicators of their success in representing and advocating the minority and minority rights. They would also not hesitate to mark their private property with a name or sign in Slovene, except for business reasons.

Conscious, but not politically active Slovenes share all these attitudes in principle, but do not engage themselves actively for achieving these goals.

These who oscillate between cultures (in German: "Kulturpendler") recognize the symbolic value of publicly represented minority place names but rank it second to good relations to the majority and would not insist on them, if they cause conflict. They would, however, differentiate between the official (administration, education), the religious, and the private sphere. While they may regard bilingual naming in the official sphere of places with a larger share of the minority as fully justified, feel emotionally affected by texts in the minority language in the religious sphere and would not like to miss this tradition, they would be afraid of being "stigmatized" by any text in Slovene on their private house or business.

'Assimilants' or 'Windische' affiliating themselves to the majority in the cultural-political sense, but recognizing their Alpine-Slavonic roots distance themselves from the public representation of minority place names, especially in their standard Slovene version arguing that there is a difference between the local Carinthian minority culture and an "imposed" national-Slovenian attitude. They are, however, in favor of minority place names in their dialect version—if not necessarily also of their public representation—and support the documentation and preservation of minority micro-toponyms like field or house names as a cultural heritage.

Radical assimilants display in principle the same attitudes but have a more offensive stance against bilingual town signs in particular as well as minority rights and 'external influence' in general.

Conscious, politically active German-Carinthians earlier opposed in principle all public representation of minority place names, especially on the very symbolic town signs. In the meantime, they (inevitably) accept relevant regulations "imposed" by federal law and would not agitate against them for the sake of "higher-ranking values" like the image of the province and its attractivity for investors and tourists. They, however, meticulously observe any motion in favor of augmenting minority place names (and minority rights in general).

Conscious, but politically non-active German-Carinthians don't differ from these principal attitudes but are not as much 'on the tenterhooks' with observing the 'opposite camp'.

The less sensitive or insensitive in cultural-political affairs do not care about minority place names at all. Legislation in this field is not important for them and they would neither react in a positive nor a negative way if their own place received a bilingual town sign. They regard activities like the collection and documentation of minority field and house names or bilingual naming on maps as academic fingering exercises.

Multiculturalists are open for the most comprehensive minority place-name regulations even exceeding the bilingual area and would appreciate a higher share of names and texts in Slovene in the linguistic landscape. They would also not oppose including more recent migrant communities into the linguistic landscape.

Advocates of a common Carinthian culture emphasize the function of minority place names as expression and recognition of the minority's share in this common Carinthian culture and of the minority's share in space-related identity. They also conceive and appreciate the public representation of minority place names as an important means of minority identity building and the preservation of the minority's language and culture—both as augmenting Carinthia's cultural wealth. They would not regard names and texts of any other linguistic group as comparable to those of the autochthonous Slovenian minority. They regard the current town-sign regulation as quite satisfying and would not risk offending local identities by expanding it. They would, however, appreciate a higher visibility of the minority in the linguistic landscape, especially in the private sphere.

Apart from this variation by subgroups of the minority and the majority our research has confirmed earlier findings and hypotheses that the minority linguistic landscape shows spatial variations mainly by three criteria: (1) share of the minority in the overall population; (2) center-periphery gradient within larger settlements with their symbolic places in the center; (3) long-term economic orientation toward industry and tourism with a detrimental impact on the minority's share in the linguistic landscape.

6.2 Těšín/Cieszyn Region (Přemysl Mácha, Uršula Obrusník, Luděk Krtička and Pavel Pilch)

6.2.1 Representation of Minority Names on Maps

As we explained in Chap. 3, our interest in the representation of minority names on maps stems from our wish to understand the normative context for people's toponymic choices and strategies. The representation of minority names on maps is a very important way of recognizing the existence of the minority and—implicitly—giving it a claim to the territory (Ormeling 1983). As a consequence, maps are always deeply political and what is not represented is often as important as what is included (Harley 1988). On the other hand, many factors including frequent errors introduced in every stage of the map creation influence its final appearance, so map analyses should not be overpoliticized.

Our research interest in minority names on maps, simple as it may appear at first glance, however, presented us with serious challenges because it was often difficult to say who constituted the minority in different moments in history due to the region's turbulent past and significant changes in its religious, linguistic, and ethnic structure as well as its administrative boundaries. As a consequence, we had to adopt a more general research approach, not focus on Polish names a priori, and rather ask what the decisions on linguistic and orthographic transcription of names on maps in different time periods tell us about the linguistic and ethnic politics of the time. We were able to focus specifically on the representation of Polish names only on the most recent maps. And as we will show, some of these maps generated heated controversies. For the subsequent analysis we select only those maps which seem to have had the largest impact on dominant cartographic representation of the region and the current toponymic discourse.

The first known map designed for the Těšín/Cieszyn region was prepared and published by Jonas Nigrin in Teschen [Cieszyn] in 1724 (see Fig. 6.75). Its approximate scale is 1:153,000. It was the first detailed map of the region which had until then appeared only on significantly less detailed maps of Silesia [Slezsko, Śląsk]. Map title and legend are rendered in Latin and the map also contains visual images of Wallachian shepherds presumably characteristic of the region. The vast majority of names indicated on the map refer to settlements (250 in the Těšín/Cieszyn region, 520 in total), a minority to other features such as hills and rivers. The linguistic and orthographic rendering of names is inconsistent. A minority of names are given only in German (i.e., *Teschen, Altstadt, Schönhoff*) and there is no necessary connection with a German-speaking population in the locality. Others are transcribed in German orthography, but the root is clearly Slavic (i.e., *Grodischt, Kotzobenz, Elgot*). The third type of names combines German and Slavic orthography (i.e., *Toschonowice, Schebischowice, Kotzurowice*). Finally, a large proportion is transcribed in an emerging Slavic orthography variously

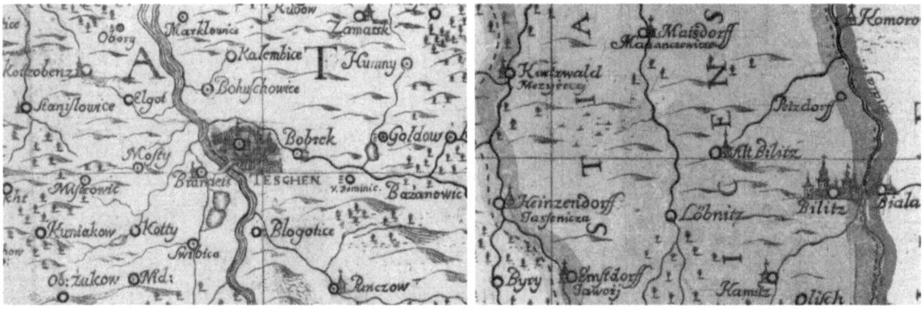

Fig. 6.75 Sections of Nigrin's map of the Těšín/Cieszyn region "Ducatus in Silesia Superiore Teschinensis cum adjacentibus regnorum vicinorum", 1724. (Source: Mollova mapova sbírka 2019)

expressing the probable Czech, Polish, or dialectal origin of the name (i.e., *Grudek, Podlessj, Lonka*). Sometimes the same names are rendered differently, for example, *Ropica/Ropicza*. Interestingly, the only instances of bilingual names appearing on the map concern the German-speaking villages in the northeastern part of the Těšín/Cieszyn region—*Kurtzwald/Mezyriczj*, *Heinzendorf/Jassenicza*, and so on. Apparently, from Nigrin's perspective, this was the minority of his times.

As we mentioned in Chap. 3, in many cases it is difficult to decide whether the names come from Czech, Polish, or the dialect. It is reasonable to presume that the vast majority of local Slavic inhabitants at this time spoke different versions of the dialect, and the dialect only, and they used corresponding dialectal forms of names. However, this is not necessarily reflected on the map, as some names where linguistic provenience can be established are often Czech (e.g., *Bludowice, Trinec*) or Polish (e.g., *Wendrina, Strumien*). Still, the map contains many dialectal names as well (e.g., *Kunska, Kopetna*), most notably using *Olsa* as the name of the principal river flowing through the region. The map is thus a linguistic and orthographic Babel, expressing both the diversity so defining for this region and the absence of clear-cut ethnic identities at this period. It was a revolutionary endeavor for its times, but it is of little use in current toponymic conflicts in which the historic antiquity and precedence of a name are often invoked. It simply fails to provide a simple toponymic picture of the region. It is therefore somewhat ironic to know that Nigrin was fined and imprisoned and almost all copies of the map were confiscated and destroyed by the state authorities because of military security reasons.

On the request of the Monarchy, in 1736 Johann Wolfgang Wieland, a military officer, published another map (see Fig. 6.76) of the Těšín/Cieszyn region in Nuremberg [Nürnberg] which contained a comparable amount of information to the Nigrin map, although rendered at a larger scale (1:116,000). The map corrected many topographical mistakes present in the Nigrin map. Nevertheless, just as the Nigrin map, the Wieland map also used several orthographies for the transcription of names. From our perspective, however, it is important to note the bilingual rendering of selected names which appear not only in the case of German-speaking villages but also in several other instances. In these cases, the main name

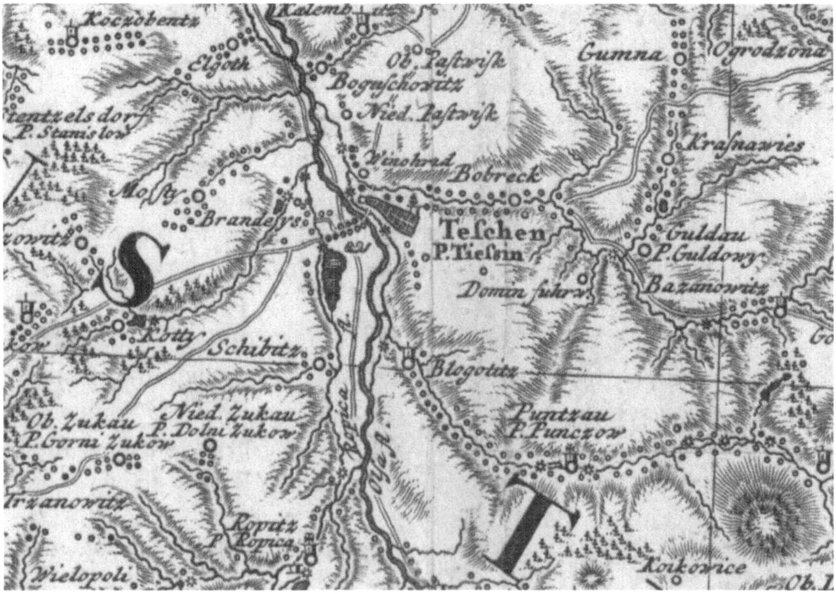

Fig. 6.76 Section of Wieland's map of the Těšín/Cieszyn region "Principatus Sielasiae Teschinensis nova et accurate", 1736. (Source: Mollova mapova sbírka 2019)

appears in larger script and the other name below in smaller font size preceded by the capital letter *P* which according to the map legend (in Latin) indicates the Polish version. Thus, we see, for example, *Teschen* and *P Tiessin*, *Dzingelau* and *P Diehilow*, and *Albersdorf* and *P Albrechtice*. Interestingly, in all three of these examples the presumed Polish version is actually the Czech one. Although this is not always the case and some names are rendered in Polish, it only further illustrates the linguistic and ethnic fluidity of the region in the 1700s.

Nigrin's map was published by a private individual with geographic curiosity (and commercial interest) and its circulation was severely limited by the authorities. The Wieland map was commissioned by the Austrian monarchy but still prepared by a private individual, albeit a military officer, for one region. We cannot therefore consider these maps as expressions of official state policy toward the territory and its inhabitants. This, however, changes with the growing awareness of the strategic importance of good maps on the part of the Austrian state which led the Monarchy to take full control of their subsequent production. The result was the state-produced Austrian land survey which became crucial for the cartographic representation and control of the territory. The military land surveys were created for the entire Monarchy and although their use was also restricted for security reasons, they expressed rather clearly the official ethnic and linguistic policies toward the regions under control. Let us briefly comment on each of these surveys as there are important differences between them.

The First or Josephinian Land Survey was a reaction to the loss of the majority of Silesia [Slezsko, Śląsk] to Prussia. The remaining Austrian part of Silesia was therefore the first to be mapped, in 1763. Unlike the maps mentioned before, title and legend of this survey (see Fig. 6.77) were in German. In spite of its large scale (1:28,800), the survey was not very exact and contained many topographic errors remedied only by later surveys. As far as the representation of names is concerned, the map contains a comparable number of names as the aforementioned but differs from them especially in the stronger presence of purely German names and orthography. At this time the German language gradually began to establish itself as the official language of state administration, replacing Czech, and thus it began to dominate cartographic works as well. Unlike on previous maps, the land survey contains no bilingual names at all. The region was thus treated as homogenous with the German element emphasized. The orthographic rendering of Slavic names continues to obscure their linguistic origin, but where it is possible to differentiate, names in Czech (e.g., *Wielopoli*), Polish, and/or dialect (e.g., *Trzytiez*), and dialect (e.g., *Kaminite*) appear without any apparent logic. Just as with the previous maps, Czech names may appear alongside Polish and dialectal ones without geographic or demographic justification. It is noteworthy that in contrast to the previous and subsequent maps up to World War I, the name of the main river flowing through the region is rendered in the form *Oelsa* which is probably an attempt to approximate the local dialectal form of the name (*Łolza*).

The Second or Franciscan Land Survey in Moravia [Morava] and Silesia [Slezsko, Śląsk] was carried out between 1836 and 1842. The scale

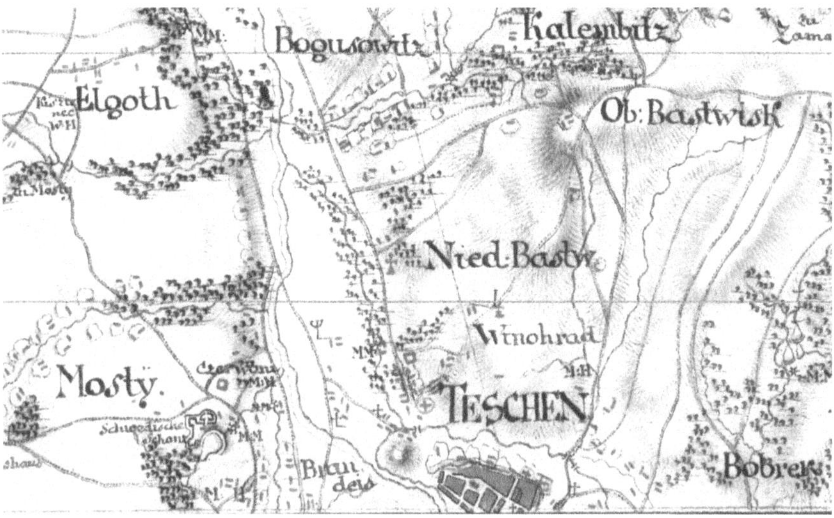

Fig. 6.77 First Austrian Land Survey, Upper Silesia [Górny Śląsk], Sections No. 7 and 8, 1763. (Source: Laboratoř geoinformatiky Fakulta životního prostředí Univerzity J.E. Purkyně 2019)

was 1:28,800 and the maps were jealously guarded by the authorities for security reasons (see Fig. 6.78). As with the previous survey, there are no bilingual names. There is also a clear movement toward the dominance of German in name use and orthography. The proportion of hybrid orthographies decreased significantly, although we still encounter it in cases when German orthography lacked letters for Slavic phonemes (e.g., *Nieder Žukau, Bažanowitz*). The Germanization of toponymy and orthography clearly expresses the Germanizing agenda of the Austrian state and the treatment of the territory as homogenous. Interestingly, where Slavic orthography was used, it was the Czech orthography of the time. It was used throughout the map, even in parts of the Těšín/Cieszyn region belonging currently to Poland. As with the First Survey, the linguistic origin of names is often obscured by German orthography, but we still can find names of discernable Polish, Czech, and dialectal origin. In 1844, generalized maps in the scale 1:144,000 based on this survey were made public. Because of the smaller scale, they contained fewer names and they were rendered in the same form as in the survey sections. It is

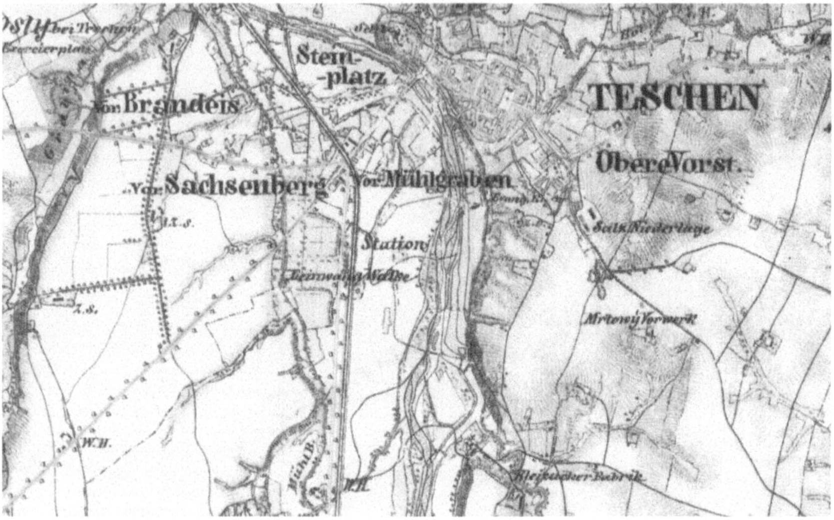

Fig. 6.78 Second Austrian Land Survey, Moravia [Morava], Sheet No. XI-6. (Source: Laboratoř geoinformatiky Fakulta životního prostředí Univerzity J.E. Purkyně 2019)

worth mentioning that some of the bilingual names for German-speaking villages in the northeastern part of the region reappeared at the time.

The Third or Francisco-Josephinian Land Survey was motivated by yet another lost war with Prussia in 1866. The mapping was carried out with modern geometric methods and the maps are significantly more exact than the previous ones. They served, with some modifications, even during Czechoslovak times as the most reliable source of geographical data until the 1950s. The recordings in the scale 1:25,000 remained unpublished, while the scale 1:75,000 appeared as Special Map (*Spezialkarte*) with elevation contour lines. Unlike previous maps, it also contained information of non-military relevance and was used for civil purposes. The Těšín/Cieszyn region was mapped in 1876 (see Fig. 6.79). As with the previous surveys, German language and orthography overwhelmingly dominated the rendering of names. The Third Survey, however, contains more geographical and toponymic information, so new names that were not mentioned in the previous surveys appear for the first time. What is also new is the use of clearly Polish orthography in purely Polish (e.g., *Czornowski,*

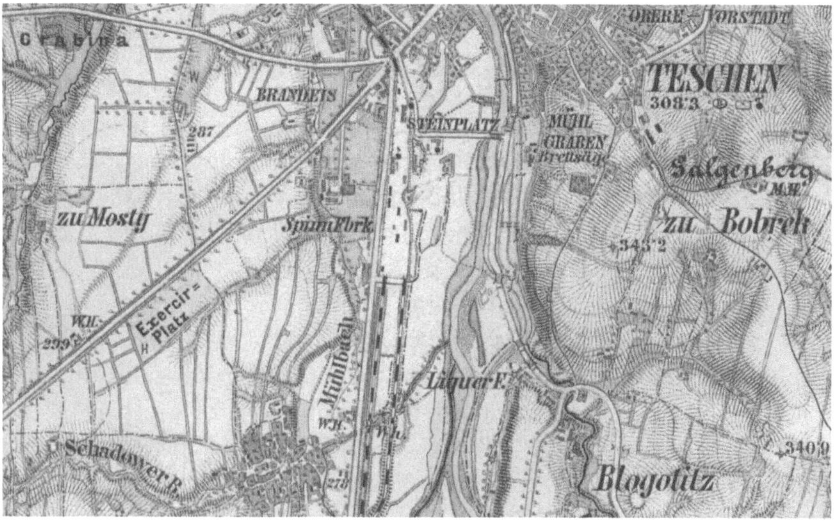

Fig. 6.79 Third Austrian Land Survey, Special Map 1:75,000, Sheet No. 4161 Teschen, Mistek und Jablunkau, 1876. (Source: Český úřad zeměměřický a katastrální 2019)

Babia góra) and many hybrid name renderings (e.g., *Bażanowitz, Mięzi lessi*), especially (but not exclusively) in those parts of the region which were inhabited by people who were classified as Polish-speaking by the later Austrian census. By contrast, in settlements with no presumed Polish-speaking population, Czech orthography was sometimes used (e.g., *Stražnica, Křižowa*). This may indicate a greater sensitivity on the part of the Austrian authorities for the linguistic preferences of the local populations. Or it may be a sign of the incipient conflict between local Czechs and Poles over the toponymic control of the region which became full-blown in the last decades of the nineteenth century. Interestingly, the only bilingual names given on the map are those for the German-speaking villages in the northeast. At this time, from the perspective of the Austrian state, German-speakers were still the only noteworthy minority.

The open toponymic war began already before World War I after the results of the Austrian population census of 1910 were seized on by Polish nationalists to demand more rights for the presumed Polish population of the region. These claims were supported also cartographically. As an example of this may serve a map from 1910 published by a Polish organization in Teschen [Cieszyn], showing the results of the census.[13] From our perspective it is important to note that map title, legend, and all names rendered on the map were in Polish language and orthography, creating thus an image of a relatively homogenous Polish space. This was by no means the last map of its kind (another one came out, for example, in 1930[14]) and it became an important source of inspiration to a recent regional map on which we comment below.

Toponymic conflict escalated after World War I, when the Těšín/Cieszyn region was claimed by both Czechoslovakia and Poland and geographical names were seen as means of supporting territorial claims. The conflict took place at different scales, from the level of street names in Cieszyn (see Mácha et al. 2018) to the representation of the entire region on state maps. An early expression of this conflict was the rendering of names on maps used by the international delimitation commission which divided the region between the two countries at the Paris peace conference in 1920. The official language of the commission was French which is also the language in which map titles and legends used by the map makers were given. While the Czechoslovak delegation argued on historical and pragmatic grounds, pointing out the industrial and transportation integration of the region into the country, the Polish representatives justified their claim to the region by the presence of the Polish majority living there. This was clearly reflected on the maps they provided to the commission. All geographical names on Czechoslovak maps are strictly in Czech language and orthography, presenting thus the entire region as a natural component of the Czechoslovak economy and transportation network.[15]

[13] Księnstwo Cieszyńskie, Stosunek narodowości według ludności z r. 1910, 1:300,000, http://maps.mapywig.org/m/Polish_maps/various/Topographic_and_tourist_maps/KSIESTWO_CIESZYNSKIE_NARODOWOSCI_WG_SPISU_Z_1910_300K_2.jpg.

[14] Ludność polska w czechosłowackiej części Śląska cieszyńskiego, 1930, no scale given http://maps.mapywig.org/m/Polish_maps/various/Topographic_and_tourist_maps/LUDNOSC_POLSKA_w_czechoslowackiej_czesci_SLASKA_CIESZYNSKIEGO_c.1930.jpg.

[15] For example, International Border Delimitation Commission Map, 1919, no scale given; http://chartae-antiquae.cz/cs/maps/28740.

Fig. 6.80 Map of Zaolzie "Śląsk Zaolziański" 1:320,000, 1938. (Source: Archivum map wojskowego instytutu geograficznego 1919–1939 2019)

In contrast, the Polish produced a map emphasizing the ethnic composition of the region and all names were rendered in Polish only.[16] It goes beyond the scope of our analysis to delve any further into post-World War I Polish cartography; it may be sufficient to say, however, that the use of maps to assert territorial claims to the Těšín/Cieszyn region did not end with the division of the region between the two countries in 1920 and resurfaced on several additional occasions, most notably in connection with the Polish occupation of the region in 1938 (see, e.g., Fig. 6.80).

We will now concentrate only on maps produced by Czechoslovakia and Czechia for the Czech part of the Těšín/Cieszyn region. As we already mentioned, Czechoslovakia adopted the Third Austrian Land Survey and used it as its basic cartographic source until the 1950s. Apart from gradually correcting different topographic mistakes and bringing the map up to date, the most visible change from our perspective was its total toponymic overhaul. Not only map titles and legends but also all names in the Těšín/Cieszyn region were rendered purely in Czech. As we see clearly

[16] Mapa orjentacyjna Śląska cieszyńskiego, 1920, no scale given;
http://maps.mapywig.org/m/Polish_maps/various/Topographic_and_tourist_maps/MAPA_ORJENTACYJNA_SLASKA_CIESZYNSKIEGO_200K_1920_1.jpg.

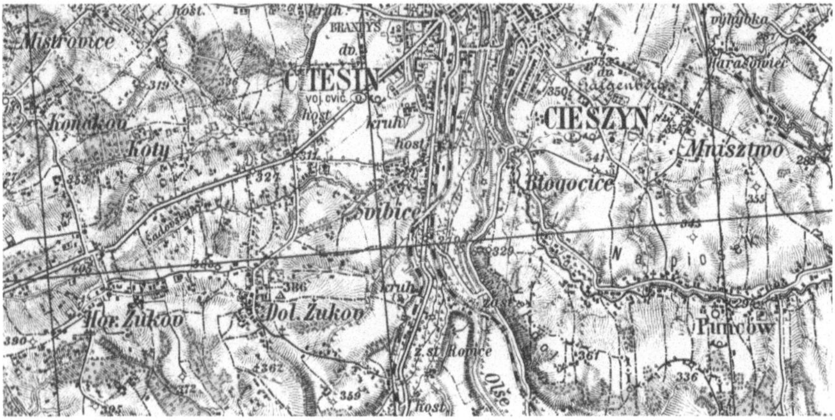

Fig. 6.81 Czechoslovak Military Map 1:75,000, Sheet 4161 Frýdek, 1935. (Source: ČÚZK 2019)

on the 1935 map (see Fig. 6.81), special care was taken to eliminate also dialectal names since they might remind of Polish names. The city of Teschen, historically appearing only as *Teschen*, now appears as *Český Těšín* (in Czechoslovakia) and *Cieszyn* (in Poland), and the river historically rendered as *Olsa* now appears only as *Olše* (in Czechoslovakia) and *Olza* (in Poland). As a consequence, the map for the Těšín/Cieszyn region does not toponymically differ in any way from a map of a Czech-only area in the center of the country.

Unlike the Germans in other parts of the country, the Poles were not recognized by bilingual names. This is undoubtedly the result of fear that a toponymic admission of the existence of Poles in the region would legitimize Poland's claim to the territory. Interestingly, this policy continued after the Polish occupation of the region in 1938 as the 1939 map edition shows,[17] during the Nazi occupation which was particularly harsh on Poles living in the region as testified by the 1943 edition which tem-

[17] Frýdek, Vojenský zeměpisný ústav v Praze, 1939, 1:75,000, https://archivnimapy.cuzk.cz/uazk/topo3v75/topo3v75_data/4161/4161_07_index.html.

porarily reintroduced *Teschen* and a Czech-German bilingual legend,[18] and after World War II as is indicated by the 1946 edition.[19]

Czech tourist maps derived from the state maps did not differ in this respect, although some did include selected dialectal names such as *Olza* or *Grudek*.[20] In this context, it is important to note that in the interwar period there were three ethnically defined tourist associations (the Czech "Klub československých turistů", the German "Beskidenverein", and the Polish "Beskid Śląski") which all published their own hiking maps of the region and promoted thereby not only their touristic activities but also their nationalist agendas.

Starting in the 1950s, a series of new state maps in the scales 1:5000, 1:10,000, 1:25,000, 1:50,000, 1:100,000, 1:200,000, and 1:500,000 was produced. As with the Austrian land surveys, access to them was limited, especially in the larger scales. Regardless of scale, however, nothing important changed for the Polish inhabitants of the region. All names are rendered in standardized Czech, eliminating thus dialectal names as well (see Fig. 6.82). The meticulousness with which the map is purged of all names even remotely Polish is actually surprising, given the generous state support for Polish cultural organizations and Polish schooling in the region. We would argue, however, that the explanation is still the same—a fear of legitimizing potential territorial claims by Poland.

The state maps censored for sensitive security information were, nevertheless, used in several public applications. There are many tourist and hiking maps from the Communist period based on the state maps which entirely conformed to the toponymic standards of the state maps, including field names.[21] The Czechoslovak Tourist Association [Klub československých turistů] was not allowed to publish its own maps during the Communist period. It only regained this authority in 1991 which

[18] Teschen, Landesvermessungsamt Böhmen u. Mähren, 1943, 1:75,000; https://archivnimapy.cuzk.cz/uazk/topo3v75/topo3v75_data/4161/4161_17_index.html.

[19] Frýdek, Zeměměřičský ústav v Praze, 1946, 1:75,000; https://archivnimapy.cuzk.cz/uazk/topo3v75/topo3v75_data/4161/4161_09_index.html.

[20] Moravskoslezské Bezkydy, not dated, approximately 1930, 1:100,000; https://www.oldmapsonline.org/map/cuni/1037879.

[21] Beskydy, Ústřední správa geodezie a kartografie, 1957, 1:75,000 Beskydy. In: Virtualní mapová sbírka Chartae-Antiquae.cz [online]. Zdiby: Výzkumný ústav geodetický, topografický a kartografický, v.v.i. [cit. 2019-12-02]. Dostupné z: http://chartae-antiquae.cz/cs/maps/46400; http://chartae-antiquae.cz/cs/maps/46400.

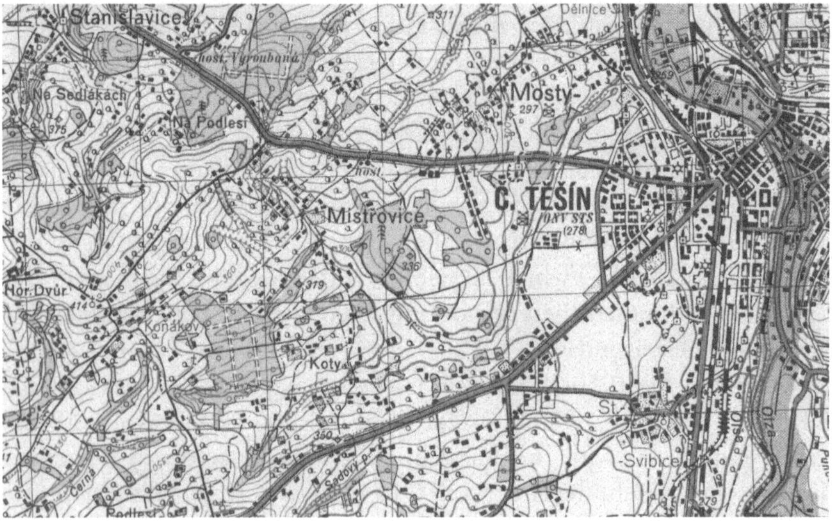

Fig. 6.82 Czechoslovak Military Map 1:50,000, Sheet M-34-74-C Český Těšín, 1956. (Source: ČÚZK 2019)

is when the cartographic (and toponymic) representation of the Těšín/Cieszyn region finally begins to change. The proliferation of maps published by the tourist association, private companies (most notably, SHOCart), and communal and regional governments democratized cartographic representation of space. However, it also brought back the linguistic and toponymic Babel we saw on the earliest maps of the region, as different name versions and different orthographic representations of their pronunciation appeared on different maps. Especially maps produced by non-state actors included a larger number of dialectal names. However, their transcription was often amateur and incorrect, causing confusion and generating criticism by the local population. As a rule, however, hardly any maps included bilingual or minority names—Polish names thus remained mostly absent. This is also true for the updated state maps which still fail to include bilingual names for populated places, but which now render train-station names bilingually and include some dialectal names (see, e.g., Fig. 6.83). This is a small victory for the Congress of Poles who have been fighting for the inclusion of bilingual names in official state maps alongside the introduction of bilingual signs. According to one Congress representative, the state resistance toward bilingual

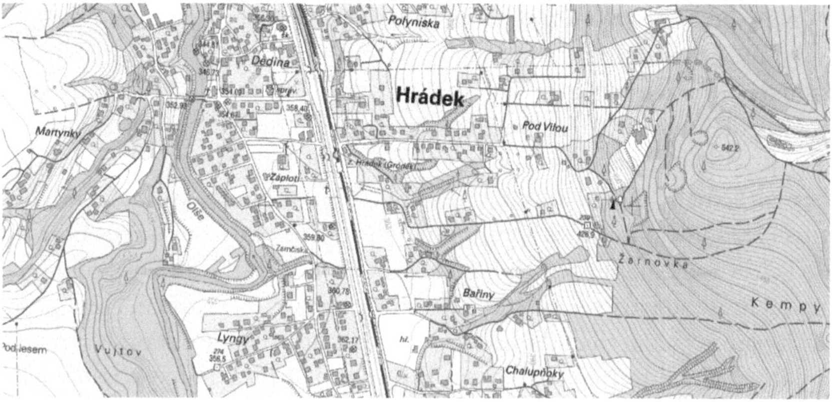

Fig. 6.83 Example of a state map combining official and dialectal names. (Source: ČÚZK 2019)

names on maps (as compared to bilingual signs) can be attributed to the facts that bilingual signs are seen only by those who travel to the region while maps are available to everyone and thus potentially weaken the Czech territorial claim to the region.[22]

Because of the large number of maps published in the last three decades and the associated cartographic cacophony, it goes beyond the scope of this text to analyze them in detail. However, two cases deserve a closer attention because they illustrate well the role of maps in ethnic relations in the region.

The first case involves a series of three local maps published by the communal government of a village at the border with Poland.[23] In the last population census, 19% of local inhabitants declared Polish nationality. There are very few bilingual signs in the village, and they have been a target of repeated vandalization. The village is an important starting point for hikes in the Silesian Beskids [Slezské Beskydy, Beskid Śląski] and thus receives many visitors. Some years ago, the commune decided to publish a detailed map of the village including field names for the use by locals and tourists. Names rendered on the map followed the standard of state maps and thus were all written in standardized Czech versions and orthography.

[22] A member of the Congress of Poles leadership, 46 years old, university professor, Pole.

[23] At request of the local representatives, we do not give the village name.

In many instances, name standardization led to a total corruption of names and even a change in their meaning. As a consequence, many inhabitants strongly criticized this map for not representing their village correctly and turning their home into a foreign space. The communal government therefore decided to print a new map which would record the local names as they are actually used. However, because Czech orthography does not have letters for some phonemes appearing in the dialect, the commune chose to use Polish orthography which is capable of expressing the dialectal pronunciation more adequately. Nevertheless, the map was met with an even more passionate criticism than the previous one because the use of Polish orthography suggested to the local Czech majority that their home was actually Polish. As a result, the map had to be scrapped once again. In the third version, the commune decided for a compromise which appears to have been accepted by the majority—the map contains local names, but they are written in Czech orthography. Its use inevitably led to misrepresentation and corruption of many dialectal names; however, national identity overrode linguistic correctness. This model has become dominant in the region since (e.g., communal map of Hrádek/Gródek).[24]

The second case involves the tourist map of Zaolzie promoted by the Congress of Poles and its partner organizations and published with public funding from the Euroregion Těšínské Slezsko—Śląsk Cieszyński in 2012 (see Fig. 6.84). No other map in recent decades has caused a comparable amount of uproar and it was widely viewed by Czechs as a reckless provocation. The polemic centered on several issues. First, the map received public funding intended for the promotion of tourism. Polish organizations presented the map as promoting the region among tourists, but the map did not contain hints at hiking trails, tourist attractions, or other tourist information—it was a general map with settlements, roads, and principal geographic features only. Additional information appearing on the map concerned details about the life of the Polish minority in individual villages, nothing was said about cultural or social life in the region as such. Second, the reference space for the map was *Zaolzie* which

[24] A similar conflict occurred in relation to a collection of proverbs published in the dialect in Jablunkov/Jabłonków. The author chose Czech orthography and faced strong criticism from members of the Polish minority for deforming the dialect. The real conflict behind this was over the national origin of the dialect—Czech or Polish. The choice of orthography was interpreted as an indication of linguistic and national ownership of the dialect (Szpyrc 2002).

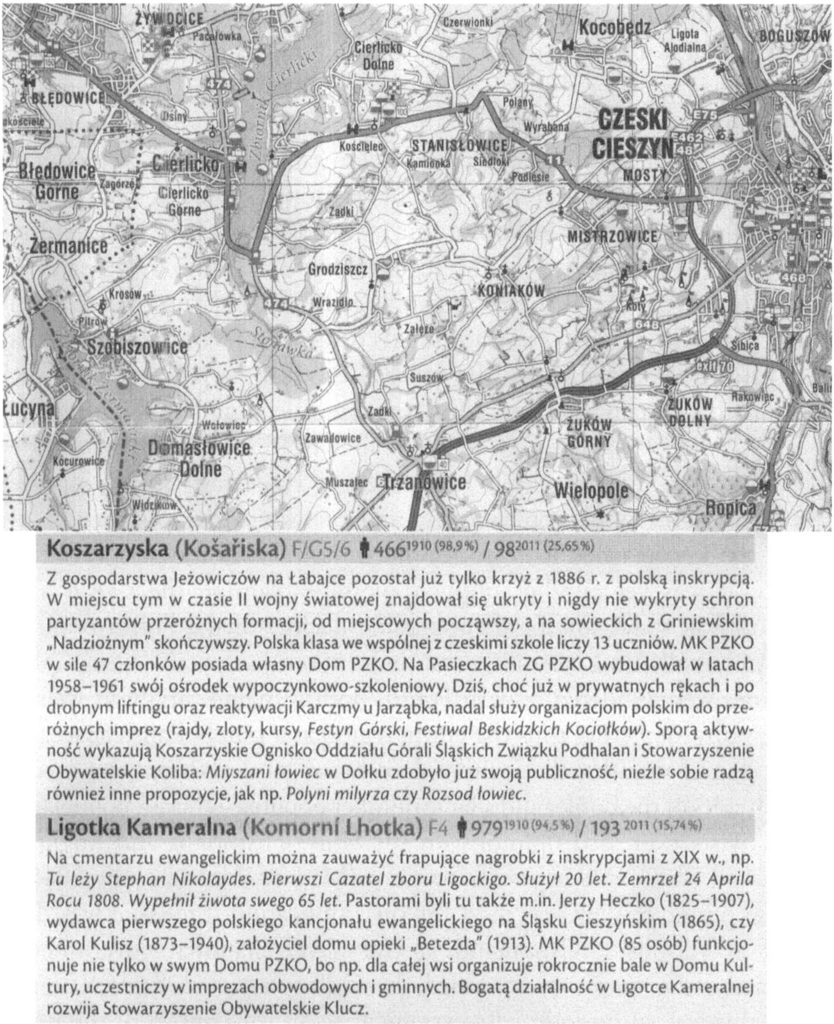

Koszarzyska (Košařiska) F/G5/6 466^{1910} (98,9 %) / 98^{2011} (25,65 %)

Z gospodarstwa Jeżowiczów na Łabajce pozostał już tylko krzyż z 1886 r. z polską inskrypcją. W miejscu tym w czasie II wojny światowej znajdował się ukryty i nigdy nie wykryty schron partyzantów przeróżnych formacji, od miejscowych począwszy, a na sowieckich z Griniewskim „Nadziożnym" skończywszy. Polska klasa we wspólnej z czeskimi szkole liczy 13 uczniów. MK PZKO w sile 47 członków posiada własny Dom PZKO. Na Pasieczkach ZG PZKO wybudował w latach 1958–1961 swój ośrodek wypoczynkowo-szkoleniowy. Dziś, choć już w prywatnych rękach i po drobnym liftingu oraz reaktywacji Karczmy u Jarząbka, nadal służy organizacjom polskim do przeróżnych imprez (rajdy, zloty, kursy, *Festyn Górski*, *Festiwal Beskidzkich Kociołków*). Sporą aktywność wykazują Koszarzyskie Ognisko Oddziału Górali Śląskich Związku Podhalan i Stowarzyszenie Obywatelskie Koliba: *Miyszani łowiec* w Dołku zdobyło już swoją publiczność, nieźle sobie radzą również inne propozycje, jak np. *Polyni milyrza* czy *Rozsod łowiec*.

Ligotka Kameralna (Komorní Lhotka) F4 979^{1910} (94,5 %) / 193^{2011} (15,74 %)

Na cmentarzu ewangelickim można zauważyć frapujące nagrobki z inskrypcjami z XIX w., np. *Tu leży Stephan Nikolaydes. Pierwszi Cazatel zboru Ligockigo. Służył 20 let. Zemrzeł 24 Aprila Rocu 1808. Wypełnił żiwota swego 65 let.* Pastorami byli tu także m.in. Jerzy Heczko (1825–1907), wydawca pierwszego polskiego kancjonału ewangelickiego na Śląsku Cieszyńskim (1865), czy Karol Kulisz (1873–1940), założyciel domu opieki „Betezda" (1913). MK PZKO (85 osób) funkcjonuje nie tylko w swym Domu PZKO, bo np. dla całej wsi organizuje rokrocznie bale w Domu Kultury, uczestniczy w imprezach obwodowych i gminnych. Bogatą działalność w Ligotce Kameralnej rozwija Stowarzyszenie Obywatelskie Klucz.

Fig. 6.84 Portions of the front and back side of the tourist map of Zaolzie ("Zaolzie—mapa turystyczna") 1:70,000, published in 2012 by Książnica Cieszyńska and Congress of Poles. (Source: Kongres Polakow w Republice Czeskij)

as we explained earlier is a politically highly charged term in the Těšín/Cieszyn region—regularly used by local Poles but fiercely contested by local Czechs because of its implicit territorial claims by Poland. The ethnic connotation of the term was emphasized by the inclusion of the anthem of the local Poles—*Płyniesz, Olzo, po dolinie*—on the back side of the map. Third, the map contained historic border lines from 1918 and 1938 associated with the Polish territorial claim to the region and its occupation before World War II. This was defended by the Polish representatives as a pragmatic geographical delimitation of *Zaolzie* which had never had any formal administrative or territorial standing. Fourth, the map was rendered only in Polish, but it was distributed in information centers in the Czech part of the region which was once again seen as undue provocation. Fifth, accompanying the title of the map was an image of two local Gorols in traditional costumes, one of them bearing a pole with the Polish flag and the other also bearing a pole but with the flag missing (apparently the other flag was Czech and it was digitally erased). Sixth, the map was accompanied by texts on the back side recounting the history of the Těšín/Cieszyn Poles and also giving the 1910 Austrian population census numbers where the proportion of Polish-speakers reached its historical maximum. From the perspective of Czechs, it was yet another implicit message blaming the Czech and Czechoslovak states for the demographic decline of the Polish population. And finally, and most importantly from our perspective, geographical names were rendered only in Polish, both in name variant as well as orthography, depicting thus the entire region as a purely Polish space.

For several months, the scandal dominated local and regional media on both sides of the border and a large number of public officials, politicians, and activists expressed opinions on the issue. Historians and linguists were asked to provide expert assessment (Map of Zaolzie 2009–2013). Because public money was used for the publication, the issue was discussed at various levels, including the Czech national government. Ultimately, the map was scrapped, becoming a rare collector's item. While this was an undoubtedly extreme case, in the context of the brief cartographic history of the Těšín/Cieszyn region we have provided above it does illustrate very well how important maps are in shaping public perception of space, how the choice of geographical names on maps influences their interpretation, and how maps may serve as proxies for endemic nationalist and ethnic tensions, historical grievances, and fears.

6.2.2 Linguistic Landscape

Another crucial source of people's ideas about place names and bilingualism is the linguistic landscape (LL). It is a significant component of the physical environment in which we live our lives and engage in everyday social relations. LL renders space more manageable and makes it possible to emphasize (or deny) ownership and belonging. It has a special influence on the relative power, status, and perception of different languages and their speakers (Cenoz and Gorter 2006). With time, LL becomes internalized and is often taken for granted as a natural state of things. When something changes—a new sign appears, an old one disappears, or an existing one is transformed—the taken-for-granted character of LL is challenged. Most changes are usually accepted, albeit with a grudge, but some may provoke a strong reaction and the eventual accommodation may turn out to be a painful and long-lasting process. This is precisely the case of the implementation of bilingual signs in the Těšín/Cieszyn region. Even after more than a decade of the presence of bilingual signs, emotions still run high and vandalization occurs regularly.

As we mentioned in Chap. 3, we mapped LL of several communes in which we subsequently carried out interviews and the questionnaire survey to understand the physical context potentially influencing people's ideas about bilingualism and place names. It was not possible to document all existing bilingual and minority signs in the entire region due to its size. By traveling around the region, however, we were able to obtain a good understanding of the number and character of signs in different settlements. This helped us to select communes for case study analyses and formulate questions for interviews.

Specifically, we documented all signs in four communes—Bystřice/Bystrzyca, Hnojník/Gnojnik, Chotěbuz/Kocobędz, and Jablunkov/Jabłonków—which represent different types of populated places in relation to the industrial and lowland areas of the region (see Fig. 6.85).[25] In addition, we included in our interview sample people from several com-

[25] We also documented all bilingual and minority signs in the center of Český Těšín/Czeski Cieszyn, but this city did not prove to be a good case study choice because of the large number of visitors it receives from Poland. It was therefore difficult to determine who was the target audience for bilingual signs and what proportion of them was put up in response to local ethnic politics rather than international tourism.

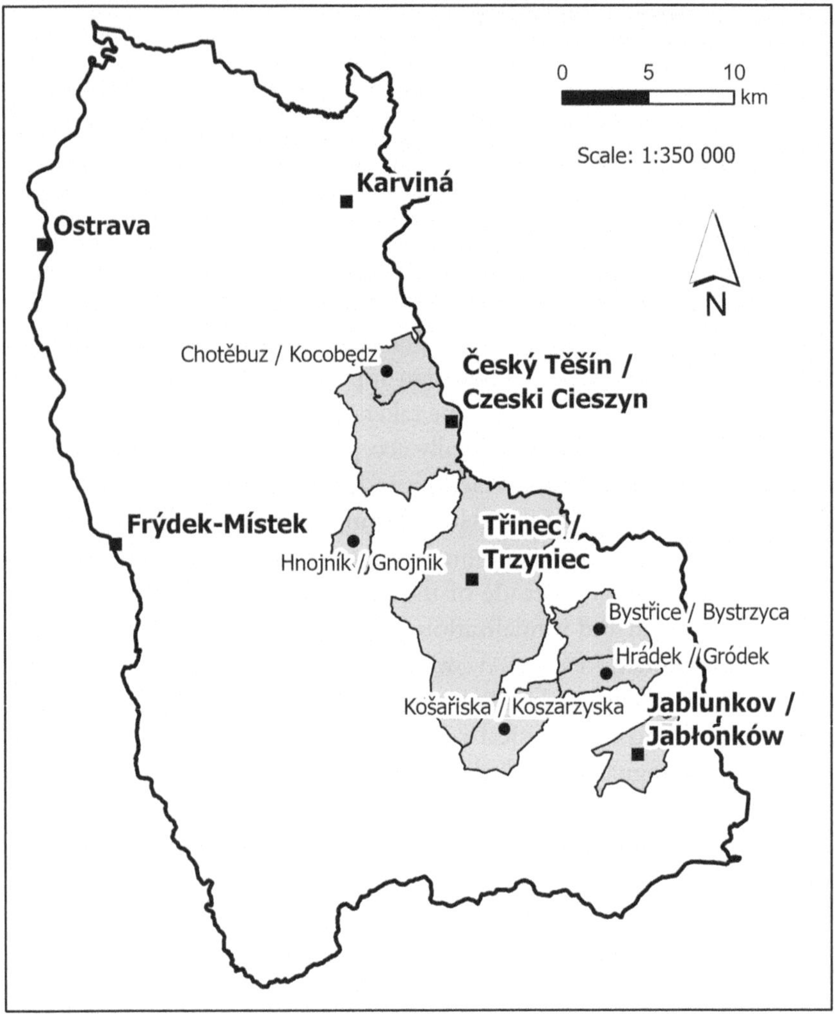

Fig. 6.85 Communes documented in the Těšín/Cieszyn region. (Cartography: Přemysl Mácha)

munes where the proportion of Poles exceeded the 10% threshold but which lacked bilingual signs (e.g., Košařiska/Koszarzyska, Nýdek/Nydek). We comment on some of those in this chapter as well. In the following paragraphs, we first outline the general characteristics of the documented signs and then we provide details on the situation in individual communes, offering some interpretation together with the raw data.

The total number of recorded signs in the four communes was 1060. It would be presumptuous to claim we successfully captured every single sign, but we are confident that we recorded at least 95% of all existing signs. The character of LL of all four communes was very similar despite their different spatial pattern, location, and size (see Table 6.8). Although the sign-per-hectare density index might suggest significant differences between, for example, Chotěbuz/Kocobędz and Jablunkov/Jabłonków, in reality most signs are concentrated in communal centers and along main access roads while the majority of the communal space often lacked any signs at all (see maps below). It would seem, therefore, that the sign-per-inhabitant density index is a more accurate metric while also indicating that the total number of signs in a commune is a function of demography rather than geography. Jablunkov/Jabłonków located in the uplands of the traditional Gorol area is officially a town with an old urban core and a clearly urban identity. The other three communes are villages to the South, Southwest, and Northwest of Český Těšín/Czeski Cieszyn. Bystřice/Bystrzyca is a large settlement including mountainous areas that also preserves Gorol identity while Chotěbuz/Kocobędz and Hnojník/Gnojnik are smaller lowland villages. All three settlements are affected by suburbanization from larger cities in the region, strengthening the Czech majority.

Most documented signs were in the majority language (75.3%) while bilingual signs and signs written in Polish only formed a minority

Table 6.8 Basic characterization of case study communes

Commune	Population (2018)	% of Poles (2011)	Number of signs	Area (ha)	Sign/person density[a]	Sign/ha density[b]
Bystřice/Bystrzyca	5334	24	423	1609	0.08	0.26
Hnojník/Gnojnik	1454	11	144	642	0.10	0.22
Chotěbuz/Kocobędz	1312	15.9	126	1061	0.10	0.12
Jablunkov/Jabłonków	5547	16	367	1039	0.07	0.35

[a]Number of signs per inhabitant
[b]Number of signs per hectare

(21.6%). As Table 6.9 shows, there are some differences between communes in this respect, notably Chotěbuz/Kocobędz stands out for its high number of bilingual signs. It is the result of the decision by the village council to implement bilingual street signs. Street signs are mandatory attributes of towns by Czech law, in villages they are voluntary, and most villages do not have them. The appearance of street names in the otherwise rather poor village LL thus dramatically changes the numbers which would otherwise be even less favorable for the minority than in the case of Hnojník/Gnojnik.

Tables 6.8 and 6.9 also cast doubt on the hypothesis that the number of bilingual and minority signs correlates with the number of minority members in the settlement as an expression of the ethnolinguistic vitality of the minority community (originally suggested by Landry and Bourhis 1997). The hypothesis seems to hold for absolute numbers only—the larger the number of Poles in a community, the higher the number of bilingual and minority signs. However, relative numbers are not consistent with this hypothesis—the proportion of bilingual and minority signs appears to be independent of the proportion of minority members. Further research on a larger sample is therefore needed to test the hypothesis with more conclusive results.

In addition to monolingual and bilingual signs, some multilingual signs can be seen as well, namely in the form of tourist information tables and welcome signs. While the former are usually small, inconspicuous, and often ignored by locals, the latter have a role to play in the communal

Table 6.9 Number of signs by commune and language (absolute/*relative*)

Commune	Majority only	Bilingual	Minority only	Multilingual	Other	Total
Bystřice/	316	71	18	7	11	423
Bystrzyca	*74.7%*	*16.8%*	*4.3%*	*1.7%*	*2.6%*	*39.9%*
Hnojník/Gnojnik	121	17	3	3	0	144
	84.0%	*11.8%*	*2.1%*	*2.1%*	*0.0%*	*13.6%*
Chotěbuz/	75	47	2	2	0	126
Kocobędz	*59.5%*	*37.3%*	*1.6%*	*1.6%*	*0.0%*	*11.9%*
Jablunkov/	286	64	7	4	6	367
Jabłonków	*77.9%*	*17.4%*	*1.9%*	*1.1%*	*1.6%*	*34.6%*
Total	798	199	30	16	17	1060
	75.3%	*18.8%*	*2.8%*	*1.5%*	*1.6%*	*100%*

and regional place-name politics. Multilingual welcome signs (featuring English and sometimes also German and Russian, in addition to Czech and Polish) are usually located at the entrance to the settlement together with bilingual signs bearing the name of the commune. Although they are few, they serve an important mitigating function, reducing the visual impact of bilingual signs by the inclusion of other languages and both watering down the potential Polish claim to the territory as well as calming down Czech nationalist sentiments. It is precisely this reason that led the communal government of Komorní Lhotka/Ligotka Kameralna to put up multilingual welcome signs after the bilingual signs at the entrance to the village had been repeatedly vandalized (Kraus-Żurová 2013).

The order of languages on bilingual (and multilingual) signs gives priority to the majority language—that is, Czech. Only in less than 1% of cases is the order reversed and the minority language appears first. It is so mostly on signs produced by private actors such as Polish cultural organizations who are also responsible for the majority of signs rendered in the minority language only (Table 6.10).

We observed 19 different sign types in all four communes (see Fig. 6.86). The most common types include company billboards, signs of stores, posters, and street signs. Together they represent one half of all signs. Other originally proposed sign types such as culture and sports signs and private information signs proved to be rather rare, so they were collapsed into a more general category of other signs, which also included, for example, signs on waste separation containers (Table 6.11).

The visual and material quality of signs differs from type to type. Some signs are standardized nationally (e.g., road signs, train stop signs, signs at the entrance to settlements, tourist signs), regionally (e.g., bus-stop signs, waste separation-container signs), or locally (e.g., street signs). As a result, there is very little variability in these signs between or within communes. Because their appearance is determined by a central authority (and often regulated by law, public notice, or an internal institutional

Table 6.10 Number of signs by order of language (absolute/*relative*)

Monolingual	Majority language first	Minority language first	Other	Total
841	201	10	8	1060
79.3%	*19.0%*	*0.9%*	*0.8%*	*100.0%*

bus-stop sign

company sign

company billboard

information table

inn or hotel sign

monument inscription

communal office sign

sign of other public institutions

Poster

Fig. 6.86 Examples of individual sign types. (Photos by Přemysl Mácha 2016–2019)

road signs

orientation signs

sacral and grave inscriptions

school sign

town signs

store signage

street signs

tourist sign

train stop sign

Fig. 6.86 (continued)

welcome sign

other signs

other signs

Fig. 6.86 (continued)

Table 6.11 Number of signs by type and language

Sign type	Majority only	Bilingual or minority	Other	Total
Bus-stop sign	68			68
Company sign	69	5	3	77
Company billboard	138		5	143
Information table	33	15		48
Inn or hotel sign	32	1	8	41
Monument inscription	4	14	1	19
Communal office sign	1	7		8
Signs of other public institutions	13	3	4	20
Other signs	40	5		45
Poster	96	31	1	128
Road sign	78	16	1	95
Inscription in/on sacral buildings and graves	3	6		9
School sign	9	9		18
Town sign	2	36		38
Store sign	112	16	3	131
Street sign	69	50		119
Tourist sign	28	8	3	39
Train-stop sign	3	6		9
Welcome sign		1	4	5
Total	798	229	33	1060

order), they are uniform and offer little to study by way of comparison. Others (namely private signs such as company signs, billboards, posters, store signs, but also some public signs put up by the commune beyond the minimum legal requirements) show a great diversity in size, color, font, and so on, but unfortunately for our purposes, they include a small number of bilingual or minority signs which makes a quantitative statistical analysis problematic. For example, we found no bilingual or minority company billboards which are otherwise the most frequent type of sign. The proportion of bilingual and minority signs among company names and store signs was also low and they in most instances emphasized the Czech language by font size (Czech name larger) and priority location (Czech name more likely on top, left, or center).

Of all signs, only posters really showed the potential of genuine creativity, challenging the prevailing norms. They often play with words which are common to both Czech and Polish (and the dialect), emphasizing thus the commonalities of the two groups. Also, they often give Polish language priority or appear in Polish language only. Indeed, of the 30 signs in the minority language that we recorded, 14 were posters. And of the 10 signs where the minority language came first, 6 were posters. Overall, almost 25% of all private posters were bilingual or in Polish, more than with any other private sign type.

On the other hand, we would run serious risks by drawing fundamental conclusions from this. It is true that posters are a very democratic form of sign making, they are the quintessential "signs from below" (Cenoz and Gorter 2006). In theory, anyone can put them up. In reality, however, only a few people actually do it and they may not necessarily represent the attitudes of the larger population. Also, posters are the most dynamic component of LL and temporal sampling errors are high. The particular proportion of majority and bilingual/minority language posters changes from day to day, and repeated documentation in the same commune over a longer period of time would be required to get a better estimate of the actual prevalence of individual languages. Overall, we therefore see posters as an expression of hope. Whether the longing becomes fulfilled will be observed in the more permanent components of LL. There are, however, important differences between individual communes, and a low proportion (or absence) of bilingual and minority posters in a given commune (e.g., Hnojník/Gnojnik, Chotěbuz/Kocobędz, Jablunkov/Jabłonków) may be an indication of the degree of popular acceptance of bilingualism in the locality (Fig. 6.87).

The vast majority of the differentiated sign types were present in all four communes and the quantitative differences between them were also minimal. What is relevant in relation to sign types, however, is that some types are significantly more likely to have bilingual rendering than others. Therefore, the bilingual LL (as compared to the entire LL) is poorer in sign types because some signs appear in the majority language either exclusively (e.g., bus-stop signs, company billboards) or predominantly (e.g., inn and hotel signs, signs of other public institutions, tourist signs). By contrast, other sign types appear primarily in bilingual form—for example, signs of communal offices, monument inscriptions, town signs, and train-stop signs. There is thus a discrepancy between the larger LL and its bilingual component. The anchor points are different and because some sign types are more spatially determined than others, the difference between majority and minority components of LL is not only qualitative and quantitative but also spatial. Bilingual and minority signs tend to concentrate near the entrance to a populated place and then in the proximity of symbolic public features (communal office, school, park). By contrast, majority signs are more determined by commercial factors, so we find them principally in the communal center and along major roads.

Fig. 6.87 Examples of bilingual and dialectal posters from the Těšín/Cieszyn region. (Photos by Přemysl Mácha 2019)

This is illustrated well by the maps below. The minority LL is a clear negative of the majority LL. We could therefore say that the number of bilingual and minority signs is not a banal function of the number of all signs but rather it is influenced by other factors, namely the symbolic importance of place and the struggle for representation and spatial control (Figs. 6.88, 6.89, 6.90, 6.91, 6.92, 6.93, 6.94, and 6.95).

The visual impact of signs is determined not only by their location but also by their size, message, and visibility. As far as size is concerned, it is important to note that although bilingual and minority signs represent 21.6% of all documented signs, only 5.6% of large signs are bilingual or minority, that is, only 3.9% of all bilingual and minority signs fall into the category of large signs. By contrast, 44.5% of all bilingual and minority signs fall into the category of small signs. It is clear that size potentially reduces the visual impact of bilingual signs. Because this is a systematic (and statistically significant) fact common to all communes, we do not see ill will behind it. Rather, we would see it as a result of the fact that certain

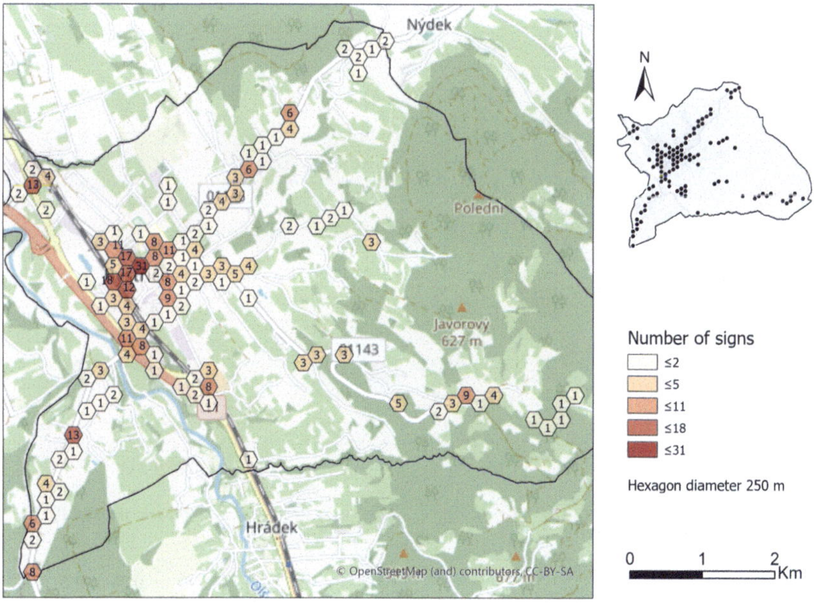

Fig. 6.88 Distribution of signs within the Commune Bystřice/Bystrzyca. (Cartography: Luděk Krtička)

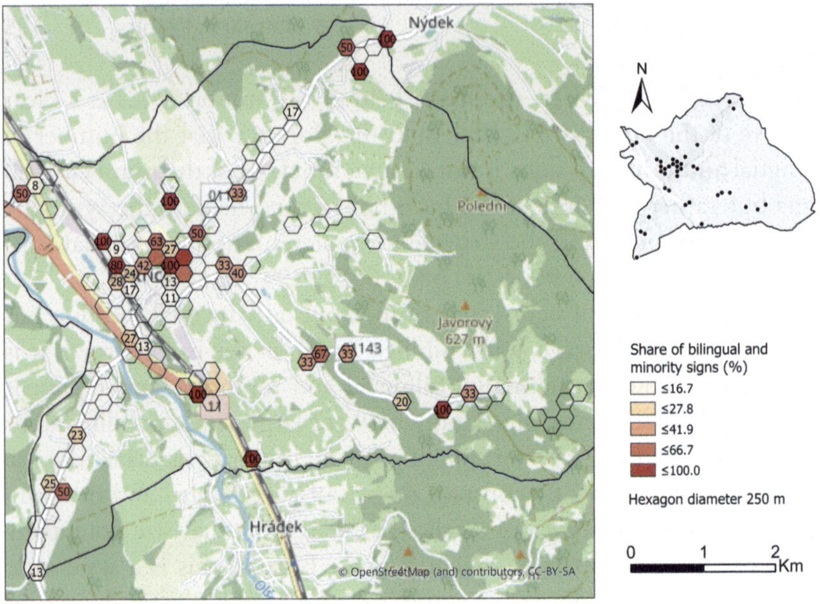

Fig. 6.89 Share of bilingual and minority signs in the Commune Bystřice/Bystrzyca. (Cartography: Luděk Krtička)

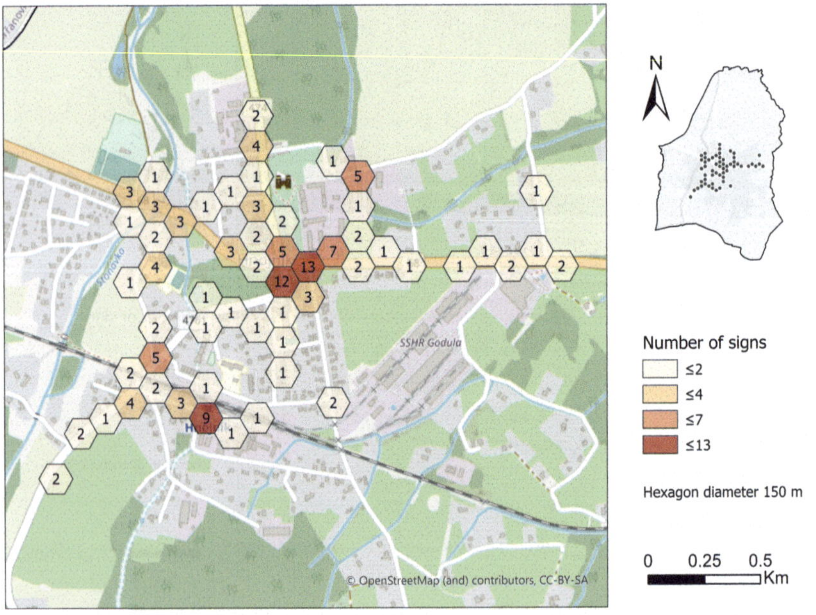

Fig. 6.90 Distribution of signs within the Commune Hnojník/Gnojnik. (Cartography: Luděk Krtička)

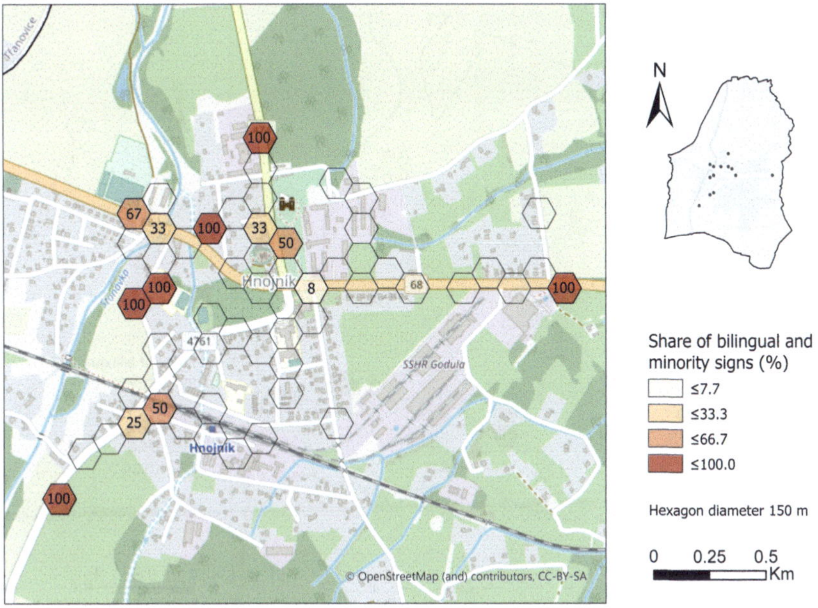

Fig. 6.91 Share of bilingual and minority signs in the Commune Hnojník/Gnojnik. (Cartography: Luděk Krtička)

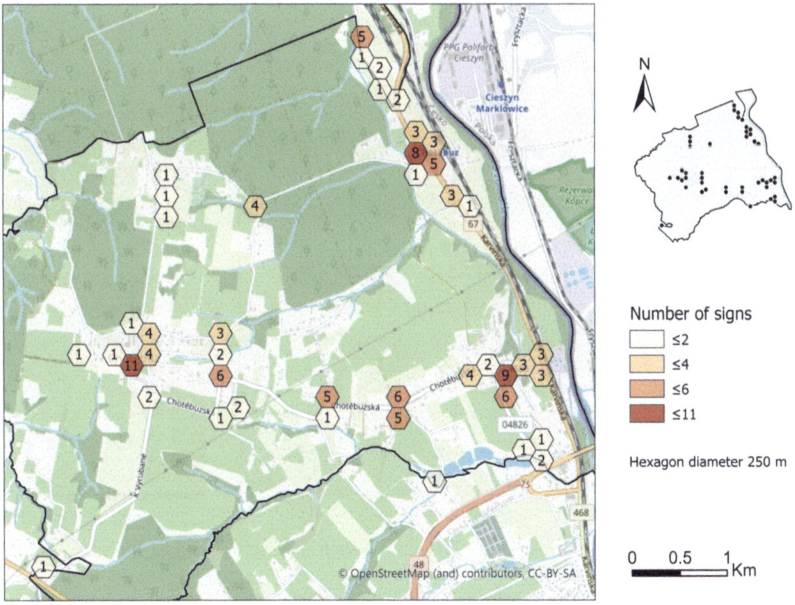

Fig. 6.92 Distribution of signs within the Commune Chotěbuz/Kocobędz. (Cartography: Luděk Krtička)

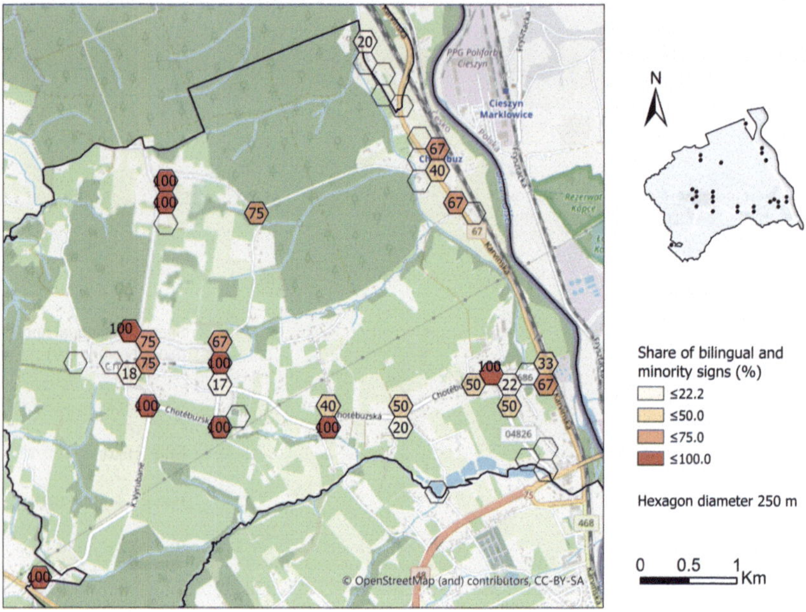

Fig. 6.93 Share of bilingual and minority signs in the Commune Chotěbuz/Kocobędz. (Cartography: Luděk Krtička)

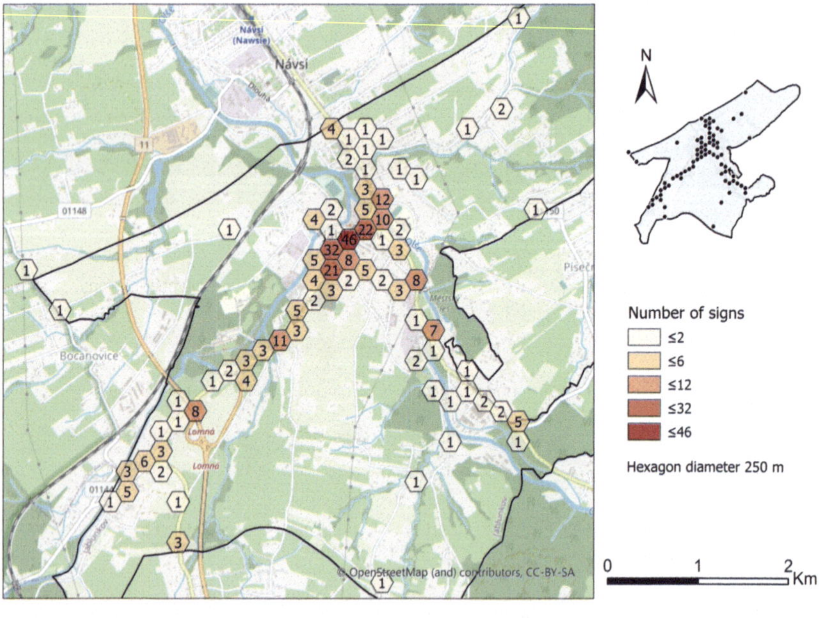

Fig. 6.94 Distribution of signs within the Commune Jablunkov/Jabłonków. (Cartography: Luděk Krtička)

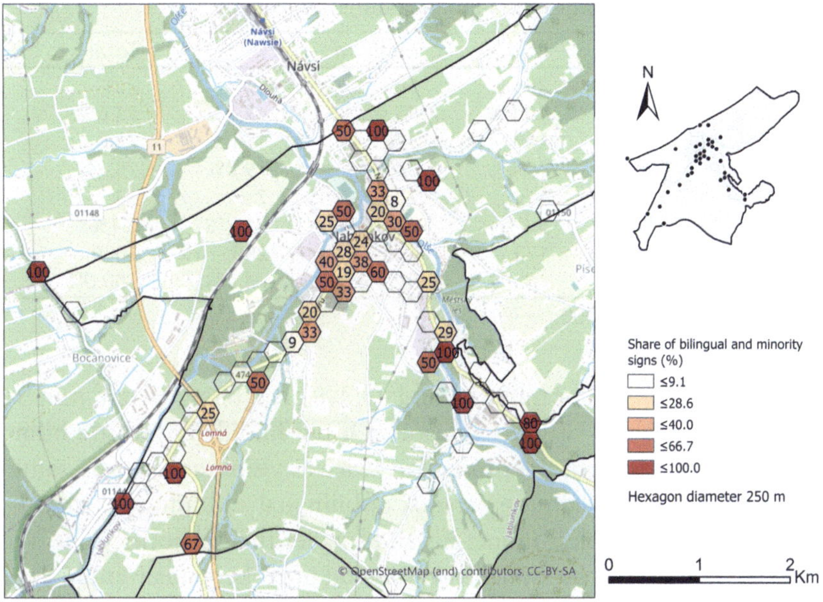

Fig. 6.95 Share of bilingual and minority signs in the Commune Jablunkov/Jabłonków. (Cartography: Luděk Krtička)

Table 6.12 Sign size by language (absolute/*relative*)

Language of sign	Small	Medium	Large	Total
Bilingual or minority	102	118	9	229
	44.5%	*51.5%*	*3.9%*	*21.6%*
Majority only	339	320	139	798
	42.5%	*40.1%*	*17.4%*	*75.3%*
Other	6	22	5	33
	18.2%	*66.7%*	*15.2%*	*3.1%*
Total	447	460	153	1060
	42.2%	*43.4%*	*14.4%*	*100.0%*

Note: Pearson χ^2, $p = 0.000$

sign types which are more likely to have bilingual or minority rendering simply appear in smaller sizes (such as street signs and posters) (Table 6.12).

On the other hand, the smaller size of bilingual and minority signs is often compensated by their visibility, that is, their visual prominence in their particular spatial setting. In our sample, bilingual and minority signs score high on visibility—38.2% of all signs in the category of high

Table 6.13 Visibility of signs by language (absolute/*relative*)

Language of sign	Low	Medium	High	Total
Bilingual or minority	25	105	99	229
	10.9%	*45.9%*	*43.2%*	*21.6%*
Majority only	147	498	153	798
	18.4%	*62.4%*	*19.2%*	*75.3%*
Other	1	25	7	33
	3.0%	*75.8%*	*21.2%*	*3.1%*
Total	173	628	259	1060
	16.3%	*59.2%*	*24.4%*	*100.0%*

Note: Pearson χ^2, $p = 0.000$

visibility are bilingual or minority and 43.2% of all bilingual and minority signs fall into this visibility category. Or put in different words: bilingual and minority signs are more than twice as visible as majority signs. This may help to explain the paradox of bilingual and minority signs—they are significantly less numerous and smaller than majority signs, yet they generate strong emotional reactions and provoke vandalization. This may be at least partially attributed to the fact that they are more visible at a statistically significant level. Together with their location in symbolic public places they acquire a great prominence which may be the key factor in generating the observed strong reactions (Table 6.13).

The message of a sign undoubtedly is another key factor influencing people's interpretation of it. However, it is not a quantitative attribute and therefore cannot be appraised on a large scale. It can be studied on individual signs or sign types, taking always into account the particular social, spatial, and historical context. The message may be encrypted into the text or made explicit by accompanying imagery. However, one of the most intriguing facts we discovered is that by far the most frequently (and often exclusively) vandalized sign type is the town sign bearing the name of the commune located at the entrance to the populated place being the center of the commune and bearing the same name. Although it represents only 16% of all bilingual and minority signs, it stands for more than 95% of all vandalization cases. The appearance of these signs is regulated by the state and is standardized nationally. No creativity is permitted, they all look identical. Yet their message is obviously far more powerful than that of any other sign. We believe they are the quintessential expressions of home in LL. As one respondent clearly put it: "You know what,

Fig. 6.96 An example of selectively vandalized in-coming town signs standing at the entrance to the communal center on opposite sides of the road. (Photos by Přemysl Mácha 2016)

when I arrive to Nýdek/Nydek, I finally say to myself, I am at home, when I pass the sign, when I see the sign Nýdek/Nydek, I feel I am returning home."[26] The metonymical character of town signs symbolizing home is also supported by the fact that the in-coming signs are more frequently vandalized, while the out-going signs are often left untouched. We suggest that this is because the Polish name of the commune on the out-going sign is already crossed out and the end of the commune may be re-interpreted as the negation of the Polish name (Fig. 6.96).

In terms of ownership, a slight majority of signs (54%) are private, usually commercial, while the rest is public, usually communal or state. It is important to note, however, that there is a statistically significant difference in the overall proportion of bilingual and minority signs on public and private signs (31% to 13%). This is an important metric in

[26] A woman from Nýdek/Nydek, 53 years old, Czech: "Viš co, když vjíždím do Nýdku, tak si tak konečně mluvím, sem doma, když projíždí tou tabulí, když vidím tu tabuli Nýdek, vracím se domů."

that it shows who are the principal actors behind the creation of the bilingual LL. Also, this metric indicates to what extent are bilingual signs promoted from the grassroots and how much they are accepted by the general public regardless of legislative stipulations. This metric does not, however, entirely cover all the motivations behind the erection of bilingual signs in the case of the public signs. Not all public signs are mandatory—many are put up by the commune as a gesture of good will well beyond the minimum standard demanded by law. In fact, of the 153 bilingual or minority public signs only 90 are mandatory, the rest is voluntary. Be it as it may, however, in most communes it is the public sector that is the main actor in the creation of the bilingual LL (Table 6.14).

Although the dominant role of the public sector is indisputable, there are important differences between individual communes in the proportion of bilingual signs among public and private signs and also in the communal government's willingness to go beyond the minimum legal requirements. These differences are important expressions of local political dynamics. For example, although in most communes the private signs are more numerous, in Chotěbuz/Kocobędz there are more public signs than private ones, due to the aforementioned decision to implement bilingual street signs (see Table 6.15). Furthermore, from Table 6.16 we see that private actors in Bystřice/Bystrzyca promote bilingualism to a greater degree than the commune, attesting to a greater acceptance of bilingualism among the local population. This stands in a sharp contrast to the situation in Hnojník/Gnojnik where only 3 out of 84 private signs included Polish. Finally, if we also take into consideration the degree of voluntarity of public signs, in Hnojník/Gnojnik the proportion of mandatory and voluntary public signs is roughly equal and in Bystřice/Bystrzyca and Jablunkov/Jabłonków the

Table 6.14 Number of signs by ownership and language (absolute/*relative*)

Ownership	Majority only	Bilingual or minority	Other	Total
Private	475	76	19	570
	83.3%	*13.3%*	*3.3%*	*53.8%*
Public	323	153	14	490
	65.9%	*31.2%*	*2.9%*	*46.2%*
Total	798	229	33	1060
	75.3%	*21.6%*	*3.1%*	*100.0%*

Note: Pearson χ^2, $p = 0.000$

Table 6.15 Ownership of signs by commune (absolute/*relative*)

Commune	Public	Private	Total
Bystřice/Bystrzyca	193 *45.6%*	230 *54.4%*	423
Hnojník/Gnojnik	60 *41.7%*	84 *58.3%*	144
Chotěbuz/Kocobędz	88 *69.8%*	38 *30.2%*	126
Jablunkov/Jabłonków	149 *40.6%*	218 *59.4%*	367

Note: Pearson χ^2, $p = 0.000$

Table 6.16 Proportion of bilingual signs by commune (absolute/*relative*)

	Public signs			Private signs		
Commune	Majority	Bilingual or minority	Other	Majority	Bilingual or minority	Other
Bystřice/ Bystrzyca	150 *77.7%*	36 *18.7%*	7 *3.6%*	166 *72.2%*	53 *23.0%*	11 *4.8%*
Hnojník/ Gnojnik	41 *68.3%*	17 *28.3%*	2 *3.3%*	80 *95.2%*	3 *3.6%*	1 *1.2%*
Chotěbuz/ Kocobędz	41 *46.6%*	45 *51.1%*	2 *2.3%*	34 *89.5%*	4 *10.5%*	0 *0.0%*
Jablunkov/ Jabłonków	91 *61.1%*	55 *36.9%*	3 *2.0%*	195 *89.4%*	16 *7.3%*	7 *3.2%*

Note: Pearson χ^2, $p = 0.000$, for both public and private signs

balance is in favor of voluntary signs, but in Chotěbuz/Kocobędz, 93% of bilingual public signs are mandatory—that is, the commune complied with its legal requirements but did no more.

These metrics illustrate three types of political arrangements that we observe throughout the Těšín/Cieszyn region. In some communes, especially but not exclusively, those, where the 10% threshold is only slightly exceeded, the commune complies with the request from the communal council for minorities to implement bilingual signs. The political negotiation tends to be very difficult and conflictual, and law is met with grinding teeth on the part of the majority. This is the most frequent case illustrated in our sample by Chotěbuz/Kocobędz and Hnojník/Gnojnik but represented by many communes in the region—Nýdek/Nydek, Střítěž/Trzycież, Mosty u Jablunkova/Mosty koło Jabłonkowa, Návsí/Nawsie, and Třinec/Trzyniec.

In other communes, political negotiations are also not easy but the gradual implementation of bilingual signs, first those legally required and later voluntary ones, leads to a wider acceptance of bilingualism. However, the public sector still plays the dominant role. In our sample, this case is represented by Jablunkov/Jabłonków, but it also applies, for example, to Hrádek/Gródek, Horní Suchá/Sucha Górna, and Stonava/Stonawa, to mention a few. The first two political arrangements support Ben-Rafael's hypothesis that LL associated with minorities may be more visible in top-down signs than bottom-up where there is an official policy toward minorities (Ben-Rafael 2009: 51). After the implementation of the European Charter of Regional or Minority Languages, the number of bilingual signs in the region increased dramatically. Many communes had private bilingual signs long even before then and their proportion significantly decreased as a result.

The third situation is illustrated by Bystřice/Bystrzyca (and also Vendryně/Węndrynia) where the public sector is active, but it is shamed by the private sector as an expression of 'popular' acceptance of bilingualism. We leave 'popular' in quotation marks because the fact remains that in all cases bilingual signs always represent only a minority of signs in a given locality. However, this last scenario illustrates Ben-Rafael's expectation that minority signs should be equally frequent in top-down and bottom-up flows where multiculturalism is taken for granted (Ben-Rafael 2009: 51). In our sample it is precisely Bystřice/Bystrzyca that manifests the highest degree of multiculturalism.

There is, however, a fourth type of political arrangement which we could not report on above because it is characterized by absence rather than presence. As discourse theory has taught us, absences are as telling as presences and there are some which are particularly noteworthy. They concern those villages in which the 10% threshold is met and even significantly exceeded but in which no bilingual signs were implemented because members of the local Polish minority decided not to ask for them in order to avoid conflict with the majority. This is, for example, the case of Ropice/Ropica, Bukovec/Bukowiec, Písek/Piosek, Bocanovice/Bocanowice, Písečná/Pioseczna, and Horní Lomná/Łomna Górna. In fact, as we already mentioned in Sect. 5.2, only 20 of 30 communes where the 10% threshold is met introduced official bilingual signs at the entrance to the commune. As a consequence, and ironically so from the perspective of LL studies, this last type of political arrangement (i.e., no signs at all) is actually the most frequent one. This stands in accordance with Ben-Rafael's hypothesis that collective identity would be absent, at least in top-down flow, where power relations work against minorities (Ben-Rafael 2009: 51).

Finally, there is an additional absence which we consider noteworthy. Despite the high vitality of the local dialect in spoken language we rarely see it on signs. There are only a few exceptions, namely Bystřice/Bystrzyca, Hrádek/Gródek, and Košařiska/Koszarzyska. In Bystřice/Bystrzyca, we recorded ten dialectal signs (on restaurants, posters). All were "signs from below", that is, signs produced by private individuals and companies. In the other two villages, the initiative came from above, that is from village representatives, but after open debates between local inhabitants and cultural organizations. In the case of Košařiska/Koszarzyska, a small mountain village with Gorol identity, where 27% of the population declared Polish nationality in the last census, representatives of the Polish minority made a decision not to ask for bilingual signs in order to avoid conflict with the Czech majority and keep the signs safe from vandalization. Instead, they opted for dialectal signs (see Fig. 6.97) as a uniting factor for the local community:

> We made the signs in the park in the dialect, in the Gorol language, and so … we do not want to provoke unnecessarily those … nationalists … who would, who might [destroy the signs] … There was a debate and we kicked the ball out of play [uses soccer terminology] …. That we don't need them … We wrote it in the Gorol language, in the dialect, and he who wants to read it, can … It is less visible and it does not pierce the hearts of some people so much … much.[27]

With a similar philosophy, the representatives of Hrádek/Gródek, a village where 31% inhabitants declared Polish nationality in the last census and where bilingual signs were implemented over a decade ago (and

[27] Representative of the Polish minority in Košařiska, man, farmer, 53 years old. In the original: "My se zrobili napisy v parku ponašimu, po gorolsku, a taki to … Ňechcymy provokovać zbytečňe ňikjerych, co by jako, co se myślum, že by to zaś … narodnostňi ty … Była tu dyskusja, a my to tak zagrali do autu bardźi. Že ňepotřebujymy … My to napisali po gorolsku, ponašimu, a kdo chce, tak se to přečito … Mjyňi to je łokate a mjyňi to žgo do serca ňikjerych ludźi."

Fig. 6.97 Signs in dialect, Košařiska/Koszarzyska. (Photo by Přemysl Mácha 2016)

vandalized one day after their appearance), chose to use some dialectal names in local administration to foster unity. The most visible step was the erection of the sign *Olza* on both sides of the bridge crossing the river in the center of the commune. It was part of a larger effort in the early 2000s to rename the river back to its original name, but other mayors eventually pulled back, afraid of a nationalist backlash. As a result, it is the only sign of its kind, generating pride among local inhabitants. The commune also succeeded in persuading the Czech Cadastral Office [Český úřad zeměměřický a katastrální] to include dialectal names in the cadastral map of the commune, contra its standardization efforts. The result is not ideal—hybrid names appear due to a lack of adequate characters in Czech orthography for some dialectal phonemes (e.g., *Lyngi* instead of *Łyngi*)—but it is still an expression of a genuine interest in fostering unity and identification with the locality through names. As the mayor of Hrádek/Gródek shared with us: "We feel that they are our names, that we have a right to them, they have been here for centuries."[28]

Although the usage of the name *Olza* in reference to the river on an official sign is unique to Hrádek/Gródek, it appears on other signs in LL

[28] In the original: "Cítíme, že to jsou naše jména, že máme na ně právo, jsou tu už po staletí."

Fig. 6.98 *Olza* on public and private signs. (Photos by Přemysl Mácha 2016)

in the region (see Fig. 6.98). For example, several companies incorporated it into their name and as a result it appears on signs such as *OlzaDent* (dentist's office) and *OlzaLogistics* (transportation company) in Český Těšín/Czeski Cieszyn and *Olza* (metal engineering) in Jablunkov/Jabłonków. Interestingly, the official Czech name of the river—*Olše*—was not adopted by any company. Because the name of the river evokes such strong emotions and the river runs through the entire region, we decided to investigate it more closely in our questionnaire survey which is summarized below. Although there are growing attempts to promote the dialect and dialectal names for use in writing, the impact of this movement on LL has so far been minimal, the aforementioned exceptions notwithstanding.

6.2.3 The "Paintbrush War" in the Těšín/Cieszyn Region: Media Discourse on Minority Place Names

In most households, place names and bilingualism are not a daily topic for conversation and they usually do not receive any media coverage either. However, when a change in the linguistic landscape is introduced, it is the media and the internet where public discussions about bilingual signs often take place (Hiss 2013). As part of our research we were keen to see how the topic of both the introduction and vandalization of bilingual signs was reported, analyzed, and represented in the local and national media. While the effects of media on the public opinion are difficult to assess (Riffe et al. 2019), we see the media coverage as another important piece of the puzzle to help us understand the themes, questions, arguments, and opinions surrounding the issue. The media shapes the public sphere, sets agendas, and influences the public discourse. It is a crucial part of the context in which people formulate, voice, and debate their opinions on bilingualism and the linguistic landscape.

As we already outlined in Chap. 3, we analyzed Czech and Polish newspaper, radio, and TV coverage of the conflict over bilingual signs and compared it with internet discussions on the same topic. In the following paragraphs, we provide both the characterization of our sample as well as some analyses of similarities and differences between Czech and Polish media and between newspapers, radios, and TVs on the one hand, and the internet debates on the other. We also offer a brief content analysis which shows how much attention was given to the process of introduction per se and how much to the cases of vandalization that followed. We present the overall tone the articles had toward the signs and we look at argument categories most commonly used in items with discernible 'for' or 'against' positions. Data for Czech and Polish media will be presented first, followed by a section presenting the results from online discussions.

The research period covered the years between 1999 and 2018 in order to include the whole length of the process from initial legal changes through the introduction of signs to its aftermath. In total we analyzed

560 media items and 572 online discussion posts. The media items were of three types: print, internet, and audio-visual. The vast majority of our data came from print media (492 items, 87.9%), followed by internet media (49 items, 8.8%), and audio-visual media (19 items, 3.4%). This difference comes from the fact that all minority media analyzed was in printed form only as well as the vast majority of Czech-language items published in regional inserts to large newspapers. Internet-based outlets reported on the issues occasionally. The Czech-language television and radio items were also relatively sparse. Regrettably, the minority audio-visual items were excluded from the analysis for reasons mentioned previously.

Out of the 560 media items 323 (57.7%) were from the Czech-language media and 237 (42.3%) were from the Polish-language media. All discussion posts were taken from comment sections of Czech newspapers and Czech blogs. In Fig. 6.99 and Table 6.17 we see the number of items identified each year with a visible increase from 2007 to 2010. This reflects the process of signs introduction which gained momentum in 2007. When looking at the graphs it is important to keep in mind that while the Czech-language media refer to a large number of outlets, the

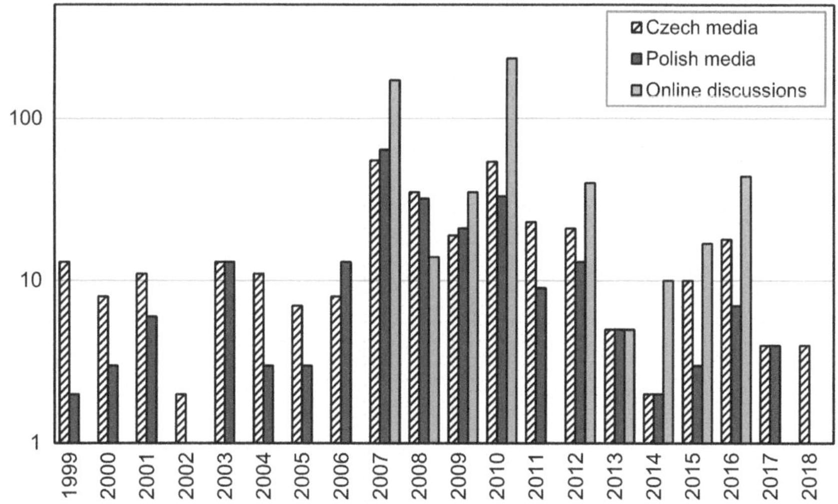

Fig. 6.99 Number of items per year. (Graphics: Uršula Obrusník)

Table 6.17 Number of items per year

	'99	'00	'01	'02	'03	'04	'05	'06	'07	'08	'09	'10	'11	'12	'13	'14	'15	'16	'17	'18
Czech	13	8	11	2	13	11	7	8	55	35	19	54	23	21	5	2	10	19	4	4
Polish	2	3	6	1	13	3	3	13	64	32	21	33	9	13	5	2	3	7	4	0
Online	0	0	0	0	0	0	0	0	172	14	35	234	1	40	5	10	17	44	0	0

Polish numbers refer to two titles only. In the years 2007 and 2008 almost every issue of *Głos Ludu* had an item about the bilingual signs.

We were interested in whether the issue was reported nationally or remained a regional concern. The majority of items were classified as having only a regional impact (365 items, 65.2%) and a smaller number was classified as having a national impact (195 items, 34.8%). This seems to suggest that in Czechia the issue of bilingual signs remained largely a regional issue despite some coverage in the media outside of the Těšín/Cieszyn region. The discussions surrounding the process often did not make it to the national discourse despite the involvement of high-ranking politicians in the most contentious years of the process.

Looking at the issues that surround the introduction of bilingual signs elsewhere in Europe (Puzey 2009; Sloboda et al. 2012), we decided to look at two possible focus themes in the items, namely introduction and vandalism. The label "introduction" was attributed to all items that offered information about the process of sign introduction, both from the legal perspective and from the practical point of view. Articles labeled with "vandalism" are concentrated on the subsequent vandalization of the signs and steps undertaken to mitigate it. A relatively small number of items covered both topics. As seen in Table 6.18 and Fig. 6.100, both Czech- and Polish-language media covered the two themes at roughly the same ratio. In both cases articles concerned with the process of sign introduction vastly outnumber articles dealing with the acts of vandalism.

We also looked at whether the concern with bilingual signs was a primary or a secondary topic in any given item. 'Primary' refers to articles where bilingual signs were the main topic of the text, while 'secondary' refers to items in which bilingual signs were mentioned but the focus was

Table 6.18 Topic of article by language (absolute/*relative*)

Language of item	Introduction	Intro/van	Vandalism	N/A	Total
Czech	224	13	63	23	323
	69.3%	*4.0%*	*19.5%*	*7.1%*	*100.0%*
Polish	180	13	44	0	237
	75.9%	*5.5%*	*18.6%*	*0.0%*	*100.0%*
Total	404	26	107	23	560
	72.1%	*4.6%*	*19.1%*	*4.1%*	*100.0%*

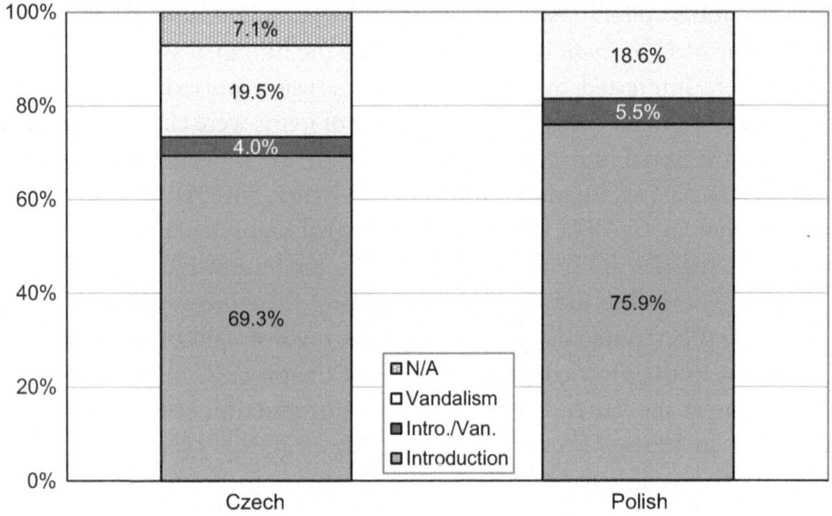

Fig. 6.100 Topic of article by language. (Graphics: Uršula Obrusník)

elsewhere. In the Czech media 243 items (75.2%) were classified as having the bilingual signs as primary and 80 (24.8%) as a secondary topic. For Polish media, the results were 182 (76.8%) for primary and 55 (23.2%) for signs as a secondary topic. Both Czech and Polish media concentrated on the process of introduction of the signs significantly more often than on the issue of vandalization.

Following existing research on bilingual signs, our interest turned into the argument categories arguing for or against minority signs (Szabó-Gilinger et al. 2012; Hiss 2013). First, we looked at the tone that the items displayed regarding the bilingual signs. We differentiated items that expressed position for the signs, against the signs, and those that remained neutral. The decision on 'tone' of the article was made based on particular arguments used within the article, linguistic choices, and framing of the issue. Items that presented a number of arguments both for and against the issue were classified based on the dominant arguments. We followed Riffe et al. (2019) in recognizing that editorial choice had to be made when selecting the presented positions. Articles that showed no discernible position or where both arguments for and against were used in the same measure were classified as 'neutral' (see Table 6.19 and Fig. 6.101).

Table 6.19 Tone of the item by language (absolute/*relative*)

Language of item	For	Neutral	Against	Total
Czech	95	210	18	323
	29.4%	*65.0%*	*5.6%*	*100.0%*
Polish	110	127	0	237
	46.4%	*53.6%*	*0.0%*	*100.0%*
Total	205	337	18	560
	36.6%	*60.2%*	*3.2%*	*100.0%*

Note: Pearson χ^2, $p = 0.000$

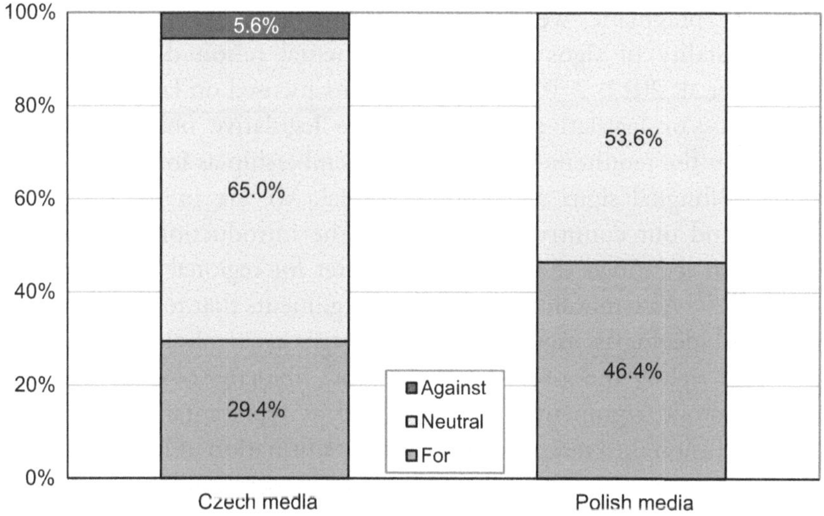

Fig. 6.101 Tone of the item by language. (Graphics: Uršula Obrusník)

This label was also applied to short informative items that displayed no charged linguistic choices or arguments.

While 'neutral' articles dominate in both Czech and Polish media, Polish media displays significantly more articles in the 'for' tone and a lack of articles whose dominating argumentative position would be 'against'. The Czech media, however, certainly did not shy away from expressing support for bilingual signs, as well, albeit to a lesser degree. We did not encounter any attempts at sensationalism so common for the tabloid press. Only occasionally catchy phrases stood out in otherwise purely informative and detached writing. A prime example is the coining of the conflict as the "Paintbrush War" [Štětcová válka], a name referring to the frequent vandalization of bilingual signs with paint (see, e.g., Horizont 2009). However, a substantial proportion of even the nominally neutral Czech media manifested an implicit educational agenda, emphasizing between the lines the importance of tolerance, pluralism, and good will while openly portraying the vandals as hooligans and xenophobes.

When clear arguments for and against signs were explicitly formulated, we classified them into one of two categories: pragmatic or symbolic. As pragmatic we had considered arguments that saw "the instrumentality of signs as their fundamental raison d'être" (Szabó-Gilinger et al. 2012: 270). These arguments focused on language retention, costs, or legislative obligation. The legislative obligation often focused on the requirements of the EU membership as in the following quote: "Bilingual signs are nothing special. We are in the European Union. And our country committed to the introduction of bilingual signs when it ratified the European Charter for regional and minority languages."[29] As symbolic we considered arguments that relied on shared values and ideologies, arguments that saw signs as "symbol of presence, a symbol of rights, and a symbol of the past" (ibidem: 274). In our case these symbolic arguments were identified as representation in public space, inclusion into mainstream society or affirmation of historical presence in the area, as in the following quote: "I would see it as evidence, that we Poles undoubtedly belong to this piece of land and that we will not be hearing the old—if you don't like it, go home to Poland. I am at home here."[30] The detailed breakdown of each category can be found in Chap. 3.

It is important to note that not every item contained an argument. Short news items stating changes in the linguistic landscape or explaining the process often lacked any explicit arguments. At the same time, some

---

[29] "Dvojjazyčné nápisy nejsou nic zvláštního. Jsme v Evropské unii. A náš stát se k zavádění dvojjazyčných nápisů zavázal, když ratifikoval evropskou chartu na ochranu menšinových a regionálních zvyků" (Sasínová 2007).

[30] "Bral bych to jako důkaz, že my Poláci do tohoto kousku země neodmyslitelně patříme a neuslyšíme už to známé—když se ti to tu nelíbí, běž si domů do Polska. Já jsem doma tady" (D.K. 2005).

items contained more than one argument category. We focused on identifying the arguments that were in line with the coded tone of the article. For example, if the article presented several arguments for and one against the signs, while maintaining the general tone as 'for', we focused on the arguments used 'for' the signs. If the item presented an equal number of arguments both for and against the signs, we classified it as 'neutral' and noted argument categories used. A total of 124 arguments were identified in Czech language media and 127 in Polish language media. In Czech media pragmatic arguments were preferred with 78 (62.9%) of those presented, while the remaining 46 (37.1%) were symbolic arguments. Polish media relied on both categories more evenly with 69 (54.3%) pragmatic arguments and 58 (45.7%) symbolic arguments identified.

During the preliminary research, we defined several possible categories for arguments that appeared both in the media and in online discussion posts. In total we worked with 11 subcategories: five for pragmatic and six for symbolic arguments. The pragmatic categories identified in the media items are as follows: language ('plan'), economic factors ('pmon'), population numbers ('pnum'), outside influence ('pout'), and other ('poth'). The six symbolic categories are representation in space ('srep'), inclusion into mainstream society ('sinc'), assimilation pressures ('sasi'), history ('shis'), impact on other minorities ('smin'), and other ('soth'). For each of these subcategories we modeled a possible argument both for and against the signs (Macnamara 2005). These 'template arguments' are presented in Chap. 3. During the research we looked for occurrences of these arguments in the studied media items. However, some categories identified during the preliminary research appeared only sporadically in the media data set ('plan', 'sasi', and 'smin'), though they do appear in larger numbers in the online discussions set. In light of this we decided to simplify the media data for presentation (see Table 6.20).

First, we present the overview of arguments used by both Czech and Polish media with a focus on their frequency. The argument that bilingual signs need to be introduced as a result of changes introduced 'from outside' (pout) was used most often in both Czech and Polish media. While there was often direct reference to legislative changes brought by the EU joining process, the idea of Europe as space where bilingual signs are normalized was also utilized: "(…) but if in whole Europe in border regions,

Table 6.20 Media argument categories (absolute/*relative*)

Argument category	Czech	Polish	Total
Pmon	16	17	33
Pragmatic monetary	*12.90%*	*13.40%*	*13.10%*
Pnum	11	8	19
Pragmatic numeric	*8.90%*	*6.30%*	*7.60%*
Pout	40	29	69
Pragmatic outside influence	*32.30%*	*22.80%*	*27.50%*
Poth	11	15	26
Pragmatic other	*8.90%*	*11.80%*	*10.40%*
Srep	17	23	40
Symbolic representation	*13.70%*	*18.10%*	*15.90%*
Sinc	9	7	16
Symbolic inclusion	*7.30%*	*5.50%*	*6.40%*
Shis	15	15	30
Symbolic history	*12.10%*	*11.80%*	*12.00%*
Soth	5	13	18
Symbolic other	*4.00%*	*10.20%*	*7.20%*
Total	124	127	251
	100.00%	*100.00%*	*100.00%*

Note: Pearson χ^2, $p = 0.000$

which are nationally diverse, there are bilingual signs and it is not a financial issue for the communes, then I think that this matter should not be seen as an issue in the Czech-Polish borderland."[31] One-third of all arguments used in the Czech media were classified as falling into this category. The argument that the signs are a way to mark the minority presence in the area (srep) was the second most used in both Czech and Polish media, though it appeared more often in Polish media. In general, the Polish media used a wider variety of arguments with a stronger emphasis on symbolic arguments, such as in this quote: "Although it is very easy to make stupid stain (i.e. vandalize the sign), Poles cannot be so easily erased from this land. They have been living here for hundreds of years, creative and active members of society, who bring forth new generations."[32]

[31] "(…) ale jeżeli w całej Europie w obszarach nadgranicznych, które są mieszane narodowo, występują nazwy dwujęzyczne i nie jest to żaden problem finansowy dla gmin, to mysle, ze ta sprawa nie powinna być postrzegana jako problem na pograniczu Czesko-Polskim" (Radłowska-Obrusnik 2007).

[32] "O ile bardzo łatwo jest walnąć głupi bohomaz, nie można wymazać Polaków z tej ziemi. Żyjących tu od setek lat, tworzących, aktywnych społeczników, wydających na świat kolejne pokolenia" (Wolff 2017).

Table 6.21 Media argument categories in relation to overall tone

Argument category	For	Against	Total
Pmon Pragmatic monetary	11	9	20
Pnum Pragmatic numeric	9	4	13
Pout Pragmatic outside influence	48	2	50
Poth Pragmatic other	12	5	17
Srep Symbolic representation	34	2	36
Sinc Symbolic inclusion	12	0	12
Shis Symbolic history	25	3	28
Soth Symbolic other	13	2	15
Total	164	27	191

Note: Pearson χ^2, $p = 0.000$

While Table 6.20 shows the frequency with which arguments of a certain type appeared in the media, Table 6.21 shows which arguments were used depending on the tone of the item. For this we only worked with items that showed a clear position for or against the signs and excluded the items that were classified as 'neutral' in their tone. It is important to keep in mind that the tone refers to an overall tone of the item, rather than just the tone of the argument. However, for practically all items with a clear tone, the argument's tone is in line with the overall tone of the item.

Both Czech and Polish media used similar categories in items that were favorable to the signs. The only notable difference was a stronger reliance on the 'pout' category in the Czech media, while Polish media made an even use of the categories across the board. We see that the argument of outside influence on the introduction process was the most commonly used argument in favor of signs, followed by an argument that signs are a way of marking minority presence in the area. As far as arguments against the signs are concerned, the monetary argument was most commonly used, claiming that the process will be expensive, and the funds are needed elsewhere. In the following example, the monetary argument is utilized to outweigh the legal rights argument; the idea that signs are pointless and often vandalized is also employed: "Even though the minorities have legal rights to the bilingual signs, I perceive the investment into these signs as wasted money, that could be used differently. I am most annoyed by Polish signs in towns and communes that are absolutely identical with the Czech ones. Czechs are welcomed by the sign *Tyra* and Poles, surprisingly, also by *Tyra*. Polish signs are also vandalized, which is yet another sign that they go against people's wishes."[33] The relatively small number of arguments identified in the items that were 'against' the signs makes generalization difficult; however, it forms an interesting starting point for comparison with arguments that were used 'against' the signs by online users.

The online discussion data supplement the data from the media analysis. By looking at discussions taking place under the articles and blogs on the topic we gained a better understanding of how the media messages were reflected and challenged by the readers (Bergmann and Jordan 2010). We identified a total of 572 individual online discussion contributions, published between 2007 and 2016, with the majority of the entries—406—posted in 2007 and 2010. This was a time when the debate around bilingual signs was at its most heated as they were becoming visible in the public space.

First, we take a look at how the attitudes toward minority signs in the online discussions compared against the attitudes in the media. From Fig. 6.102 we see that unlike the media, the online discussions show a large number of negative attitudes, both in terms of absolute numbers and in ratio to positive and neutral positions. More than a half of identified online entries were coded as being against the bilingual signs. It is telling that the 'neutral' space, so prevalent in both the majority and minority media has been significantly diminished in the online discussions. Sloboda et al. (2012: 65) reached a similar conclusion, stating: "Polish language signs receive indiscriminate negative evaluations partic-

[33] "I když mají na dvojjazyčná označení národnostní menšiny ze zákona právo, vnímám investice do těchto označení jako vyhozené peníze, které by mohly být využity jinak. Nejvíce mě zvedají ze židle polská označení měst a obcí, která jsou naprosto totožná s těmi českými. Čecha vítá cedule *Tyra* a Poláka, překvapivě, také označení *Tyra*. Polské nápisy bývají navíc ničeny vandaly, což je jen další známka toho, že jsou lidem proti chuti" (S.P. 2012).

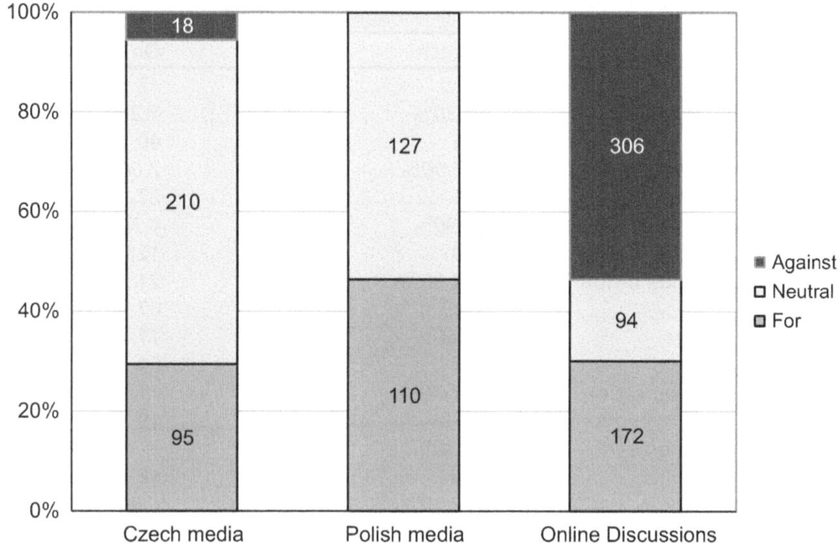

Fig. 6.102 Media and online discussions by tone. (Graphics: Uršula Obrusník)

ularly in anonymous Internet discussions." These results are statistically significant (Pearson χ^2, $p = 0.000$).

We also wanted to compare the frequency with which argument categories were used in the online discussions. This is presented first in Table 6.22 and then in Fig. 6.103 where we can see the usage of argument categories in online debates compared to both Czech and Polish media.

We can see that categories which rarely appeared in the media figure more prominently in the online discussions. This is especially the case for the argument on language use ('plan'), assimilation ('sasi'), and possible growing demands of other minorities ('smin'). In online discussions the 'pragmatic other' category contains a number of arguments such as, that there are no Czech signs on the Polish side of the border or that the signs will either serve to confuse tourists or, conversely, that they will make the region more attractive for tourists from Poland.

Table 6.22 Argument categories by media (absolute/*relative*)

Argument category	Czech	Polish	Online
Plan	9	2	42
Pragmatic language	*7.30%*	*1.60%*	*7.30%*
Pmon	16	17	40
Pragmatic monetary	*12.90%*	*13.40%*	*7.00%*
Pnum	11	8	27
Pragmatic numeric	*8.90%*	*6.30%*	*4.70%*
Pout	40	29	125
Pragmatic outside influence	*32.30%*	*22.80%*	*21.90%*
Poth	2	13	67
Pragmatic other	*1.60%*	*10.20%*	*11.70%*
Srep	17	23	64
Symbolic representation	*13.70%*	*18.10%*	*11.20%*
Sinc	9	7	19
Symbolic inclusion	*7.30%*	*5.50%*	*3.30%*
Shis	15	15	46
Symbolic history	*12.10%*	*11.80%*	*8.00%*
Sasi	1	5	58
Symbolic assimilation	*0.80%*	*3.90%*	*10.10%*
Smin	2	3	53
Symbolic other minorities	*1.60%*	*2.40%*	*9.30%*
Soth	2	5	31
Symbolic other	*1.60%*	*3.90%*	*5.40%*
Total	124	127	572
	100.00%	*100.00%*	*100.00%*

Note: Pearson χ^2, $p = 0.000$

Finally, Table 6.23 presents the breakdown of argument categories used in the online discussions by the tone of the entry. As with the media before, the 'neutral' entries were excluded, and we only look at posts with a clearly stated 'for' or 'against' position.

In the online discussions the most commonly used arguments 'for' were pragmatic arguments, similarly to the media. The argument that bilingual signs should be introduced because of outside pressure, whether from national or supranational level, was again the most commonly used one by supporters of the signs. Those who were against them in online discussions relied much more on the assimilation argument, that is that the minority should adapt and should not expect special treatment. A significant number of entries saw bilingual signs as an undue provocation from the minority trying to make its mark in the public space ('srep') as

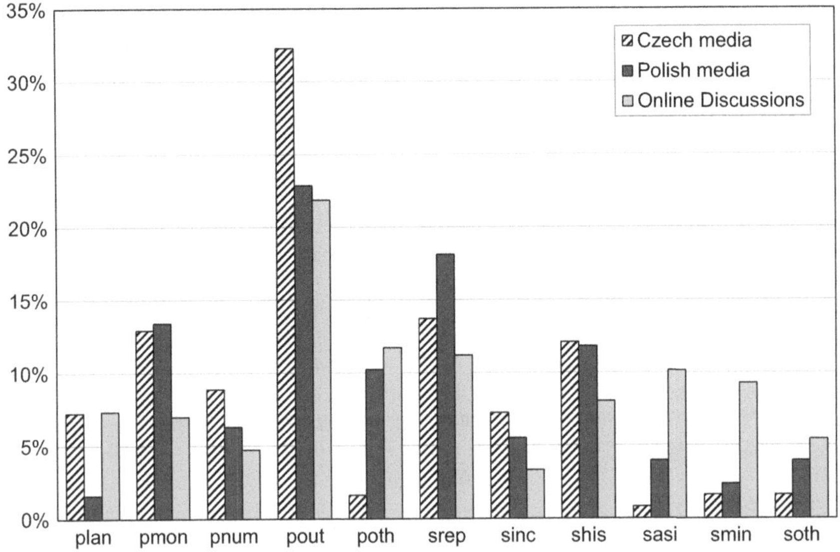

Fig. 6.103 Argument categories by media. (Graphics: Uršula Obrusník)

well as an unfair demand, considering there were no Czech signs on the Polish side of the border ('poth').

To summarize, both minority and majority media in the Těšín/Cieszyn region presented the issue of bilingual signs in mostly neutral terms. Polish media often displayed a positive position, although neutral stance was the most common as well. A small number of negative articles were found in the Czech media. The character of online discussions was remarkably different, with the majority of comments being against the signs, while neutral positions were the least common. Pragmatic arguments were favored across the board, although both Polish media and internet users made a significant use of symbolic arguments, despite their opposing positions.

The discursive framework reproduced by the media was predominantly in favor of signs, relating the Těšín/Cieszyn situation to the European and national legislation and the life of minorities in other EU countries. Effectively, it rendered the introduction of signs as an inevitable and desirable fact. Very little meaningful space was given to the opponents of

Table 6.23 Online argument categories by tone (absolute/*relative*)

Argument category	For	Against	Total
Plan	11	24	35
Pragmatic language	*6.40%*	*7.80%*	*7.30%*
Pmon	9	29	38
Pragmatic monetary	*5.20%*	*9.50%*	*7.90%*
Pnum	7	12	19
Pragmatic numeric	*4.10%*	*9.50%*	*4.00%*
Pout	68	38	106
Pragmatic outside influence	*39.50%*	*12.40%*	*22.20%*
Poth	19	40	59
Pragmatic other	*11.10%*	*13.10%*	*12.40%*
Srep	9	40	49
Symbolic representation	*5.20%*	*13.10%*	*10.30%*
Sinc	14	2	16
Symbolic inclusion	*8.10%*	*0.70%*	*3.30%*
Shis	16	20	36
Symbolic history	*9.30%*	*6.50%*	*7.50%*
Sasi	2	54	56
Symbolic assimilation	*1.20%*	*17.60%*	*11.70%*
Smin	11	28	39
Symbolic other minorities	*6.40%*	*9.20%*	*8.20%*
Soth	6	19	25
Symbolic other	*3.50%*	*6.20%*	*5.20%*
Total	172	306	478
	100.00%	*100.00%*	*100.00%*

Note: Pearson χ^2, $p = 0.000$

signs. It would be therefore an exaggeration to speak of a true public debate, let alone about public reconciliation. As a consequence, the conflict boiled at the local level and the Polish minority had to fight many private battles with a differing degree of success. By contrast, the opponents of signs felt marginalized and vented their frustration by other means, including vandalization.

6.2.4 Interviews

Having documented the spatial and discursive contexts in which the conflict over bilingual signs takes place, we interviewed 105 local inhabitants, politicians, and minority representatives to find out how this context

Table 6.24 Interview sample in the Těšín/Cieszyn region

Characteristic	Structure of respondents
Gender	50.5% men and 49.5% women
Age	5.7% (<19), 21.9% (19–39), 34.3% (40–59), 38.1% (>60)
Education	1% elementary, 46% secondary, 53% tertiary
Mother tongue	34% Czech, 27% Polish, 35% dialect, 4% combination

influences and interacts with their values, opinions, and attitudes related to minority names and bilingual signs (see Table 6.24).

The interviews took place in several villages and towns in the Těšín/Cieszyn region, namely (but not exclusively) in Bystřice/Bystrzyca, Nýdek/Nydek, Košařiska/Koszarzyska, Vendryně/Wędrynia, Třinec/Trzyniec, and Český Těšín/Czeski Cieszyn. The interview consisted of four thematic blocks covering the linguistic and national identity of the respondent, his/her relationship to local place names, and his/her attitudes toward bilingual signs. The interviews often had a narrative character and responses to individual questions tended to become part of a larger personal biography linking family, history, schooling, community, place, language, and identity. Implicit echoes or explicit reflections of the wider cartographic, linguistic landscape, and media context, in addition to local and regional politics, surfaced in all interviews. Nearly all interviews were recorded, amounting to some 150 hours of recordings. In this section we therefore inevitably present only a small fraction of the wealth of personal stories and opinions shared with us by the respondents. We focus on findings which capture prevailing trends, illustrating them with typical responses voiced by our respondents. We include the translations in the main text, while the original responses are transcribed in footnotes.

The interviews usually started with questions related to linguistic preferences and territorial and national identities. We were interested to what extent a person's self-identification influenced his/her attitudes toward bilingual signs. Most respondents identified themselves as Czech or Polish (occasionally also Silesian) but a few people refused to identify nationally and either chose a supranational (e.g., European), regional (e.g., Gorol, Zaolziak), or local identity (e.g., Bystřičan, stela),[34] or their combination.

[34] 'Bystřičan'—a native of Bystřice, 'stela'—'from here' in the dialect.

We found out that national identification among our respondents was most commonly determined by upbringing (family, schooling), often without the subjective emotional embracement one might expect: "Well, (I am) Polish, I didn't choose it, I was born with it."[35] Nationality was thus predominantly perceived as a given rather than something one can choose.

However, those who declared identities other than Czech and Polish often provided interesting reflexive explanations indicating that at least for some of our respondents group identity was a choice and not an incontrovertible fact of nature: "It is not important for me very much because if you asked me which nationality I was then I would tell you that I am Bystřičan. I am neither Czech nor Pole, I am simply Bystřičan, I am 'tutajši' (i.e. from here)."[36] According to this woman, differentiating between Czechs and Poles never brought any good to the community. The mayor of Hrádek/Gródek, a village with a high percentage of Poles, expressed a similar support for the transcendence of Czech-Polish dichotomies: "Silesian nationality. Or better said—'tu stela' (i.e. 'from here'). People can also have such identity, it is not mandatory today (to have Czech or Polish nationality) … they (i.e. the government) did not recognize such identity before."[37]

While the aforementioned respondents prioritized sub-national identities as a way to transcend national dichotomies, European nationality also lends itself to such a re-interpretation of local belonging as was voiced by another respondent who commented on the last population census: "Listen, I made fun of it. I wrote 'European nationality'."[38] The reasons for this attitude included both the aforementioned effort to avoid Czech-Polish dichotomies as well as the life experiences and educational background of the respondent who had traveled a lot and who considered

[35] A man from Nýdek/Nydek, 28 years old, Pole: "No polskou, ja sem si to nevybral, ja sem se s tim narodil."

[36] A woman from Bystřice/Bystrzyca, 44 years old, teacher, Czech: "Není to tak pro mě důležité, protože kdyby ses mě zeptal, jaká jsem národnost, tak ti řeknu, že jsem Bystřičan. Nejsem ani Čech ani Polák, já jsem prostě Bystřičan, já jsem „tutajši."

[37] Mayor of Hrádek/Gródek, man, Pole: "Slezská národnost, nebo lépe řečeno ‚tu stela'. To taky může mít člověk takovou identitu, není to dneska povinné mít …, protože kdysi to oni neuznávali."

[38] A man from Bystřice/Bystrzyca, 80 years old, retired, Pole: "Posłuchej, jo z tego mňoł srande, jo tam napisoł narodnost: evropska."

traditional nationalities as unnecessary complications. A different opinion on European nationality was voiced by a woman from Český Těšín/Czeski Cieszyn: "Here in Těšín, I see it as more logical because the city is divided between two countries but in fact it is a single city. It is simply Europe."[39]

A special type of identification is represented by the Gorol identity which turned out to be equally salient as the Czech-Polish axis. The Gorol identity is an important component of the personal identity of those who grew up in the region and kept in close touch with certain traits associated with the Gorol culture—traditional agriculture and sheep-herding, residence in the mountains, preference for folklore traditions, or aspiring to a romantic image of the Gorols as hard-working, good-hearted people living in harmony with nature (although they may occasionally be looked down upon by the city folk for being conservative, poor, and backward). One respondent characterized the Gorols as "a way of thinking and living. The Gorol is his own, he has to be independent, on the one hand he can't rely on the help of others, on the other hand he has to look out for others because mutual help is necessary, so he is somewhat different, he has to think things through in advance and he has to be a jack of all trades in order to survive in these hills and villages which are buried in snow or suffer from droughts."[40] Some respondents mentioned the possible pejorative association of the Gorols but even those who did not see themselves as Gorols expressed certain sympathies for them. Not only does the Gorol identity transcend the Czech-Polish divide, but many Gorols also feel a cultural proximity to other Carpathian pastoral groups, making thus the Gorol identity a supranational one.

However, some respondents see the Gorol culture either as Polish or as something that Poles are trying to monopolize. One respondent voiced a

[39] A woman from Český Těšín/Czeski Cieszyn, 40 years old, journalist, Pole: "Tady v Těšíně mi to připadá nejvíc logické, tím, že to město je rozděleno na dva státy a při tom je to jedno město. Je to proste Evropa."

[40] A woman from Košařiska/Koszarzyska, 52 years old, social worker, Czech: "Zpusob myšlyňo a żyćo. Gorol je taki svojski, muši być taki samostatny, na jednum strune ňimože spolygać na to že mu kdośi s čimši pomože, ale na drugum strune zaś se muśi dźivać co by pumug tym drugim bo jak by było třeja že by se navzajem pomogać, tak je taki kapke inačy, je taki muśi rozmyślać dopřodku a muśi być taki na vjyncej strun naučuny aby był schopny tu na tych kopcach a tu na tych dźedźinach jak to śňegym zafujo albo zaś jak vody ňima, aby był schopny tu přežyć."

common doubt: "So this really is a question, why the Gorol always speaks Polish. Or why most of them speak Polish and in the dialect, why the name 'Gorol' is associated with Polish speakers here in Silesia. I have to look into it because I fail to see why there are few or no Czech-speaking Gorols. And if they are here, why they do not have information stands in the Gorol festivals, etc. Why is it always associated with Polish folklore, Polish songs and so on."[41] This can be attributed to the fact that the maintenance and promotion of the Gorol folklore is commonly done by Polish minority associations which are fundamental in the social and cultural life of local communities even though Poles represent only a minority in all of them. As a consequence, the Polish language and the regional dialect are used during Gorol cultural events and festivals more often than Czech. Czechs interested in the Gorol folklore prefer to join these Polish minority organizations rather than establishing their own. As a result, the 'nationalization' of the Gorol folklore is rather a strategy for some militants but most see it as irrelevant, preferring to view it as a uniting factor unique to the region.

There is, however, a substantial number of respondents who declared their national identity very strongly, occasionally even expressing nationalist sentiments and antagonisms. For example, one respondent felt the need to emphasize that "this territory always belonged—from the Luxemburgs till 1918—historically as a fief to the Czech[42] kings. Until 1918, everyone spoke only Czech here and it was a Czech territory."[43] Such statements are common, and they show how historic memories still shape current debates. On the other hand, most respondents eventually chose a conciliatory tone, considering Czech-Polish relations as good and seeing the observed conflicts in the region as the work of a few extremists.

[41] A man from Bystřice/Bystrzyca, 43 years old, property caretaker, Czech: "Takže je vyloženě otázka, proč ten Gorol je vždycky polsky mluvící. Respektive proč většina z nich mluví spíše polsky a ponašimu, proč je slovo Gorol spojeno jenom s polsky mluvícími občany tady na Slezsku. Tuhletu věc si musím doplnit, protože mi trošku uniká, proč nejsou, nebo jich je málo, čeští Goroli. A pokud jsou, proč nemají stánky na slavnostech gorolských a tak dál. Proč je to vždycky většinou spojeno s tím polským folklórem, polskými písněmi a tak dále."

[42] *Czech* is here used instead of correctly *Bohemian*.

[43] A man from Bystřice/Bystrzyca, 77 years old, retired, Czech: "Protože tato země patřila vždycky, od Lucemburků až po rok 1918, patřila historicky, jako … léno českých králů. Tady se vlastně až do roku 1918 mluvilo česky a byla to země česká."

From a linguistic perspective, we can say that the dialect serves as a regional lingua franca, although some Czechs do not speak it actively. Many of our respondents considered the dialect their mother tongue and refused to classify it as simply Polish or Czech, pointing out the numerous German lexemes and the dialect's transitional character. The dialect seems to play a special role in many people's personal identity and some feel very strongly about it as was demonstrated by one Czech respondent who otherwise refuses everything Polish: "Many people speak the dialect there, although some become pissed off nowadays (expletive). Mainly one engineer (expletive) keeps saying 'We are in the Czech land, so we will speak Czech!' (expletive), and so I told her that I could speak the dialect even in courts (expletive)."[44] Where national identity divides people, the dialect brings them together.

In the subsequent part of our interviews we tried to find out the meaning of place names for our respondents. This proved to be a challenging task—although all of us use names on a regular basis, we rarely do it reflexively and consciously. Place names typically do not figure as explicit subjects of everyday conversations and the knowledge of toponymic lore appears to be dramatically less frequent than the works of Basso (1988, 1996) might suggest. It was therefore difficult to explain to our respondents what we were interested in and to prompt them to start thinking about names consciously. After initial explanations, however, we usually found a common ground and were able to explore our respondents' personal toponomies.

We specifically asked about places of special importance. For the most part, people mentioned places associated with their family and personal experiences. In many cases, these names referred to people who lived or used to live there (i.e., *U Cupka, U Hetmaňoka*), so personal, communal, and landscape histories merged into one. At times, we had to overcome the authority of maps because many people doubted that the names, they used, were the correct names or even names at all because they did not appear on maps. These names often conserved an older layer of top-

[44] A man from Bystřice/Bystrzyca, 32 years old, steel-mill worker, Czech: "… no tam ponašimu muvi dość ludźi, sice teraz już to ňikjerych sere, ku*va. Bo głuvňe jednum takum inženyrke, ku*va, bo ‚Sme v česke zemi, tak budeme mluvit česky!', kur*a, tak jo ji muvił że ponašimu mogym muvić aji u sunduv, kur*a."

onymic knowledge, mostly forgotten now: "But there were names such as *Na Čvurke* because it was No. 4, for example, where we live, as you go up the hill, there is an old house and dilapidated barns, that is my birth place and it is the first brick house in Bystřice. And they started to give numbers to the individual houses built from brick and stone, but not to the wooden ones. Na Čvurke or Na Jedynke (No. 1), our place was called *Do Parlamentu* ('to the parliament') because my grandfather was the mayor."[45]

In addition to these very personal names, our respondents often mentioned names of landscape dominants such as hills, streams, fields, settlements, and neighborhoods. These function both as orientation points and as identity anchors, participating in the conformation of home. Places in the mountains are very important and many people return there regularly. Some were even created by them: "I founded the first chapel of St. Hubertus in the Moravian-Silesian Beskids, I founded a cemetery for hunting dogs, I erected a bear stone because the last bear was killed here, so I put a stone there."[46]

The situation in cities was similar. Respondents recalled places associated with city dominants such as parks, squares, and pubs: "I spent my entire childhood between Pjonierak and Sikora Park. And so it is even now when I go for a walk with my children, we always go along Olza, now we have the opportunity to cross the sports bridge to Polish Těšín and walk there. These are places connected with my childhood and which I especially like. I would also like to mention that we plan to build a house and it will stand by Olza, near to the Sikora Park."[47]

[45] A woman from Bystřice/Bystrzyca, 85 years old, retired, Pole: "Ale tuž to były taki nazvy, *na Čvurke*, bo był nume štyry, napřiłlad u nas, jak do kopječka pan vyjdźe, tak je tako staro chałpa, rozbite stodoły, tak to je muj rodny dum a to je pjyršy murovany dum v Bystřicy. A začyło śe to numerovać, jak śe začyły budovać dumy ňi drevňanne, ale kamjynne, a murovane. Na čvurke albo na jedynke, albo do nas śe nazyvało do Parlamentu. Bo muj dźadźo był převodňićuncy gminy."

[46] A man from Bystřice/Bystrzyca, 77 years old, retired, Czech: "Já jsem tady založil první kapličku svatého Huberta v Moravskoslezských Beskydách, založil jsem hřbitov horských loveckých psů, posadil jsem medvědí kámen, protože tady se v roce 1984 nebo pět ulovil takový poslední medvěd, čili tam jsem dal kámen."

[47] A man from Český Těšín/Czeski Cieszyn, 42 years old, doctor, Pole: "Celé dětství jsem trávil mezi Pjonierakiem a Parkiem Sikory. A to trvá dodnes, když jdeme s dětmi na procházku, tak vždycky jdeme nad Olze, teď už máme tu možnost, že můžeme klidně jít přes ten sportovní most do polského Těšína a tam pokračovat. To jsou fakt místa, která jsou spojena s dětstvím a která mám obzvlášť rád. Ve spojitosti s tím se taky musím pochlubit, že chceme stavět dům a ten bude taky u Olzy, kousek u parku Sikory."

One's relationship to place, however, was not necessarily determined by, or tied to, the name itself but rather to the events and memories associated with it. Sometimes the name even seemed unimportant, and places were mentioned in their generic form—school, parish house, church, football stadium, playground, river, campground, and so on. Commonly, they were childhood places as explained by one respondent: "Well, look, I have lived there for 50 years, in Brandys, so I just know everything there. As a child I covered the whole area, so you know everything, you do not need to take the road, you know the side ways and you will get to where you want to go. It is close to me, your birthplace is always closest to you, it applies to the entire surroundings, you simply know every stone there, as they say."[48] In such responses, we hear the echoes of Ingold's processual approach to place names as stories (Ingold 2011: 165–175).

Others, however, were more explicit about the link between their identity and place names: "In Polední where the monument to the Czechoslovak legionnaires is, I think this is now one such place when for me as a Czech it is a place where the mound of Czech statehood is being created, organized tourist processions go there every year, we commemorate the traditions of the creation of Czechoslovakia in 1918–1920, this is probably the place I have the most heartfelt relationship with."[49]

We have, nevertheless, found it as mostly unproductive to try collecting stories related to place names that would be shared in the community. Actually, most respondents did not know any stories at all. Only a few offered folk etymologies and occasional historical explanations. For example, one respondent explained the etymology of the name *Koňské pleso* ('horse pond') thus: "It is called this way because the water there used to be so deep that two horses with a carriage drowned there, and

[48] A man from Nýdek/Nydek, 50 years old, Czech: "No podivej se, tam bydlim padesat roku, tak prostě, ten Brandys, to máš, všecko to tam znám no. Jako za děcka to mám všecko sbíhané, tak to tam znáš, že nemusíš chodit po cestě, jak se pustíš, tak dojdeš tam no. To mi je blízké, rodný kraj ti je nejbližší no, takže to se vztahuje i na to nejbližší okolí, že tam vlastně znáš, jak se říká, každý kámen no."

[49] A man from Bystřice/Bystrzyca, 57 years old, unemployed, Czech: "Na Polední, kde je památník československým legionářům a tam si myslím, že to je teďka pro mě takové místo, kdy pro Čecha je to pro mě místo, kde vzniká i mohyla české státnosti, každý rok tam chodí turistické pochody k tomu, připomínáme si tradice vzniku Československa v roce 1918–1920, že toto je možná tím místem, kde bych řekl, že k tomu mám nějaký srdečnější vztah."

that is why it is called *Horse Pond*. Even though the water is only ankle-deep now."[50] One of the few examples of local legends is the story of knights sleeping in the Čantoryja mountain: "Under Čantoryja below the last beech forest, as you stand below the beech forest, it is to your left. And there, under Čantoryja, where the rocks are, people say there are sleeping knights."[51]

Overall, based on our material, we cannot but conclude that Basso's research among the Cibecue Apache can hardly be applied to Central Europe. Actually, Basso admits that place name lore is poorly known even among Apaches themselves. Yet, as the conflict over bilingual signs attests, names are hardly losing their power because of it. In other words, the poetics of place-name lore is not required for names to possess a high emotional value.

What we found as crucial, however, was the importance of language for the perception of names. The transcription and pronunciation of names clearly emerged as a fundamental aspect of the name itself, more important than its etymology (Kearns and Berg 2002). To most respondents it mattered a great deal which linguistic variant of a name was used in the written or spoken form. A particularly strong criticism was directed at the standardization of dialectal names which often led to a change in their meaning: "(Researcher: Is it important how a name is written or pronounced?) Well, certainly. Some people czechisize the names to such a degree that you do not know where you are."[52] Another respondent provided a concrete example: "I think that if they wrote *Zarumbek* (instead of *Zárubek*), it would be written in a way that everyone understands as it should be. But they corrupt it to the extent that it is strange. It can even have a different meaning, can't it? For example, there is a place

[50] A woman from Bystřice/Bystrzyca, 55 years old, engineer, Czech: "Koňske pleso? Tam se tomu říká, že tam kdysi byla hluboká voda, že se tam utopili koně dva i s vozem, a proto se mu tak říká, Koňské pleso. I když tam už teďkom voda je po kotníky třeba."

[51] A man from Nýdek/Nydek, 33 years old, Pole: "Pod Čantoryjum niž tyn ostatni bukovy las a jak stojiš pod tym bukovym lasym, tak to je vlevo. A tam pod tum Čantoryjum, jak sum ty skały, tak se muvi, že tam sum ti spici rytíři."

[52] A woman from Bystřice/Bystrzyca, 49 years old, Czech: "Tak určiće. Tuž ňikjety do tak počešćujum fest, že to potym ňevjyš kaj žeš je."

called *Na Břyžkach* but (on maps) they write *Na Břízkách*[53] or something like that which is also strange."[54]

The problem of standardization is not unique to Czech, it applies equally to the Polish rendering of names in the case of communes and neighborhoods which often appear on bilingual signs. One of the most frequently cited examples was the commune Návsí/Nawsie which is rendered *Nawsie* in Polish but pronounced *Nowsi* in the dialect. To add insult to injury, when upcoming train stops are announced on the train and the train station, even the Polish name *Nawsie* is mispronounced to *Nawsze*, the meaning changing from 'in the village' to 'in the lice'.[55] Overall, most people expressed support for the use of dialectal names, often presenting them as their cultural heritage and implying a moral dimension to names: "I think our grandparents deserve that dialectal names are used and not corrupted. This, I think, would not be fair to them."[56] Interestingly, this strongly echoes the Apache elders who strongly reprimanded Basso when he repeatedly failed to pronounce the place names correctly (see Basso 1996: 10).

However, people's support for dialectal names cannot be easily translated into positive action. First, many names exist in different variants, so their potential use on maps and signs would present a challenge. The selection of any single variant would undoubtedly generate criticism by those who prefer other variants. More importantly, there is little agreement on how the names should be transcribed. As we commented in the section on maps, most speakers of Czech prefer the use of Czech orthography which, however, lacks characters for some dialectal phonemes and inevitably leads to their corruption even with the best of intentions. The use of Polish orthography would mostly (but not entirely) solve this

[53] The dialectal name is derived from the appellative *břyžek*, that is small bank or slope, the standardized name is derived from *bříza*, that is the Czech name for the tree *Betula pendula*.

[54] A woman from Košařiska/Koszarzyska, 66 years old, retired, Pole: "Jo se myšlym, že jak by tam był tym *Zarumbek* a pisane to tak jak to každy pochopi, že to mo być. A uni to tam tak překomolum, že to je potym taki dźivne. To je, no inši aji, že. Tu jak je třebas *na Břyžkach*, a tež tam pišum *na Břízkách* albo cośi takigo inšego, to tež je taki dźivne."

[55] For transcription of dialectal names into the international phonetic alphabet, see Appendix F.

[56] A woman from Bystřice/Bystrzyca, 53 years old, Pole: "Jo śe myšlym, že aji by śe to ty dźadki zasłužyły, kdyby to śe dźeržało tum gvarum a ňezkomolało śe. To, śe myšlym že to zaś jakśi by ňebyło spravedlive."

problem, but it would render dialectal names as Polish in the eyes of most Czechs which, too, is politically not acceptable. Interestingly, some Czech respondents who otherwise expressed a reluctance to accept bilingual names, had, nevertheless, no problem with the transcription of dialectal names on maps using the Polish orthography: "It seems normal to me, simply as it should be. It is something we are used to. No problem. (When written in Czech) it is very strange. Once it is a dialectal name, it should be written this way."[57] Such cases notwithstanding, the use of dialectal names does not have good prospects and with a few exceptions (such as the communes of Hrádek/Gródek or Košařiska/Koszarzyska mentioned above) they are seldom used on official maps and signs. However, they are well and alive in the spoken language, their vitality is high, and a conscious pride in their usage is rather unique in the Czech context.

No name demonstrates this better than *Olza*, the dialectal name of the river officially called *Olše* flowing through the entire Těšín/Cieszyn region. This issue was often brought up by respondents themselves as an example of a blatant discrepancy between popular and official use. It is common knowledge that most people in the region call the river *Olza*, not *Olše*, and this name is used even by people who otherwise do not speak the dialect and who often express little sympathy for the Polish minority. Typically, most respondents would say something like this: "I think that everyone here has simply learned to say *Olza*. I do not know anyone who would use *Olše*. At all."[58] As we explained in Sect. 5.2, some respondents, however, associate *Olza* with Poland, fear that its use would legitimize the Polish claim to the territory, and argue for the use of *Olše*, sometimes on presumed historic grounds: "We know from history that Jiří Borek from Rostropice and Vendryně ordered Janek Slanina to build and clear-cut there beyond the river Olša (cites from archival sources), this record is historic, we even found it with the mayor of Vendryně, it is from 1596 and it clearly states *Olša*, and old maps never say *Zaolzi*, but

[57] A woman from Bystřice/Bystrzyca, 18 years old, student, Czech: "Mě to přijde takové normální, prostě takové v pořádku. Je to něco, na co jsme zvyklí. Takže úplně v pohodě … Ono je to strašně divné, to jsou takové patvary. Když už jednou po naszymu, tak to napsat tím způsobem."

[58] A woman from Bystřice/Bystrzyca, 19 years old, accountant, Czech: "Myslím, že tady už se to prostě všichni naučili říkat *Olza*, nesetkávám se, že by někdo říkal *Olše*. Vůbec."

always *Zaolsi*. So it is clear that the name *Olza* is the product of sulking people who wish the name was *Olza*. But it is very doubtful."[59]

Interestingly, the support for the official use of *Olše* was often voiced by people who otherwise use *Olza* in daily life: "(Researcher: Is it important for you how a name is written or pronounced?) Yes, *Bystřica nad Łolzum*, or *nad Olší*, I would prefer the Czech name. I don't know how it is correctly, based on history, but I would prefer *nad Olší*. (Researcher: Why would you prefer the Czech form?) Well, we live in the Czech Republic, so it is the Czech name. (Researcher: And what about other names such as *Guchufka* (dialect) versus *Hluchová* (Czech)?) *Guchufka* is po našimu ('in our language'), it is a completely different thing!"[60] As we demonstrate in the following section summarizing the results of our questionnaire survey, this is a widespread phenomenon which clearly shows the importance of names in anchoring national identity. Dialectal names do not have a carte-blanche support of their users—they seem to be acceptable only to the extent that they do not threaten the dominant discourse of the Czech national character of the region. It also shows how important it is to take into account the discursive context and historical narratives of belonging which shape one's perception and emotional interpretation of names.

Only in the last part of our interviews did we ask explicitly about bilingual signs. Sometimes they were mentioned by the respondents spontaneously earlier, but we did not press the topic until we have learned about the person's life and a basic rapport and trust has been established. The topic of bilingual signs is highly sensitive, and we wanted our respondents to feel comfortable expressing their true opinions. A few people had even refused the interview when we told them that it would include

[59] A man from Bystřice/Bystrzyca, 77 years old, retired, Czech: "My z historie víme, že Jiří Borek z Rostropic a na Vendryni poručil Janku Slaninovi, aby sobě zbudoval a vypasejčil tam za vodú Olšú [cituje z archivního materiálu pro svou knihu], ten doklad je historický, dokonce jsme ho našli s bývalým starostou ve Vendryni, z roku 1596 a je tam jasně napsáno *Olšú*, a pohledem do starých map, nikde není *Zaolzi*, ale Zaolsi. Čili je vidět, že ten název *Olza* je víceméně tak nějak vytrucovaný lidmi, kteří by si přáli, aby se jmenovala *Olza*. Ale je to velmi sporné."

[60] A woman from Bystřice/Bystrzyca, 55 years old, Czech: "(Je pro vás důležité, jak se dané jméno píše nebo vyslovuje?) Jo, *Bystřica nad Łolzum*, anebo *nad Olší*, to jako bych to preferovala česky. *Nad Olší*. Nevím, jak to je správně, podle historických těch, ale preferovala bych *nad Olší*. (Proč byste preferovala ten český tvar?) No, tak bydlíme v České republice, tak to je český název. (U jiných jmen vám to nevadí?) *Guchuvka/Hluchová* apod. Tak to se říká ponašemu, tak to mi nevadí vůbec."

this topic, so we gave the respondents freedom to talk about it and did not press them for an answer. Most, however, shared their views and this helped us to understand better how their identity, family ties, residence, language use, and communal experiences influenced their views.

The vast majority of our respondents knew the reasons why bilingual signs were introduced in the region although they may not be entirely familiar with the legislative details. They were aware of the national and/or European legal requirements and understood that it was not a devious pastime of local Poles as some opponents of bilingual signs had suggested. However, some respondents, namely mayors and minority representatives, emphasized that ordinary people often forgot that it was the Czech national government and not the Polish minority that legally bound itself to introduce bilingual signs, so the Polish minority still attracts much criticism. One's opinion on bilingual signs depends to a large extent on his/her role in their introduction. Mayors face the greatest challenges because they are the ones who have to implement the law and introduce the signs (should the local council on minorities ask for them), but they then have to explain it to the community and face public criticism and attacks from the opponents. The mayor of Stonava/Stonawa shared with us his amazement at the degree of antagonism the introduction of bilingual signs generated in his commune: "It was difficult for me to understand why there sometimes was such a problem from the Czech side. I don't understand it, but it may be a certain xenophobia typical for smaller nations, it is even natural, we also were … I also had a problem with my own ethnic group … (Researcher: With Poles?) With those who were here (in the communal council) before us. (Researcher: In what sense?) Well, there was a certain xenophobia towards the Czech majority as well."[61]

As the preceding quote shows, although most people may be aware of the legislative framework, they do not necessarily accept it. Some Czechs either support or tolerate the bilingual signs but most believe that these

[61] Mayor of Stonava/Stonawa, man, Pole: "Bylo pro mě těžké pochopit, proč až tak někdy … byl problém z české strany. Nechápu to, měl bych, ale nechápu, ale to je určitá xenofobie menších národů, to je přirozená věc dokonce, my jsme taky.. Měl jsem určitý problém s naším etnikem … [Jako s Poláky?] S těmi našimi předchůdci. [V jakém smyslu?] No taky určitá xenofobie vůči české většině."

signs do not have a place in the commune. They argue, for example, that signs are needless expenses as are the repairs of vandalized signs; that the vandalized signs are ugly; that there are no bilingual signs in Poland; that signs are used by Poles to create tension even though they speak Czech well; and so on: "It is stupid. There is no reason for it, it is an insult for Czechs (Researcher: And do you know why they were introduced?) No, I don't and I don't care, they simply should not be here. There are no Czech signs in Poland either, we are in Czechia, so the name should be in Czech."[62] Some go even further and reject not only bilingual signs but the Polish minority as well; fortunately, however, they are a small minority.

Poles, by contrast, mostly expressed support for the introduction of bilingual signs. They argued, for example, that signs were an expression of respect by the Czech majority for the Polish minority; that they were a signal that Poles still lived there; that they were a symbol of Polish historic presence in the region; that the law required them and the law had to be obeyed; that they gave the region an attractive and unique character. Supporters of bilingual signs were well aware of the problems that signs generate, including vandalization, but that was not a reason to halt their introduction. However, doubts sometimes remained: "At the beginning, I was extremely pleased that they were here but later when they started to vandalize them, I became very disappointed and now I don't know if they even fulfill their function."[63]

Some mentioned that in several communes (see the case of Košařiska/Koszarzyska in Sect. 6.2.1) Poles preferred to give up bilingual signs to avoid conflicts with the Czech majority. By contrast, in other communes, bilingual signs were surprisingly put in place even though there was no demand for them, presumably for purely pragmatic reasons: "In some places it was the Czechs who introduced the signs when they wanted the political support of Poles. In Hnojník, in Střítež, nobody asked for them, but the mayor when he saw that he was losing control, there even were

[62] A man from Nýdek/Nydek, 26 years old, Czech: "Blbost. Nemá to opodstatnění, důvod, to je urážka pro Čechy. [A víš, proč se to vůbec dělalo?] Nevím, nezajímá mě to, nemá to tady co dělat. V Polsku taky nejsou české názvy, bohužel, jsme v Česku, tak musí být český název."

[63] A man from Nýdek/Nydek, 33 years old, steel-mill worker, Pole: "Z počuntku mě to strašně potěšiuo, že to tu było, akorat potym jak to začli likvidovat, devastovat, tak mě to začło strašně mrzet a teraz nievjym ani či to robi jako spełňo svojum funkce."

lawsuits against him, he had many problems, but he needed every vote, so there are (bilingual signs) in Hnojník."[64]

The vandalization of bilingual signs is paradoxically perceived negatively even by their critics. For some, it is precisely the reason why they should not be introduced in the first place; others clearly differentiate between the rejection of signs and their vandalization which they criticize even more: "Well, sure, I don't like (the vandalization). Personally, I also don't like the graffiti painters who immediately destroy (expletive in the original) a newly built bridge etc. (Researcher: So, you don't like the Polish signs, but you also don't like their vandalization.) No, I don't like that. It shouldn't be done. If they are once introduced, then let them stay, but I don't like that."[65] Many respondents recalled concrete examples of vandalization, most often by paint or breakage. They also remembered that the reaction of their surroundings was mostly critical of such acts although some admitted they knew a few people who supported vandalization and that Poles and the commune should not be surprised.[66]

Both supporters and opponents of bilingual signs agree that the Czech language should be visually dominant or preferred. As we showed in the section on the linguistic landscape, the majority of signs give Czech a clear priority if not by size, at least by order. In some cases, however, Czech and Polish signs are visually identical and discursively equal while on a minority of signs Polish comes first. This bothers many respondents who already have a difficult time accepting bilingual signs in any form. None of the opponents would allow a bilingual sign on his/her house or fence, those who see signs favorably would not mind this.

[64] A man from Třinec/Trzyniec, 47 years old, teacher, Pole: "Někde dokonce to zavedli i Češi, když chtěli podporu Poláků, v Hnojníku, na Stříteži, tam se o to nikdo nehlásil, ale ten starosta, když viděl, ze se mu to tam všechno sere, ho tam i po soudech někde smykali, měl tam nějaké svoje věci, ale potřeboval každý hlas, no tak v Hnojníku jsou [dvojjazyčné cedule]."

[65] A man from Bystřice/Bystrzyca, 68 years old, retired, Czech: "No jasne. Jako mje to vadźi. Mje osobňe vadźi aji ći sprejeři, co, uňi postavjum novy most nebo co, a momentalňe to je po … to … Takže na jednu stranu vám vadí polské nápisy, ale jejich ničení také. No to mi tež vadźi. To by śe ňimjało. Ješi to už tu je roz, tak až to je, ale tak mje to vadźi."

[66] During an interview with one man from Bystřice/Bystrzyca it became clear that he himself vandalized several signs and that the vandalization was a planned and well-organized collective act during which all Polish signs at the entrance to the commune disappeared. This information was provided off the record.

An older bilingual layer of signs preceding the recent introduction of official signs represent an interesting class of signs. They usually appear on private businesses and shops. For example, in the center of Bystřice/ Bystrzyca we find a butcher's shop owned by a local Pole who put Czech and Polish signs on the shop in the early 1990s. He considered it a natural decision which followed upon an even earlier tradition common during Communism. A nearby grocery store, by contrast, a long-time herald of bilingualism and a central point in the village, underwent total remodeling in 2018 which led to the disappearance of Polish signs entirely. However, a sweets shop next door belonging to the same owner continues to preserve its bilingual signs. From the interview with this owner it became clear that the decisions by store owners on bilingualism are not necessarily strategically planned, and graphic studios and construction companies bear their share of influence on the final appearance of a commercial building (as was the case with the grocery store).

In the neighboring village of Vendryně/Węndrynia there is also a group of shops, most of them marked with bilingual signs. Local respondents viewed them positively and did not recall any protests against them or any vandalization cases. This was also confirmed to us by the shop owners themselves. We consider this as an important finding. The public debate about signs is, in fact, a debate about public signs, that is, about the use of public funds for public signs in and for public spaces, and on and for public institutions. Private signs albeit in public spaces appear to be respected, and perhaps even appreciated. While private owners enjoy a full right to freely use their private property, public signs are implicitly an expression of the community in which the opponents of signs do not feel represented. Rather, they feel their community is being hijacked by Polish militants and their Czech sympathizers.

We have so far commented on the perception of bilingual signs from within the local community. The arguments for/against bilingual signs, however, sometimes include external views, as well. For the supporters, bilingual signs express a unique character of the region and its friendly and respectful approach to minorities which would be appreciated by tourists and encourage them to explore the richness of the local culture. One respondent related her experience from Bautzen/Budyšin located in the Sorabian region of Germany to the Těšín/Cieszyn situation: "I confirmed this to myself in Germany, everything was marked (with bilingual signs) there and the city immediately became more transparent for me and it was a very positive experience. So I now view the situation here also positively, even though Czech and Polish are not that different."[67] By contrast, the opponents of bilingual signs argue that tourists will be confused when they see signs in different languages which he/she will not be able to read anyway, as one respondent told us: "A Pole from Poland will never know where *Dworcowa* (Polish name of a street in Český Těšín/Czeski Cieszyn) is, he will only use Czech names. Yes, certainly. And he will most likely go to *náměstí* ('square' in Czech) and not *rynek* ('main square' in Polish), because for him, *rynek* is on the Polish side."[68] As we commented in Sect. 5.2, the region receives little tourism from the outside, so these arguments are intriguing in that they seek external sanctions for internal affairs.

Our interviews showed that people perceive the debate on bilingual signs through the prism of their own identification with the region. Local and regional identities sometimes overshadowed the national ones, although respondent's nationality was a key factor determining the person's attitude. This nationality was in turn most commonly perceived as a given, inherited within the family, and influenced by schooling. The uniting factors cutting across ethnic boundaries included the local dialect but also a warm link to nature and the mountains whose names maintain a strong presence in everyday life. Dialectal names in particular elicit a strong emotional response underlining the importance of names in establishing and maintaining an intimate link to place perceived as home. People generally voiced conciliatory opinions on the use of Polish names on signs and maps, although there were sharply nationalistic exceptions on both sides of the debate. The accommodation to the introduction of bilingual signs in the region appears to be progressing mostly successfully,

[67] A woman from Český Těšín/Czeski Cieszyn, 40 years old, journalist, Pole: "Třeba jsem si to potvrdila v tom Německu, tam taky bylo označeno úplně všechno a pro mě se tím to město stalo mnohem přehlednější a pro mě to bylo velice pozitivní. Takže dneska na to hledím taky pozitivně, i když ta čeština a polština zase není natolik odlišná."

[68] A man from Český Těšín/Czeski Cieszyn, 46 years old, pastor, Pole: "Polák z Polska ten už vůbec nebude vědět, kde je *Dworcowa*, ten bude používat jenom české názvy. Ano, určitě. A dokonce nejspíš půjde na *náměstí* a ne na *rynek*, protože pro něj *rynek* je na polské straně a tady je *náměstí*."

at least in those communes where signs have been introduced. Continuing resistance to bilingual signs, however, is still strong and the accommodation process will not end in the next few years.

6.2.4.1 Conflict over Place Names

The process of the introduction of bilingual signs on the ground was far from straightforward. While in some communes it went smoothly, in others was a stark resistance to the process followed by long delays, arguments, and negotiations. While we had initially gained understanding of the conflict from both media and interviews with members of the public, we were interested to hear the perspectives of the political representatives as well. In this short section, we will look at how the conflict was reflected by some of the actors during the research interviews that took place as part of this project.

It is important to note that the situation often varied from commune to commune, with some acting fairly swiftly and others taking more time. Initially, however, several communes decided to wait with the introduction of signs. Władysław Niedoba, one of the proponents of bilingual signs in the commune of Chotěbuz/Kocobędz highlighted the difference between the planned and the actual process of implementation: "Although we were one of the firsts who initiated it, when it came to realization we were somewhere in the middle. There were lots of delays as some people were hoping that the law would change and there would be no need to introduce the signs at all."[69] The initial approach of 'wait and see' in hopes that bilingual signs could be somehow avoided has also been expressed by Ladislav Olšar, mayor of Bystřice/Bystrzyca at the time: "And even when I knew that the decree is in effect, already at the time when I was the mayor, I ignored it, because I thought: We have fulfilled [our obligation], we have the signs in blue and yellow, no one minds that, it is clear and transparent that the official name is one and the other is the name in the

[69] "Byť jsme byli jedni z prvních, kteří to iniciovali, tak co se realizace týče, byli jsme někde uprostřed; hodně průtahů s tím, že někteří doufali, že se zákon změní a nebude potřeba ty cedule vůbec zavádět."

language of the national minority"[70] While Bystřice/Bystrzyca did have an older version of signs where a Polish sign, much smaller and in different color combination was placed under the Czech sign, the mayor was reluctant to initiate the introduction of signs that would, at least visually, put the Czech and Polish version of the name on a more equal footing, both being rendered in the same font and color scheme.

The fact that bilingual signs in a way made the minority visible in public space was also reflected in the interviews. Józef Szymeczek, former director of the KP (Congress of Poles) compares the presence of bilingual signs with another marker of minority presence in the region, namely minority schools: "Introduction of signs causes the majority to feel that Poles gained something above standard new, even though in reality they were only symbolically acknowledged in the public sphere. No one in here had any issues with Polish schools, but they did with the signs, those were new."[71] In some places, the signs at the entry to the communal center were seen as a natural extension to the bilingualism that existed in public space during Communism as the mayor of Stonava/Stonawa Ondřej Feber pointed out "(…) and that we just continued with it, there were communes where after the [Velvet] revolution the council even was helping to destroy bilingualism, but that was not our case, maybe because the long-serving mayor was a Pole, hard to say. The only new signs that were introduced were the entry/exit signs, all other signs in the village had been preserved from the past, even doors in the town hall, mayor, officials, all was there."[72] However, while bilingual signs were present in public space during Communism, Stonava/Stonawa was indeed one of the few communes that maintained them after regime change.

[70] "A i když jsem věděl, že ta vyhláška je platná, už možná ještě v době, kdy jsem byl starostou, tak jsem to ignoroval, protože jsem si řekl: My jsme splnili, máme to označení v tom modrém žlutém, nikomu to nevadí, je to jasné, transparentní, že oficiální název je jedno a název obce v jazyce národnostním je jiný."

[71] "Zavedeni nápisů budí u majority pocity, že Poláci něco vybojovali nadstandardního, noveho, když ve skutečnosti byli pouze symbolicky uznání ve veřejném prostoru, nikdo neměl problém s polským školstvím, ale nápisy ano, ty byly nové."

[72] "(…) a to, že jsme v tom pokračovali, byly obce, kde po revoluci zastupitelstvo obce dokonce pomáhalo likvidovat tu dvojjazyčnost a u nás to nebylo, možná tím, že tady po mnoho let byl starostou Polák, těžko říct. Jediné cedule které byly zavedené navíc byly ty vjezdove/vyjezdove, všechny jiné ještě ve vesnici byly zachované z minulosti, měli jsme označené i dveře na obecním úřadě—úředníci, starosta, všechno bylo."

The arguments politicians remembered being used against the signs were along the lines of arguments appearing in the press of the time. They were either based on the sense of wastefulness or, where a more conciliatory tone was to be maintained, they were based on the idea that signs were "not needed" as the local Polish minority spoke and understood Czechs easily. The supporters of signs argued that the funds for bilingual signs came from specially dedicated funds of the state and as such could not be used for other purposes. Bogusław Raszka, then vice-mayor of Vendryně/Wędrynia commented on the arguments used during the process of sign implementation in the commune: "We saw it as a representative thing for those who come and want to orientate themselves a bit on how the life goes on here, who lives here. In some sense it was a recognition of these traditions, which were made visible. The arguments against were that everyone understands [Czech] after all, it is unnecessary fuss, money wasted. When it came to the money, the argumentation was quite baseless, because the amounts were not terribly high. I have to say openly that these arguments appeared also among the people with Polish nationality."[73]

When asked for their reflection on why the introduction of signs was problematic in some communes and not in others, the representatives often mentioned that the pre-existing relationship between the majority and minority played an important role as well as the perceived 'pressure' the Polish minority exerted in particular contexts. For example, in the town of Český Těšín/Czeski Cieszyn, bilingual signs were introduced over a number of years in a phased installation and they faced a relatively small number of incidents of vandalism. Currently, all streets in the town carry bilingual signs, making thus the minority presence in the area very visible to both locals and visitors.

On the other hand, Třinec/Trzyniec, another town with a similar percentage of Polish minority population at the time of the debate over the

[73] "Viděli jsme to jako věc reprezentační pro ty, co sem přijedou a chtějí se trochu zorientovat, jake je tady soužití, kdo tu bydlí. Bylo to v určitém smyslu vzdání holdu těm tradicím, které byly tímto zviditelněny. Argumenty proti: všichni tomu přece rozumíme, jsou to zbytečné cavyky, zbytečné peníze. Pokud šlo o ty peníze, tak ta argumentace neměla moc zaklad, protože to nebyly nějaké strašné částky, které na to šly. Musím otevřeně říci, že se tyto argumenty objevovaly i u lidi polské národnosti."

signs, became one of the most contentious cases. First, the Committee for National Minorities, advisory body to the communal council set up in communes where at least one minority reaches 10% population threshold, was reluctant to propose the installation of bilingual signs. This was largely due to the fact that the Committee consisted of representatives of numerous smaller minorities living in the town who did not necessarily share the desire for Polish signs and together with the Czech members of the Committee had a numerical majority. Secondly, the council kept postponing the decision for several years using various arguments ranging from a presumed lack of analysis across forgotten paperwork to avoiding difficult decisions before and after local elections. Bilingual signs were eventually introduced in a much more limited scope (entry and exit to the town but no street signs) once new representatives of the Polish minority were selected for the Committee and decided to strike a more compromising tone. Those who pushed for the signs in the years before were subsequently labeled as "radical", "militant", and "thinking about their own political gain". "Nationality-based clash is one of the ways of getting into higher politics. The conflict had always escalated before elections. The national topic was misused here for political ambitions"[74] commented the mayor of Třinec/Trzyniec Věra Palkovská, presenting thus the legitimate demands of minority representatives as attempts to appropriate the issue for political gain.

Cases of vandalism were also discussed at the time. Some newspapers used imaginative metaphors for the conflict (e.g., "paintbrush war" as it was dubbed by the local newspaper "Horizont"). Others considered whether the defacing of signs was simple vandalism or an act of xenophobia. When discussing the tensions with the benefit of hindsight, political representatives in general tried to avoid any such labeling. In the interviews, those who were politically active at the time stressed that they followed the appropriate legal process by informing the police at the time. Alternatively, they avoided giving these events more publicity and opted for simply cleaning up or renewing the signs themselves.

[74]"Národnostní střet je jedním ze způsobů, jak se dostat do vyšší politiky. Konflikt vždycky eskaloval před volbami. Národnostní téma tady bylo zneužito pro politické ambice."

When looking at the quotes provided, we need to take into account the context of the interviews. The majority of responses do come from interviews in which the nationality of the researcher matched the nationality of the respondent, which we hoped would lead to a degree of openness from our respondents. Two of the interviews took place during a group meeting of the whole research team with the politician, and thus we should take into account the way the matter is presented to the international audience. At the end, however, many of the interviews were very conciliatory in tone, stressing the good cooperation between the local government and the Polish minority while minimizing the friction areas and cases of vandalism. The mayor of Třinec/Trzyniec, the town with perhaps the most complex road to bilingual signs, Věra Palkovská, stressed in 2017 that "now there are the best relationships with Poles that there ever were"[75] while also explaining that the signs should be useful for both groups. "[We wanted] signs to be practical and functional, so that both Czechs and Poles have a good feeling about it."[76] This can be read as a way of minimizing the difference through an appeal to a shared use. One, however, has to wonder why it is necessary to reference the majority's use for the signs which are rooted in minority presence in the area.

Several political actors framed the conflict as a matter of a few radical Polish activists trying too hard to push for overly extensive application of the law. These opinions could be heard from both the majority and minority politicians. We can see this as a way of diminishing the validity of uncomfortable demands that were, after all, based solely on existing legislation. As Stanisław Gawlik, a retired local minority politician, stated: "The problems stemmed from the fact that the majority society was not informed that this is not our idea but that [it is something] the Czech Republic agreed to. It is not a matter of rights but of obligation. Rights are one thing but the obligation has to be fulfilled. We did not make this up, we are only pointing out that if the state committed to it, why isn't it doing it? Many local representatives, Poles, don't want to provoke their Czech neighbors with the demands, in order to maintain good relationships. They say 'we can't just keep demanding things' but we are not

[75] "Teď jsou nejhezčí vztahy s Poláky jaké kdy byly."

[76] "Nápisy—aby to bylo praktické a funkční, aby z toho Češi i Poláci měli dobrý pocit."

'demanding', we are only pointing out that someone committed to something so that they should do it. It is not our idea, we did not sign the agreements."[77] While many argued that the demand for bilingual signs is not something made up by the Polish minority but that it was required by the European integration process, rarely did we hear such a clear statement highlighting the fact that the state took certain obligations upon itself and that, in effect, all that the minority did was to hold the state accountable to its own obligations.

The legal provisions surrounding the implementation process were also often subjected to critique. First, the legislation pertaining to bilingual signs offered, according to our respondents, too much space for interpretation and made the process overly dependent on the good will of local politicians, as the then-director of PZKO (Polish Cultural and Educational Association) Jan Ryłko surmised: "We do not have laws that would force people to have goodwill."[78] Even in places where the good will existed, the legal process was seen as complicated. The former deputy mayor of Vendryně/Wędrynia Bogusław Raszka acknowledged that the council felt a sense of obligation to install the signs but despite the assurances it did not find the process particularly clear or well managed: "We felt a sense of obligation that we should fulfill. Though as I said, it could have been solved differently from the legislative point of view. If the state ordered it 'from above' and offered a clear funding system, it would have been better. Also, if the communal obligations were clarified; it was very unclear at the beginning."[79]

Second, the 10% threshold minority was also criticized for being, in a way, too short-sighted a solution, especially considering future demo-

[77] "Problémy byly jen s tím, že většinová společnost nebyla informována o tom, že to není náš výmysl, ale že s tim souhlasila Česká Republika. Nejde o právo, tady jde o závazek, ne? Právo je jedna věc, ale závazek se musí realizovat. My jsme si to nevymysleli, my se jen díváme na to, že pokud se k něčemu stát zavázal, tak proč to nedělá. Hodne místních zastupitelů, Poláků, nechce, v rámci zachování dobrých vztahů dráždit své české sousedy nějakými požadavky, říkají, že nemůžeme pořád něco požadovat, my ale 'nepožadujeme', my pouze poukazujeme na to, že někdo se k něčemu zavázal a tedy by to měl dělat. To není náš výmysl, my jsme tu smlouvu nepodepsali."

[78] "Nemáme zákony, které by donutily lidi k dobré vůli."

[79] "Měli jsme pocit povinnosti, že to máme splnit, byť jak už jsem říkal, mohlo to být legislativně jinak řešené. Kdyby to stát nařídil zvrchu a dal jasné financování bylo by to lepší. Taky kdyby byly jasněji vymezené povinnosti obcí; to bylo ze začátku velmi nejasné."

graphic changes. The then-director of PZKO Jan Ryłko put it thus: "The signs are not as important, the law might change regarding the percentage threshold and then the number of bilingual signs will decrease and even if that will not happen, we will have demographic changes and the number of Poles will go down. So the majority will be able to change it back to Czech signs only. The percentual threshold should not be determinative, the history of this place should be taken into account as well."[80] Others suggested that a fixed list of communes where bilingual signs have to be installed would have helped with both the issues of interpretation as well as with preventing the disappearance of signs should the minority numbers go down.

Finally, many considered the long-term impact that the process of introduction of bilingual signs had had on local communities. The balance was not entirely positive. Some minority politicians, even years later, reflected on their surprise that certain tensions came forth that they had thought long gone. As Roman Wróbel, the current mayor of Bystřice/Bystrzyca contemplated: "Although it was a lesson for me that one has to go in-depth with the issue and then sometimes unwillingly negative emotions and tendencies come up to surface and sometimes not much is needed for them to boil over and then one is left thinking 'man, I thought that everything is alright and yet there is something hateful asleep here; that hurts most of all.'"[81]

The notion that issues from the past were brought forth during this process is also present in the following statement from Józef Szymeczek, the former director of the KP, who, however, also brings a more pragmatic aspect to the discussion: "The matter is for the shadow of history not to limit the real life in our present, so that it does not haunt us anymore. And the question is, whether we helped to improve the quality [of life in the region] or not. I would risk saying that it was slightly positive,

[80] "Nápisy nejsou zas tak důležité, zákon se může změnit pokud jde o procentuální limit a pak počet dvojjazyčných obcí půjde dolů a i když k tomu nedojde, tak budeme mit demografické změny a Poláků bude čím dál méně, takže ta většina to bude moci zase změnit zpátky na české nápisy. Procentuální kritérium by nemělo být rozhodující, měl by se brát ohled na historii toho místa."

[81] "Byť to pro mě byla lekce v tom, že člověk musí jít do hloubky problematiky a pak občas nechtěně se na povrch dostanou takove negativní emoce a tendence, a občas stačí málo k tomu, aby ony vybublaly, a pak si člověk říká, 'kurde, myslel jsem si že je všechno v pořádku a přece jen tu něco nenávistného spí, to bolí nejvíce.'"

a little bit yes."[82] He goes on to say that the implementation of bilingual signs helped local tourism and infrastructure and thus improved the life in the region in tangible ways for both the majority and the minority.

6.2.5 Questionnaires

Interviews allowed us to gain insight into the personal links between language, landscape, and identity. The semi-structured nature of the interviews provided sufficient flexibility and space to explore topics in detail and learn how family histories, personal experiences, schooling, and communal social life impact one's attitudes toward bilingual signs and minority names. The number of interviews was relatively high, and the wealth of information shared with us by the respondents shed light on many issues while pointing out problems to be explored further. For all the benefits of personal, semi-structured interviews, they still faced a definite limit of representativity. In these interviews, we learned a lot about specific individuals, but we would step on risky ground if we drew general conclusions about the larger population from them. For this reason, we decided to complete our investigation by a questionnaire survey on a large population sample (see Table 6.25) and test our hypotheses with statistical methods.

Of course, resorting to a questionnaire inevitably required a simplification of the problems and high selectiveness in the choice of questions to

Table 6.25 Questionnaire sample in the Těšín/Cieszyn region

Characteristic	Structure of respondents
Usable questionnaires	1804, 1454 in Czech and 350 in Polish
Gender	44.4% men and 55.6% women
Age	32.3% (<19), 14.8% (19–39), 39.7% (40–59), 13.2% (>60)
Residence	43.3% town residents, 56.7% village residents
Education	31.8% elementary, 47.9% secondary, 20.3% tertiary
Ethnicity	69.7% Czech, 25.4% Polish, 4.9% other

[82] Protože jde o to, aby nám ty stíny dějin nevymezovaly hranice reálného života naší současnosti. Aby nás to pořád nestrašilo. A teď je otázka, jestli jsme pomohli tu kvalitu [života v regionu] zlepšit, nebo ne. Já bych riskoval tvrzení, že to bylo lehce pozitivní, trošku ano.

be explored. Compared to the personal interviews, the questionnaire therefore may appear superficial and partial. However, we believe the survey results are extremely important because they have allowed us to test the influence of key factors shaping people's attitudes toward bilingual signs and minority names by differentiating strong and weak relationships between different variables. The survey made it possible to generalize our findings to the entire population while leaving space for the exploration of many topics in future research.

As we mentioned in Chap. 3, the Těšín/Cieszyn questionnaire had 19 questions, which were mostly identical in both language versions (see Appendix E). The only exceptions were the first two questions asking for one's relationship to the Těšín/Cieszyn region where the results showed that we had failed to translate them adequately for the two language versions to have the same meaning. In consequence, we had to disregard answers to these questions and could not, unfortunately, test the influence of one's ties to the region on the person's attitudes toward bilingual signs.[83] The first analyzed section of the questionnaire (Questions 3–7) therefore asked about language use and preference in different social situations, including toponymic practice. The second part (Questions 8–10) concerned the use and acceptability of the name *Olza* as a specific example of a controversial name which most respondents use regularly. The third part (Questions 11–14) focused on attitudes toward bilingual signs and majority-minority relations. The last section (Questions 15–19) asked for basic socio-demographic information and nationality.

As for language use and preference, the results were rather interesting and in many ways surprising. Question 3 asked which language was predominantly used in six social situations—at home, with friends, with neighbors, at school or work, in shops, and at public institutions. This question was included in our questionnaire to see how widespread was the use of the three linguistic codes spoken in the region and how true

[83] With a grain of salt, we could, however, interpret the confusion about the meaning of *Těšínsko* as an indication of the weakening of regional identity among the Czech population, additionally manifested by the small number of people declaring Silesian identity in our sample. This interpretation is also supported by our analysis of street name changes in Český Těšín and Cieszyn during the twentieth century. We clearly found that regional identity was much more heavily accented in Cieszyn whereas in Český Těšín, the national identity expressed through names strongly overshadowed all other levels of identification (Mácha et al. 2018).

Table 6.26 Prevailing language use in different social situations (absolute/*relative*)

Language	Home	Friends	Neighbors	School/work	Shops	Communal office
Dialect	817	749	825	592	339	139
	42.8%	*39.1%*	*41.8%*	*31.3%*	*17.6%*	*7.4%*
Czech	1075	1161	1142	1285	1579	1711
	56.3%	*60.6%*	*57.9%*	*67.9%*	*82.0%*	*91.9%*
Polish	18	7	7	17	8	11
	0.9%	*0.4%*	*0.4%*	*0.9%*	*0.4%*	*0.6%*
Total	1910	1917	1974	1894	1926	1861

Note: Pearson χ^2, $p = 0.000$, the number of responses is higher than the number of respondents because some indicated a combination of languages

was the stereotypical observation often heard locally that "everyone speaks the dialect". In addition, we were also interested if language practices had any effect on attitudes toward bilingual signs.

The results summarized in Table 6.26 indicate significant differences in language use in different social situations. To begin with, we are forced to discard the common wisdom that "everyone speaks the dialect". The predominant language of communication in all situations is Czech. Or, more precisely, the predominant language of communication in all situations is a linguistic code considered 'Czech' by the respondents—in most cases probably a colloquial form of Czech mixed with dialectal words and grammatical forms.

Second, we observe a decrease in the use of the dialect with social distance and an analogical increase in the use of Czech. The relationship is statistically significant for all situations (Pearson χ^2, $p = 0.000$). Strategic code switching is a well-known phenomenon (Scotton and Urry 1977). There is, however, one caveat in these summary results. The second most common situation in which the dialect is used is communication with neighbors rather than friends. This is in congruence with Williams's findings on the use of minority languages in different social situations across Europe (see Williams 2005: 160). The explanation in our case lies in nationality, as we will show below.

Finally, we see an almost total lack of the use of Polish even in the most intimate social situation—at home. This puts the struggle over Czech-Polish bilingual signs into a special perspective—the conflict is over signs rendered in a language few people use on a daily basis. The appearance of

Table 6.27 Prevailing language use among Czech respondents ($n = 1255$)

	Home	Friends	Neighbors	School/work	Shops	Communal office
Dialect	28.7%	26.5%	31.2%	20.7%	8.4%	2.4%
Czech	71.3%	73.5%	68.7%	79.0%	91.6%	97.6%
Polish	0.1%	0.0%	0.1%	0.3%	0.0%	0.1%

Note: Pearson χ^2, $p = 0.000$

Table 6.28 Prevailing language use among Polish respondents ($n = 454$)

	Home	Friends	Neighbors	School/work	Shops	Communal office
Dialect	86.0%	79.1%	69.7%	63.5%	40.6%	20.5%
Czech	10.8%	19.6%	29.6%	34.0%	58.3%	77.9%
Polish	3.3%	1.4%	0.8%	2.6%	1.1%	1.6%

Note: Pearson χ^2, $p = 0.000$

Polish on bilingual signs is therefore mainly symbolic, not functional. This was well expressed by Józef Szymeczek, the president of the Congress of Poles at the time, in the early debates about bilingual signs: "We do not want to introduce them because we could not speak Czech. Rather, they are a sign of our identity and historical belonging to this region"[84] (Třinecký hutník 2005).

When we differentiate the responses by nationality (Tables 6.27 and 6.28), we continue to see clear and statistically significant differences in language use in different social situations, but we observe additional differences which were hidden in the summary results. The most striking difference is that between Czechs and Poles. Most importantly, dialect use is significantly and systematically higher among Poles than Czechs. In fact, for the Polish population in the Těšín/Cieszyn region, the dialect functions as a substitute for the Polish language. In terms of practice, it is one of the defining features of the regional Polish identity. We also see a consistent decrease in dialect use with increasing social distance. In the Czech population, by contrast, the most common situation in which dialect is used is communication with neighbors. It appears that for Czechs the dialect serves as a lingua franca and a regional identity marker ("jo je tu z tela"—"I am from here") whereas Czech national identity is expressed

[84] In the original: "Nechceme je prosazovat proto, že bychom nerozuměli česky. Jedná se spíše o znak naší identity a dějinné příslušnosti k tomuto území."

by the usage of the Czech language at home. As we will show by many other examples below, although we tested the influence of different variables, declared national identity was by far the most important variable in explaining variation within our sample. The influence of age, for example, in this case was observable but not statistically significant.

In Question 4, we asked the respondents if they saw the dialect as Czech, Polish, or mixed. As we explained in Sect. 5.2, the boundaries of spoken language are blurred in the region as is the understanding of the linguistic and national origin of the dialect. The dialect may be variously perceived as Czech, Polish, or mixed. For some people the Polish language and the dialect are even synonyms. While the previous question concerned linguistic practice, in this question we move to the politics of language. Table 6.29 shows that the majority of respondents consider the dialect as mixed. This holds true for both Czechs and Poles although significantly more Poles see the dialect as Polish.

In Question 5 we asked about language preference in toponymic use. Table 6.30 indicates that in total slightly more people prefer Czech toponyms than dialectal ones and that very few people declare using Polish toponyms. Once again, we see a statistically significant difference between Czechs and Poles. What we also see is that the percentage of Czechs declaring the use of dialectal toponyms is higher than the percent-

Table 6.29 National character of the dialect

	Czech		Polish		Mixed		Total
	abs.	rel.	abs.	rel.	abs.	rel.	
Czechs	102	8.1%	159	12.7%	994	79.2%	1255
Poles	13	2.9%	173	38.1%	268	59.0%	454
All	115	6.7%	332	19.4%	1262	73.8%	1709

Note: Pearson χ^2, $p = 0.000$

Table 6.30 Linguistic preference in toponymy ($n = 1784$)

	Czechs	Poles	Total
Dialect	34.3%	73.8%	45.6%
Czech	65.0%	13.7%	50.3%
Polish	0.6%	12.5%	4.0%
Total	71.4%	28.6%	100.0%

Note: Pearson χ^2, $p = 0.000$

age of Czechs using the dialect in any social situation. In other words, even Czechs who do not speak the dialect use dialectal place names.

In Question 6 we asked the respondents to write up to five names for places in their surroundings that had a special meaning for them and to explain their choice. The purpose for this question was threefold. First, we wanted to see how the linguistic forms of names mentioned in people's responses compared to their linguistic preferences indicated in Question 5. Second, we wanted to know if some names were systematically mentioned more frequently than others. Third, we were interested in how people transcribed dialectal names and how large was the toponymic plurality of the region. And finally, we wanted to see what types of names with special meaning were mentioned and whether there was any trend in their selection (e.g., proximity, visual prominence, historic symbolism, personal experiences). We will only comment on the first two aspects, leaving the other two for another text due to the scope of the question.

As for the linguistic form chosen, the overall result resembles answers to Question 5, but the difference between Czechs and Poles is even more pronounced. Many Czech respondents who declared the use of dialectal names chose to write Czech names instead. This can be, undoubtedly, attributed to the normative authority of the questionnaire as well as the lack of experience in writing in the dialect. By contrast, some Polish respondents who declared using Polish names actually wrote down dialectal names which once again confirms the importance of paying attention to the nationalist politics of dialect classification (Table 6.31).

A total of 2385 responses were given with 851 unique names in some 1346 linguistic and orthographic variants. By far the most frequently mentioned name was *Olza*—115 cases—while *Olše* was mentioned only

Table 6.31 Linguistic forms of names (n = 2385)

	Czechs	Poles	Total
Dialect	23.0%	79.7%	43.9%
Czech	76.0%	13.4%	53.0%
Polish	0.9%	6.8%	3.1%
Total	63.1%	36.9%	100.0%

Note: Pearson χ^2, p = 0.000

21 times. As we will show below, this corresponds well with the usage of the name in everyday life.

The last question from the section on linguistic preference in communication and toponymy asked whether the respondent would welcome the use of dialectal names on official signs and maps. The results are rather surprising. First, both Czechs and Poles oppose dialectal names on signs. Second, Czechs are much less in favor of dialectal names on signs than Poles. And third, significantly fewer respondents want dialectal names on official signs and maps than those who actually use them. In other words, many people who use dialectal names do not wish them to appear on official signs and maps. This, once again, holds true for both Czech and Poles, although they differ significantly in their preference (Table 6.32).

These results appear to go against what we heard in the majority of personal interviews in which people expressed strong feelings for dialectal names. It seems that either our interview sample was schewed or the interview format gave people more time to think about names and appreciate them more. We would explain the questionnaire results once again by the fear that the official use of the dialect may weaken the ethnic claims to the territory by either side and thus the use of the national languages is preferred. This interpretation has support in interviews with Congress of Poles representatives who rejected the use of the dialect on official signs in spite of its unifying potential for the fear that it would undermine the status of Polish and the legitimacy of the Polish national movement in Czechia.[85] Such interpretation is also supported by the

Table 6.32 The use of dialect on official signs (n = 1719)

	Czechs	Poles	Total
Yes	6.2%	27.0%	11.8%
No	70.1%	57.3%	66.7%
Does not know/care	23.7%	15.7%	21.6%
Total	1259	459	100.0%

Note: Pearson χ^2, $p = 0.000$

[85] For example, interview with Dariusz Branny, February 2, 2017, Karviná, and interview with Józef Szymeczek, October 18, 2017, Český Těšín.

statistically significant relationship between the support for dialectal names on signs and the perception of bilingual signs (Questions 11 and 12). The results of both tests are summarized in Tables 6.33 and 6.34. In both cases, those who opposed dialectal names on signs also more probably opposed bilingual signs (and vice versa, with the neutral respondents standing exactly in the middle). Last but not least, the answers to the next three questions provide further support for this interpretation.

In Questions 8, 9, and 10 we inquired about the use of the name *Olza*, its perception in terms of the national language, and its potential promotion as the official name of the river. As already mentioned, we thought this name was sufficiently well-known to give all respondents a very concrete example of a bilingual name, rather than asking about bilingual names in general. We were correct in assuming strong emotions would be associated with the name; however, further tests showed that attitudes toward this name do not necessarily correlate with attitudes toward bilingual names (more below).

What did we find? As Table 6.35 shows, the majority of respondents use exclusively or predominantly *Olza* rather than *Olše*. This holds true

Table 6.33 The use of dialect on official signs and perception of bilingual signs (n = 1656)

	Perception of bilingual signs					
Dialectal names on signs	Very negative	Negative	Neutral	Positive	Very positive	Total
Yes	4.2%	5.2%	19.4%	39.8%	31.4%	191
No	36.2%	21.8%	25.0%	11.6%	5.4%	357
Does not know/ care	10.1%	17.9%	41.2%	25.2%	5.6%	1108

Note: Pearson χ^2, $p = 0.000$

Table 6.34 The use of dialect on official signs and argument preference (n = 1656)

	Arguments on bilingual signs		
Dialectal names on signs	Against	For	Total
Yes	18.8%	81.2%	191
No	76.2%	23.8%	1108
Does not know/care	51.5%	48.5%	357

Note: Pearson χ^2, $p = 0.000$

Table 6.35 Usage of the name *Olza*

	Only *Olza*	Mostly *Olza*	Both *Olza* and *Olše*	Mostly *Olše*	Only *Olše*	Total
All	754 *41.70%*	468 *25.90%*	277 *15.30%*	101 *5.60%*	118 *6.50%*	1718
Czechs	399 *31.70%*	388 *30.80%*	258 *20.50%*	98 *7.80%*	116 *9.20%*	1259
Poles	355 *77.30%*	80 *17.40%*	19 *4.10%*	3 *0.70%*	2 *0.40%*	459

Note: Pearson χ^2, $p = 0.000$

Table 6.36 *Olza* as a Czech, Polish, or dialectal name

	Czech	Polish	Dialectal	Other	Total
All	237 *13.8%*	631 *36.7%*	734 *42.7%*	116 *6.8%*	1718
Czechs	223 *17.7%*	334 *26.5%*	640 *50.8%*	62 *4.9%*	1259
Poles	14 *3.1%*	297 *64.7%*	94 *20.5%*	54 *11.8%*	459

Note: Pearson χ^2, $p = 0.000$

for both Czechs and Poles, although Poles use *Olza* significantly more often. This confirms the generally held view in the region that *Olza* is by far the preferred linguistic choice. Responses to Question 6 discussed above support this claim.

In view of the aforementioned politics of language classification, we also asked in Question 9 whether the respondents saw the name *Olza* as Czech, Polish, dialectal, or other (Table 6.36). Overall, most people considered the name as dialectal; however, the breakdown by nationality shows that there are, once again, statistically significant differences between Czechs and Poles in the perception of the name. In the context of the politics of language and identity, we see an indication of an effort on the part of Czechs to justify their usage (and ownership) of the name by ascribing it a dialectal origin and therefore neutral or potentially even Czech nationality while Poles are largely claiming it as Polish. And, put into association with the previous question, those Czechs who use exclusively or predominantly *Olše* rather than *Olza* are significantly more likely to see the name as Polish, and vice versa.

Table 6.37 *Olza* as the official name of the river

	For	Against	Doesn't know	Total
All	553	622	543	1718
	32.2%	*36.2%*	*31.6%*	
Czechs	241	575	443	1259
	19.1%	*45.7%*	*35.2%*	
Poles	312	47	100	459
	68.0%	*10.2%*	*21.8%*	

Note: Pearson χ^2, $p = 0.000$

The most interesting results in this section, however, were brought to light by Question 10. In this question we asked whether *Olza* should be the official name of the river and why or why not. As Table 6.37 demonstrates, the population is divided roughly in thirds which is in itself interesting, given the percentage of the population that declares using exclusively or predominantly the name *Olza*. The situation becomes even more pronounced when we break down the sample by nationality. We see then that although 62% of Czechs use exclusively or predominantly *Olza* (see Question 8), only 19% wish for *Olza* to be the official name of the river. Poles are significantly more in favor of *Olza* as official and also significantly less against it. Still, even a part of the Polish population does not wish for *Olza* to be the official name. In other words, we have a large number of Czech respondents and some Poles who use the name *Olza* but do not wish it as the official name of the river. This is a very intriguing paradox which deserves further exploration as it touches directly on the main issues with which our text is concerned—namely, the relationship between language, identity, and landscape.

Respondents themselves offered an explanation of this paradox in the second part of Question 10. They were encouraged to provide reasons for their position in an unstructured manner. Almost 700 respondents did so and after limiting responses to Czechs and Poles and coding them into categories we were able to understand (and statistically test) differences in respondents' reasoning. The categories we decided to use to categorize answers to this question were similar but not identical to those used in our media analysis, and they were less numerous simply because fewer types of arguments were voiced. This stems from what we discovered to

Table 6.38 Categories and typical responses

Category type	Typical responses
Utilitarian	Everyone uses *Olza*
Ideological	This is Czechia and we should only use Czech names, *Olza* is Polish
Aesthetic	*Olza* sounds better
Historic	It has always been called *Olza*
Pragmatic	A change would cost a lot of money
Other	Why? Why not! It doesn't matter to me

be an important difference in the perception of the two issues—bilingual signs on the one hand and *Olza/Olše* on the other. For the same reasons, these categories are also partly different from categories used in Question 12 (see below).

We used six categories expressing typical responses. Table 6.38 provides an overview of individual categories and characteristic answers. Table 6.39 summarizes responses and breaks them down by nationality. There are two notable findings. First, as with the previous questions, there is a statistically significant difference between Czech and Polish respondents in their choice of arguments. And two, the choice of arguments also depends on the position toward the name change—respondents in favor of the change argue differently than those opposing it. Once again, these differences are statistically significant. What does it mean? Let us first summarize the results. Czechs in favor of the name change prefer utilitarian arguments—that is, "Most/all people use the name". By this argument they strategically avoid the question of the national/linguistic origin of the name, making it irrelevant in the debate. This position is also supported by a minority of Poles. By contrast, Czechs opposing the change prefer ideological/nationalistic arguments—that is, "This is the Czech Republic and we should only use Czech names", implying thus that *Olza* is a Polish name. Here, the usage of the name (as documented by answers to Question 8) is discarded as irrelevant and the (incorrectly) perceived national origin of the name is emphasized. This position is also supported by a small minority of Poles opposing the change. Finally, Poles supporting the change favor historic arguments—that is, "*Olza* is the original name of the river". This goes well together

Table 6.39 Arguments for/against *Olza* as official (Q3, Q4)

	Argument categories						
	Utilitarian	Ideological	Aesthetic	Historical	Pragmatic	Other	
All respondents	104	214	49	181	39	41	628
	16.6%	*34.1%*	*8.2%*	*28.8%*	*6.2%*	*6.5%*	
Czechs for	80	6	23	36	6	4	155
	51.6%	*3.9%*	*14.8%*	*23.2%*	*3.9%*	*2.6%*	
Czechs against	5	179	7	4	24	23	242
	2.1%	*74.0%*	*2.9%*	*1.7%*	*9.9%*	*9.5%*	
Poles for	19	15	18	141	8	12	213
	8.9%	*7.0%*	*8.5%*	*66.2%*	*3.8%*	*5.6%*	
Poles against	0	14	1	0	1	2	18
	0.0%	*77.8%*	*5.6%*	*0.0%*	*5.6%*	*11.1%*	

Note: The value of all Pearson χ^2 tests for the individual relationships was $p = 0.000$

with the larger discourse promoted by the Polish community about the historic roots of the Polish population in the region (in reaction to claims of recent migration promoted by some Czech nationalists). One name, multiple interpretations, disjointed worlds. What this shows clearly, however, is that names definitely are not arbitrary labels and that emotions attached to them may be so strong that they may even override everyday toponymic practice.

An interesting category standing somewhat aside from the others is the aesthetic argument. Not many respondents chose it and when they did, it usually was to argue in favor of the change: for example, "*Olza* sounds better" or "*Olše* sounds strange". The aesthetic quality of a name is often overshadowed by its ideological message and thus underestimated, but as Kearns and Berg (2002) pointed out, name's sound and pronunciation are as important to its meaning and acceptance as its written form. Ainala (2016) and Berezkina (2016) found the same for Helsinki/Helsingfors and Oslo, respectively. To be sure, there undoubtedly is a politics implicit in aesthetic appreciation, but beauty may at times stand at the same level of importance, if not higher than, truth and goodness. Names should not be corrupted and that includes their aesthetic quality.

The last research section of our questionnaire concerned the central problem of our investigation—the attitudes toward bilingual signs (Questions 11–13) and the perception of Czech-Polish relations (Question 14). As Table 6.40 indicates, overall, more people are against bilingual signs than in favor of them. As with all previous tables, however, the summary results hide significant differences by nationality (and other

Table 6.40 Perception of bilingual signs

	Strongly against	Against	Neutral	In favor	Strongly in favor	Total
All	453	321	491	308	145	1718
	26,4%	*18,7%*	*28,6%*	*17,9%*	*8,4%*	
Czechs	423	280	372	147	37	1259
	33,6%	*22,2%*	*29,5%*	*11,7%*	*2,9%*	
Poles	30	41	119	161	108	459
	6,5%	*8,9%*	*25,9%*	*35,1%*	*23,5%*	

Note: Pearson χ^2, $p = 0.000$

factors—on those see below). Notably, the figures for Czechs and Poles are reversed—while the majority of Czechs oppose bilingual signs, the majority of Poles support them. Approximately one-fourth of both groups remain neutral. The differences are once again statistically significant.

In Question 11 we asked respondents to choose from predefined options that argument with which they identified the most. We formulated the arguments (see Table 6.41) on the basis of the media analysis and on the basis of arguments which most frequently appeared in interviews. Following Szabó-Gilinger et al. (2012), we chose two pragmatic and two symbolic arguments, always one in favor and one against bilingual signs. Although respondents were asked to select one argument only, some chose two, so the number of answers in Table 6.42 is slightly higher than the number of respondents. When two arguments were selected, in all cases it was a consistent choice—both against or both in favor.

Table 6.41 Arguments used in Question 12

Argument type	Argument
Pragmatic against	The introduction of bilingual signs is a waste of public money, the Těšín/Cieszyn Poles speak Czech
Pragmatic for	The Polish minority has a legal right to bilingual signs
Symbolic for	The introduction of bilingual signs is a recognition of the existence of the minority, that it is also at home here
Symbolic against	We are in Czechia, signs should only be in Czech, there are no Czech signs in Poland either
Other	All other arguments

Table 6.42 Arguments for and against bilingual signs

	Pragmatic against	Symbolic against	Pragmatic for	Symbolic for	Other	Total
All	368 *19.7%*	805 *43.1%*	183 *9.8%*	472 *25.3%*	40 *2.1%*	1868
Czechs	300 *22.3%*	740 *54.9%*	71 *5.3%*	206 *15.3%*	30 *2.2%*	1347
Poles	68 *13.1%*	65 *12.5%*	112 *21.5%*	266 *51.1%*	10 *1.9%*	521

Note: Pearson χ^2, $p = 0.000$

As the results show, once again we see an overall prevalence of negative attitudes and a statistically significant difference between arguments preferred by Czechs and Poles. While almost 70% of Czechs chose arguments against bilingual signs, a slightly higher percentage of Poles preferred arguments in favor. This is entirely consistent with answers to previous questions in our questionnaire. However, what deserves further exploration are those Czechs who support bilingual signs and those Poles who oppose them. It is clear that other factors, besides nationality, influence one's attitudes and we comment on those below.

The third question in this section asked which signs should be bilingual. Respondents could choose up to ten types of signs or they could choose no sign at all. Interestingly, most Czechs did not choose any signs while most Poles chose at least one (see Table 6.43). This is entirely consistent with the results mentioned before and cannot be interpreted as a simple lack of interest in responding. In fact, many respondents opposed to bilingual signs felt it necessary to emphasize they did not want any at all and therefore wrote it in the questionnaire often in large capital letters, underlined and/or followed by exclamation point(s), and so on.

Furthermore, when we compare responses to Questions 11, 12, and 13, we see a counterintuitive increase in support for bilingual signs among Czech respondents—while only less than 15% are in favor or strongly in favor of bilingual signs (Question 11) and only 21% chose a symbolic or a pragmatic argument in favor of signs (Question 12), almost 28% indicated at least one bilingual sign (Question 13). In other words, some Czechs oppose bilingual signs, but concede that Poles may have legiti-

Table 6.43 Preference for different types of bilingual signs by nationality

	At least one sign type	No sign type at all	
All	701	1017	1718
	40.8%	*59.2%*	
Czechs	348	911	1259
	27.6%	*72.4%*	
Poles	353	106	459
	76.9%	*23.1%*	

Note: Pearson χ^2, $p = 0.000$

mate demands (or simply wish to avoid further conflict), so with a great reluctance they agree to the introduction of some bilingual signs. This interpretation is supported by further findings. On average, Czechs chose only 3.6 signs while Poles chose 5. Finally, only 18.7% of those Czechs who chose at least one sign (i.e., 5.2% of all Czechs) selected more than five. By contrast, 39.4% of Poles who chose at least one sign (i.e., 30.3% of all Poles) selected more than five. In other words, the support for bilingual signs among Czechs sympathetic with the Polish cause is lukewarm, at best.

As far as individual types of signs are concerned, by far the strongest support among all respondents is for the introduction of signs indicating the name of the commune which are typically located at the entrance to the communal center. Incidentally, this is also by far the most frequently vandalized type of sign. This may be attributed to the fact that these signs in our research area already are mostly bilingual, so a support for their bilingual rendering may be interpreted as a simple acknowledgment or acceptance of the status quo. It appears that the less frequent the individual types of signs in linguistic landscape in the area are, the smaller is the support for their bilingual rendering. We have not tested this 'habituation' hypothesis though; hopefully, future researchers will. Differences in support for other types of signs are smaller and not statistically significant as a whole set, although differences in support for some signs would probably be significant. (We did not test individual combinations.) However, regardless of the chosen metric, support for the bilingual rendering of all signs is significantly higher among Poles (Pearson χ^2, $p = 0.000$), and this holds true for both their support among those who responded to the question as well as among all respondents.

Two additional findings are noteworthy. First, we observe a high difference between Czechs and Poles in support for bilingual signs on schools and cultural institutions as compared to other signs. It would appear that for Czech respondents a bilingual sign on a school may present a greater threat than most other signs. This is probably caused by the close association between language, Czech nationalism, and schooling in the Czech language. And second, the majority of all respondents opposed the mandatory introduction of bilingual signs on (private) shops and businesses. As our interviews already indicated, the conflict over bilingual signs is a

Table 6.44 Preference of bilingual signs by type and nationality

	a	b	c	d	e	f	g	h	i	J
All	583	386	308	401	219	256	219	299	72	289
% of responded	*83.2*	*55.1*	*43.9*	*57.2*	*31.2*	*36.5*	*31.2*	*42.7*	*10.3*	*41.2*
% of all	*33.9*	*22.5*	*17.9*	*23.3*	*12.7*	*14.9*	*12.7*	*17.4*	*4.2*	*16.8*
Czechs	266	140	91	182	99	103	89	136	21	132
% of responded	*76.4*	*40.2*	*26.1*	*52.3*	*28.4*	*29.6*	*25.6*	*39.1*	*6.0*	*37.9*
% of all	*21.1*	*11.1*	*7.2*	*14.5*	*7.9*	*8.2*	*7.1*	*10.8*	*1.7*	*10.5*
Poles	317	246	217	219	120	153	130	163	51	157
% of responded	*89.8*	*69.7*	*61.5*	*62.0*	*34.0*	*43.3*	*36.8*	*46.2*	*14.4*	*44.5*
% of all	*69.1*	*53.6*	*47.3*	*47.7*	*26.1*	*33.3*	*28.3*	*35.5*	*11.1*	*34.2*

Note: a—entrance to village; b—communal office; c—library, school; d—train stops; e—bus stops; f—parts of communes; g—street signs; h—orientation signs; i—shops; j—maps

Table 6.45 The quality of Czech-Polish relations in the region

	Very bad	Rather bad	Neither bad nor good	Rather good	Very good	Total
All	47	151	862	648	99	1807
	2.6%	*8.4%*	*47.7%*	*35.9%*	*5.5%*	
Czechs	35	98	590	468	68	1259
	2.8%	*7.8%*	*46.9%*	*37.2%*	*5.4%*	
Poles	9	44	237	149	20	459
	2.0%	*9.6%*	*51.6%*	*32.5%*	*4.4%*	

Note: Pearson χ^2, $p = 0.159$

struggle over the control of the public sphere, leaving the private sphere the freedom to choose the language of commerce as it sees fit (Table 6.44).

The last question in this section asked about the perception of the quality of Czech-Polish relations in the region. We expected the Poles as the disadvantaged group confronted with a backlash from the Czech majority to view the quality of Czech-Polish relations as worse than Czechs. The numbers may even seem to nominally confirm that (see Table 6.45). However, the observed differences were small, and they were not statistically significant. This is actually good news for further development in the region—in spite of occasional conflicts and clear differences in attitudes there seems to be good will on both sides and a generally positive or neutral view of the other group. We may only wish that this will not change in the future.

In the preceding paragraphs, we have presented the overall findings of our survey. Initially, we only planned to give summary results for the entire sample and then provide explanations for the observed differences. However, once we identified the large and statistically significant differences between Czechs and Poles, we concluded that to present only the summary results for the whole sample would be highly misleading. Nevertheless, we have thus inadvertently disclosed the principal finding of our survey: the key factor in explaining variance in our sample is self-declared nationality. As we demonstrate below, no other variable was even close to nationality in its explanatory power.

At the end of the questionnaire, we asked—in addition to nationality—for basic socio-demographic characteristics (place of residence, age, gender, and educational level). We tested the influence of these variables to compare them with the effect of nationality. We also ran tests between other variables to see if answers to some questions were related. Some expected relationships turned out not statistically significant, others passed the significance level threshold but the actually observed differences were small. Let us address them one by one.

The effect of age on attitudes toward bilingual signs (Questions 11 and 12) was partially statistically significant and—considering the large number of young people in our sample—rather weak. Our expectation based on Franssen et al. (2013) was that younger people would be more tolerant and cosmopolitan. We ran ANOVA (Bonferroni) analysis to compare age means between people holding negative and positive opinions toward signs as measured by a scale in Question 11 and grouping of arguments (for/against) in Question 12. In the first case, there is a consistent mean age decrease as we move from the most negative to the most positive stances; however, only some of the observed mean age differences were statistically significant. The largest mean age difference was observed between people with a "very negative" and "positive" attitude: 5.5 ($p = 0.000$). However, although there also was a consistent mean age difference between the "very negative" and "negative" position (2), it was not statistically significant ($p = 1.000$). In the case of Question 12, the mean age difference between people who chose arguments in favor of bilingual signs as compared to those who chose arguments against them was 3 and it was statistically significant ($p = 0.004$). We must therefore conclude that age has an effect on attitudes toward bilingual signs, but it is not entirely straightforward, and the overall influence is mediated by other factors.

Place of residence was primarily included in our questionnaire in order to ascertain the geographical origin of our sample. Once this data was available and we knew our sample corresponded quite well with the geographical scope of our other research methods, we were interested if geographical origin had any effect on attitudes toward bilingual signs. Because we did not have sufficient respondents from any given locality, we could not test, for example, whether the percentage of Poles living in the commune influences attitudes toward signs among Czech respondents—a hypothesis based on the generally accepted fact that we fear most that which we do not know (Carleton 2016). However, we tried to approximate this test by grouping respondents into urban and rural dwellers. As already pointed out in Chap. 5, the proportion of Polish inhabitants in our research area is generally higher in villages than in towns. The test, nevertheless, did not find any statistically significant differences between urban and rural dwellers on either Question 11 or Question 12. In other words, Czechs living in urban and rural areas seem to hold similar views on bilingual signs.

By contrast, the effect of education was observable and statistically significant. It is generally accepted that educational attainment is negatively correlated with ethnic prejudice (Coenders and Scheepers 2003). People with higher education tend to be more tolerant toward ethnic outgroups. This was also confirmed by our data which showed a stronger support for bilingual signs among university-educated than among high-school educated respondents. We only included people over the age of 27 to minimize the number of students with incomplete education. The relationship between education and attitudes toward bilingual signs holds true for both the perception of bilingual signs (Question 11, Kendall's tau c, $p = 0.000$) as well as preference for arguments for/against bilingual signs (Question 12, Pearson χ^2, $p = 0.000$). In both instances, however, the actual proportional difference between educational groups was not

Table 6.46 Preference of arguments on bilingual signs by education (28 years and older, $n = 1108$)

	Educational level			
	Elementary	High school w/o certificate	High school with certificate	University
Arguments against bilingual signs	17 *73.9%*	212 *76.5%*	360 *73.9%*	160 *49.8%*
Arguments in favor of bilingual signs	6 *26.1%*	65 *23.5%*	127 *26.1%*	161 *50.2%*
Total	23	277	487	321

Note: Pearson χ^2, $p = 0.000$

large (see Table 6.46), nothing comparable to the 'mirror image' reversals of attitudes among Czechs and Poles. The majority of respondents were clearly against bilingual signs and even among the university-educated only one-half supported them. However, when we break down the numbers by nationality, the picture becomes even bleaker. Among the university-educated Czechs, only 24.3% support bilingual signs which is almost twice as many as high-school-educated Czechs but still significantly below the sample average raised to 50% by university-educated Poles (see Table 6.46).

The effect of gender in our sample was, like age, observable and statistically significant only partially. The relationship between gender and preference for arguments for/against bilingual signs was not statistically significant (Pearson χ^2, $p = 0.327$). The relationship between gender and perception of bilingual signs was significant but we were not able to confirm the results of other researchers indicating that women manifested more positive attitudes toward outgroups (see, e.g., Hoxter and Lester 1994). In our sample, women were actually more likely to hold the middle-ground positions while men were more likely to support the extreme positions on both ends of the scale. Once again, however, the differences were rather small, and we also lack any non-essentializing explanation for this finding (Table 6.47).

Since most of our variables were nominal or ordinal, we were not able to run any multi-factor analyses and multiple regression models. From the aforementioned findings, it is, however, clear that nationality accounts

Table 6.47 Differences in attitudes to bilingual signs by gender

	Perception of bilingual signs					
	Very negative	Negative	Neutral	Positive	Very positive	Total
Men	233	136	198	124	72	763
	30,5%	*17,8%*	*26,0%*	*16,3%*	*9,4%*	
Women	230	190	293	183	79	975
	23,6%	*19,5%*	*30,1%*	*18,8%*	*8,1%*	
Total	463	326	491	307	151	1738

Note: Pearson χ^2, $p = 0.010$

for most variance in the sample. In no instance did this variable explain all variation though. For example, when we tested the influence of nationality on attitudes toward bilingual signs, we found that the majority of Czechs opposed them, and the majority of Poles supported them. Still, some 21% of Czechs were in favor of bilingual signs and some 26% of Poles were against them. How do we account for these 'dissenters'? In our opinion, educational level and age may be additional factors to be taken into account. For example, when we break down attitudes toward bilingual signs as expressed by preference for arguments for/against these signs (Question 12) by nationality and age groups—27 years and younger versus 28 and older, we see clear differences (see Table 6.48). Young people of both nationalities support bilingual signs to a significantly larger degree than older people.

The same holds true for the effect of education (see Table 6.49)—both Czechs and Poles with university education declare a stronger support for bilingual signs than Czechs and Poles with only high-school education (with or without the certificate). We may conclude that nationality together with age and education are the best predictors of attitudes toward bilingual signs in our research area.

In addition to nationality and selected socio-demographic characteristics, we also tested the significance of relationships between some questions. Namely, we were interested if the use of dialect, preference for the official use of dialectal names, and the perception of the quality of Czech-Polish relations in the region could be related to attitudes toward bilingual signs. As we hinted above, the use of the dialect is a tricky variable. We already demonstrated that its use is significantly related to nationality—

Table 6.48 Differences in support for bilingual signs by age group and nationality ($n = 1656$)

	Under 28			28 and older		
	Against	For	Total	Against	For	Total
Total	338	240	578	726	352	1078
	58.5%	41.5%	100%	67.3%	32.7%	100%
Czechs	326	152	478	622	116	738
	68.2%	*31.8%*	*82.7%*	*84.3%*	*15.7%*	*68.5%*
Poles	12	88	100	104	236	340
	12.0%	*88.0%*	*100.0*	*30.6%*	*69.4%*	*31.5%*

Note: Pearson χ^2, $p = 0.000$—the results are identical for both the total as well as for Czechs and Poles

Table 6.49 Support for bilingual signs by education and nationality (28 and older, $n = 1075$)

	High school educated			University educated		
	Against	For	Total	Against	For	Total
Total	570	194	764	154	157	311
	74.6%	*25.4%*		*49.5%*	*50.5%*	
Czechs	492	74	566	128	41	169
	86.9%	*13.1%*		*75.7%*	*24.3%*	
Poles	78	120	198	26	116	142
	39.4%	*60.6%*		*18.3%*	*81.7%*	

Note: Pearson χ^2, $p = 0.000$—the results are identical for both the total as well as for Czechs and Poles

most Poles use it at home, while only a minority of Czechs do. As a consequence, testing the influence of dialect use risks testing the influence of nationality. However, when we tested the influence of dialect use at home for Czech respondents only, we found a statistically significant relationship (Pearson χ^2, $p = 0.005$). Czechs who speak dialect at home express a stronger support for bilingual signs. Our data does not make it possible to explain the attenuating influence of dialect use. Dialect itself may be the factor because of the phonological proximity of many dialectal and Polish names or—and we consider this more likely—dialect may only be a proxy for other factors such as family history or Polish schooling.

When we discussed these findings with representatives of the Congress of Poles, some suggested that Czechs speaking the dialect at home in fact were former Poles. There is an emphatic dialectal term for such people—*"škopyrtok"* ('someone who tripped and lost his/her way'). While there is a clear political agenda on the part of the Polish representation behind their claim of assimilated Poles lost to the Polish cause, family histories are indeed very complex in the Těšín/Cieszyn region and Czech and Polish identities intermix in a multiplicity of ways. Be it as it may, it is clear, however, that dialect use among Czech respondents does relate with positive attitudes toward bilingual signs and it may be considered an additional factor explaining observed variation in our sample (Table 6.50).

In addition to dialect use, we hypothesized that the support of official dialectal use in toponymy may also correlate with attitudes toward bilingual signs. We chose the name *Olza* as an example of such name. Given the strong and emotional responses people gave to Question 10, we expected there to be a significant and strong relationship with attitudes toward bilingual signs. As Table 6.51 demonstrates, the difference

Table 6.50 Effect of dialect use at home on attitudes toward bilingual signs among Czechs

	Against	For	Total
Use Czech	754	183	937
	80.5%	*19.5%*	*77.4%*
Use dialect	189	85	274
	69.0%	*31.0%*	*22.6%*
Total	943	268	1211
	77.9%	*22.1%*	*100.0%*

Note: Pearson χ^2, $p = 0.005$

Table 6.51 Effect of support for *Olza* as official on attitudes toward bilingual signs

	Against bilingual signs	For bilingual signs	Total
Olza as official	212	316	528
	40.2%	*59.8%*	*46.6%*
Olza as not official	509	95	604
	84.3%	*15.7%*	*53.4%*
Total	721	411	1132
	63.7%	*36.3%*	*100.0%*

Note: Pearson χ^2, $p = 0.000$

between supporters and opponents of *Olza* in their view of bilingual signs is rather stark and statistically highly significant. This is further corroborated by the statistically equally significant relationship between the general support for dialectal names on signs (Question 7) and attitudes toward bilingual signs (Questions 11 and 12) that we mentioned earlier. We may thus conclude that people who support the official use of dialectal names more likely agree with bilingual signs in the region.

Finally, we tested the influence of the perception of Czech-Polish relations in the region on attitudes toward bilingual signs. We assumed that people who viewed these relations more favorably would also be more supportive of bilingual signs. The correlation test between Questions 11 and 14 formally supported this hypothesis but the relationship—although significant—was extremely weak (Kendall tau-$b = 0.066$, $p = 0.010$). The relationship between Questions 12 and 14 was not significant at all. We therefore dismissed this variable as a useful explanatory factor.

To summarize this section, nationality, age, education, dialectal use, and support for the official use of dialectal names are related to attitudes toward bilingual signs. Nationality is by far the strongest and together they explain the majority of the variation within our sample. On the other hand, several tested factors such as residence, gender, or perception of Czech-Polish relations proved weak or irrelevant. There undoubtedly are other potentially important factors which we did not test at all (e.g., strength of ties to the region, family histories, effect of Polish schooling, personal contacts with members of the Polish community, etc.). Future research will have to amend that.

6.2.6 Conclusion

Our research in the Těšín/Cieszyn region succeeded in combining different methods to provide a complex picture of the debate on bilingual signs and factors influencing people's attitudes toward them. It is clear that the individual levels/contexts—maps, linguistic landscape, media, communal politics, national, regional and local identities, and personal life histories—significantly influence each other. None of these levels is simple and straightforward in itself, let alone in combination with the others.

Nevertheless, across these levels we see common topics, problems, and challenges, all arising from the complex ethnic, linguistic, and political history of the region. National loyalties and regional ties continue to clash, as do the associated linguistic and toponymic practices. People are often torn between conflicting demands and loyalties and it is not always easy for them to maneuver them safely. As the results of the questionnaire survey show, the road to a region-wide reconciliation is still long and open. The demographic development of the region works against this process and the next population census in 2021 is undoubtedly going to reveal yet another significant decrease in the proportion of the Polish minority. This will in turn negatively impact its negotiating position in debates over the introduction of bilingual signs as well as in other areas such as schooling, cultural activities, or Polish journalism. Some hard-won gains may even be reversed, and new ones will be more difficult to achieve. The Commune Těrlicko/Cierlicko may stand as a warning—after the last population census in 2011, the proportion of the Polish minority in the Commune fell to 9.6% (i.e., just under the minimum 10% threshold), bilingual town signs were immediately vandalized and they were never renewed. After the results of the next census are announced, we will quickly learn how far the national reconciliation in the Těšín/Cieszyn region has progressed and what new challenges it will bring. We can be sure, however, that place names will continue to play a prominent role in this process.

References

Ainala, T. (2016). Attitudes to street names in Helsinki. In G. Puzey & L. Kostanski (Eds.), *Names and naming. People, places, perceptions and power* (pp. 106–119). Bristol: Chanell View Publications/Multilingual Matters.

Anonymous. (s.a. [post 1945]). *Maps relating to the ethnical structure of Slovenian Carinthia*. s.l.

Anton Melik Geographical Institute & Institute of Geography (Ed.). (2001). *National atlas of Slovenia*. Ljubljana: Rokus Publishing House.

Archivum map wojskowego instytutu geograficznego 1919–1939. (2019). Retrieved May 26, 2020, from http://maps.mapywig.org.

Arnberger, E., & Kretschmer, I. (1975). *Wesen und Aufgaben der Kartographie. Topographische Karten*. Wien: Deuticke.

Austrian Board on Geographical Names (AKO) (Ed.). (2017, August 8–17). *Recommendation for the naming of urban traffic areas*. 11th United Nations Conference on the Standardization of Geographical Names, New York, E/CONF.105/21/CRP.21. Retrieved March 24, 2020, from https://unstats.un.org/unsd/geoinfo/UNGEGN/docs/11th-uncsgn-docs/E_Conf.105_21_CRP.21_9_Recommendations%20.pdf.

Austria Presse Agentur (APA) (Ed.) (2002). APA defacto. Retrieved April 25, 2020, from www.apa-defacto.at.

Austrian State Archive. Retrieved from https://www.oesta.gv.at/.

Basso, K. H. (1988). Speaking with names: Language and landscape among the Western apache. *Cultural Anthropology*, III(2), 99–130.

Basso, K. H. (1996). *Wisdom Sits in Places. Landscape and Language among the Western Apache*. Albuquerque: University of New Mexico Press.

Beimrohr, W. (2012). Die Erste und die Zweite Landesaufnahme von Tirol. Retrieved August 28, 2020, from https://www.tirol.gv.at/fileadmin/themen/kunstkultur/landesarchiv/downloads/Erste_Zweite_Landesaufnahme-Tirol.pdf.

Ben-Rafael, E. (2009). A sociological approach to the study of the linguistic landscapes. In E. Shohamy & D. Gorter (Eds.), *Linguistic landscape: Expanding the scenery* (pp. 40–54). New York: Routledge.

Berezkina, M. (2016). Linguistic landscape and inhabitants' attitudes to place names in multicultural Oslo. In G. Puzey & L. Konstanski (Eds.), *Names and naming: People, places, perceptions and power* (pp. 106–119). Bristol, Buffalo, Toronto: Multilingual Matters.

Bergmann, H., & Jordan, P. (Eds.). (2010). *Geographische Namen—Vielfalt und Norm. 40 Jahre institutionalisierte Ortsnamenforschung und -standardisierung in Österreich, 65. Geburtstag von Isolde Hausner*. Wien: Praesens Verlag.

Bildungsdirektion Kärnten (Ed.) (2019). *Standorte zweisprachiger Schulen*. Retrieved May 1, 2020, June 15, 2020 from Bildung-ktn.gv.at/Minderheitenschulwesen/Standorte.html.

Breu, J. (1970). Die Behandlung der geographischen Namen in der österreichischen staatlichen Kartographie. In Bundesamt für Eich- und Vermessungswesen (Ed.), *Die amtliche Kartographie Österreichs* (pp. 153–174). Wien, Bundesamt für Eich- und Vermessungswesen.

Bundesamt für Eich- und Vermessungswesen (BEV) (Ed.). (2020). Austrian Map. Retrieved April 2, 2020, from http://www.austrianmap.at/.

Bundeskanzleramt der Republik Österreich (BKA). Retrieved March 17, 2020., from https://www.ris.bka.gv.at/.

Bundesministerium des Inneren (Ed.). (1999). *Erster Bericht der Bundesrepublik Deutschland gemäß Artikel 25 Absatz 1 des Rahmenübereinkommens des Europarates zum Schutz Nationaler Minderheiten*. Berlin.

Carleton, R. N. (2016). Fear of the unknown: One fear to rule them all? *Journal of Anxiety Disorders*, 41, 5–21.

Cenoz, J., & Gorter, D. (2006). Linguistic landscape and minority languages. *International Journal of Multilingualism*, 3(1), 67–80.

Český úřad zeměměřický a katastrální (ČÚZK) (Ed.). (2019). Retrieved May 26, 2020, from https://archivnimapy.cuzk.cz.

Coenders, M., & Scheepers, P. (2003). The effect of education on nationalism and ethnic exclusionism: An international comparison. *Political Psychology*, 24(2), 313–343.

D.K. (2005). Trinec-Trzyniec, obec - gmina, ne - tak? Trinecký Hutník/Hutnik *Trzyniecki*, 2(3), 2.

Dörflinger, J. (1986). Militärgeographisches Institut Wien. In I. Kretschmer & F. Wawrik (Eds.), *Lexikon zur Geschichte der Kartographie* (p. 93). Deuticke: Wien.

Eberhardt, P. (2015). *Ethnic groups and population changes in twentieth-century Central-Eastern Europe: History, data and analysis*. London, New York: Routledge.

Franssen, V., Dhont, K., & Hiel, A. V. (2013). Age-related differences in ethnic prejudice: Evidence of the mediating effect of right-wing attitudes. *Journal of Community & Applied Social Psychology*, 23(3), 252–257.

Grdina, I., & Stabej, M. (2002). Slowenisch. In M. Okuka & G. Krenn (Eds.), *Lexikon der Sprachen des europäischen Ostens (= Wieser Enzyklopädie des europäischen Ostens, 10)* (pp. 495–508). Wieser Verlag: Klagenfurt/Celovec, Wien, Ljubljana.

Harley, J. B. (1988). Maps, knowledge, and power. In D. Cosgrove & S. Daniels (Eds.), *The iconography of landscape* (pp. 277–312). Cambridge: Cambridge University Press.

Hiss, F. (2013). Tromsø as a "Sámi town"? Language ideologies, attitudes, and debates surrounding bilingual language policies. *Language Policy*, 12(2), 177–196.

Horizont. (2009). Štětcovou válku proti polským názvům čeští nacionalisté pořád nevzdávají. *Horizont*, 48(12), 1–2.

Hoxter, A. L., & Lester, D. (1994). Gender differences in prejudice. *Perceptual and Motor Skills*, 79(3 suppl), 1666.

Hugelmann, K. G. (1934). *Das Nationalitätenrecht des alten Österreich*. Wien, Leipzig: Braumüller.
Ingold, T. (2011). *Being alive. Essays on movement, knowledge, and description*. London, New York: Routledge.
Jordan, P. (1988). *Möglichkeiten einer stärkeren Berücksichtigung slowenischer Ortsnamen in den heutigen amtlichen topgraphischen Karten Österreichs*. Wien: Österreichische Akademie der Wissenschaften, Institut für Kartographie.
Jordan, P. (2005). The concept of the cultural region and the importance of coincidence between administrative and cultural regions. *Romanian Review of Regional Studies*, 1, 13–18.
Jordan, P., Kocsis, K., Eberhardt, P., Kahl, T., Klemenčić, M., Kicošev, S., Kowalski, M., Lozovanu, D., & Zupančič, J. (2006). Ethnic consciousness in central and Southeast Europe around 2000. In P. Jordan (Ed.), *Atlas of eastern and Southeastern Europe* (no. 2.9–2G9). Stuttgart: Borntraeger.
Jurič, M. (1987). Implikacija večinsko-manjšinske problematike na Koroškem glede na specifičnost po spolu. *Razprave in gradivo*, 20, 201–221.
K.k. statistische Central-Commission (Ed.). (1883). *Special-Orts-Repertorium von Kärnten/Specijalni Repertorij Krajev na Koroškem*. Alfred Hölder: Wien.
K.k. statistische Central-Commission (Ed.). (1894). *Special-Orts-Repertorium von Kärnten. Neubearbeitung auf Grund der Ergebnisse der Volkszählung vom 31. Dezember 1890/Specijalni Repertorij Krajev na Koroškem. Na novo predelan po rezultatih popisa ljudstva dne 31. decembra 1890*. Alfred Hölder: Wien.
Kearns, R. A., & Berg, L. D. (2002). Proclaiming place: Towards a geography of place name pronunciation. *Social and Cultural Geography*, 3(3), 283–302.
Kongres Polakow w Republice Czeskije. Retrieved May 15, 2020., from www.polonica.cz.
Köstler, H. J. (1986). Das Elektrostahlwerk in Ferlach (1909–1940). *Carinthia II*, 176(96), 43–56.
Kozler, P. (1853). *Zemljovid Slovenske dežele in pokrajin* [1:576,000]. S.l.
Kranzmayer, E. (1958). *Ortsnamenbuch von Kärnten. II. Teil: Alphabetisches Kärntner Siedlungsnamenbuch*. Klagenfurt: Verlag des Geschichtsvereines für Kärnten.
Kraus-Żurová, I. (2013, January 30). Nové tabule čtyřjazyčně. *Třinecký Hutník / Hutnik Trzyniecky*, p. 3.
Kronsteiner, O. (1974). *Die slowenischen Namen Kärntens in Geschichte und Gegenwart (= Österreichische Namenforschung, Sonderreihe 1)*. Wien: Österreichische Gesellschaft für Namenforschung.
Laboratoř geoinformatiky Fakulta životního prostředí Univerzity J.E. Purkyně. (2019). Retrieved May 26, 2020, from http://oldmaps.geolab.cz.

Landry, R., & Bourhis, R. Y. (1997). Linguistic landscape and ethnolinguistic vitality: An empirical study. *Journal of Language and Social Psychology*, 16(1), 23–49.

Lukan, W. (1978). Die slowenischen Wörterbücher. Ein historischer Abriß. *Österreichische Osthefte*, 20(1), 193–216.

Mácha, P., Lassak, H., & Krtička, L. (2018). City divided: Place names and nationalism in the Czech-polish borderlands. *Mitteilungen der Österreichischen Geographischen Gesellschaft*, 160, 303–329.

Macnamara, J. R. (2005). Media content analysis: Its uses, benefits and best practice methodology. *Asia Pacific Public Relations Journal*, 6(1), 1–34.

Map of Zaolzie (2009–2013). *Archive file with official communication regarding the map, expert opinions and newspaper articles covering the debate*. Congress of Poles, Český Těšín.

Marí Mayans, I. (2011). Policies governing the use of languages in relations between the authorities and the public. In M. Strubell & E. Boix-Fuster (Eds.), *Democratic policies for language revitalisation: The case of Catalan* (pp. 84–118). London: Palgrave Macmillan.

Mollova mapova sbírka. (2019). Retrieved May 26, 2020, from http://mapy.mzk.cz.

Nigrin, J. (1724). *Dvcatvs in Silesia Svperiore Teschinensis cum adjacentibus regnorum vicinorum, Hungariae videlicet et Poloniae nec non Marchionatus Moraviae etc.* [ca. 1:158,660]. Teschen. Retrieved June 20, 2020, from http://mapy.mzk.cz/mzk03/001/029/372/2619321452,sign.Moll-0001.876a.

Ormeling, F. (1983). *Minority Toponyms on maps. The rendering of linguistic minority Toponyms on topographic maps of Western Europe*. Utrecht: University of Utrecht.

Österreichische Rektorenkonferenz (Ed.). (1988). *Bericht der Arbeitsgruppe „Lage und Perspektiven der österreichischen Volksgruppen"*. Wien.

Piko-Rustia, M. (2016, April 25-29). *Slovene field and house names in Carinthia [Kärnten]*. Intangible cultural heritage of the Austrian Commission for UNESCO. Working Paper No. 11/16 of the 29th UNGEGN Session, Bangkok.

Pohl, H.-D. (2004). Sprache und Politik, gezeigt am Glottonym *Windisch*. In T. Krisch, T. Lindner, & U. Müller (Eds.), *Analecta homini universali dicata. Festschrift Oswald Panagl zum 65. Geburtstag (= Stuttgarter Arbeiten zur Germanistik, 421)* (pp. 625–636). Stuttgart: Hans-Dieter Heinz—Akademischer Verlag.

Puzey, G. (2009). Opportunity or Threat? The Role of Minority Toponyms in the Linguistic Landscape. In W. Ahrens, S. Embleton, & A. Lapierre (Eds.), *Names in multi-lingual, multi-cultural and multi-ethnic contact. Proceedings of the 23rd international congress of onomastic sciences August 17-22, 2008, York University, Toronto, Canada*. Toronto, York University. Retrieved from http://yorkspace.library.yorku.ca/xmlui/handle/10315/4022.

Radłowska-Obrusník, M. (2007). Pomoze swiadectwo starszej generacji. *Glos Ludu*, 23(6), 3.

Rajšp, V. (1998). *Slovenija na vojaškem zemljevidu 1763-1787, Karte, 4. zveszek*. Ljubljana: Znanstvenoraziskovalni center Slovenske akademije znanosti in umetnosti, Arhiv Republike Slovenije.

Ramovš, F. (1936). *Kratka zgodovina slovenskega jezika*. Ljubljana.

Riffe, D., Lacy, S., Fico, F., & Watson, B. (2019). *Analyzing media messages: Using quantitative content analysis in research*. London, New York: Routledge.

Rumpler, H. (2005). Die nationale Frage im Spannungsfeld von kärntnerischem Landespatriotismus, österreichischem Staatsbewusstsein und völkischem Nationalismus 1918-1938. In C. Fräss-Ehrfeld & H. Rumpler (Eds.), *Kärnten und Wien. Zwischen Staatsidee und Landesbewusstsein (= Kärnten und die nationale Frage, 4)* (pp. 9–82). Klagenfurt/Celovec, Ljubljana/Laibach, Wien/Dunaj: Hermagoras/Mohorjeva.

S.P. (2012). Názory pro a proti: proti. *Horizont*, 8(7), 2.

Sasínová, P. (2007). Vztahy jsou lepsí nez dríve. *Mladá Fronta Dnes*, 22(8), 3.

Schmiedt, F., Schmiedt, G., & Martin, F. (2020). *100 Jahre Unterkärnten in historischen Fotografien*. Erfurt: Sutton Verlag.

Scotton, C. M., & Urry, W. (1977). Bilingual strategies: The social functions of code-switching. *International Journal of the Sociology of Language*, 13, 5–20.

Sloboda, M., Simicic, L., Szabo-Gillinger, E., & Vigers, D. (2012). The policies on public signage in minority languages and their reception in four traditionally bilingual European locations. *Media and Communication Studies*, 63, 51–88.

Statistische Zentralkommission (Ed.). (1918). *Spezialortsrepertorium von Kärnten. Bearbeitet auf Grund der Ergebnisse der Volkszählung vom 31. Dezember 1910*. Wien.

Stranj, P. (1992). *The submerged community: An a to Ž of the Slovenes in Italy*. Trieste: Slovenski raziskovalni inštitut.

Svane, G. O. (1958). *Grammatik der slowenischen Schriftsprache*. Kopenhagen: Rosenkilde & Bagger.

Szabó-Gilinger, E., et al. (2012). Discourse coalitions for and against minority languages on signs: Linguistic landscape as a social issue. In D. Gorter, H. Marten, & L. Van Mensel (Eds.), *Minority languages in the linguistic landscape* (pp. 263–280). New York: Palgrave Macmillan.

Szpyrc, D. (2002). V Jablunkově se rozhořel spor o přepis nářečí. *Horizont*, 10(14), 3.

Třinecký hutník. (2005). Retrieved May 20, 2020, from https://www.yumpu.com/xx/document/view/6922380/trinecky-hutnik-c2005-29.

Vilfan, S. (2001). Historische Stereotype in der Alpen-Adria-Region. In A. Moritsch (Ed.), *Alpen-Adria. Zur Geschichte einer Region* (pp. 37–49). Klagenfurt/Celovec, Ljubljana/Laibach, Wien/Dunaj: Hermagoras Verlag/Mohorjeva založba.

Wieland, J. W. (1736). *Principatvs Sielasiae Teschinensis nova et accurata: Delinatio distincte instimmul exhibens Status minores vulgo Burglehn, dictos, Freystadt, Roy Reichwaldav, Bielitz, Friedeck, Devtschlevthen et Oderberg* [ca. 1:116,000]. Norimbergae: Excusa. ab Homannianis Heredibus. Retrieved June 20, 2020, from http://mapy.mzk.cz/mzk03/001/028/158/2619322319,sign.Moll-0001.815,20.

Williams, G. (2005). *Sustaining Language Diversity in Europe: Evidence from the Euromosaic Project (= Palgrave studies in minority languages and communities)*. London: Palgrave Macmillan.

Wolff, T. (2017). Nikt tego za nas nie zetrze. *Głos Ludu*, 27(7), 1.

Zdovc, P. (1974). Einige Aspekte zu Ortsnamenfragen in Kärnten. *Carinthia*, I(164), 283–303.

Zupančič, J. (1999). *Slovenci v Avstriji (= Geographica Slovenica, 32)*. Ljubljana: Inštitut za geografijo.

7

Comparative Interpretation of Research Results

Contributed by Peter Jordan and Přemysl Mácha

In addition to a long common history as parts of the Habsburg empire, striking similarities such as place-name conflicts and historical burdens such as military occupations by the mother country of the minority had been a major motivation for comparing the two minority situations. The longer, however, the research proceeded, the more it became obvious that similarities are balanced out by differences and that even similarities differ in detail in both regions. This chapter will discuss at first similarities and then differences, both in a sequence from historical, political, cultural, and geographical backgrounds toward place-name and more specifically minority place-name issues proceeding from politics across the legislative frameworks to place-name use, and the perceptions of and attitudes toward minority place names.

P. Jordan et al., *Place-Name Politics in Multilingual Areas*, Palgrave Studies in Minority Languages and Communities, https://doi.org/10.1007/978-3-030-69488-3_7

Similarities

1. Both minorities were parts of the Habsburg realm from the High Middle Ages and belonged to the 'Austrian half' of the Habsburg empire ('Cisleithania') from 1867 until the end of World War I. In consequence, both minorities were also subjected to similar minority policies and legislation. After the rise of national ideas in the course of the nineteenth century, the Habsburg empire conceived itself as a supranational state. Following the Austro-Hungarian Compromise in 1867, the 'Austrian half' in contrast to the Hungarian perpetuated this idea, administratively structuring the territory as a federation of rather autonomous crownlands. In this federation, Germans had the relative population majority and German was the dominant official language, but national and regional languages coexisted also enjoying official status, much in contrast to contemporary nation states like Hungary, Germany, Italy, or France. Minority rights including the right to official place names were substantial.
2. Industrialization of the nineteenth and twentieth centuries, more significant in the Czech case than in Carinthia, had an assimilating effect in the direction of the majority in both regions.
3. The interwar period saw Austria as well as Czechoslovakia as nation states conceived to be the possession of their dominant nation (Germans in Austria, Czecho-Slovaks in Czechoslovakia) justified in their efforts of national homogenization. Although required by the Paris peace treaties and the League of Nations, minority protection in both countries was not a priority and the implementation of minority legislation left much to be desired.
4. Nazi occupation and Germanization efforts with all kinds of suppressions affected both minority situations with the significant difference that the majority population in Carinthia and Austria (at least the largest part of it) cooperated with the regime, while in the Czech case the majority was also its victim.
5. Both minority regions were subject to claims and occupations by the 'mother country' of the minority. In Carinthia this occurred after World War I and after World War II, in the Těšín/Cieszyn region

after World War I and in the later interwar period. These events last as burdens on the two minority situations up to the present day, affecting relations between the local groups as well as making minority legislation a delicate political affair.

6. EU integration had a strong impact on both situations. On the Czech by at first stimulating/exerting positive pressure on minority legislation and later by open borders facilitating exchange between the minority and their mother country (Poland); on the Austrian mainly by invalidating the argument that minority rights and most specifically bilingual place names would mark a territory that could be claimed by a hostile neighbor (Yugoslavia, Slovenia).
7. Religion has played an important role in the rise and vitality of the minority's national consciousness in both regions. Churches were important public and social spaces where minority identity was expressed and affirmed. Church representatives were key members of the intellectual and political elite of the minority and they were instrumental as leaders of the social and cultural life of the minority while representing minority demands at regional and state levels. In the Carinthian case, it was and is the Roman-Catholic Church that served/serves as a minority 'haven' while in the Těšín/Cieszyn case it was the Lutheran Church.
8. Dialect(s) is/are the autochthonous minority language(s), while the minority's standard language (Slovene in Carinthia, Polish in the Těšín/Cieszyn region) is introduced, but represented in public space, including place names. Some minority members criticize that, others do not wish to see their (intimate) dialect names publicly exposed.
9. Neither minority is internally homogenous. Furthermore, other identities compete for the minority members' loyalty, so the situational mix of identities in any given moment is very fluid and often even contradictory. In the Carinthian case, the 'Windische' consider themselves not Slovene or even Carinthian Slovene. For them, their objective Slavic cultural roots do not necessarily imply a political loyalty toward, and a subjective identification with the Slovene minority in Carinthia. It therefore comes as no surprise that the 'Windische' do not support bilingual signs, and if they do, these had to be in the local dialect. In the Těšín/Cieszyn case, by contrast,

supraregional and supranational identities may fracture the Polish minority internally. The Gorol identity is a case in point. Shared by people in Czechia, Poland, and Slovakia and supported by 'objective' linguistic and cultural affinity arising from the traditions of mountain pastoralism, it crosscuts national identities in the Těšín/Cieszyn region while bringing together Czechs and Poles from the mountainous part of the region for common festivals and lived pastoral practices. At times, some Polish groups attempt to appropriate the Gorol traditions and present them as "Polish", but this is resisted by local Czechs who also feel Gorol. By contrast, Poles living in the lowland part of the region look down upon Gorols of any nationality.

10. In both cases the minority in the most comprehensive sense (including all sub-identities) is just defined by the dialect, not by a standard language. In Carinthia, all minority members—including the 'Windische'—would accept to speak one of the Carinthian Slovenian dialects, while at least the 'Windische' refer to German as their standard language, and in fact German is the standard language also for many more minority members. In the Těšín/Cieszyn region the local dialect "po naszymu" is the identity marker of the minority, while the minority is at the standard language level divided into Czech and Polish.
11. The minorities differ by group identity from their mother nations (although the latter claim that the minorities belonged to them): in Carinthia by dialect, regional (Carinthian) and national (Austrian) consciousness; in the Těšín/Cieszyn region by dialect, religious orientation (a strong presence of Lutheran denominations), and Czech national consciousness in many cases.
12. Silesian regional identity offers a certain way out of the national Czech-Polish dichotomy as Carinthian identity does related to the German-Carinthian/Carinthian Slovene, but contrary to Carinthian Silesian regional identity is much weaker due to frequent boundary changes and re-interpretations of the spatial concept.
13. Both minorities are politically divided, and competition between political factions has both positive and negative impacts on minority life. On the one hand, internal competition may weaken the negotiating position of the minority with the majority; on the other,

competition generates a larger offer of social and cultural events for the minority.

14. Both minorities are small by numbers and demographically declining. In both cases, one of the reasons for the decline is assimilation toward the majority. Additional reasons differ by region. In Carinthia, it is mainly out-migration from a predominantly rural and peripheric area which leads to the erosion of social networks and local dialects. In the Těšín/Cieszyn region, it is rather the historic in-migration of Czechs and other groups to a heavily industrialized region which reduced the singularity and coherence of the autochthonous minority.
15. The number/share of the minority was always a salient and intensively discussed criterion for the majority as well as for the minority—the minority questioning official population censuses, the majority demanding a "minority assessment" as a precondition for granting minority rights. Contrary to other minority situations (e.g., the Kanal Valley [Val Canale/Val Cjanâl/Kanalska dolina/Kanaltal] in Italy) hardly anybody ever raised the idea that a regional identity drawing from several sources and symbolized by a bi- or multilingual namescape could have an added value augmenting regional identity and that numbers or shares were not essential in this respect.
16. Traditional toponymy conforms to the local dialect. This causes problems insofar as standard language toponymy in public space as well as on official topographical maps prevails in Austria and Czechia. A part of the minority then feels not represented by minority place names converted to standard language versions.
17. Place-name conflicts including the vandalization of bilingual signs ('place-name storm' in Carinthia, 'paintbrush war' in the Těšín/Cieszyn region, to mention just the peak events) repeatedly popped up and were in the focus of public debates on minority and minority legislation.
18. Minority names on town signs are the politically most critical and contested, while they play a rather subordinate role as street names or on maps.
19. A significant share of the minority regards minority language including minority names as a private (intimate) matter, a matter of the family and between neighbors, and does not wish to go public with it.

20. Power relations today (not in the past and even not in the near past) work in favor of the minority's visibility in the linguistic landscape. The top-down direction (public sector) is by far dominating the bottom-up direction (private sector). Official policy today promotes the minority.
21. The public sector is the promotor of minority signs in public space, while the private sector lags behind. In general, minority signage is not a grassroots movement.
22. Public-sector minority signs tend to concentrate in the vicinity of symbolic places such as communal offices, schools, monuments, and at entrances to a populated place (town signs). Therefore, centers of populated places have both the highest density and the highest share of public-sector minority signs in the linguistic landscape.
23. Media from outside the region prefer distanced, pragmatic arguments when they discuss minority problems and minority rights, while media from the region—whether they are close to the majority or to the minority—prefer emotional, symbolic arguments.
24. Symbolic arguments are more commonly used by minority media which emphasize the emotional value of names and belonging, whereas majority media prefer pragmatic arguments.
25. Most media express either neutral or favorable opinions on bilingual signs.

Differences

1. While Carinthia is a very traditional historical-cultural region (supported by geomorphological conditions) with a strong regional consciousness shared by majority and minority, regional consciousness in the Czech case refers not to the minority region (the Těšín/Cieszyn region), but to Silesia in the wider sense. And even the concept of Silesia is faint, since it underwent several changes, including more recent history (1742, 1919). Thus, in the Carinthian case applying the ethnic principle on drawing political boundaries would contradict the regional principle much more than in the Czech case.

2. While in Carinthia a historical social stratification between current majority and minority has its lasting effects up to the present day, social differences between the current majority and minority are not significant in the Czech case. (The earlier German population of the region would be a different case.)
3. While the German language backed by social stratification in favor of German-speakers was the dominant in Carinthia at least from the end of the Middle Ages resulting in linguistic assimilation with social ascent, in the Czech situation no linguistic assimilation toward one of the current standard languages occurred up to 1919, since a third language (German) was dominant then, and Czech and Polish were equal in status in Austrian Silesia from 1867.
4. In the Czech case, the majority and the minority partly differ in their cultural-national identity. Most Poles, however, see themselves as equally Czech and Polish, so the identity difference is often situationally negotiated and becomes manifest only under certain circumstances. In Carinthia, by contrast, the majority as well as the minority affiliate themselves today to the Austrian cultural nation defined not by language, but by common history, and thus offering the inclusion of ethnic groups. Slovene-speakers in Austria are Austrians; Polish-speakers in the Těšín/Cieszyn region are both Czech and Polish.
5. In the decades after World War II up to 1989, inter-ethnic relations and minority affairs in Czechoslovakia were regarded under the auspices of 'Communist internationalism' conceiving ethnic and national consciousness as resulting from social inequalities to be overcome by egalitarian Socialism. The unfortunate consequence of this ideology was the neglect of minority problems and slow implementation of minority legislation. In contrast, postwar democratic Austria—up to 1955 under occupation, later neutral in military terms, but otherwise West-oriented—roughly adopted Western minority protection standards.
6. While in Carinthia the bilingual area was (and to some extent still is) characterized by an urban-rural ethnic dichotomy throughout the region, there is no such ethno-spatial differentiation present in the Těšín/Cieszyn region.

7. Already since the later nineteenth century, but even more so since World War II, tourism has been a much more important socio-economic factor in Carinthia than in the Těšín/Cieszyn region. It has significantly contributed to the assimilation of the minority and to German monolinguality in the linguistic landscape.
8. While the minority in the Těšín/Cieszyn region resides relatively compactly, the Carinthian minority extends over a wider area, today frequently interrupted by places without any minority population.
9. While the minority situation in Carinthia was the most prominent and most discussed in the First and Second Austrian Republic, the situation of the Polish minority in the Těšín/Cieszyn region remained in the shadow of the much more prominent situations of Germans and Hungarians in interwar and Slovaks and Hungarians in postwar Czechoslovakia.
10. While minority legislation started in postwar Austria immediately, although mainly due to external pressure and implemented rather reluctantly, in Czechia it was implemented to a remarkable extent only in the context of EU accession.
11. Group identities are more flexible and shifting in Carinthia than in the Těšín/Cieszyn region. There is a wide transition zone between "Carinthian Slovene" and "German-Carinthian" with additional choices like "Windisch", "Carinthian", or "Austrian", while in the Těšín/Cieszyn region just "Czech", "Pole", and eventually also "Silesian" are the alternatives.
12. While dialect use by the minority is declining in Carinthia and is substituted by the Slovenian standard language as an identity marker, the local dialect in the Těšín/Cieszyn region is widely used not only by the minority.
13. In the Czech case, various ethnic/linguistic groups played the role of the regional minority throughout the region's history, while this role has exclusively been associated with the Carinthian Slovenes at least since the later Middle Ages in Carinthia.
14. While the Carinthian minority situation and the conflicts around it tend to be widely reflected in the media and at times blow up to nationwide affairs, the Czech case has mostly remained of regional relevance, receiving little nationwide attention.

15. In Carinthia, no administrative reforms at the level of communes or districts were undertaken to weaken (at least as a hidden agenda) the minority. On the contrary, the partition of the Political District Klagenfurt-Land into Feldkirchen and Klagenfurt-Land with practically all minority members of the former Political District Klagenfurt-Land remaining in the new one let the minority's share rise significantly in the latter. It can also be assumed that the district courts Ferlach, Eisenkappel, and Bleiburg would have been closed down already, if they were not bilingual. In the Czech case, the partition of the District Český Těšín covering the entire minority region into the districts Frýdek-Místek and Karviná in 1960 was perceived by the minority as a deliberate attempt to weaken its demographic position and political representation.
16. While the Těšín/Cieszyn region had mainly due to industrialization from the later nineteenth century well into the Communist period a positive migration balance with not only Czechs, but also Slovaks and Roma migrating in, the migration balance of southern Carinthia is negative since times immemorial—both with ambivalent effects on the minorities. While economic prosperity resulting in in-migration to the Těšín/Cieszyn region meant on the one hand stable living conditions and no motivation for out-migration for the minority, their regional social standing, social coherence, and homogeneity were undermined by the newcomers on the other. In the Carinthian case, continued out-migration meant on the one hand a loss of minority members and an erosion of compact rural social and cultural structures, on the other less intrusion of external factors as it could have been in the opposite case. (Intrusion of external elements by tourism, second homes of urban dwellers a.s.o. was, however, strong enough.)
17. While Polish as the standard language of a larger part of the minority in the Těšín/Cieszyn region is rather the larger and more important language compared to Czech on the European and global scene, in the Carinthian situation, it is the opposite: The majority language is by far the larger and more important with the consequence that the motivation of majority members to learn the standard language of the minority, Slovene, has to draw from other motives.

18. While in Czechia place-name standardization is completely a top-down process controlled by central authorities, in Austria, with the exception of minority names of populated places, place-name standardization is in principle a bottom-up process. This opens in Austria, for example, opportunities for bilingual street naming by communes, bilingual commune naming by communes with the consent of the federal state, and even monolingual minority naming of individual buildings and properties up to monolingual minority naming of fields, mountains, or waters.
19. In the Czech case, official bilingual names of communes, populated places, streets, and public buildings may be introduced by decision of the communal assembly in communes with a minority share of 10% or more. However, the process is not automatic. The commune is required to introduce bilingual signs only when it is asked to do it by a resolution of the local council on national minorities or when it receives a petition from a local minority organization. Even when this condition is met, it is ultimately up to the commune to decide which signs and where signs will be rendered bilingually. By contrast, in Austria, federal law defines the list of officially bilingual names of populated places on the basis of a 17.5% share of the minority in the respective populated place without providing space for local participation in either direction. The Memorandum of 2011 opens, however, limited opportunities for communes to define additional names of populated places as bilingual.
20. In contrast to Austria, in Czechia the right on bilingual names is not fixed once and for ever, but it is tied to the results of the last two population censuses and whether the minority surpasses then the 10% threshold. In face of a decreasing minority and trends toward non-declaration of nationality and regional—for example, Silesian—affiliation this can mean an erosion of the right.
21. While in Carinthia majority and minority names are in most cases linguistically distinct enough to let their bilingual representation in public space or on maps appear reasonable, in the Těšín/Cieszyn region Czech, Polish and dialectal names often look and sound very similar differentiated only by orthographies or pronunciation. This

may let their bilingual representation appear as a waste of money and an academic fingering exercise.

22. While in Carinthia a significant share of the minority regards minority language including minority names as a private (intimate) matter, a matter of the family and between neighbors and does not wish to go public with it, no empirical evidence for such an attitude could be found in the Czech case.
23. Polish tourists are an economically valid argument for bilingual Czech-Polish signage in the Těšín/Cieszyn region due to intensive Polish shopping tourism already since the fall of Communism, but even more intensive after Czech and Polish EU accession in 2004 and the complete opening of the border. In the Carinthian case, generating markets of tourism are in the first line German-speaking countries (Germany, Austria) and other countries not speaking Slovene, while Slovenia plays only a small role in this respect. Thus, there is from this perspective hardly a motivation for German-Slovene bilingualism in public space. If, however, a multicultural milieu is understood as touristically attractive in its own right (also for non-Slovene speakers), bilingualism in public space in Carinthia could also be motivated economically.

8

Conclusions

Contributed by Peter Jordan and Přemysl Mácha

Our principal research questions were:

1. What do place names mean for the identity of human communities in general and more specifically for linguistic minorities? What is the relationship between language, place, and identity, and how do we make ourselves at home through place names?
2. What toponymic strategies have been employed by different actors in establishing, maintaining, and subverting ethnic/national boundaries, and what are the principal social forces structuring the contemporary toponymic landscape and everyday toponymic practice?
3. How are the multilingual/multiethnic city-text and linguistic landscape produced, performed, interpreted, and contested?
4. When we speak of minority rights and cultural preservation, what role do place names play in this discussion? Why, how, by what means and procedures, by whom, and for whom should place names be protected?

P. Jordan et al., *Place-Name Politics in Multilingual Areas*, Palgrave Studies in Minority Languages and Communities, https://doi.org/10.1007/978-3-030-69488-3_8

The following conclusions drawn from our research may provide some answers with the ambition to be applicable to some other minority situations and exceptionally also to be generalized. They are presented in a thematically logical sequence proceeding from the general to the specific and ending with what may also be read as recommendations for minority policies and legislation. They may repeat some of the "similarities" and "differences" outlined before but do this, because they are conceived as a selection of most relevant issues worth to be finally highlighted even if they don't apply to both situations.

1. **Place names play essential roles in relating humans and geographical space and are in this respect specifically important for linguistic minorities.** General roles of place names in relating humans and geographical space are: (1) Place names often highlight characteristics of space important for a certain community, and thus they reflect a human community's perception of space. (2) Place names mark the territory of a community. (3) Place names structure space mentally. (4) Place names support emotional ties between people and place and thus promote space-related identity building. Two of these roles are specifically important for linguistic minorities: The role of marking the territory of a community and the role of supporting emotional ties.
2. **Place-name conflicts in minority areas are—as place-name conflicts in general—just the symbolic surface of deeper societal problems.** In the Těšín/Cieszyn region, the antagonists in this respect are national-minded Czechs and the local minority mostly speaking a specific Slavonic dialect and conceiving themselves as an identity group. The deeper reason for this conflict is that Czechs are looking at this minority as Poles claiming disproportionate cultural rights while feeling reminded of territorial claims made by Poland in the interwar period. In Carinthia, the antagonists are the majority of German-Carinthians and the minority of Carinthian Slovenes. The deeper reasons for this conflict are that Carinthian Slovenes don't feel adequately recognized by the majority as a traditional autochthonous community and as having essentially contributed to common Carinthian culture as well as that German-Carinthians are afraid of

a historical land's partition and of losing control over a part of it ("Carinthian primeval fear").

3. **Industrialization, tourism, and suburbanization have a detrimental impact on autochthonous, usually rural cultural minorities and in consequence also on the preservation of minority place names.** This may in the first line be attributed to the fact that they mean the intrusion of others into so far closed local societies, to external orientation of locals and—voluntary or not—acceptance of external power.
4. **Minority place names on town signs are politically the most sensitive and the most frequent target of vandalization.** This is a finding that contradicts Jordan (1988) who argued that state resistance toward bilingual names on maps (even as compared to bilingual signs in public space) and consequently an inclination toward avoiding them on official topographical maps is based on the ground that minority town signs are noticed only by those who travel to the place while maps convey an impression of a minority area at a glance and thus potentially weaken the state's territorial claim to this area. However, our media analysis, interviews, and questionnaires clearly demonstrated that rather the opposite is true: Town signs are visible for everybody, are perceived as very symbolic and marking the identity of a place and are therefore also in the focus of political action, media and public discourse, and vandalization. By comparison, maps mostly attract relatively little attention, usually just among a rather limited, better-educated fraction of the society. However, if in particular official topographical maps are not so much regarded as means of daily communication (as reflected by our respondents), but in their other function as official documents of political relevance, sensitivity of minority place-name representation on them may indeed be higher.
5. **Signs of the private sector in public space tend to be respected as expressions of personal attitudes, while signs of the public sector are implicitly an expression of the community represented by this public authority and arouse opposition, when a part of the community does not feel to be represented by these signs.**

6. **Minority names in public space have a high symbolic charging and concentrate therefore in symbolic sites like centers of populated places, places of worship, or representative buildings.**
7. **The main promoters of minority names in public space are public authorities, while private initiatives—even from the minority—remain less common.** This can be explained by the intention not to arouse conflict bearing in mind that the minority is in the weaker position. It also shows that the conflict over bilingual signs is a struggle over the control of the public sphere leaving the private sphere the freedom to choose the language as it sees fit.
8. **The younger and the better educated have a more positive attitude toward bilingual town signs and bilingual signage in general.**
9. **In historically burdened situations it is difficult to conduct interviews about cultural characteristics and identity.** The interviewer is in danger of being categorized by the respondent and attributed to one of the 'camps' resulting in receiving responses the interviewer is presumed to expect. Respondents tend either to repeat one of the standard opinions circulating in the public instead of reflecting on the issue themselves or to refrain from answering out of fear of being politically categorized.
10. **Even when dialect names are vital elements of the spoken language, speakers do not necessarily want to see them represented in public space.** This is equally true for the minority and the majority and can stem from several reasons: the intimate character of dialect names, the assumption that the public sphere requires standard language, and the lack of practice in writing dialect in general.
11. **Etymologies and stories associated with places are not needed to achieve a high emotional value of place names.** Meanings and 'stories' behind place names very often fall into oblivion and the name continues to function just as a label—with only a few people knowing what it means, how it came into being, and so on. Nevertheless, its meaning can become relevant and heavily contested overnight. A case in point is commemorative street names: We use them frequently without wasting a thought on what they mean (or of whom or what they should remind us), but a political change can cause their exchange almost overnight.

12. **Pronunciation, intonation, choice of the dialect, or standard language variant of a place name has a strong impact on its perception.** Pronouncing an endonym, for example, in a strange way not in accordance with local use can result in wider emotional distance than using an exonym. While the dialect name is the real endonym in the sense of the name that supports emotional ties with the place, the name in the standard language is frequently conceived as already alienated, rather as an exonym than an endonym.
13. **The existence of dialectal names may create additional challenges, especially for dialects which do not have standard orthography.** Writing down a dialectal name may in itself become a political matter as the choice of orthography can be interpreted as signaling group membership. This commonly happens with dialectal names in the Těšín/Cieszyn region, both in their writing as well as in their pronunciation. The poetics of place names includes all these aspects which often generate stronger emotional reaction than the name itself. This also applies to regions with languages written in different alphabets or writing systems. Special sensitivity is needed to accommodate the wishes of minorities in such cases.
14. **Neither the local majority nor the local minority regard minority names in public space as an asset for tourism.** The idea that multicultural structure and corresponding visual appearance could be assets for tourism sounds strange even to most minority members.
15. **Indicators of an adequate minority place-name representation on topographical maps are the following:**
    (a) Relative share of minority names and their spelling according to the orthography of the minority or majority language.
    (b) Consistency across feature categories: Does a certain mode of minority-name rendering only affect certain feature categories or all of them?
    (c) Consistency across map scales: Do minority names only or predominantly occur in the largest scale or are they—in line with generalization—also reflected in derived scales?
    (d) Visual representation of minority in relation to majority names: Are both represented in the same font size and type? Are they separated by a slash or is the minority name set in brackets?

16. **Private maps such as tourist and hiking maps or maps published by communes frequently substitute the role of the state in supporting the representation of minority names on maps.** Notable and highly inspirational in this respect are the efforts of the Urban Jarnik Institute in Klagenfurt, which has in cooperation with communes published several bilingual maps of the minority area in Carinthia, or the communal activities in the Těšín/Cieszyn region incorporating traditional names into official maps. Civil society, private actors, and communes can thus contribute greatly to the promotion of tolerance and understanding at the local and regional levels.
17. **Strong regional, historical-cultural space-related identities have the potential of mitigating (antagonistic) national identities.** Common landscape/region-related, that is, umbrella identities crosscut ethnic and national identities and work thus in favor of interethnic relations and peace in the region.
18. **In minority situations burdened by (historical) social stratification and historical conflict minority members hesitate to expose their identity by minority names for their place or house for various reasons:** (a) They regard their language as secondary to the dominant language or not as adequate for official communication. (b) They regard their names as intimate. (c) They don't wish to have their names in local dialect transformed to the standard language versions as it would be conventional with signs in public space. (d) They do not want to stir up conflict (with neighbors). (e) They do not want to be "stigmatized".
19. **Historical situations of dominance and subordination can be perpetuated by estimating one's own language or place names as "less important", "less attractive", or "less suitable to be exposed to the public".**
20. **There is a gap between exponents and political representatives of a minority and the wider group of minority members as far as minority rights and place-name representation in the public sphere are concerned.** While minority representatives display a kind of activism (in competition among themselves and minority organizations), 'normal' minority members frequently regard their identity (expressed also by place names) rather as private.

21. **Attributing the responsibility for minority legislation in general and for minority place names in particular to the state level has the advantage of exempting them from local and regional disputes. But this needs careful preparations, the involvement of all relevant regional and local actors, and a special assessment of local acceptance.**
22. **Minority rights (including the right to minority place names) dependent on a flexible share of the minority in the total population and subject to shifts from census to census have the potential of keeping a conflict alive instead of solving the question once and for ever.** Regulations of this kind attribute also too much weight to shares and thresholds for minority legislation while disregarding the contribution of the minority as such (independent of numbers and shares) to the culture of a region.

Our research uncovered very intimate links between names, place, and identity. We observed these links consistently in archival sources, media, linguistic landscapes, interviews, and questionnaires in both of our study regions. It is clear now that names are everything but arbitrary labels. They anchor and orient, connect and divide, and claim and share. Paradoxically, however, we rarely think about them. They are so deeply woven into the fabric of our everyday lives that we only notice them when something about them suddenly changes. An example of such a change is the appearance of minority signs in a previously monolingual linguistic landscape. Suddenly all that had been taken for granted has been put into doubt and the identity of the place as well as the identity of the observing individual is profoundly shaken.

In such a situation, the gut reaction of many members of the majority often is: "They are taking my home away from me!" And the response may be the vandalization of the sign. From the perspective of the minority, however, the gut reaction may be the exact opposite: "I am finally at home here. No one can tell me anymore that I am a stranger to this place!" Both reactions are deeply emotional and formally rational arguments put forth for or against minority signs often only mask these basic emotions. There is nothing illegitimate or shameful about them, and they have to be figured in and built upon in any negotiations that take place between representatives of the majority and the minority. And as with any emotions, they can be harnessed constructively or destructively, depending on the good will of the people who participate in the negotiations.

Our book cannot provide a manual for such negotiations because every situation is unique if only due to the uniqueness of the people taking part in it. What our book can, and has attempted to do, however, is to offer experiences from two areas which have gone through years of painful conflicts and which have finally arrived at alternative solutions. Much can be learned from these results and even more from the process itself. We express hope that our book will serve as an inspiration for researchers and practitioners searching for solutions to similar conflicts in other countries and regions. And we look forward to learning from their experiences when they decide to share them.

Reference

Jordan, P. (1988). *Möglichkeiten einer stärkeren Berücksichtigung slowenischer Ortsnamen in den heutigen amtlichen topographischen Karten Österreichs*. Wien: Österreichische Akademie der Wissenschaften.

Appendices

Appendix A: List of Bilingual German-Slovene Settlements [Ortschaften] in the Federal Province of Carinthia According to Law No. 46/2011 (the First Name Is the German, the Second Behind the Slash the Slovene)

This list is structured by political districts and communes in alphabetical order. The long name versions of political districts and communes are first presented in English and then in German (in rectangular brackets).

Political District Hermagor [Politischer Bezirk Hermagor]

Urban Commune Hermagor-Pressegger See [Stadtgemeinde Hermagor-Pressegger See]

Dellach/Dole
Potschach/Potoče

P. Jordan et al., *Place-Name Politics in Multilingual Areas*, Palgrave Studies in Minority Languages and Communities, https://doi.org/10.1007/978-3-030-69488-3

Political District Klagenfurt-Land [Politischer Bezirk Klagenfurt-Land]

Market Commune Ebenthal in Kärnten [Marktgemeinde Ebenthal in Kärnten]

Kossiach/Kozje
Kreuth/Rute
Lipizach/Lipica
Radsberg/Radiše
Schwarz/Dvorec
Tutzach/Tuce
Werouzach/Verovce

Market Commune Feistritz im Rosental [Marktgemeinde Feistritz im Rosental]

Hundsdorf/Podsinja vas
Sankt Johann im Rosental/Šentjanž v Rožu

Urban Commune Ferlach [Stadtgemeinde Ferlach]

Bodental/Poden
  Loibltal/Brodi
  Strugarjach/Strugarje
  Tratten/Trata
  Waidisch/Bajdiše
  Windisch Bleiberg/Slovenji Plajberk

Commune Köttmannsdorf [Gemeinde Köttmannsdorf]

Neusaß/Vesava
Plöschenberg/Plešivec

Commune Ludmannsdorf [Gemeinde Ludmannsdorf]

Bach/Potok
Edling/Kajzaze
Fellersdorf/Bilnjovs
Franzendorf/Branča vas
Großkleinberg/Mala Gora
Ludmannsdorf/Bilčovs
Lukowitz/Koviče
Moschenitzen/Moščenica
Muschkau/Muškava
Niederdörfl/Spodnja vesca
Oberdörfl/Zgornja vesca
Pugrad/Podgrad
Rupertiberg/Na Gori
Selkach/Želuče
Strein/Stranje
Wellersdorf/Velinja vas
Zedras/Sodraževa

Market Commune Schiefling [Marktgemeinde Schiefling]

Techelweg/Holbiče

Commune Sankt Margareten im Rosental [Gemeinde Sankt Margareten im Rosental]

Trieblach/Treblje

Commune Zell [Gemeinde Zell]

Zell-Freibach/Sele-Borovnica
Zell-Homölisch/Sele-Homeliše
Zell-Koschuta/Sele-Košuta

Zell-Mitterwinkel/Sele-Srednji Kot
Zell-Oberwinkel/Sele-Zvrhnji Kot
Zell-Pfarre/Sele-Cerkev
Zell-Schaida/Sele-Šajda

Political District Villach-Land [Politischer Bezirk Villach-Land]

Market Commune Arnoldstein [Marktgemeinde Arnoldstein]

Hart/Ločilo

Market Commune Finkenstein am Faakersee [Marktgemeinde Finkenstein am Faakersee]

Goritschach/Zagoriče
Oberferlach/Zgornje Borovlje
Petschnitzen/Pečnica
Sigmontitsch/Zmotiče
Susalitsch/Žužalče
Unterferlach/Spodnje Borovlje
Untergreuth/Spodnje Rute

Commune Hohenthurn [Gemeinde Hohenthurn]

Achomitz/Zahomec

Market Commune Rosegg [Marktgemeinde Rosegg]

Frög/Breg
Raun/Ravne

Market Commune Sankt Jakob im Rosental [Marktgemeinde Sankt Jakob im Rosental]

Frießnitz/Breznica
Greuth/Rute
Kanin/Hodnina
Lessach/Leše
Maria Elend/Podgorje
Mühlbach/Reka
Sankt Jakob im Rosental/Šentjakob v Rožu
Sankt Peter/Šentpeter
Srajach/Sreje
Tösching/Tešinja

Market Commune Velden am Wörthersee [Marktgemeinde Velden am Wörthersee]

Pulpitsch/Pulpače
Treffen/Trebinja

Political District Völkermarkt [Politischer Bezirk Völkermarkt]

Urban Commune Bleiburg [Stadtgemeinde Bleiburg]

Aich/Dob
Bleiburg/Pliberk
Dobrowa/Dobrova
Draurain/Brege
Ebersdorf/Drveša vas
Einersdorf/Nonča vas
Kömmel/Komelj
Kömmelgupf/Komeljski Vrh
Loibach/Libuče
Moos/Blato
Replach/Replje

Rinkenberg/Vogrče
Rinkolach/Rinkole
Ruttach/Rute
Sankt Georgen/Šentjur
Sankt Margarethen/Šmarjeta
Schilterndorf/Čirkovče
Wiederndorf/Vidra vas
Woroujach/Borovje

Market Commune Eberndorf [Marktgemeinde Eberndorf]

Buchbrunn/Bukovje
Eberndorf/Dobrla vas
Edling/Kazaze
Gablern/Lovanke
Gösselsdorf/Goselna vas
Hof/Dvor
Mökriach/Mokrije

Market Commune Eisenkappel-Vellach [Marktgemeinde Eisenkappel-Vellach]

Bad Eisenkappel/Železna Kapla
Blasnitzen/Plaznica
Ebriach/Obirsko
Koprein Petzen/Pod Peco
Koprein Sonnseite/Koprivna
Leppen/Lepena
Lobnig/Lobnik
Rechberg/Rebrca
Remschenig/Remšenik
Trögern/Korte
Unterort/Podkraj
Vellach/Bela
Weißenbach/Bela
Zauchen/Suha

Market Commune Feistritz ob Bleiburg [Marktgemeinde Feistritz ob Bleiburg]

Dolintschitschach/Dolinčiče
Feistritz ob Bleiburg/Bistrica pri Pliberku
Gonowetz/Konovece
Hinterlibitsch/Suha
Hof/Dvor
Lettenstätten/Letina
Penk/Ponikva
Pirkdorf/Breška vas
Ruttach-Schmelz/Rute
Sankt Michael ob Bleiburg/Šmihel pri Pliberku
Tscherberg/Črgoviče
Unterlibitsch/Podlibič
Unterort/Podkraj
Winkel/Kot

Commune Gallizien [Gemeinde Gallizien]

Drabunaschach/Drabunaže
Enzelsdorf/Encelna vas
Freibach/Borovnica

Commune Globasnitz [Gemeinde Globasnitz]

Globasnitz/Globasnica
Jaunstein/Podjuna
Kleindorf/Mala vas
Podrain/Podroje
Slovenjach/Slovenje
Sankt Stefan/Šteben
Traundorf/Strpna vas
Tschepitschach/Čepiče
Unterbergen/Podgora
Wackendorf/Večna vas

Commune Neuhaus [Gemeinde Neuhaus]

Graditschach/Gradiče
Hart/Breg
Heiligenstadt/Sveto mesto
Kogelnigberg/Kogelska Gora
Oberdorf/Gornja vas
Schwabegg/Žvabek
Unterdorf/Dolnja vas

Commune Sankt Kanzian am Klopeiner See [Gemeinde Sankt Kanzian am Klopeiner See]

Grabelsdorf/Grabalja vas
Horzach I/Horce I
Horzach II/Horce II
Lauchenholz/Gluhi Les
Mökriach/Mokrije
Nageltschach/Nagelče
Obersammelsdorf/Žamanje
Sankt Primus/Šentprimož
Sankt Veit im Jauntal/Šentvid v Podjuni
Unternarrach/Spodnje Vinare
Vesielach/Vesele

Commune Sittersdorf [Gemeinde Sittersdorf]

Goritschach/Goriče
Kleinzapfen/Malčape
Kristendorf/Kršna vas
Müllnern/Mlinče
Obernarrach/Zgornje Vinare
Pogerschitzen/Pogrče
Rückersdorf/Rikarja vas
Sagerberg/Zagorje

Sittersdorf/Žitara vas
Sonnegg/Ženek
Tichoja/Tihoj

Appendix B: Comparison of the Four Regulations for Official Bilingual Names of Populated Places in Carinthia 1972–2011

This list is structured by political districts and communes in alphabetical order. The long name versions of political districts and communes are first presented in English and then in German (in rectangular brackets).

Columns from left to right:

(1) 1972: Official bilingual place names according to the Law of 1972 (later withdrawn) with a threshold of 20% according to population census 1971
(2) 1977: Official bilingual place names according to the Decree on Topography for Carinthia as of 1977 (BGBl. 306/1977) with a threshold of 25% according to population census 1951
(3) 2001: Official bilingual place names according to the decision of the Constitutional Court of Justice as of 2001 recommending a threshold of 10% according to recent population censuses (not effectuated according to this threshold). If the requirement of 10% is not met by all population censuses after World War II, the last population census complying to this criterion is mentioned (e.g., 1961).
(4) 2011: Official bilingual place names according to Law No. 46/2011 with a threshold of 17.5% according to the population census 2001 (see Appendix A)

Sources

Arbeitsgruppe des Rates der Kärntner Slowenen zur „Zweisprachigen Topographie" (ed.). (2011). *10 Jahre Ortstafelerkenntnis. Die zweisprachigen Aufschriften in Kärnten/Koroška—eine Information.*

Klagenfurt/Celovec: Narodni svet koroških Slovencev/Rat der Kärntner Slowenen.
Bundeskanzleramt der Republik Österreich (BKA). Retrieved from https://www.ris.bka.gv.at/ Law No. 46/2011

Political District Hermagor [Politischer Bezirk Hermagor]

Urban Commune Hermagor-Pressegger See [Stadtgemeinde Hermagor-Pressegger See]

Name of the populated place	1972 (20%)	1977 (25%)	2001 (10%)	2011 (17.5%)
Brugg/Moste	x		x	
Dellach/Dole			x	x
Fritzendorf/Limarče	x		1971	
Latschach/Loče			x	
Mellach/Mele	x		1961	
Mellweg/Melviče			x	
Nampolach/Napole	x		1961	
Potschach/Potoče	x		x	x

Political District Klagenfurt-Land [Politischer Bezirk Klagenfurt-Land]

Market Commune Ebenthal in Kärnten [Marktgemeinde Ebenthal in Kärnten]

Name of the populated place	1972 (20%)	1977 (25%)	2001 (10%)	2011 (17.5%)
Berg/Rute pri Medgorjah			x	
Kosasmojach/Kozasmoje			x	
Kossiach/Kozje	x	X	x	x
Kreuth/Rute	x	X	x	x
Lipizach/Lipica	x	X	x	x
Moosberg/Kajže	x		x	
Radsberg/Radiše	x	X	x	x
Schwarz/Dvorec	x	X	x	x
Tutzach/Tuce	x	X	x	x
Werouzach/Verovce	x	X	x	x

Market Commune Feistritz im Rosental [Marktgemeinde Feistritz im Rosental]

Name of the populated place	1972 (20%)	1977 (25%)	2001 (10%)	2011 (17.5%)
Hundsdorf/Podsinja vas			x	x
Matschach/Mače			x	
Rabenberg/Šentjanške Rute			x	
St. Johann im Rosental/Šentjanž v Rožu	x		x	x
Suetschach/Sveče			x	

Urban Commune Ferlach [Stadtgemeinde Ferlach]

Name of the populated place	1972 (20%)	1977 (25%)	2001 (10%)	2011 (17.5%)
Bodental/Poden		X	x	x
Dörfl/Kajže			x	
Glainach/Glinje	x		x	
Laak/Loka	x		x	
Loibltal/Brodi	x	X	x	x
Otrouza/Otrovca	x		1961	
Rauth/Rute	x		x	
Seidolach/Ždovlje			x	
Singerberg/Žingarca			x	
Strugarjach/Strugarje		X	x	x
Tratten/Trata			x	x
Unterglainach/Vesca	x		x	
Waidisch/Bajdiše	x		x	x
Windisch Bleiberg/Slovenji Plajberk	x	X	x	x

Commune Grafenstein [Gemeinde Grafenstein]

Name of the populated place	1972 (20%)	1977 (25%)	2001 (10%)	2011 (17.5%)
Sand/Prod	x		1961	

Commune Keutschach [Gemeinde Keutschach]

Name of the populated place	1972 (20%)	1977 (25%)	2001 (10%)	2011 (17.5%)
Dobein/Dobajna	x		x	
Höflein/Dvorec	x		x	
Höhe/Gora	x		1961	
Keutschach/Hodiše	x		1971	
Pertitschach/Prtiče			x	
Plaschischen/Plašišče	x		x	
Plescherken/Plešerka	x		x	
Rauth/Rut	x		1991	
Reauz/Rjavec	x		1961	
Sankt Margarethen/ Šmarjeta	x		x	
Sankt Nikolai/Šmiklavž			x	

Commune Köttmannsdorf [Gemeinde Köttmannsdorf]

Name of the populated place	1972 (20%)	1977 (25%)	2001 (10%)	2011 (17.5%)
Gaisach/Čežava	x		x	
Göriach/Gorje	x		1971	
Hollenburg/Humberk			x	
Neusaß/Vesava	x		x	x
Plöschenberg/Plešivec	x		x	x
Preliebl/Preblje	x		1981	
Sankt Kandolf/Šent Kandolf	x		x	
Sankt Margarethen/ Šmarjeta			x	
Schwanein/Zvonina	x		1961	
Trabesing/Trabesinje	x		x	
Tschachoritsch/Čahorče	x		x	
Wurdach/Vrdi	x		1991	

Commune Ludmannsdorf [Gemeinde Ludmannsdorf]

Name of the populated place	1972 (20%)	1977 (25%)	2001 (10%)	2011 (17.5%)
Bach/Potok	x	X	x	x
Edling/Kajzaze	x	X	x	x
Fellersdorf/Bilnjovs	x	X	x	x
Franzendorf/Branča vas	x	X	x	x
Großkleinberg/Mala Gora		X	x	x

Name of the populated place	1972 (20%)	1977 (25%)	2001 (10%)	2011 (17.5%)
Ludmannsdorf/Bilčovs	x	X	x	x
Lukowitz/Koviče	x	X	1991	x
Moschenitzen/Moščenica	x	X	x	x
Muschkau/Muškava	x	X	x	x
Niederdörfl/Spodnja vesca	x	X	x	x
Oberdörfl/Zgornja vesca	x	X	x	x
Pugrad/Podgrad		X	x	x
Rupertiberg/Na Gori	x	X	x	x
Selkach/Želuče	x	X	x	x
Strein/Stranje	x	X	x	x
Wellersdorf/Velinja vas	x	X	x	x
Zedras/Sodraževa	x	X	x	x

Commune Maria Rain [Gemeinde Maria Rain]

Name of the populated place	1972 (20%)	1977 (25%)	2001 (10%)	2011 (17.5%)
Obertöllern/Zgornje Dole			x	
Saberda/Zabrda			x	

Commune Poggersdorf [Gemeinde Poggersdorf]

Name of the populated place	1972 (20%)	1977 (25%)	2001 (10%)	2011 (17.5%)
Eibelhof/Ovčjak	x		1961	

Commune Sankt Margareten im Rosental [Gemeinde Sankt Margareten im Rosental]

Name of the populated place	1972 (20%)	1977 (25%)	2001 (10%)	2011 (17.5%)
Dobrowa/Dobrava	x		1991	
Gupf/Vrh			x	
Homölisch/Homeliše	x		x	
Niederdörfl/Dolnja vas			x	
Sabosach/Zavoze			x	
Sankt Margareten im Rosental/ Šmarjeta v Rožu			x	
Seel/Selo			x	
Trieblach/Treblje			x	x

Market Commune Schiefling [Marktgemeinde Schiefling]

Name of the populated place	1972 (20%)	1977 (25%)	2001 (10%)	2011 (17.5%)
Ottosch/Otož	x		x	
Raunach/Ravne	x		1961	
Sankt Katrein/Podjerberk	x		x	
Techelweg/Holbiče	x		x	x

Commune Zell [Gemeinde Zell]

Name of the populated place	1972 (20%)	1977 (25%)	2001 (10%)	2011 (17.5%)
Zell-Freibach/Sele-Borovnica	x	X	x	x
Zell-Homölisch/Sele-Homeliše	x	X	x	x
Zell-Koschuta/Sele-Košuta		X	?	x
Zell-Mitterwinkel/Sele-Srednji Kot	x	X	x	x
Zell-Oberwinkel/Sele-Zvrhnji Kot	x	X	x	x
Zell-Pfarre/Sele-Cerkev	x	X	x	x
Zell-Schaida/Sele-Šajda	x	X	x	x

Political District Villach-Land [Politischer Bezirk Villach-Land]

Market Commune Arnoldstein [Marktgemeinde Arnoldstein]

Name of the populated place	1972 (20%)	1977 (25%)	2001 (10%)	2011 (17.5%)
Hart/Ločilo	x		x	x
Krainberg/Strmec	x		1971	
Sankt Leonhard bei Siebenbrunn/Šent Lenart pri Sedmih studencih	x		1991	

Market Commune Finkenstein am Faakersee [Marktgemeinde Finkenstein am Faakersee]

Name of the populated place	1972 (20%)	1977 (25%)	2001 (10%)	2011 (17.5%)
Altfinkenstein/Stari Grad			x	
Goritschach/Zagoriče			x	x
Kopein/Kopanje	x		1961	
Latschach/Loče			x	
Oberferlach/Zgornje Borovlje	x		1971	x
Obertechanting/Zgornje Teharče	x		1971	
Petschnitzen/Pečnica	x		x	x
Ratnitz/Ratenče			x	
Sigmontitsch/Zmotiče			x	x
Susalitsch/Žužalče	x		x	x
Unteraichwald//Spodnje Dobje	x		x	
Unterferlach/Spodnje Borovlje	x		x	x
Untergreuth/Spodnje Rute			x	x

Commune Hohenthurn [Gemeinde Hohenthurn]

Name of the populated place	1972 (20%)	1977 (25%)	2001 (10%)	2011 (17.5%)
Achomitz/Zahomec	x		x	x
Dreulach/Drevlje			x	

Market Commune Rosegg [Marktgemeinde Rosegg]

Name of the populated place	1972 (20%)	1977 (25%)	2001 (10%)	2011 (17.5%)
Dolintschach/Dolinčič			x	
Duel/Dole	x		1981	
Frög/Breg			x	x
Frojach/Broje			x	
Obergoritschach/Zgornje Goriče			x	
Prik/Brezje	x		1971	
Raun/Ravne	x		x	x
Sankt Johann/Ščedem	x		x	

Market Commune Sankt Jakob im Rosental [Marktgemeinde Sankt Jakob im Rosental]

Name of the populated place	1972 (20%)	1977 (25%)	2001 (10%)	2011 (17.5%)
Dreilach/Dravlje			x	
Feistritz/Bistrica	x		x	
Frießnitz/Breznica	x		x	x
Gorintschach/Gorinčiče	x		x	
Greuth/Rute	x		x	x
Kanin/Hodnina	x		x	x
Längdorf/Velika vas	x		x	
Lessach/Leše	x		x	x
Maria Elend/Podgorje			x	x
Mühlbach/Reka	x		x	x
Sankt Jakob im Rosental/ Šentjakob v Rožu	x		x	x
Sankt Peter/Šentpeter	x		x	x
Schlatten/Svatne	x		x	
Srajach/Sreje	x		x	x
Tösching/Tešinja	x		x	x
Winkl/Kot	x		x	

Market Commune Velden am Wörthersee [Marktgemeinde Velden am Wörthersee]

Name of the populated place	1972 (20%)	1977 (25%)	2001 (10%)	2011 (17.5%)
Augsdorf/Loga vas	x		1971	
Dieschitz/Deščice	x		x	
Latschach/Loče	x		x	
Pulpitsch/Pulpače	x		x	x
Treffen/Trebinja	x		x	x

Commune Wernberg [Gemeinde Wernberg]

Name of the populated place	1972 (20%)	1977 (25%)	2001 (10%)	2011 (17.5%)
Wudmath/Vudmat	x		x	

Political District Völkermarkt [Politischer Bezirk Völkermarkt]

Urban Commune Bleiburg [Stadtgemeinde Bleiburg]

Name of the populated place	1972 (20%)	1977 (25%)	2001 (10%)	2011 (17.5%)
Aich/Dob	x	X	x	x
Bleiburg/Pliberk			x	x
Dobrowa/Dobrova		X	x	x
Draurain/Brege	x	X	x	x
Ebersdorf/Drveša vas			x	x
Einersdorf/Nonča vas	x	X	x	x
Grablach/Grablje	x		x	
Kömmel/Komelj	x	X	x	x
Kömmelgupf/Komeljski Vrh	x	X	x	x
Loibach/Libuče	x		x	x
Lokowitzen/Lokovica	x		x	
Moos/Blato	x	X	x	x
Replach/Replje	x	X	x	x
Rinkenberg/Vogrče	x	X	x	x
Rinkolach/Rinkole	x	X	x	x
Ruttach/Rute	x	X	x	x
Sankt Georgen/Šentjur	x		x	x
Sankt Margarethen/ Šmarjeta	x		x	x
Schattenberg/Senčni kraj	x		x	
Schilterndorf/Čirkovče	x	X	x	x
Weißenstein/Belšak	x		x	
Wiederndorf/Vidra vas	x	X	x	x
Woroujach/Borovje	x		x	x

Commune Diex [Gemeinde Diex]

Name of the populated place	1972 (20%)	1977 (25%)	2001 (10%)	2011 (17.5%)
Grafenbach/Kneža	x		1971	
Bösenort/Hudi kraj			x	
Großenegg/Tolsti Vrh			x	
Haimburgerberg/Vovbrske gore			x	

Market Commune Eberndorf [Marktgemeinde Eberndorf]

Name of the populated place	1972 (20%)	1977 (25%)	2001 (10%)	2011 (17.5%)
Buchbrunn/Bukovje			x	x
Eberndorf/Dobrla vas			x	x
Edling/Kazaze			x	x
Gablern/Lovanke	x		x	x
Gösselsdorf/Goselna vas			x	x
Graben/Graben	x		x	
Hof/Dvor	x		x	x
Homitzberg/Homec			x	
Köcking/Kokje			x	
Loibegg/Belovče			x	
Mökriach/Mokrije	x		x	x
Pudab/Pudab	x		x	
Sankt Marxen/Šmarkež			x	

Market Commune Eisenkappel-Vellach [Marktgemeinde Eisenkappel-Vellach]

Name of the populated place	1972 (20%)	1977 (25%)	2001 (10%)	2011 (17.5%)
Bad Eisenkappel/Železna Kapla			x	x
Blasnitzen/Plaznica	x	X	x	x
Ebriach/Obirsko	x	X	x	x
Koprein Petzen/Pod Peco	x	X	x	x
Koprein Sonnseite/Koprivna	x	X	x	x
Leppen/Lepena	x	X	x	x
Lobnig/Lobnik	x	X	x	x
Rechberg/Rebrca		X	x	x
Remschenig/Remšenik	x	X	x	x
Trögern/Korte	x	X	x	x
Unterort/Podkraj		X	1981	x
Vellach/Bela	x	X	x	x
Weißenbach/Bela		X	?	x
Zauchen/Suha		X	x	x

Market Commune Feistritz ob Bleiburg [Marktgemeinde Feistritz ob Bleiburg]

Name of the populated place	1972 (20%)	1977 (25%)	2001 (10%)	2011 (17.5%)
Dolintschitschach/Dolinčiče	x	X	x	x
Feistritz ob Bleiburg/Bistrica pri Pliberku	x	X	x	x
Gonowetz/Konovece	x	X	x	x
Hinterlibitsch/Suha	x	X	x	x
Hof/Dvor	x	X	x	x
Lettenstätten/Letina	x	X	x	x
Penk/Ponikva		X	x	x
Pirkdorf/Breška vas	x	X	x	x
Rischberg/Rižberk	x	X	?	x
Ruttach-Schmelz/Rute	x	X	x	x
Sankt Michael ob Bleiburg/Šmihel pri Pliberku	x	X	x	x
Tscherberg/Črgoviče	x	X	x	x
Unterlibitsch/Podlibič		X	x	x
Unterort/Podkraj	x	X	x	x
Winkel/Kot	x	X	x	x

Commune Gallizien [Gemeinde Gallizien]

Name of the populated place	1972 (20%)	1977 (25%)	2001 (10%)	2011 (17.5%)
Abtei/Apače	x		x	
Dolintschach/Dolinče			x	
Drabunaschach/Drabunaže			x	x
Enzelsdorf/Encelna vas			x	x
Freibach/Borovnica	x		x	x
Gallizien/Galicija			1991	
Pirk/Brezje			x	
Pölzing/Pecelj			x	
Robesch/Robež	x		x	
Unterkrain/Podkrinj			x	

Commune Globasnitz [Gemeinde Globasnitz]

Name of the populated place	1972 (20%)	1977 (25%)	2001 (10%)	2011 (17.5%)
Globasnitz/Globasnica	x	X	x	x
Jaunstein/Podjuna	x	X	x	x
Kleindorf/Mala vas	x	X	x	x
Podrain/Podroje			x	x
Sankt Stefan/Šteben	x	X	x	x
Slovenjach/Slovenje	x	X	x	x
Traundorf/Strpna vas	x	X	x	x
Tschepitschach/Čepiče	x	X	x	x
Unterbergen/Podgora	x	X	x	x
Wackendorf/Večna vas	x	X	x	x

Market Commune Griffen [Marktgemeinde Griffen]

Name of the populated place	1972 (20%)	1977 (25%)	2001 (10%)	2011 (17.5%)
Enzelsdorf/Encelna vas	x		1961	
Großenegg/Tolsti Vrh			x	
Grutschen/Krčanje	x		1961	
Obere Gemeinde/Zgornja Gmajna	x		1961	
Stift Griffen/Grebinjski Klošter			x	
Untergreutschach/Spodnje Krčanje	x		x	

Commune Neuhaus [Gemeinde Neuhaus]

Name of the populated place	1972 (20%)	1977 (25%)	2001 (10%)	2011 (17.5%)
Bach/Potoče	x		1981	
Draugegend/Pri Dravi		X	?	
Graditschach/Gradiče			x	x
Hart/Breg		X	x	x
Heiligenstadt/Sveto mesto		X	x	x
Illmitzen/Ivnik	x		x	
Kogelnigberg/Kogelska Gora			x	x
Oberdorf/Gornja vas	x	X	x	x
Schwabegg/Žvabek	x	X	x	x
Unterdorf/Dolnja vas	x	X	x	x
Wesnitzen/Beznica			x	

Commune Ruden [Gemeinde Ruden]

Name of the populated place	1972 (20%)	1977 (25%)	2001 (10%)	2011 (17.5%)
Eis/Led	x		1961	
Grutschen/Gruča			x	
Kanaren/Kanarn			x	
Sankt Martin/Šmartin	x		1991	
Sankt Nikolai/Šmiklavž	x		x	
Sankt Radegund/Šent Radegunda	x		1971	
Untermitterdorf/Srednja vas	x		x	

Commune Sankt Kanzian am Klopeiner See [Gemeinde Sankt Kanzian am Klopeiner See]

Name of the populated place	1972 (20%)	1977 (25%)	2001 (10%)	2011 (17.5%)
Grabelsdorf/Grabalja vas	x		x	x
Horzach I/Horce I	x		x	x
Horzach II/Horce II	x		x	x
Kleindorf I/Mala vas I			x	
Kleindorf II/Mala vas II			x	
Klopein/Klopinj			x	
Lanzendorf/Lancova	x			
Lauchenholz/Gluhi Les	x		x	x
Mökriach/Mokrije	x		x	x
Nageltschach/Nagelče	x		x	x
Obersammelsdorf/Žamanje	x		x	x
Sankt Kanzian/Škocjan			1991	
Sankt Primus/Šentprimož			x	x
Sankt Veit im Jauntal/Šentvid v Podjuni	x		x	x
Srejach/Sreje			x	
Unterburg/Podgrad			x	
Unternarrach/Spodnje Vinare	x		x	x
Untersammelsdorf/Samožna vas	x		1981	
Vesielach/Vesele	x		x	x

Commune Sittersdorf [Gemeinde Sittersdorf]

Name of the populated place	1972 (20%)	1977 (25%)	2001 (10%)	2011 (17.5%)
Altendorf/Stara vas			x	
Blasnitzenberg/Plaznica			x	
Dullach/Dule	x		x	
Goritschach/Goriče			x	x
Kleinzapfen/Malčape	x		x	x
Kristendorf/Kršna vas			x	x
Müllnern/Mlinče	x		x	x
Obernarrach/Zgornje Vinare			x	x
Pfannsdorf/Banja vas			x	
Pogerschitzen/Pogrče	x		x	x
Polena/Polane	x		x	
Proboj/Proboj			x	
Rückersdorf/Rikarja vas			x	x
Sagerberg/Zagorje			x	x
Sielach/Sele			x	
Sittersdorf/Žitara vas	x		x	x
Sonnegg/Ženek			x	x
Tichoja/Tihoja			x	x
Wigasnitz/Vijasce			x	
Winkel/Kot			x	
Wrießnitz/Breznica			x	

Urban Commune Völkermarkt [Stadtgemeinde Völkermarkt]

Name of the populated place	1972 (20%)	1977 (25%)	2001 (10%)	2011 (17.5%)
Arlsdorf/Orlača vas			x	
Attendorf/Vata vas	x		1981	
Bach/Potok			x	
Gattersdorf/Štriholče			x	
Krenobitsch/Hrenovče	x		1961	
Kulm/Hom	x		1961	
Penk/Klopče	x		x	
Rammersdorf/Ramovča vas			x	
Sankt Jakob/Šent Jakob	x		1961	
Wenzach/Brnce	x		1981	

Appendix C: Interview Guidelines Carinthia

Personal characteristics:

(1) What is your relation to the place, where you are living today? Are you born here? Have you moved in later?
(2) Age
(3) Gender
(4) Highest level of education completed
(5) Did you attend a bilingual kindergarten/school?
(6) Occupation
(7) Denomination

Group affiliation:

(8) To which group do you feel to belong? (Carinthian Slovenes, Slovenian-speaking Carinthians, Slovenes, 'Windische', German-Carinthians, Germans, Austrians, "This is not important for me", "I rather feel a local/regional affiliation", "I am just a Carinthian", "I am just an Austrian, I am just a European")
(9) How would you rank your affiliations in this respect? (e.g., first Carinthian Slovene, second Austrian, third Inhabitant of the valley)
(10) What is it that gives you the identity of a Carinthian Slovene/ Carinthian/German-Carinthian?
(11) Were you led to this identity from childhood or have you adopted it later?
(12) Are you active in an organization representing your group identity or do you support such an organization? If yes, which one?
(13) Do you read/hear/watch regional media in Slovene or German?

Language:

(14) What is your mother tongue?
(15) Which language do you use at home (in the family)?

(16) Which language do/did you use at school, when you talk/talked with friends sharing your language?
(17) Which language do you use when talking with neighbors, who share your language?
(18) Do you use in all these situations rather dialect or standard language?
(19) When do you use standard language?
(20) Do you regard your (Slovenian or German) Carinthian dialect as something very specific and distinct from the standard language?
(21) How important is it to learn Slovene as a second language in Carinthia?
(22) Is speaking Slovene an advantage in the professional context?
(23) Does German/Slovene bilingualism make southern Carinthia a more attractive tourism destination?

Toponyms:

(24) How do you call places in your surroundings? (Up to ten examples from near to far)
(25) Do you also know the Slovenian names of these places (in cases, when the respondent just mentioned the German name)?
(26) Do you also know some places in Slovenia?
(27) Do you also know the German/Slovenian names of these places? (Depending on whether the respondent has earlier mentioned the Slovenian or German name)
(28) Was the Carinthian place-name conflict settled to your satisfaction?
(29) What should still be done?
(30) Has the official settling of the conflict improved the relation between the various groups in Carinthia?
(31) What do bilingual town signs in Carinthia mean for you?
(32) When and where are bilingual town signs appropriate and when/where not?

(33) How do you/would you perceive bilingual signs in your village/town?
(34) What do/would these signs in your village/town express?
(35) Have the bilingual signs in your village/town (earlier) been vandalized?
(36) What was your reaction to this vandalization?
(37) What was the reaction of your neighbors?
(38) Should also street plates be bilingual?
(39) Would you find it appropriate to reflect Slovenian toponyms also on tourism/hiking/biking maps?
(40) Are also the official Austrian topographical maps to reflect official Slovenian toponyms?
(41) Would you as a Carinthian Slovene (also if you are not a member of this group) sign your own house with a Slovenian name? If yes, why; if not, why?

Appendix D: Interview Topics and Questions in the Těšín/Cieszyn Region

Personal characteristics:

(1) Relationship to place (born there/moved in)
(2) Age
(3) Gender
(4) Education (Czech/Polish school)
(5) Occupation
(6) Religion (denomination—Catholic, Evangelical, Lutheran, none)

Nationality:

(1) To which nationality do you belong?
(2) What makes you Czech/Polish/Silesian? Were you led to this identity from childhood? Or have you adopted it in adulthood?

(3) (For Poles) Are you active in a Polish organization or do you support such an organization?
(4) Do you read Czech/Polish/regional newspapers? Is the representation of Czech/Polish language sufficient?
(5) Do you consider yourself a Gorol? Is it more a national or a regional or other term?

Language:

(1) What is your native language? (In case he/she responds “Polish”, is it standard Polish?)
(2) Which language do you use at home?
(3) Which language did you speak at school with friends?
(4) Which language do you use with your neighbors?
(5) When do you use standard Czech/Polish?
(6) Do you see po naszymu as a Czech or a Polish dialect?

Toponyms:

(1) Tell me names for some places in your surroundings/How do you call places in your surroundings?
(2) What is the origin of these names?
(3) Are any stories associated with them?
(4) Which places/names are important for you and why?
(5) Is it important how a name is spelled and pronounced?
(6) (Over a map) How do you like the transcription of this specific name? How would you write it down, if you were the author of the map?
(7) How do you perceive the double name for Olza/Olše river? Have you noticed any conflicts related to it? Are they real or artificially created?

Bilingualism:

(1) How do you perceive the appearance of bilingual signs in your village?
(2) When are bilingual signs appropriate and when not?
(3) Should all toponyms have their official Czech and Polish equivalents?
(4) Did you put a bilingual sign on your house? Would you agree with the placement of a bilingual sign on your street?
(5) What is your perception of the current look of the bilingual signs (placement, appearance, order of names, size, etc.)?
(6) Which names should appear on maps? Is the current form of names appearing on tourist maps appropriate?
(7) Have the bilingual signs in your village been vandalized?
(8) What was your reaction to this vandalization? What was the reaction of your neighbors?
(9) Do you think that the presence of bilingual signs affects the perception of the region by tourists?

Appendix E: Questionnaire Survey—Geographical Names in the Těšín/Cieszyn Region

Dear Madam/Sir,
We would like to ask you to take part in this brief survey the goal of which is to learn how you as an inhabitant of the Těšín/Cieszyn region perceive the bilingual character of the region as regards geographical names, their use in everyday life, and their significance from your relationship to your home and the region. Responding to the following questions will take a maximum of ten minutes of your time and the survey is completely **anonymous**. Your participation is **very important** for us. Return the filled-out questionnaires to our local collaborators as soon as possible. As the survey coordinator will gladly answer your questions.
Dr. Přemysl Mácha, Ph.D.

How would you characterize your feelings toward the Těšín/Cieszyn region?

How would you characterize your relation toward the Těšín/Cieszyn region?

a) Born in this region
b) Attended school here
c) Have relatives here
d) Have married someone from here
e) Have migrated to this region
f) Work here
g) Spend vacations or leisure time here

Which language do you *mostly* use in the following situations? Circle your answer.

At home	Dialect	Czech	Polish
With friends	Dialect	Czech	Polish
At work/school	Dialect	Czech	Polish
With neighbors	Dialect	Czech	Polish
In the shop	Dialect	Czech	Polish
At the municipal office	Dialect	Czech	Polish

Is the Tesin dialect for you rather a Czech, Polish, or mixed dialect?
a) Czech
b) Polish
c) Mixed

Imagine you are talking about places in your surroundings (village part, stream, hill, etc.), which form do you *preferably* use?
a) Dialect
b) Czech
c) Polish

Which places in the Tesin region (settlements, village parts, streams, hills, etc.) are *particularly* important to you? Write down their names (max. 5) in the form in which you use them (e.g., in the dialect, Czech, or Polish) and briefly describe what type of place it is and why it is important for you (e.g., Na Kympě—neighborhood—I live there).

Place name	Type of place	Why it is important

Would you appreciate if dialectal names were used on official signs and maps?

a) Yes, I would like that
b) I don't know, it does not matter to me
c) No, dialectal names should not be used on official signs and maps

Which form of the river name Olza/Olše do you use daily?
a) Only Olza
b) Mostly Olza
c) Equally Olza and Olše
d) Mostly Olše
e) Only Olše

Is Olza for you
a) a Czech name?
b) a Polish name?
c) a dialectal name?

Should the name Olza become the official name of the river? Briefly explain why yes or no.

a) Yes b) No
Reason:

What are your feelings about bilingual signs?

With which of the following arguments do you identify the most?

a) Implementation of bilingual signs is a waste of public funds, Poles in the Tesin region speak Czech.

b) The Polish minority has a legal right to bilingual signs.
c) Implementation of bilingual signs expresses a public recognition of the existence of the Polish minority that it is also at home here.
d) We are in the Czech Republic, signs should be only in the Czech language, there are no Czech signs in Poland either.
e) Others (specify):

If you agree with the implementation of bilingual signs, in which cases should they be implemented? Indicate with a cross all cases in which the bilingual signs should appear.

Entrance to village	Municipal office	Library, school, etc.	Train stops	Bus stops
Village parts	Street signs	Orientation signs inside villages	Shops, companies, pubs	Local and regional maps

How do you see the quality of the relationship between Czechs and Poles in the Tesin region?

Place of residence (name of municipality where you live):

____

Gender: Male/Female **Age:** ____________

Education:
a) Elementary
b) Vocational high school
c) High school
d) University

Nationality:
a) Czech
b) Polish
c) Slovak
d) Silezian
e) other: __________

Thank you for filling out the questionnaire!

Appendix F: Place Names from the Těšín/Cieszyn Region Transcribed into the International Phonetic Alphabet

Czech	IPA	Polish	IPA	Dialect	IPA
Albrechtice	albrɛxcɪt͡sɛ	Olbrachcice	ɔlbraxt͡ɕit͡sɛ	Albrechcice	albrɛxt͡ɕit͡sɛ
Beskydy	bɛskɪdɪ	Beskidy	bɛskidɨ	Beskidy	bɛskidɨ
Bílsko	biːlskɔ	Bielsko	bjɛlskɔ	Bielsko	bjɛlskɔ
Bocanovice	bɔt͡sanɔvɪt͡sɛ	Boconowice	bɔt͡sɔnɔvit͡sɛ	Bocanowice	bɔt͡sanɔvit͡sɛ
Bohumín	bɔɦumiːn	Bogumin	bɔgumin	Bogumin	bɔgumin
Brandýs	brandiːs	Brandys	brandɨs	Brandys	brandɨs
Bukovec	bukɔvɛt͡s	Bukowiec	bukɔvjɛt͡s	Bukowiec	bukɔvjɛt͡s
Bystřice	bɪstr̝̊ɪt͡sɛ	Bystrzyca	bɨstʂɨt͡sa	Bystrzica	bɨstr̝̊it͡sa
Čantoryje	t͡ʃantɔrɪjɛ	Czantoria	t͡ʂantɔrija	Czantoryja	t͡ʂantɔrɨja
Česko	t͡ʃɛskɔ	Czechy	t͡ʂɛxɨ	Czechi	t͡ʂɛxi
Český Těšín	t͡ʃɛskiː cɛʃiːn	Czeski Cieszyn	t͡ʂɛski t͡ɕɛʂɨn	Czeski Cieszyn	t͡ʂɛski t͡ɕɛʂɨn
Darkov	darkɔf	Darków	Darkuf	Darków	darkʊf
Dětmarovice	ɟɛtmarɔvɪt͡sɛ	Dziećmorowice	d͡ʑɛt͡ɕmɔrɔvit͡sɛ	Dzietmarowice	d͡ʑɛtmarɔvit͡sɛ
Do Parlamentu	dɔ parlamɛntu	Do Parlamentu	dɔ parlamɛntu	Do Parlamentu	dɔ parlamɛntu
Dolní Líštná	dolɲiː liːʃtnaː	Leszna Dolna	lɛʂna dɔlna	Dolnio Łyszno	dɔlɲɔ wɨʂnɔ
Dolní Lomná	dolɲiː lɔmnaː	Łomna Dolna	wɔmna dɔlna	Dolnio Łómno	dɔlɲɔ wʊmnɔ
Dolní Lutyně	dolɲiː lutɪɲe	Lutynia Dolna	lutɨɲa dɔlna	Dolnio Lutynia	dɔlɲɔ lutɨɲa
Dolní Marklovice	dolɲiː marklɔvɪt͡sɛ	Marklowice Dolne	marklɔvit͡sɛ dɔlnɛ	Dolni Marklowice	dɔlɲi marklɔvit͡sɛ
Dolní Suchá	dolɲiː suxaː	Sucha Dolna	suxa dɔlna	Dolnio Sucho	dɔlɲɔ suxɔ
Dolní Těrlicko	dolɲiː cɛrlɪt͡skɔ	Cierlicko Dolne	t͡ɕɛrlit͡skɔ dɔlnɛ	Dolni Cierlicko	dɔlɲi t͡ɕɛrlit͡skɔ
Dolní Žukov	dolɲiː ʒukɔf	Żuków Dolny	ʐukuf dɔlnɨ	Dolni Żuków	dɔlɲi ʐukʊf
Doubrava	dɔu̯brava	Dąbrowa	dɔ̃brova	Dómbrowo	dʊmbrɔvɔ
Frýdek	friːdɛk	Frydek	frɨdɛk	Frydek	frɨdɛk
Fryštát	frɪʃtaːt	Frysztat	frɨʂtat	Frysztat	frɨʂtat

(*continued*)

(continued)

Czech	IPA	Polish	IPA	Dialect	IPA
Godula	gɔdula	Godula	gɔdula	Godula	gɔdula
Guty	gutɪ	Guty	gutɨ	Guty	gutɨ
Havířov	ɦaviːr̝ɒf	Hawierzów	xavjɛʐ̧uf	Hawiyrzów	ɦavjɨr̝ʊf
Hluchová	ɦluxɔvaː	Głuchówka	Gwuxufka	Guchówka	guxʊfka
Hnojník	ɦnɔjɲiːk	Gnojnik	gnɔjɲik	Gnojnik	gnɔjɲik
Horní Líštná	ɦɔrɲiː liːʃtnaː	Leszna Górna	lɛʂna gurna	Górnio Łyszno	gʊrɲɔ lɨʂnɔ
Horní Lomná	ɦɔrɲiː lɔmnaː	Łomna Górna	wɔmna gurna	Górnio Łómno	gʊrɲɔ wʊmnɔ
Horní Lutyně	ɦɔrɲiː lutɪɲɛ	Lutynia Górna	lutɨɲa gurna	Górnio Lutynia	gʊrɲɔ lutɨɲa
Horní Suchá	ɦɔrɲiː suxaː	Sucha Górna	suxa gurna	Górnio Sucho	gʊrɲɔ suxɔ
Horní Těrlicko	ɦɔrɲiː cɛrlɪt͡skɔ	Cierlicko Górne	t͡ɕɛrlikɔ gurnɛ	Górni Cierlicko	gʊrɲi t͡ɕɛrlikɔ
Horní Tošanovice	ɦɔrɲiː tɔʃanɔvɪt͡sɛ	Toszonowice Górne	tɔʂɔnɔvit͡sɛ gurnɛ	Górni Toszanowice	gʊrɲi tɔʂanɔvit͡sɛ
Horní Žukov	ɦɔrɲiː ʒukɔv	Żuków Górny	ʐ̧ukuf gurnɨ	Górni Żuków	gʊrɲi ʐ̧ukʊf
Hrádek	ɦraːdɛk	Gródek	grudɛk	Gródek	grʊdɛk
Hrčava	ɦrt͡ʃava	Herczawa	xɛrt͡ʂava	Hyrczawa	ɦɨrt͡ʂava
Chotěbuz	xɔcɛbus	Kocobędz	kɔt͡sɔbɛ̃t͡s	Kocobync	kɔt͡sɔbɨnt͡s
Jablunkov	jabluŋkɔf	Jabłonków	jabwɔŋkuf	Jabónków	jabʊŋkʊf
Karpentná	karpɛntnaː	Karpętna	karpɛ̃tna	Karpyntno	karpɨntnɔ
Karviná	karvɪnaː	Karwina	Karvina	Karwino	karvinɔ
Kojkovice	kɔjkɔvɪt͡sɛ	Kojkowice	kɔjkɔvit͡sɛ	Kojkowice	kɔjkɔvit͡sɛ
Komorní Lhotka	kɔmɔrɲiː lɦɔtka	Ligotka Kameralna	ligɔtka kamɛralna	Ligotka	ligɔtka
Konská	kɔnskaː	Końska	kɔ̃jska	Kóńsko	kʊɲskɔ
Koňské pleso	kɔɲskɛː plɛsɔ	Jezioro końskie[a]	jɛʑɔrɔ kɔ̃jsk’ɛ	Kóński ploso	kʊɲski plɔsɔ
Košařiska	kɔʃar̝ɪska	Koszarzyska	kɔʂaʐ̧ɨska	Koszarziska	kɔʂar̝iska
Kozubová	kɔzubɔvaː	Kozubowa	kɔzubɔva	Kozubowo	kɔzubɔvɔ
Kyčera	kɪt͡ʃɛra	Kiczory	kit͡ʂɔrɨ	Kiczera	kit͡ʂɛra
Louky nad Olší	lɔukɪ nat ɔlʃiː	Łąki nad Olzą	wɔ̃ki nat ɔlzɔ̃	Łónki nad Olzóm	wʊŋki nat wɔlʊm
Luhy	luhɪ	Łęgi[a]	wɛ̃gi	Łyngi	wɨŋgi

Lyžbice	lɪʒbɪt͡sɛ	Łyżbice	wɨz̨bit͡sɛ	Łyżbice	wɨz̨bit͡sɛ
Milíkov	mɪliːkɔf	Milików	Milikuf	Milików	milikʊf
Mistřovice	mɪstr̝̊ɔvɪt͡sɛ	Mistrzowice	mistʂɔvit͡sɛ	Mistrzowice	mistr̝̊ɔvit͡sɛ
Moravskoslezské Beskydy	mɔrafskɔslɛskɛː bɛskɪdɪ	Beskid Morawsko-Śląski	bɛskit mɔrafskɔ ɕlɔ̃ski	Morawsko-ślónski Beskidy	mɔrafskɔ ɕlʊnski bɛskidɨ
Mosty u Jablunkova	mɔstɪ ʔu jabluŋkɔva	Mosty koło Jabłonkowa	mɔstɨ kɔwɔ jabwɔŋkɔva	Mosty u Jabónkowa	mɔstɨ ʔu jabʊŋkɔva
Na Břízkách	na br̝iːskaːx	Na brzyżkach	na bz̨ɨʂkax	Na brzyżkach	na br̝ɨʂkax
Na čtyřku[a]	na t͡ʃtɪr̝̊ku	Na czwórkę[a]	na t͡ʂfurkɛ̃	Na czwórke	na t͡ʂfʊrkɛ
Na jedničku[a]	na jɛdɲɪt͡ʃku	Na jedynkę[a]	na jɛdɨŋkɛ̃	Na jedynke	na jɛdɨŋkɛ
Na milíři	na mɪliːr̝ɪ	Na mielerzu[a]	na mjɛlɛz̨u	Na milyrzu	na milɨr̝u
Nádražní	naːdraʒɲiː	Dworcowa	dvɔrcɔva	Nadrażnio	nadraz̨ɲɔ
Návsí	naːfsiː	Nawsie	nafɕɛ	Nowsi	nɔfɕi
Nebory	ɲɛbɔrɪ	Niebory	ɲɛbɔrɨ	Niebory	ɲɛbɔrɨ
Nýdek	niːdɛk	Nydek	nɨdɛk	Nydek	nɨdɛk
Odra	ɔdra	Odra	ɔdra	Odra	wɔdra
Oldřichovice	ɔldr̝ɪxɔvɪt͡sɛ	Oldrzychowice	ɔldz̨ɨxɔvit͡sɛ	Oldrzichowice	wɔldr̝ixɔvit͡sɛ
Olše	ɔlʃɛ	Olza	ɔlza	Olza	wɔlza
Orlová	ɔrlɔvaː	Orłowa	ɔrwɔva	Orłowo	wɔrwɔvɔ
Ostrava ?	ɔstrava	Ostrawa	ɔstrava	Ostrawa	wɔstrava
Ostravice	ɔstravɪt͡sɛ	Ostrawica	ɔstravit͡sa	Ostrawica	wɔstravit͡sa
Ostrý	ɔstriː	Ostry	ɔstrɨ	Ostry	wɔstrɨ
Pasečky	pasɛt͡ʃkɪ	Pasieczki	paɕɛt͡ʂki	Pasiyczki	paɕɨt͡ʂki
Paseky	pasɛkɪ	Pasieki	paɕɛki	Pasieki	paɕɛki
Petrovice	pɛtrɔvɪt͡sɛ	Piotrowice	pjɔtrɔvit͡sɛ	Petrowice	pɛtrɔvit͡sɛ
Petřvald	pɛtr̝valt	Pietwałd	pjɛtfawt	Pietwółd	pjɛtfʊwt
Pionýrák	pɪjɔniːrak	Pionierak[a]	pjɔɲɛrak	Pjonierak	pjɔɲɛrak
Písečná	piːsɛt͡ʃnaː	Pioseczna	pjɔsɛna	Pioseczno	pjɔsɛnɔ
Písek	piːsɛk	Piosek	pjɔsɛk	Piosek	pjɔsɛk

(*continued*)

(continued)

Czech	IPA	Polish	IPA	Dialect	IPA
Polední	pɔlɛdniː	Połedna	pɔwɛdna	Połednio	pɔwɛdɲɔ
Polsko	pɔlskɔ	Polska	pɔlska	Polska	pɔlska
Prstná	pr̩stnaː	Pierstna	pjɛrstna	Piersno	pjɛrsnɔ
Ráj	raːj	Raj	Raj	Raj	raj
Ropice	rɔpɪt͡sɛ	Ropica	rɔpit͡sa	Ropica	rɔpit͡sa
Rychvald	rɪɣvalt	Rychwałd	rɨxvawt	Rychwółd	rɨxvʊwt
Řeka	r̝ɛka	Rzeka	z̨ɛka	Rzeka	r̝ɛka
Sikorův Park	sɪkɔruːf par̩k	Park Sikory	park ɕikɔrɨ	Sikorów park	ɕikɔrʊf park
Slezská Ostrava	slɛskaː ɔstrava	Śląska Ostrawa	ɕlɔ̃ska ɔstrava	Ślónsko Ostrawa	ɕlʊnskɔ ɔstrava
Slezské Beskydy	slɛskɛː bɛskɪdɪ	Beskid Śląski	bɛskit ɕlɔ̃ski	Ślónski Beskidy	ɕlʊnski bɛskidɨ
Slezsko	slɛskɔ	Śląsk	ɕlɔ̃sk	Ślónsk	ɕlʊnsk
Smilovice	smɪlɔvɪt͡sɛ	Śmiłowice	ɕmiwɔvit͡sɛ	Śmiłowice	ɕmiwɔvit͡sɛ
Stanislavice	staɲɪslavɪt͡sɛ	Stanisłowice	staɲiswɔvit͡sɛ	Stanisławice	staɲiswavit͡sɛ
Stonava	stɔnava	Stonawa	stɔnava	Stonawa	stɔnava
Střítež	str̝̊iːtɛʃ	Trzycież	tʂɨt͡ɕɛʂ	Trzycież	tʂɨt͡ɕɛʂ
Svibice	svɪbɪt͡sɛ	Sibica	ɕibica	Świbica	ɕfibica
Šenov	ʃɛnɔf	Szonów	ʂɔnuf	Szenów	ʂɛnʊf
Těrlicko	cɛrlɪt͡skɔ	Cierlicko	t͡ɕɛrlit͡skɔ	Cierlicko	t͡ɕɛrlit͡skɔ
Těšín	cɛʃiːn	Cieszyn	t͡ɕɛʂɨn	Cieszyn	t͡ɕɛʂɨn
Třanovice	tr̝̊anɔvɪt͡sɛ	Trzanowice	tʂanɔvit͡sɛ	Trzanowice	tr̝̊anɔvit͡sɛ
Třinec	tr̝̊ɪnɛt͡s	Trzyniec	tʂɨɲɛt͡s	Trzyniec	tr̝̊ɨɲɛt͡s
Tyra	tɪra	Tyra	tɨra	Tyra	tɨra
U Cupka	ʔu t͡supka	Przy Cupkowi[a]	pʂɨ t͡supkɔvi	U Cupka	ʔu t͡supka
U Hetmaňoka	ʔu ɦɛtmaɲɔka	Przy Hetmaniokowi[a]	pʂɨ xɛtmaɲɔkɔvi	U Hetmanioka	ʔu ɦɛtmaɲɔka
Vělopolí	vjɛlɔpɔliː	Wielopole	vjɛlɔpɔlɛ	Wielopole	vjɛlɔpɔlɛ

Vendryně	vɛndrɪɲɛ	Wędrynia	vɛ̃drɨɲa	Wyndrynia	vɨndrɨɲa
Věřnovice	vjɛr̝ɲɔvɪt͡sɛ	Wierzniowice	vjɛʐɲɔvit͡sɛ	Wierzniowice	vjɛrɲɔvit͡sɛ
Visla	vɪsla	Wisła	Viswa	Wisła	viswa
Záolší[b]	zaːɔlʃiː	Zaolsze[a]	zaɔlʂɛ	Zołolzi	zɔwɔlʑi
Záolží	zaːɔlʒiː	Zaolzie	zaɔlʑɛ	Zaolzie[a]	zawɔlʑɛ
Zárubek	zaːrubɛk	Zarębek	zarɛ̃bɛk	Zorymbek	zɔrɨmbɛk
Závada	zaːvada	Zawada	Zavada	Zowada	zɔvada

Note: Dialectal names are given in the dialect version spoken in the south-central part of the minority area and they are transcribed using Polish orthography for lack of a standard dialectal alphabet and an absence of adequate characters in the Czech alphabet. The orthographic choice should not be interpreted as a statement on the Polishness of the dialect. We consider it a transitional Slavic dialect without national membership

[a]Reconstructed names not documented in actual use

[b]Záolší is a part of the village and commune of Vendryně/Wędrynia, while Záolží refers to the minority homeland, Zaolzie

Name Index[1]

[1] Note: Page numbers followed by 'n' refer to notes.

P. Jordan et al., *Place-Name Politics in Multilingual Areas*, Palgrave Studies in Minority Languages and Communities, https://doi.org/10.1007/978-3-030-69488-3

B

L

M

Subject Index[1]

[1] Note: Page numbers followed by 'n' refer to notes.

P. Jordan et al., *Place-Name Politics in Multilingual Areas*, Palgrave Studies in Minority Languages and Communities, https://doi.org/10.1007/978-3-030-69488-3

G

M

Q

R

S

The manufacturer's authorised representative in the EU is Springer Nature Customer Service Centre GmbH, Europaplatz 3, 69115 Heidelberg, Germany. If you have any concerns regarding our products, please contact ProductSafety@springernature.com

Printed and bound by CPI Group (UK) Ltd, Croydon, CR0 4YY
30/09/2026
02225443-0012